Sustainable Control Strategies of Plant Pathogens in Horticulture

Sustainable Control Strategies of Plant Pathogens in Horticulture

Editors

Hillary Righini
Roberta Roberti
Stefania Galletti

Basel • Beijing • Wuhan • Barcelona • Belgrade • Novi Sad • Cluj • Manchester

Editors

Hillary Righini
University of Bologna
Bologna
Italy

Roberta Roberti
University of Bologna
Bologna
Italy

Stefania Galletti
Council for Agricultural
Research and Economics
(CREA)
Bologna
Italy

Editorial Office
MDPI
St. Alban-Anlage 66
4052 Basel, Switzerland

This is a reprint of articles from the Special Issue published online in the open access journal *Horticulturae* (ISSN 2311-7524) (available at: https://www.mdpi.com/journal/horticulturae/special_issues/Plant_Pathogens_Horticulture).

For citation purposes, cite each article independently as indicated on the article page online and as indicated below:

Lastname, A.A.; Lastname, B.B. Article Title. *Journal Name* **Year**, *Volume Number*, Page Range.

ISBN 978-3-7258-0479-5 (Hbk)
ISBN 978-3-7258-0480-1 (PDF)
doi.org/10.3390/books978-3-7258-0480-1

© 2024 by the authors. Articles in this book are Open Access and distributed under the Creative Commons Attribution (CC BY) license. The book as a whole is distributed by MDPI under the terms and conditions of the Creative Commons Attribution-NonCommercial-NoDerivs (CC BY-NC-ND) license.

Contents

About the Editors vii

Hillary Righini, Roberta Roberti and Stefania Galletti
Special Issue "Sustainable Control Strategies of Plant Pathogens in Horticulture"
Reprinted from: *Horticulturae* **2024**, *10*, 146, doi:10.3390/horticulturae10020146 1

Conrado Parraguirre Lezama, Omar Romero-Arenas, Maria De Los Angeles Valencia de Ita, Antonio Rivera, Dora M. Sangerman Jarquín and Manuel Huerta-Lara
In Vitro Study of the Compatibility of Four Species of *Trichoderma* with Three Fungicides and Their Antagonistic Activity against *Fusarium solani*
Reprinted from: *Horticulturae* **2023**, *9*, 905, doi:10.3390/horticulturae9080905 6

Mansour M. El-Fawy, Kamal A. M. Abo-Elyousr, Nashwa M. A. Sallam, Rafeek M. I. El-Sharkawy and Yasser Eid Ibrahim
Fungicidal Effect of Guava Wood Vinegar against *Colletotrichum coccodes* Causing Black Dot Disease of Potatoes
Reprinted from: *Horticulturae* **2023**, *9*, 710, doi:10.3390/horticulturae9060710 20

Norma Hortensia Alvarez, María Inés Stegmayer, Gisela Marisol Seimandi, José Francisco Pensiero, Juan Marcelo Zabala, María Alejandra Favaro and Marcos Gabriel Derita
Natural Products Obtained from Argentinean Native Plants Are Fungicidal against Citrus Postharvest Diseases
Reprinted from: *Horticulturae* **2022**, *8*, 562, doi:10.3390/horticulturae9050562 33

Hela Chikh-Rouhou, Ana Garcés-Claver, Lydia Kienbaum, Abdelmonem Ben Belgacem and Maria Luisa Gómez-Guillamón
Resistance of Tunisian Melon Landraces to *Podosphaera xanthii*
Reprinted from: *Horticulturae* **2022**, *8*, 1172, doi:10.3390/horticulturae8121172 49

Rromir Koçi, Fabrice Dupuy, Salim Lebbar, Vincent Gloaguen and Céline Faugeron Girard
A New Promising Plant Defense Stimulator Derived from a By-Product of Agar Extraction from *Gelidium sesquipedale*
Reprinted from: *Horticulturae* **2022**, *8*, 958, doi:10.3390/horticulturae8100958 61

Hillary Righini, Roberta Roberti, Silvia Cetrullo, Flavio Flamigni, Antera Martel Quintana, et al.
Jania adhaerens Primes Tomato Seed against Soil-Borne Pathogens
Reprinted from: *Horticulturae* **2022**, *8*, 746, doi:10.3390/horticulturae8080746 76

Caterina Morcia, Isabella Piazza, Roberta Ghizzoni, Stefano Delbono, Barbara Felici, Simona Baima, et al.
In Search of Antifungals from the Plant World: The Potential of Saponins and Brassica Species against *Verticillium dahliae* Kleb.
Reprinted from: *Horticulturae* **2022**, *8*, 729, doi:10.3390/horticulturae8080729 94

Catello Pane, Riccardo Spaccini, Michele Caputo, Enrica De Falco and Massimo Zaccardelli
Multi-Parameter Characterization of Disease-Suppressive Bio-composts from Aromatic Plant Residues Evaluated for Garden Cress (*Lepidium sativum* L.) Cultivation
Reprinted from: *Horticulturae* **2022**, *8*, 632, doi:10.3390/horticulturae8070632 110

Stefania Galletti, Stefano Cianchetta, Hillary Righini and Roberta Roberti
A Lignin-Rich Extract of Giant Reed (*Arundo donax* L.) as a Possible Tool to Manage Soilborne Pathogens in Horticulture: A Preliminary Study on a Model Pathosystem
Reprinted from: *Horticulturae* **2022**, *8*, 589, doi:10.3390/horticulturae8070589 . . . . **129**

Hamada El-Gendi, Abdulaziz A. Al-Askar, Lóránt Király, Marwa A. Samy, Hassan Moawad and Ahmed Abdelkhalek
Foliar Applications of *Bacillus subtilis* HA1 Culture Filtrate Enhance Tomato Growth and Induce Systemic Resistance against *Tobacco mosaic virus* Infection
Reprinted from: *Horticulturae* **2022**, *8*, 301, doi:10.3390/horticulturae8040301 . . . . **150**

Simona Chrapačienė, Neringa Rasiukevičiūtė and Alma Valiuškaitė
Control of Seed-Borne Fungi by Selected Essential Oils
Reprinted from: *Horticulturae* **2022**, *8*, 220, doi:10.3390/horticulturae8030220 . . . . **171**

Yanan Duan, Yifan Zhou, Zhao Li, Xuesen Chen, Chengmiao Yin and Zhiquan Mao
Effects of *Bacillus amyloliquefaciens* QSB-6 on the Growth of Replanted Apple Trees and the Soil Microbial Environment
Reprinted from: *Horticulturae* **2022**, *8*, 83, doi:10.3390/horticulturae8010083 . . . . **181**

Laura Orzali, Mohamed Bechir Allagui, Clemencia Chaves-Lopez, Junior Bernardo Molina-Hernandez, Marwa Moumni, Monica Mezzalama and Gianfranco Romanazzi
Basic Substances and Potential Basic Substances: Key Compounds for a Sustainable Management of Seedborne Pathogens
Reprinted from: *Horticulturae* **2023**, *9*, 1220, doi:10.3390/horticulturae9111220 . . . . **198**

Marco Scortichini
Sustainable Management of Diseases in Horticulture: Conventional and New Options
Reprinted from: *Horticulturae* **2022**, *8*, 517, doi:10.3390/horticulturae8060517 . . . . **214**

About the Editors

Hillary Righini

She is a postdoc at the Department and Food Sciences of the University of Bologna. Her research focuses on the use of natural substances extracted from algae and cyanobacteria for the sustainable management of fungal diseases in horticultural plants. Currently, she works on polysaccharides and phycobiliproteins for their plant biostimulant properties and their protective role in triggering plant defense responses. She collaborates with national and international centers and been involved in national and international projects. She has published articles in indexed journals and participated in national and international conferences.

Roberta Roberti

She is a professor at the Department and Food Sciences of the University of Bologna, and she teaches programs on plant pathology and crop protection in integrated and organic cultivation. She studies the mechanisms of action of fungal microorganisms and the bioactivity of extracts and compounds from cyanobacteria and macroalgae for disease management under laboratory and greenhouse conditions. She collaborates with national and international centers and has been involved in national and international projects. She has published more than 100 publications as an author or co-author in national and international scientific journals. She has participated in national and international conferences.

Stefania Galletti

She works for the Council for Research in Agriculture and Economics, Research Centre for Agriculture and Environment, Bologna, Italy. She studies the biological activity of microorganisms, natural compounds, and extracts from micro- and macroalgae for crop protection purposes at laboratory, greenhouse, and field levels. Her other interests concern the use of microorganisms for applications in the sector of renewable energy from lignocellulosic biomass. She has participated in several projects funded by the Italian Ministry of Agriculture and is the author or co-author of more than 100 publications in national and international scientific journals.

Editorial

Special Issue "Sustainable Control Strategies of Plant Pathogens in Horticulture"

Hillary Righini [1,*], Roberta Roberti [1,*] and Stefania Galletti [2]

[1] Department of Agriculture and Food Sciences, Alma Mater Studiorum, University of Bologna, 40127 Bologna, Italy
[2] Research Centre for Agriculture and Environment, Council for Agricultural Research and Economics, 40128 Bologna, Italy; stefania.galletti@crea.gov.it
* Correspondence: hillary.righini2@unibo.it (H.R.); roberta.roberti@unibo.it (R.R.)

1. Introduction

European Regulation No. 1107/2009 [1] recommends the adoption of alternatives to synthetic products among plant protection products, thereby repealing Council Directives 79/117/EEC [2] and 91/414/EEC [3]. This recommendation is driven by the potential adverse effects of synthetic products on the environment, as well as on human and animal health [4,5]. Furthermore, the European "Green Deal" has introduced various initiatives that aim to facilitate a green transition to counteract climate change and safeguard the environment.

One of the key objectives outlined in the "Green Deal" is to substantially reduce the use of chemical pesticides in agriculture by 2030 [6,7], to reduce the associated risks and to address potential challenges in the management of plant pathogens. This strategy also outlines a set of measures to be achieved by 2030, including the promotion of 25% of organic agriculture. This ambitious target highlights the need for innovative research to identify alternative solutions to chemical pesticides. Over the past ten years, most of the research has focused on the use of beneficial microorganisms such as fungi and bacteria [8–10], natural substances [11,12], resistant varieties [13,14], RNAi gene silencing that targets specific pathogens [15,16] and organic cultivation systems [17].

2. Special Issue Contents

The Special Issue on "Sustainable Control Strategies of Plant Pathogens in Horticulture" features research articles and two reviews. These contributions present potential alternatives to synthetic pesticides, showing innovative results related to the use of natural substances, beneficial microorganisms and resistant varieties in the management of several pathogens affecting plants.

The majority of the articles deal with the antifungal activity of natural substances derived from plants (Table 1), such as essential oils from thyme, common juniper and hyssop (1); giant reed extract (*Arundo donax*) (2) against *Alternaria* spp. and *Phytium ultimum*; and extracts from Argentinian plant species against *Penicillium* spp. and *Geotrichum citri-aurantii* (3). On the same topic, two articles demonstrate the in vitro antifungal activity of saponins from *Medicago* species; oat grains and homogenates from sprouts of Brassica species against *Verticillium dahliae* (4); and the *in planta* disease reduction of *Rhizoctonia solani*, *P. ultimum* and *Fusarium oxysporum* following tomato seed treatment with water-soluble polysaccharides from *Jania adhaerens* (5). Three more articles focus on the potential of by-products and bio-composts to reduce disease symptoms and increase plant resistance to pathogens. An alkaline residue from *Gelidium sesquipedale* agar production elicits resistance in tomato and reduces *Plasmopara viticola* symptoms in the vineyard (6), and the guava wood vinegar by-product of charcoal production effectively reduces potato black dot disease by *Colletotrichum coccodes* (7). Several bio-composts from aromatic plant residues controlled damping-off by *R. solani* and *Sclerotinia sclerotiorum* on garden cress (8).

Citation: Righini, H.; Roberti, R.; Galletti, S. Special Issue "Sustainable Control Strategies of Plant Pathogens in Horticulture". *Horticulturae* **2024**, *10*, 146. https://doi.org/10.3390/horticulturae10020146

Received: 4 January 2024
Revised: 29 January 2024
Accepted: 1 February 2024
Published: 4 February 2024

Copyright: © 2024 by the authors. Licensee MDPI, Basel, Switzerland. This article is an open access article distributed under the terms and conditions of the Creative Commons Attribution (CC BY) license (https://creativecommons.org/licenses/by/4.0/).

Table 1. Main results obtained with natural substances in this Special Issue.

Natural Substances	Plant/Pathogen/Method	Activity	Ctrb.
Essential oils of thyme, common juniper, hyssop	Carrot (C), tomato (T) and onion (O) seeds naturally infected by *Alternaria* spp. Agar plate assay amended with essential oils.	Thyme and common juniper: 40–100% long-lasting antifungal activity for C, T and O. Hyssop: no activity for C; 20–60% long-lasting antifungal activity for T and O.	1
Giant reed extract	Zucchini. Plant growth substrate inoculated with *Pythium ultimum* and treated with the extract.	Disease reduction up to 73% and pathogen growth (colony forming units) in the substrate by 90%.	2
Forty extracts from 20 Argentinian plant species	*Penicillium digitatum*, *P. italicum* and *Geotrichum citri-aurantii*. Agar plate diffusion assay.	Inhibition of *G. citri-aurantii* growth of more than 50% by most of the extracts. Inhibition of *P. digitatum* and *P. talicum* by some extracts.	3
Saponins from *Medicago* species and oat grains and homogenates from sprouts of *Brassica* species	*Verticillium dahliae*. Agar plate assay amended with saponins and homogenates. Maize and tomato seeds treated and sown on filter paper.	Reduction in mycelium growth and conidium formation. No phytotoxic effect on seed germination.	4
Water-soluble polysaccharides from *Jania adhaerens*	Tomato. Seeds treated with polysaccharides. Plant growth substrate inoculated with *Rhizoctonia solani* and *P. ultimum* before seeding or with *Fusarium oxysporum* before transplant.	Disease reduction of *R. solani*, *P. ultimum* and *F. oxysporum* up to 58%, 53% and 29%, respectively. Increase in seedling emergence and plant development. Up-regulation of HQT, HCT, PR1 PAL and PR2 genes. Increase in β-1,3-glucanase activity.	5
Gelidium sesquipedale by-product (alkaline residue)	Tomato. Greenhouse experiments. Grapevine. *Plasmopara viticola* in field trials.	Increase in peroxidase and PAL activities and up-regulation of PR9 genes in tomato plants. Reduction in downy mildew symptoms in grapevine.	6
Guava wood vinegar by-product of charcoal production	Potato. *Colletotrichum coccodes*. Agar plate assay amended with the by-product. Pot experiments—stem/soil inoculation with the pathogen.	Inhibition of pathogen mycelial growth. Black dot disease reduction by an average of 23% (stem colonization), 20% (roots covered with sclerotia) and 30% (wilted plants) in the two seasons of experiments.	7
Bio-composts from aromatic plant residues	Garden cress. *R. solani*, *Sclerotinia sclerotiorum*.	Reduction in *S. sclerotiorum* damping-off by all of the raw composts. Reduction in *R. solani* damping-off by 7 composts. Overall, 2 composts showed suppression levels up to 60%.	8

Three articles deal with beneficial microorganisms to control fungal and viral diseases (Table 2). Among these, *Bacillus amyloliquefaciens* as a soil amendment and *B. subltilis* culture filtrate that was sprayed on leaves were effective in reducing apple replant disease (9) and TMV accumulation in tomato (10). One contribution showed the mycoparasitic activity of *Trichoderma* species against *Fusarium solani*, the compatibility of *T. asperellum* with captan and mancozeb and the incompatibility of all *Trichoderma* species with chlorothalonil in vitro (11).

Table 2. Main results obtained with beneficial microorganisms and resistant varieties in this Special Issue.

	Plant/Pathogen/Method	Activity	Ctrb.
Microorganisms			
Bacillus amyloliquefaciens QSB-6	Apple replant disease. Soil amendment. Field conditions.	Increase in plant growth parameters (i.e., plant height), soil bacteria population (i.e., *Actinomycetes*) and soil enzymatic activity. Reduction in soil phenolic acid content and *Fusarium* spp. population.	9
Bacillus subtilis HA1 culture filtrate	Tomato. TMV. Foliar treatment. Pot experiments.	Increase in plant growth (root and shoot parameters). Increase in total phenolic and flavonoid content up to 27 and 50%, respectively, and in the activity of ROS-scavenging enzymes. Reduction in TMV accumulation up to 91%. Up-regulation of PR1, PAL, CHS and HQT genes.	10
Trichoderma asperellum *T. hamatum* *T. harzianum* *T. koningiopsis*	*Fusarium solani*. Dual plate assay on agar medium not amended or amended with the fungicides captan, chlorothalonil and mancozeb.	*Trichoderma* species inhibited *F. solani* up to 67%. High compatibility of *T. asperellum* with captan and mancozeb. No compatibility of *Trichoderma* species with chlorothalonil.	11
Resistant varieties			
Fourteen Tunisian melon landraces	*Podosphaera xanthii*, 3 races (2, 3.5 and 5). Artificial infection in a growth chamber. Natural infection in a greenhouse.	Susceptibility of all landraces to the 3.5 and 5 races and resistance of several landraces to race 2, in the growth chamber. The resistance of three landraces to *P. xanthii* race 2 was confirmed under natural conditions.	12

Table 2 also reports an article on the possibility of using melon landraces to counteract powdery mildew caused by three races of *Podosphaera xanthii* (2, 3.5 and 5) under both artificial and natural infection, showing that the resistance of the three melon landraces to race 2 of the pathogen was confirmed under natural conditions (12).

The topic of this Special issue is complemented by two reviews (13, 14). One review (13) focuses on sustainable options for the management of diseases in horticulture, such as the use of biocontrol agents, natural products, forecasting models, precision farming, nanotechnology, endotherapy, systemic resistance inducers and gene silencing. The second review (14) deals with the use of basic substances against several seed-borne pathogens, fungi, oomycetes, phytoplasma, bacteria and viruses. The basic substances are active, non-toxic substances which fulfil the criteria of a "foodstuff" as defined in Article 2 of Regulation (EC) No 178/2002 [18]. For their use as plant protection products, basic substances are regulated in the EU according to criteria presented in Article 23 of Regulation (EC) No 1107/2009 [1]. The basic substances examined in this review (14) are those already approved in Europe and some of those that are still under evaluation.

3. Conclusions

This Special Issue comprises articles that aimed to identify alternative, sustainable and effective strategies for the management of important plant diseases. The extensive use of synthetic fungicides in managing plant diseases has led to the development of resistance in fungi and oomycetes. Additionally, in recent years, there has been growing consumer demand for food devoid of residues and produced in an environmentally friendly manner. Following this trend, the potential strategies outlined in this Special Issue offer viable alternatives for adoption in large-scale trials.

All the articles contribute valuable insights that enhance understanding in these research fields and have the potential to facilitate future sustainable practical solutions for plant disease management.

The Guest Editors express their gratitude to all authors for delivering interesting research findings in this Special Issue and for sharing their expertise. Furthermore, they extend appreciation to the Journal "*Horticulturae*" for its helpful support, enabling the realization of this Special Issue.

Author Contributions: Writing—Original Draft Preparation, Writing—Review and Editing, Visualization, Supervision, H.R., R.R. and S.G. All authors have read and agreed to the published version of the manuscript.

Funding: This research received no external funding.

Conflicts of Interest: The authors declare no conflicts of interest.

List of Contributions

1. Chrapačienė, S.; Rasiukevičiūtė, N.; Valiuškaitė, A. Control of Seed-Borne Fungi by Selected Essential Oils. *Horticulturae* **2022**, *8*, 220. https://doi.org/10.3390/horticulturae8030220.
2. Galletti, S.; Cianchetta, S.; Righini, H.; Roberti, R. A Lignin-Rich Extract of Giant Reed (*Arundo donax* L.) as a Possible Tool to Manage Soilborne Pathogens in Horticulture: A Preliminary Study on a Model Pathosystem. *Horticulturae* **2022**, *8*, 589. https://doi.org/10.3390/horticulturae8070589.
3. Alvarez, N.H.; Stegmayer, M.I.; Seimandi, G.M.; Pensiero, J.F.; Zabala, J.M.; Favaro, M.A.; Derita, M.G. Natural Products Obtained from Argentinean Native Plants Are Fungicidal against Citrus Postharvest Diseases. *Horticulturae* **2023**, *9*, 562. https://doi.org/10.3390/horticulturae9050562.
4. Morcia, C.; Piazza, I.; Ghizzoni, R.; Delbono, S.; Felici, B.; Baima, S.; Scossa, F.; Biazzi, E.; Tava, A.; Terzi, V.; Finocchiaro, F. In Search of Antifungals from the Plant World: The Potential of Saponins and *Brassica* Species against *Verticillium dahliae* Kleb. *Horticulturae* **2022**, *8*, 729. https://doi.org/10.3390/horticulturae8080729.
5. Righini, H.; Roberti, R.; Cetrullo, S.; Flamigni, F.; Quintana, A.M.; Francioso, O.; Panichi, V.; Cianchetta, S.; Galletti, S. *Jania adhaerens* Primes Tomato Seed against Soil-Borne Pathogens. *Horticulturae* **2022**, *8*, 746. https://doi.org/10.3390/horticulturae8080746.
6. Koçi, R.; Dupuy, F.; Lebbar, S.; Gloaguen, V.; Faugeron Girard, C. A New Promising Plant Defense Stimulator Derived from a By-Product of Agar Extraction from *Gelidium sesquipedale*. *Horticulturae* **2022**, *8*, 958. https://doi.org/10.3390/horticulturae8100958.
7. El-Fawy, M.M.; Abo-Elyousr, K.A.M.; Sallam, N.M.A.; El-Sharkawy, R.M.I.; Ibrahim, Y.E. Fungicidal Effect of Guava Wood Vinegar against *Colletotrichum coccodes* Causing Black Dot Disease of Potatoes. *Horticulturae* **2023**, *9*, 710. https://doi.org/10.3390/horticulturae9060710.
8. Pane, C.; Spaccini, R.; Caputo, M.; De Falco, E.; Zaccardelli, M. Multi-Parameter Characterization of Disease-Suppressive Bio-composts from Aromatic Plant Residues Evaluated for Garden Cress (*Lepidium sativum* L.) Cultivation. *Horticulturae* **2022**, *8*, 632. https://doi.org/10.3390/horticulturae8070632.
9. Duan, Y.; Zhou, Y.; Li, Z.; Chen, X.; Yin, C.; Mao, Z. Effects of *Bacillus amyloliquefaciens* QSB-6 on the Growth of Replanted Apple Trees and the Soil Microbial Environment. *Horticulturae* **2022**, *8*, 83. https://doi.org/10.3390/horticulturae8010083.
10. El-Gendi, H.; Al-Askar, A.A.; Király, L.; Samy, M.A.; Moawad, H.; Abdelkhalek, A. Foliar Applications of *Bacillus subtilis* HA1 Culture Filtrate Enhance Tomato Growth and Induce Systemic Resistance against *Tobacco mosaic virus* Infection. *Horticulturae* **2022**, *8*, 301. https://doi.org/10.3390/horticulturae8040301.
11. Parraguirre Lezama, C.; Romero-Arenas, O.; Valencia de Ita, M.D.L.A.; Rivera, A.; Sangerman Jarquín, D.M.; Huerta-Lara, M. In Vitro Study of the Compatibility of Four Species of *Trichoderma* with Three Fungicides and Their Antagonistic Activity against *Fusarium solani*. *Horticulturae* **2023**, *9*, 905. https://doi.org/10.3390/horticulturae9080905.
12. Chikh-Rouhou, H.; Garcés-Claver, A.; Kienbaum, L.; Ben Belgacem, A.; Gómez-Guillamón, M.L. Resistance of Tunisian Melon Landraces to *Podosphaera xanthii*. *Horticulturae* **2022**, *8*, 1172. https://doi.org/10.3390/horticulturae8121172
13. Scortichini, M. Sustainable Management of Diseases in Horticulture: Conventional and New Options. *Horticulturae* **2022**, *8*, 517. https://doi.org/10.3390/horticulturae8060517.
14. Orzali, L.; Allagui, M.B.; Chaves-Lopez, C.; Molina-Hernandez, J.B.; Moumni, M.; Mezzalama, M.; Romanazzi, G. Basic Substances and Potential Basic Substances: Key Compounds for a Sustainable Management of Seedborne Pathogens. *Horticulturae* **2023**, *9*, 1220. https://doi.org/10.3390/horticulturae9111220.

References

1. Regulation (EC) No 1107/2009 of the European Parliament and of the Council of 21 October 2009 Concerning the Placing of Plant Protection Products on the Market and Repealing Council Directives 79/117/EEC and 91/414/EEC. Available online: https://eur-lex.europa.eu/eli/reg/2009/1107/oj (accessed on 1 January 2024).
2. Council Directive 79/117/EEC of 21 December 1978 Prohibiting the Placing on the Market and Use of Plant Protection Products Containing Certain Active Substances. Available online: https://eur-lex.europa.eu/legal-content/EN/TXT/?uri=CELEX:31979L0117 (accessed on 1 January 2024).
3. Council Directive 91/414/EEC of 15 July 1991 Concerning the Placing of Plant Protection Products on the Market. Available online: https://eur-lex.europa.eu/legal-content/EN/ALL/?uri=CELEX:31991L0414 (accessed on 1 January 2024).
4. Geiger, F.; Bengtsson, J.; Berendse, F.; Weisser, W.W.; Emmerson, M.; Morales, M.B.; Ceryngier, P.; Liira, J.; Tscharntke, T.; Winqvist, C.; et al. Persistent negative effects of pesticides on biodiversity and biological control potential on European farmland. *Basic Appl. Ecol.* **2010**, *11*, 97–105. [CrossRef]
5. Pathak, V.M.; Verma, V.K.; Rawat, B.S.; Kaur, B.; Babu, N.; Sharma, A.; Dewali, S.; Yadav, M.; Kumari, R.; Singh, S.; et al. Current status of pesticide effects on environment, human health and it's eco-friendly management as bioremediation: A comprehensive review. *Front. Microbiol.* **2022**, *17*, 962619. [CrossRef] [PubMed]
6. Guyomard, H.; Soler, L.G.; Détang-Dessendre, C.; Réquillart, V. The European Green Deal improves the sustainability of food systems but has uneven economic impacts on consumers and farmers. *Commun. Earth Environ.* **2023**, *4*, 358. [CrossRef]
7. Montanarella, L.; Panagos, P. The relevance of sustainable soil management within the European Green Deal. *Land Use Policy* **2021**, *100*, 104950. [CrossRef]
8. Vieira, M.E.O.; Nunes, V.V.; Calazans, C.C.; Silva-Mann, R. Unlocking plant defenses: Harnessing the power of beneficial microorganisms for induced systemic resistance in vegetables—A systematic review. *Biol. Control* **2023**, *188*, 105428.
9. Pastor, N.; Palacios, S.; Torres, A.M. Microbial consortia containing fungal biocontrol agents, with emphasis on *Trichoderma* spp.: Current applications for plant protection and effects on soil microbial communities. *Eur. J. Plant Pathol.* **2023**, *167*, 593–620. [CrossRef]
10. Ünlü, E.; Çalış, Ö.; Say, A.; Karim, A.A.; Yetişir, H.; Yılmaz, S. Investigation of the effects of *Bacillus subtilis* and *Bacillus thuringiensis* as bio-agents against powdery mildew (*Podosphaera xanthii*) disease in zucchini (*Cucurbita pepo* L.). *Microb. Pathog.* **2023**, *185*, 106430. [CrossRef] [PubMed]
11. Righini, H.; Somma, A.; Cetrullo, S.; D'Adamo, S.; Flamigni, F.; Quintana, A.M.; Roberti, R. Inhibitory activity of aqueous extracts from *Anabaena minutissima*, *Ecklonia maxima* and *Jania adhaerens* on the cucumber powdery mildew pathogen in vitro and in vivo. *J. Appl. Phycol.* **2020**, *32*, 3363–3375. [CrossRef]
12. Davari, M.; Ezazi, R. Mycelial inhibitory effects of antagonistic fungi, plant essential oils and propolis against five phytopathogenic *Fusarium* species. *Arch. Microbiol.* **2022**, *204*, 480. [CrossRef] [PubMed]
13. Zhang, Y.; Wang, J.; Xiao, Y.; Jiang, C.; Cheng, L.; Guo, S.; Luo, C.; Wang, Y.; Jia, H. Proteomics analysis of a tobacco variety resistant to brown spot disease and functional characterization of NbMLP423 in *Nicotiana benthamiana*. *Mol. Biol. Rep.* **2023**, *50*, 4395–4409. [CrossRef] [PubMed]
14. Mbinda, W.; Masaki, H. Breeding strategies and challenges in the improvement of blast disease resistance in finger millet. A current review. *Front. Plant Sci.* **2021**, *11*, 602882. [CrossRef] [PubMed]
15. Biedenkopf, D.; Will, T.; Knauer, T.; Jelonek, L.; Furch, A.C.U.; Busche, T.; Koch, A. Systemic spreading of exogenous applied RNA biopesticides in the crop plant *Hordeum vulgare*. *ExRNA* **2020**, *2*, 12. [CrossRef]
16. Haile, Z.M.; Gebremichael, D.E.; Capriotti, L.; Molesini, B.; Negrini, F.; Collina, M.; Sabbadini, S.; Mezzetti, B.; Baraldi, E. Double-stranded RNA targeting dicer-like genes compromises the pathogenicity of *Plasmopara viticola* on grapevine. *Front. Plant Sci.* **2021**, *12*, 667539. [CrossRef] [PubMed]
17. Van Bruggen, A.H.C.; Finckh, M.R. Plant diseases and management approaches in organic farming systems. *Annu. Rev. Phytopathol.* **2016**, *54*, 25–54. [CrossRef] [PubMed]
18. Regulation (EC) No 178/2002 of the European Parliament and of the Council of 28 January 2002 Laying Down the General Principles and Requirements of Food Law, Establishing the European Food Safety Authority and Laying Down Procedures in Matters of Food Safety. Available online: https://www.legislation.gov.uk/eur/2002/178/contents (accessed on 1 January 2024).

Disclaimer/Publisher's Note: The statements, opinions and data contained in all publications are solely those of the individual author(s) and contributor(s) and not of MDPI and/or the editor(s). MDPI and/or the editor(s) disclaim responsibility for any injury to people or property resulting from any ideas, methods, instructions or products referred to in the content.

Article

In Vitro Study of the Compatibility of Four Species of *Trichoderma* with Three Fungicides and Their Antagonistic Activity against *Fusarium solani*

Conrado Parraguirre Lezama [1], Omar Romero-Arenas [1,*], Maria De Los Angeles Valencia de Ita [1], Antonio Rivera [2], Dora M. Sangerman Jarquín [3] and Manuel Huerta-Lara [4,*]

[1] Centro de Agroecología, Instituto de Ciencias, Benemérita Universidad Autónoma de Puebla, Edificio VAL 1, Km 1.7, Carretera a San Baltazar Tetela, San Pedro Zacachimalpa 72960, Puebla, Mexico; conrado.parraguirre@correo.buap.mx (C.P.L.); mavi1179@outlook.es (M.D.L.A.V.d.I.)
[2] Centro de Investigaciones en Ciencias Microbiológicas, Instituto de Ciencias, Benemérita Universidad Autónoma de Puebla, Ciudad Universitaria, Puebla 72570, Puebla, Mexico; jart70@yahoo.com
[3] Campo Experimental Valle de México-INIFAP, Carretera Los Reyes-Texcoco Km 13.5, Coatlinchán 56250, Texcoco, Mexico; sangerman.dora@inifap.gob.mx
[4] Departamento de Desarrollo Sustentable, Instituto de Ciencias, Benemérita Universidad Autónoma de Puebla, Ciudad Universitaria, Puebla 72570, Puebla, Mexico
* Correspondence: biol.ora@hotmail.com (O.R.-A.); batprofessor@hotmail.com (M.H.-L.)

Abstract: Strawberry wilt is a disease caused by *Fusarium solani*, which it provokes the death of the plant. Farmers mainly use chemical methods for its control, which has a negative impact on the environment and human health. Given the growing demand for organic agricultural products, compatible alternatives must be sought for disease management that can reduce the doses of fungicides. A combination of pesticides and biological control agents could be an alternative for the management of *F. solani*. Consequently, investigations on fungicide compatibility and synergistic effects are recommended in relation to the biological control of strawberry wilt. In this study, potential antagonism was calculated according to the class of mycoparasitism and the percentage inhibition of radial growth in order to later design a compatibility model of the different species of *Trichoderma* with three protective fungicides at different concentrations. The potential antagonism showed that *Trichoderma asperellum* presented high compatibility with the fungicides Captan and Mancozeb added in concentrations of 450, 900, and 1350 mg L^{-1}. The use of antagonistic strains together with the fungicide Chlorothalonil in its three concentrations showed a negative effect on the growth of *Trichoderma* species, which caused low and null compatibility against the MA-FC120 strain of *F. solani* in vitro.

Keywords: potential antagonism; *F. solani*; compatibility; strawberry wilt; antagonist agents

Citation: Parraguirre Lezama, C.; Romero-Arenas, O.; Valencia de Ita, M.D.L.A.; Rivera, A.; Sangerman Jarquín, D.M.; Huerta-Lara, M. In Vitro Study of the Compatibility of Four Species of *Trichoderma* with Three Fungicides and Their Antagonistic Activity against *Fusarium solani*. *Horticulturae* **2023**, *9*, 905. https://doi.org/10.3390/horticulturae9080905

Academic Editors: Hillary Righini, Stefania Galletti, Roberta Roberti and Boqiang Li

Received: 15 June 2023
Revised: 1 August 2023
Accepted: 2 August 2023
Published: 8 August 2023

Copyright: © 2023 by the authors. Licensee MDPI, Basel, Switzerland. This article is an open access article distributed under the terms and conditions of the Creative Commons Attribution (CC BY) license (https://creativecommons.org/licenses/by/4.0/).

1. Introduction

Crown and root rot in strawberry cultivation is the most important and destructive disease of the crop worldwide [1]. The disease limits plant growth and fruit production, causing large economic losses and low yields [2,3], especially when heat and humidity stress is induced [4].

Strawberry wilt is a disease induced by *Fusarium solani*, which produces stunted plant growth, followed by wilting, and finally plant death [5–7]. Lately, in Spain, it has been reported in strawberry crops, both in nurseries and production fields [8]. Likewise, it has been described as a strawberry pathogen in Italy [9], Iran, and Pakistan [10]. However, in Mexico there have been few reports [11].

To reduce losses caused by *F. solani* in strawberry production, farmers use chemical control methods, which have become an essential part of agriculture [12,13]. Consequently, chemical fungicides are still used recklessly as the primary means of disease control throughout the world. Isolates of *F. solani* species have recently been reported to show reduced susceptibility to azoles in agriculture, an attention-grabbing fact [14,15]. In addition, the intensive use of fungicides has an adverse effect on human health and a negative impact on the environment [16,17].

There is currently awareness and concern among fruit and vegetable consumers regarding pesticide residues in food, with an emphasis on minimizing their use and application without increasing risks in agricultural production [18]. It is crucial to intensify efforts to develop alternative farm management practices that can reduce the use of chemicals to control *F. solani* in a sustainable and environmentally friendly manner. A combination of pesticides and biological control agents (BCAs) could be an effective tool and improve the fight against plant pathogens in a more reliable way [19]. Likewise, environmental impacts could be reduced by using low concentrations of fungicides, which would help to minimize the danger associated with the use of chemical pesticides [20–22].

The possibility of generating a synergistic effect with the application of fungicides and biological antagonists, such as *Trichoderma* spp. has been reported by several authors [23]. Terrero et al. [24] reported the compatibility of *Trichoderma* species with azoystrobin and copper hydroxide fungicides, and Ruano et al. [25] implemented the use of *Trichoderma* spp. and the fungicide fluazinam (in concentrations of 0.01 and 0.05 mg L^{-1}) for the control of *Rosellinia necatrix* in avocado plants. In a study conducted by Wang et al. [26], a greater control of root rot was reported as well as elevated survival of coneflower seedlings (*Echinacea* spp.) with the combination of *Trichoderma* and fludioxonil against *Fusarium* sp. in greenhouses. However, the control of wilting of strawberry crops remains a challenge. Therefore, the objective of this investigation was to evaluate the in vitro compatibility of four species of *Trichoderma* with three broad-spectrum fungicides through potential antagonism (PA) against the MA-FC120 strain of *F. solani*.

2. Materials and Methods

2.1. Biological Material

The strains used were T-H4 of *Trichoderma harzianum*, T-K11 of *Trichoderma koningiopsis*, T-AS1 of *Trichoderma asperellum*, and T-A12 of *Trichoderma hamatum*, isolated from the root of *Persea americana*; the sequences of these strains were deposited in the National Center for Biological Information (NCBI) database with accession numbers MK779064, MK791648, MK778890, and MK791650, respectively. The MA-FC120 strain of *F. solani* was used, which is pathogenic for strawberry cultivation [3], with accession number OM616884, characterized by the Phytopathology Laboratory 204 of the Agroecology Center of the Institute of Sciences, BUAP.

2.2. Characterization of the Rate of Development and Growth

For the evaluation of development rate, 10-day-old fragments of 5 mm in diameter with active growth of *F. solani* (MA-FC120), T-H4 of *T. harzianum*, T-K11 of *T. koningiopsis*, T-AS1 of *T. asperellum*, and T-A12 of *T. hamatum* were seeded individually in Petri dishes (9 cm in diameter) containing 20 mL of PDA (Bioxon, Becton Dickinson, Cdad., Mexico, Mexico) and incubated at 28 °C for 10 days in the dark. Every 12 h, the diameter of the mycelium was measured with a digital vernier (CD-6 Mitutoyo, Naucalpan de Juarez, Mexico) to assess the growth rate (cm d^{-1}), which was calculated with the linear growth function [27] as per Equation (1):

$$y = mx + b \tag{1}$$

where

y = Distance;
m = Slope;
x = Time;
b = Constant factor.

The experiment was repeated in duplicate in a completely randomized statistical design, with three replicates for each treatment.

2.3. Dual Test on Poisoned Culture Medium In Vitro

The controlled poisoning technique was performed [28] on potato dextrose agar (PDA, Bioxon, Becton Dickinson and Company, Querétaro, Mexico) using three protective fungicides (Chlorothalonil, Mancozeb, and Captan) at different concentrations (Table 1). To determine the percentage inhibition of radial growth (PIRG), we used PDA discs (5 mm in diameter) with the mycelium of *Trichoderma* spp. and *F. solani*. They were placed at the ends of the poisoned Petri dishes with 7.5 cm between them (antagonist–phytopathogen). All plates were sealed with Parafilm® and incubated in the dark at 28 °C for 10 days.

Table 1. Fungicides used at different concentrations evaluated.

Fungicide (Tradename)	Active Ingredient	Molecular Formula	Concentration (mg L^{-1})		
			Low	Recommended	High
Control	Water	H_2O	-	-	-
Captan 50®	Captan	$C_9H_8Cl_3NO_2S$			
Mancosol 80®	Mancozeb	$C_4H_6MnN_2S_4$	450	900	1350
Talonil 75®	Chlorothalonil	$C_8Cl_4N_2$			

Radial growth of the fungal colony was evaluated every 12 h until the first contact between the mycelia of each antagonist with *F. solani* occurred. The percentage inhibition of radial growth (PIRG) was calculated based on the formula of Equation (2) [29]:

$$PIRG\% = (R1 - R2)/R1 \times 100 \qquad (2)$$

where

PIRG = Percentage inhibition of radial growth;
R1 = Radial growth (mm) of *F. solani* without *Trichoderma* spp.;
R2 = Radial growth (mm) of *F. solani* with *Trichoderma* spp.

Mycoparasitism ability due to the invasion of the antagonist or colonization on the mycelial surface of *F. solani* was evaluated according to the scale proposed by Bell et al. [30] (Table 2).

Table 2. Mycoparasitism class using Bell scale [30] in vitro.

Class	Mycoparasitism (%)	Characteristics
I	100	*Trichoderma* grew completely over *F. solani* and covered the entire mid-surface.
II	75	*Trichoderma* grew over at least two-thirds of the mid-surface.
III	50	*Trichoderma* and *F. solani* colonized approximately half of the mid-surface.
IV	25	*F. solani* colonized at least two-thirds of the mid-surface.
V	0	*F. solani* grew completely over *Trichoderma* and occupied the entire mid-surface.

The experiment was repeated in duplicate in a completely randomized statistical design, with four replicates for each treatment.

2.4. Potential Antagonism

Potential antagonism (PA) was calculated by averaging the results of the mycoparasitism class (Bell) and percentage inhibition of the radial growth rate (PIRG), as proposed by Reyes-Figueroa et al. [31] (3):

$$PA\% = (Bell + PIRG)/2 \qquad (3)$$

where

PA = Potential antagonism expressed as a percentage;

Bell = *Trichoderma* mycoparasitism against *F. solani*;
PIRG = Inhibition of radial growth of *F. solani*.

The design of the model of compatibility (C) of *Trichoderma* with different fungicides included the identification of the highest and lowest potential antagonism (PA). Then, the difference between them was calculated and divided by four [32]. Finally, the quotient was added to the lower performance progressively until the formation of four groups was achieved (Table 3).

Table 3. Compatibility of *Trichoderma* strains under different protective fungicides.

Grade	Ranges (%)	Compatibility
I	23 to 42	Null
II	43 to 59	Low
III	60 to 75	Medium
IV	76 to 93	High

2.5. Statistical Analysis

Data were analyzed using ANOVA (two-way) in the statistical package SPSS Statistics version 17 for Windows. Growth rate and development rate were response variables. Data were subjected to Bartlett's test of homogeneity, and subsequently a Tukey–Kramer comparison of means test was performed with a probability level of $p \leq 0.05$.

The mycoparasitism class (Bell), percentage inhibition of radial growth (PIRG), and potential antagonism (PA) of the strains were expressed in percentages and transformed with angular arccosine $\sqrt{x+1}$. Subsequently, the variable compatibility (C) was analyzed under the analysis of variance (multivariate ANOVA) using a quadratic response model to determine significant differences between treatments, under the following mathematical model (4):

$$C\gamma j = \mu + t_i + \varepsilon_{ij} \quad (4)$$

where

$C\gamma j$ = Value of the response variable of the experimental unit associated with the γ-th treatment and the j-th repetition;
μ = Corresponds to the overall mean of the response variable in the experiment;
t_i = Effect of the γ-th treatment;
$\varepsilon\gamma j$ = Error of the experimental unit associated with the γ-th treatment;
j = The j-th repetition;
$\gamma = 1, 2, 3, 4, \ldots, 40$;
j = 1, 2, 3, 4;
C = Compatibility (C%).

Finally, a Tukey–Kramer mean comparison test was performed with a probability level of $p \leq 0.05$.

3. Results

The rate of development and growth had highly significant differences ($p \leq 0.05$); *T. koningiopsis* (T-K11) obtained the highest value with 1.17 ± 0.02 mm/h^{-1} and 26.82 ± 0.6 cm d^{-1}, respectively (Table 4). *F. solani* showed the lowest growth rate (6.71 ± 0.08 cm d^{-1}).

The percentage inhibition of radial growth (PIRG) in the control group did not show significant differences ($p = 0.087$). Confrontation of the different *Trichoderma* species against *F. solani* showed an inhibition between 60 and 70% from the tenth day (Table 5). However, the highest percentage inhibition was obtained with *T. harzianum* (DTHR0), achieving 67.31%. Similarly, *T. hamatum* (DTH0) presented the second-best inhibition with 66.95%; both antagonistic strains presented a class II classification (Figure 1) on the scale established by Bell et al. [30]. *T. konigiopsis* presented the lowest percentage inhibition of radial growth of 64.34%.

Table 4. Development and growth rate of 10-day-old colony of *Trichoderma* strains and *F. solani* on PDA medium.

Code	Species	Strains	* Development Rate (mm h^{-1})	* Growth Rate (cm d^{-1})
DTF0	*F. solani*	MA-FC120	0.28 ± 0.08 [e]	6.71 ± 0.08 [e]
DTH0	*T. hamatum*	T-A12	1.02 ± 0.02 [d]	26.19 ± 0.53 [b]
DTA0	*T. asperellum*	T-AS1	1.07 ± 0.05 [c]	24.92 ± 1.34 [d]
DTK0	*T. konigiopsis*	T-K11	1.17 ± 0.02 [a]	26.82 ± 0.6 [a]
DTHR0	*T. harzianum*	T-H4	1.14 ± 0.5 [b]	25.20 ± 0.02 [c]

* Means with different letters indicate statistically significant differences by ANOVA–Tukey test ($p < 0.05$).

Table 5. Percentage inhibition of radial growth (PIRG) and mycoparasitism on the Bell scale [30].

Code	Fungicide	Concentration (mg L^{-1})	Species	* PIRG (%)	SE	Bell Scale
DTH0	Control (PDA)	-	*T. hamatum*	66.95 [bcde]	± 0.34	II
DTA0			*T. asperellum*	66.91 [bcde]	± 0.26	II
DTK0			*T. konigiopsis*	64.34 [def]	± 0.59	II
DTHR0			*T. harzianum*	67.31 [bcde]	± 1.16	II
DTH1	Captan	450	*T. hamatum*	71.11 [bcde]	± 1.63	II
DTA1			*T. asperellum*	76.33 [a]	± 1.16	I
DTK1			*T. konigiopsis*	67.37 [bcde]	± 1.31	II
DTHR1			*T. harzianum*	74.82 [b]	± 1.01	I
DTH2		900	*T. hamatum*	69.19 [bcde]	± 3.16	II
DTA2			*T. asperellum*	73.68 [bc]	± 1.34	I
DTK2			*T. konigiopsis*	66.30 [bcde]	± 0.32	II
DTHR2			*T. harzianum*	74.58 [b]	± 1.11	I
DTH3		1350	*T. hamatum*	68.37 [bcde]	± 2.19	II
DTA3			*T. asperellum*	72.84 [bcd]	± 1.08	I
DTK3			*T. konigiopsis*	64.84 [cde]	± 2.32	II
DTHR3			*T. harzianum*	72.15 [bcd]	± 2.51	I
DTH4	Mancozeb	450	*T. hamatum*	67.96 [bcde]	± 0.85	II
DTA4			*T. asperellum*	68.56 [bcde]	± 1.40	I
DTK4			*T. konigiopsis*	64.76 [cde]	± 1.09	II
DTHR4			*T. harzianum*	69.60 [bcde]	± 1.27	II
DTH5		900	*T. hamatum*	65.88 [bcde]	± 1.00	II
DTA5			*T. asperellum*	67.93 [bcde]	± 1.14	I
DTK5			*T. konigiopsis*	62.82 [efg]	± 0.82	II
DTHR5			*T. harzianum*	68.21 [bcde]	± 1.80	II
DTH6		1350	*T. hamatum*	65.72 [bcde]	± 0.82	II
DTA6			*T. asperellum*	66.16 [bcde]	± 0.91	I
DTK6			*T. konigiopsis*	62.47 [efg]	± 0.15	II
DTHR6			*T. harzianum*	67.29 [bcde]	± 0.60	II
DTH7	Chlorothalonil	450	*T. hamatum*	49.52 [h]	± 1.22	III
DTA7			*T. asperellum*	28.71 [i]	± 2.24	IV
DTK7			*T. konigiopsis*	55.38 [fgh]	± 1.88	III
DTHR7			*T. harzianum*	46.33 [h]	± 2.36	III
DTH8		900	*T. hamatum*	48.38 [h]	± 0.68	III
DTA8			*T. asperellum*	28.62 [i]	± 2.94	IV
DTK8			*T. konigiopsis*	50.91 [gh]	± 1.96	III
DTHR8			*T. harzianum*	28.15 [i]	± 1.83	IV
DTH9		1350	*T. hamatum*	48.05 [h]	± 3.65	III
DTA9			*T. asperellum*	24.35 [i]	± 0.97	IV
DTK9			*T. konigiopsis*	49.62 [h]	± 1.26	III
DTHR9			*T. harzianum*	26.72 [i]	± 1.19	IV

* Means with different letters indicate statistically significant differences by ANOVA–Tukey test ($p < 0.05$); SE = standard error.

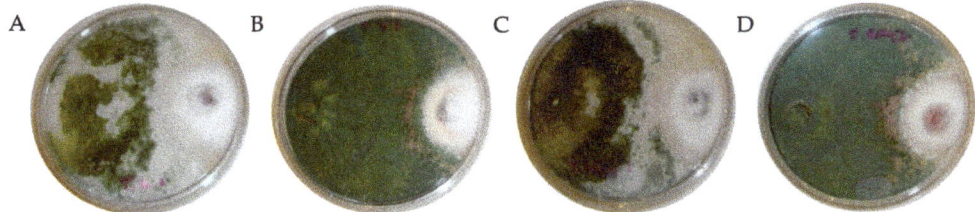

Figure 1. In vitro mycoparasitism of 10-day-old species of *Trichoderma* against *F. solani* in PDA. (**A**) *T. konigiopsis*; (**B**) *T. asperellum*; (**C**) *T. hamatum*; and (**D**) *T. harzianum*.

In general, the class of mycoparasitism found in this study at 10 days was similar in all *Trichoderma* strains. In the present investigation, the *Trichoderma* strains demonstrated a class II classification (Figure 1) according to the scale of Bell et al. [30].

The treatment that obtained the highest percentage inhibition of radial growth (PIRG) in the poisoned culture medium was DTA1 (76.33%), in which the antagonist strain corresponded to *T. asperellum* in potato dextrose agar dishes with 450 mg L^{-1} of Captan fungicide added. This difference was statistically significant with respect to the other treatments (p = 0.007). Similarly, *T. harzianum* (DTHR1) showed the second-best inhibition at 74.82% with the same concentration of Captan fungicide.

T. konigiopsis presented the lowest percentage inhibition of radial growth (PIRG), 64.84%, in the PDA culture medium with 1350 mg L^{-1} of Captan fungicide added (Table 5). The antagonistic strains *T. asperellum* and *T. harzianum* demonstrated a class I classification (Figure 2) according to the scale of Bell et al. [30]. This was dissimilar to *T. konigiopsis* and *T. hamatum* which had a classification of mycoparasitism II.

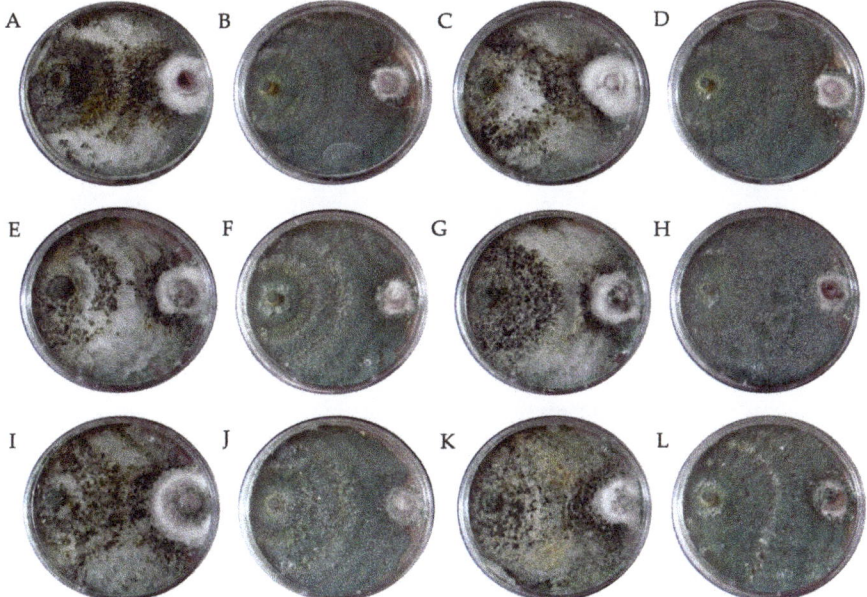

Figure 2. In vitro mycoparasitism of different species of *Trichoderma* against *F. solani* in PDA medium with Captan added at different concentrations. (**A–D**) Captan at 450 mg L^{-1}; (**E–H**) Captan at 900 mg L^{-1}; (**I–L**) Captan at 1350 mg L^{-1}; (**A,E,I**) *T. konigiopsis*; (**B,F,J**) *T. asperellum*; (**C,G,K**) *T. hamatum*; (**D,H,L**) *T. harzianum*.

The behavior of the Mancozeb fungicide showed significant differences (p = 0.002). The highest percentage of radial growth inhibition (PIRG) was obtained in the DTHR4 treatment (69.60%), corresponding to *T. harzianum* in the PDA culture medium with 450 mg L^{-1} of Mancozeb fungicide. Similarly, *T. asperellum* (DTA4) demonstrated the second-best inhibition at 68.56%, classified as class I on the scale established by Bell et al. [30], of the three concentrations used. *T. konigiopsis* presented the lowest percentage inhibition of radial growth (PIRG), between 62.47 and 64.84%, in the potato dextrose agar dishes with 450, 900, and 1350 mg L^{-1} of Mancozeb fungicide (Table 5). The antagonistic strains *T. konigiopsis*, *T. hamatum*, and *T. harzianum* demonstrated a class II classification (Figure 3) according to the scale of Bell et al. [30].

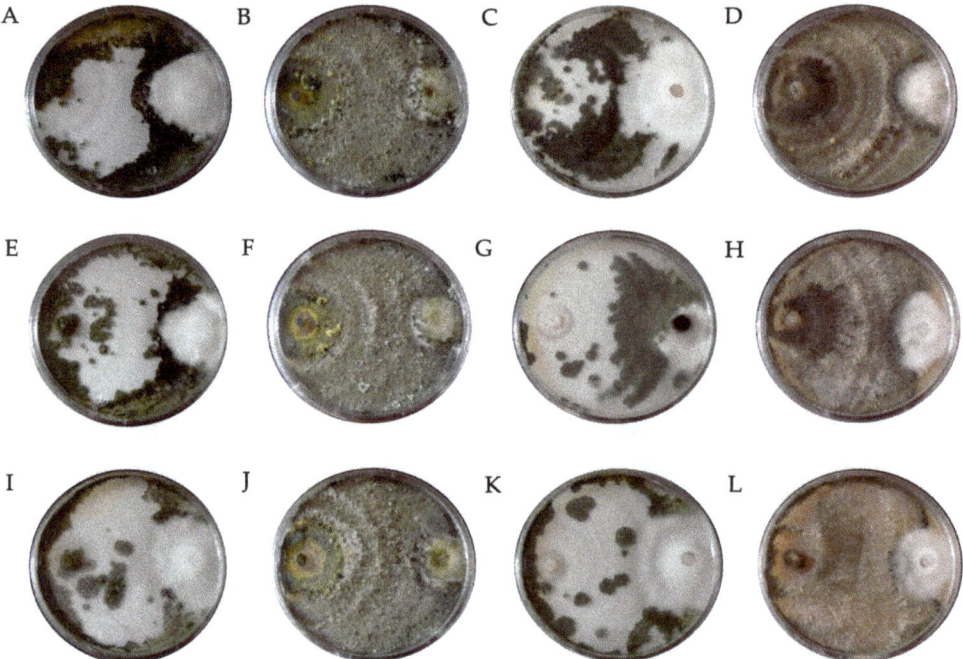

Figure 3. In vitro mycoparasitism of different species of *Trichoderma* against *F. solani* in PDA medium with Mancozeb added at different concentrations. (**A–D**) Mancozeb at 450 mg L^{-1}; (**E–H**) Mancozeb at 900 mg L^{-1}; (**I–L**) Mancozeb at 1350 mg L^{-1}; (**A,E,I**) *T. konigiopsis*; (**B,F,J**) *T. asperellum*; (**C,G,K**) *T. hamatum*; (**D,H,L**) *T. harzianum*.

The treatments in which the Chlorothalonil fungicide was not added presented a higher PIRG; however, we observed that *T. konigiopsis* (DTK8) demonstrated the best inhibition (55.38%), classified as class III on the scale established by Bell et al. [30]. This difference was statistically significant with respect to the other treatments (p = 0.001).

T. hamatum in the treatments with 450, 900, and 1350 mg L^{-1} of the fungicide Chlorothalonil (Table 4) in the PDA culture medium presented the second-best percentage inhibition of radial growth (PIRG). The results obtained were between 48.05 and 49.52%. The strain that presented the least antagonism was *T. asperellum*, reaching a PIRG of 24.35% in the potato dextrose agar dishes with 450 mg L^{-1} of the Chlorothalonil fungicide, classified as grade IV mycoparasitism (Figure 4). This was followed by the *T. harzianum* strain with the same concentration.

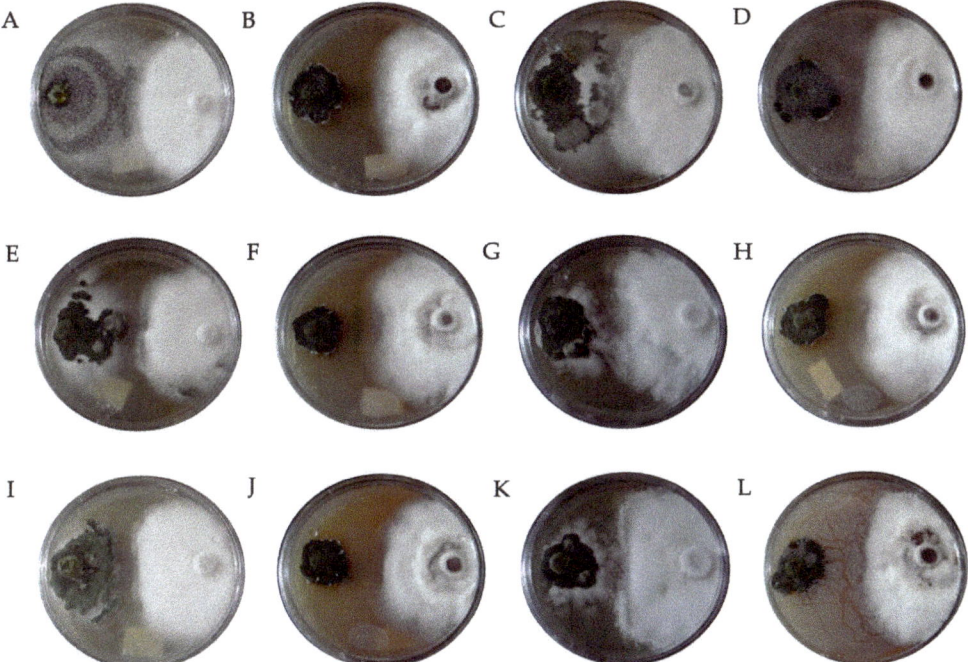

Figure 4. In vitro mycoparasitism of different species of *Trichoderma* against *F. solani* in PDA medium with Chlorothalonil added at different concentrations. (**A–D**) Chlorothalonil at 450 mg L^{-1}; (**E–H**) Chlorothalonil at 900 mg L^{-1}; (**I–L**) Chlorothalonil at 1350 mg L^{-1}; (**A,E,I**) *T. konigiopsis*; (**B,F,J**) *T. asperellum*; (**C,G,K**) *T. hamatum*; (**D,H,L**) *T. harzianum*.

The MA-FC120 strain of *F. solani* presented greater radial growth in the treatments with 450, 900, and 1350 mg L^{-1} of the fungicide Chlorothalonil (Table 4) in the PDA medium in vitro. Likewise, less inhibition exerted by the four *Trichoderma* strains was observed, classified as grade III and IV mycoparasitism on the scale established by Bell [30] (Figure 4). Furthermore, a 70% decrease in the development rate of the four *Trichoderma* strains was observed in comparison with the control group. The *T. asperellum* strain T-AS1 was the most affected of the three concentrations evaluated.

Potential antagonism (PA), in terms of the fungicide and the *Trichoderma* species evaluated in the present study, revealed a biologically important scenario. The strains of *T. asperellum* and *T. harzianum* presented higher levels of PA than *T. konigiopsis* and *T. hamatum* (Figure 5) with the fungicides Captan and Mancozeb at the three concentrations evaluated. We might expect that the known intraspecific compatibility variability of different *Trichoderma* species and phylogenetic affinities towards PA against *F. solani* would not apply to all species tested. However, when we analyzed the data by species and compatibility, we surprisingly found that the Chlorothalonil fungicide at concentrations of 450, 900, and 1350 mg L^{-1} presented a low PA in the PDA culture medium when added to the four species of *Trichoderma*. For the most part, this was below 50%.

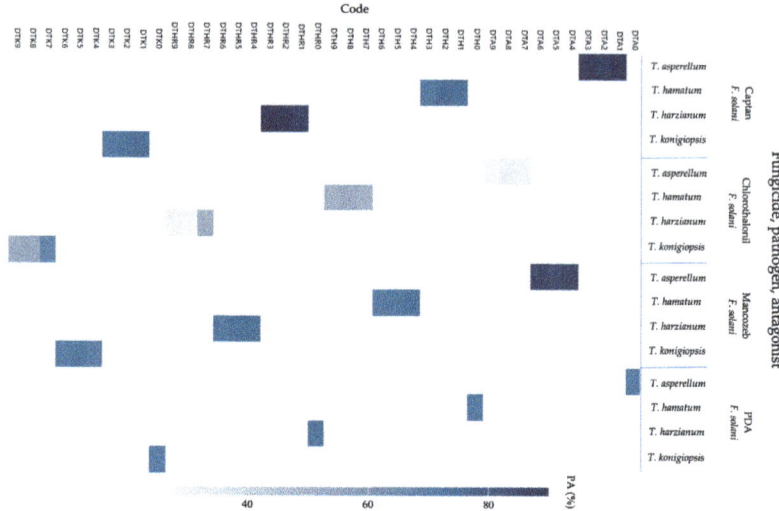

Figure 5. Potential antagonism (PA) of *Trichoderma* species against *F. solani* at different concentrations of fungicides.

In all cases, we observed that treatments with 450, 900, and 1350 mg L^{-1} of the fungicide Chlorothalonil added to the strain of *T. konigiopsis* demonstrated the best antagonism potential (58.94%). Statistically significant differences were observed with respect to the other treatments ($p = 0.001$). *T. hamatum* in the treatments with 450, 900, and 1350 mg L^{-1} of the Chlorothalonil fungicide (Figure 5) showed the second-best PA at 49.19%. The strain that showed the lowest PA was *T. harzianum*, reaching 27.36% in the treatment with 900 mg L^{-1} of the fungicide Chlorothalonil in the PDA medium in vitro.

Table 6 shows the summary of the multivariate ANOVA analysis for the quadratic response surface model, which revealed highly significant statistical differences. The model values of F = 64.55 (PIRG), 157.538 (Bell), and 204.495 (PA) implied a highly significant model for compatibility (C) in vitro.

Table 6. Quadratic model of surface response for compatibility (C).

Origin	Dependent Variable	Sum of Squares (Type III)	gL	Mean Square	F	Sig.
	X_1 = PIRG	27,481.902 a	39	704.664	64.55	<0.001
Corrected mode	X_2 = Bell	64,000.000 b	39	1641.026	157.538	<0.001
	X_3 = PA	41,743.713 c	39	1070.352	204.495	<0.001
	X_1	589,547.5	1	589,547.5	54,004.82	<0.001
Intersection	X_2	702,250	1	702,250	67,416	<0.001
	X_3	644,671.1	1	644,671.1	123,166.8	<0.001
	X_1	27,481.9	39	704.664	64.55	<0.001
Compatibility (C%)	X_2	64,000	39	1641.026	157.538	<0.001
	X_3	41,743.71	39	1070.352	204.495	<0.001
	X_1	1309.989	120	10.917		
Error	X_2	1250	120	10.417		
	X_3	628.095	120	5.234		
	X_1	618,339.4	160			
Total	X_2	767,500	160			
	X_3	687,043	160			
	X_1	28,791.89	159			
Total corrected	X_2	65,250	159			
	X_3	42,371.81	159			

(a) R-squared = 0.960 (R-squared adjusted = 0.947); (b) R-squared = 0.989 (R-squared adjusted = 0.985); (c) R-squared = 0.986 (R-squared adjusted = 0.981).

The compatibility association analysis considering mycoparasitism, PIRG, and potential antagonism (PA) showed that *T. asperellum* presented high compatibility (100%) with the fungicides Captan and Mancozeb added at concentrations of 450, 900, and 1350 mg L^{-1}. Compatibility was low–null (25–50%) with the fungicide Chlorothalonil. *T. hamatum* showed high compatibility (100%) with the fungicide Captan at 450 and 900 mg L^{-1}. Compatibility was medium (75%) with the fungicide Mancozeb in its three concentrations and with Captan at 1350 mg L^{-1}, and low (25%) with the fungicide Chlorothalonil in its three concentrations (Figure 6).

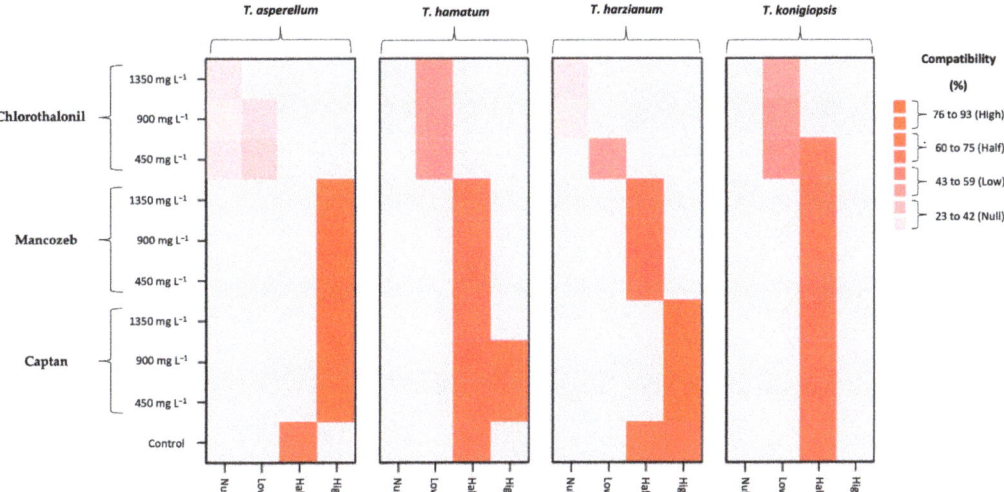

Figure 6. Compatibility analysis of *Trichoderma* strains based on multivariate ANOVA analysis with the data of mycoparasitism (Bell), percentage inhibition of radial growth (PIRG), and potential antagonism (PA). The color intensity represents the degree of compatibility of *Trichoderma* strains under the effects of three fungicides at different concentrations (450, 900, and 1350 mg L^{-1}).

T. harzianum showed high compatibility (100%) with the fungicide Captan at 450, 900, and 1350 mg L^{-1}; medium compatibility (75%) with the fungicide Mancozeb at 450, 900, and 1350 mg L^{-1}, and low compatibility (50%) with the fungicide Chlorothalonil at 450 mg L^{-1}. Compatibility was null (25%) at the 900 and 1350 mg L^{-1} concentrations of Chlorothalonil (Figure 6). Finally, *T. konigiopsis* showed medium compatibility (75%) with Captan fungicide in its three concentrations, as well as with Mancozeb fungicide in all cases, and low compatibility (50%) with 450, 900, and 1350 mg L^{-1} of Chlorothalonil fungicide (Figure 6).

4. Discussion

The genus *Fusarium* develops in the vascular tissues of plants, causing necrosis in the xylem, which limits water transport and brings about wilting of the plant [33]. Chemical pesticides are still commonly used in strawberry cultivation to suppress *F. solani* [34,35]. The scant research available limits the ability of strawberry growers to choose effective products and increases the risk of generating strains resistant to the active principles used in fungicides [36]; therefore, this study examined the compatibility between biological and chemical control.

The rapid growth and development observed in the present study may explain the competitive ability of the four *Trichoderma* species against the MA-FC120 strain of *F. solani*. Morales et al. [37] found a lower growth rate for the TH-4 strain (1.86 ± 0.22 cm d^{-1}) and a higher development rate (1.67 ± 0.01 mm h^{-1}) in comparison with the present investigation. *F. solani* presented a slower growth rate (0.4718 ± 0.00063 mm h^{-1}). The results obtained in the present investigation were consistent with a study carried out by Miguel Ferrer et al. [38], where different species of *Trichoderma* had superior mycelial growth compared with *F. solani* (4.71 cm d^{-1}).

A reduction in growth rate in dual cultures is an indicator of the antagonistic capacity of *Trichoderma* [39]. Suárez et al. [40] studied 12 isolates of *T. harzianum* confronted with *F. solani*, and they obtained a PIRG of between 60 and 70% antagonism. This was similar to the results obtained in the present investigation for the PDA culture medium without the addition of fungicides. This indicates that *T. harzianum* has a higher ingestion and metabolism rate than *F. solani*, as well as different hydrolytic enzyme induction mechanisms that may be involved in fungicide degradation processes [41]. In addition, *T. harzianum* can inhibit the growth of *F. oxysporum* because it produces numerous antibiotics such as tricodermin, suzukacillin, alamethicin, dermadin, penicillin, trichothecenes, and trichorzianins, among others [42].

The integrated use of *T. asperellum* T8a and a low dose of Captan (0.1 g L^{-1}) led to greater in vitro growth inhibition of *C. gloeosporioides* ATCC MYA 454, a pathogenic strain that causes anthracnose in mango [43]. Similarly, Ruocco et al. [44] explained that the ability of *Trichoderma* to resist relatively high concentrations of a variety of synthetic and natural toxic compounds depends on a complex system of membrane pumps through which efficient cellular detoxification mechanisms are carried out. In this regard, Singh and Varma [45] reported that the fungicide Mancozeb was the most effective in reducing the mycelial growth of *F. solani* and was compatible with *T. harzianum* and *T. viride* at concentrations of 0.05 and 0.1%.

In the present study, the fungicides Captan and Mancozeb demonstrated high compatibility with the four strains of *Trichoderma* spp. and inhibited the growth of the pathogen. González et al. [46] reported that the use of the C2A strain of *T. reesei* in combination with Mancozeb at a concentration of 0.1 mg L^{-1} improved the mycoparasitic capacity against *F. oxysporum*. This is in agreement with the results obtained in the present investigation, where the strain of *T. asperellum* was able to overgrow on *Fusarium solani*. Of note, Huilgol et al. [47] observed the compatibility of *T. harzianum* with Mancozeb (71.80%). Similarly, Maheshwary et al. [48] concluded that COC and Mancozeb at 500 ppm favor the growth of *T. asperellum*. This could explain the greater potential antagonism with respect to the control group in the present investigation.

There are reports of many factors causing the tolerance of *Trichoderma* strains to pesticides, such as the change in function of oxidoreductase genes and ABC transporter genes resulting in the tolerance of *Trichoderma* spp. to dichlorvos, Mancozeb, thiram, tebuconazole, and carbendazim [49–51].

Finally, it was found that the four strains of *Trichoderma* spp. were incompatible with the fungicide Chlorothalonil added at concentrations of 450, 900, and 1350 mg L^{-1}. This was similar to the results obtained by Elshahawy et al. [52]. Likewise, Gangopadhyay et al. [53] reported that the fungicide Chlorothalonil is highly toxic for *T. viride* and *T. harzianum*, as was also observed in the present investigation, demonstrating no compatibility with *T. harzianum*. Chlorothalonil is a non-systemic organochlorine foliar fungicide that is widely used throughout the world [54]. Specifically, it is a broad-spectrum polychlorinated aromatic component that delays mycelial growth and inhibits spore germination. It acts mainly on the respiration of fungal cells; that is, it affects the Krebs cycle by reducing ATP synthesis, causing cell death [55]. For this reason, it may be incompatible with biological control agents, as is the case for different species of *Trichodermas* used in the present investigation.

5. Conclusions

The evaluation of different strains of Trichoderma showed a medium potential antagonism (PA) against the MA-FC120 strain of *F. solani* in a PDA medium without the addition of fungicides; however, *T. harzianum* was associated with the highest inhibition in vitro.

T. asperellum demonstrated the highest potential antagonism (PA) against *F. solani* in the poisoned culture medium, with the fungicides Captan and Mancozeb at concentrations of 450, 900, and 1350 mg L^{-1} showing greater compatibility. *T. harzianum* achieved the second highest, demonstrating a high potential antagonism (PA) against *F. solani* in the PDA medium with the fungicide Captan in the three concentrations evaluated. This was followed by *T. hamatum* at concentrations of 450 and 900 mg L^{-1}.

T. koningiopsis demonstrated a medium potential antagonism (PA), inhibiting the growth of *F. solani* by between 60 and 75% in the PDA medium with the fungicides Captan and Mancozeb at the three concentrations evaluated in this study.

The use of *Trichoderma* strains, together with the fungicide Chlorothalonil at 450, 900, and 1350 mg L^{-1} in the PDA culture medium, showed an incompatible adverse effect on potential antagonism (PA) given the less than 59% inhibition against *F. solani*.

F. solani demonstrated better development and adaptation than the *Trichoderma* strains in the fungicide Chlorothalonil at concentrations of 450, 900, and 1350 mg L^{-1} in the PDA medium in vitro.

6. Recommendations

The results obtained in the present investigation are important, and we believe that they can be taken to the greenhouse and field for strawberry cultivation. It is recommended to use *T. asperellum* in combination with the fungicides Captan and Mancozeb at a concentration of 450 mg L^{-1}. In addition, *T. harzianum* and *T. hamatum* should be used in combination with the fungicide Captan at a concentration of 450 mg L^{-1}.

Finally, the use of the fungicide Chlorothalonil in combination with any strain of *Trichoderma* is not recommended to control crown and root rot in strawberry cultivation.

Author Contributions: Conceptualization, O.R.-A., C.P.L. and D.M.S.J.; methodology, A.R., D.M.S.J. and O.R.-A.; software, M.H.-L. and O.R.-A.; validation, M.D.L.A.V.d.I., A.R. and O.R.-A.; formal analysis, D.M.S.J., C.P.L. and O.R.-A.; resources, O.R.-A. and C.P.L.; original—draft preparation, M.D.L.A.V.d.I., M.H.-L. and O.R.-A.; writing—review and editing, A.R. and O.R.-A.; visualization, O.R.-A. and D.M.S.J.; supervision, M.H.-L.; project administration, O.R.-A.; funding acquisition, O.R.-A. All authors have read and agreed to the published version of the manuscript.

Funding: This research was supported by the Consejo Nacional de Ciencia y Tecnología (CONACyT), number 47406; and Benémerita Universidad Autónoma of Puebla, number 100420500.

Data Availability Statement: Informed consent was obtained from all subjects involved in the study.

Acknowledgments: The authors are grateful to Consejo Nacional de Ciencia y Tecnología (CONACyT) and Laboratory 204 of the Center for Agroecology at Benémerita Universidad Autónomaof Puebla.

Conflicts of Interest: The authors declare no conflict of interest.

References

1. Fang, X.L.; Phillips, D.; Li, H.; Sivasithamparam, K.; Barbetti, M.J. Severity of strawberry crown and root diseases and associated fungal and oomycete pathogenesis in Western Australia. *Australas Plant Pathol.* **2011**, *40*, 109–119. [CrossRef]
2. Hassan, O.; Chang, T. Morphological and molecular characteristics of fungal species associated with strawberry crown rot in South Korea. *Mol. Biol. Rep.* **2022**, *49*, 51–62. [CrossRef] [PubMed]
3. Martínez, K.; Ortiz, M.; Albis, A.; Gilma Gutiérrez Castañeda, C.; Valencia, M.E.; Grande Tovar, C.D. The effect of edible chitosan coatings incorporated with *Thymus capitatus* essential oil on the shelf-life of strawberry (*Fragaria x ananassa*) during cold storage. *Biomolecules* **2018**, *8*, 155. [CrossRef] [PubMed]
4. Hong, S.; Kim, T.Y.; Won, S.-J.; Moon, J.-H.; Ajuna, H.B.; Kim, K.Y.; Ahn, Y.S. Control of fungal diseases and fruit yield improvement of strawberry using *Bacillus velezensis* CE 100. *Microorganisms* **2022**, *10*, 365. [CrossRef] [PubMed]
5. Coronel, A.C.; Parraguirre-Lezama, C.; Pacheco-Hernández, Y.; Santiago-Trinidad, O.; Rivera-Tapia, A.; Romero-Arenas, O. Efficacy of four In vitro fungicides for control of wilting of strawberry crops in Puebla-Mexico. *Appl. Sci.* **2022**, *12*, 3213. [CrossRef]
6. Ruiz-Romero, P.; Valdez-Salas, B.; González-Mendoza, D.; Mendez-Trujillo, V. Antifungal effects of silver phyto-nanoparticles from *Yucca shilerifera* against strawberry soil-borne pathogens: *Fusarium solani* and *Macrophomina phaseolina*. *Mycobiology* **2018**, *46*, 47–51. [CrossRef]

7. De la Lastra, E.; Villarino, M.; Astacio, J.D.; Larena, I.; De Cal, A.; Capote, N. Genetic diversity, and vegetative compatibility of *Fusarium solani* species complex of strawberry in Spain. *Phytopathology* **2019**, *109*, 2142–2151. [CrossRef]
8. Villarino, M.; De la Lastra, E.; Basallote-Ureba, M.J.; Capote, N.; Larena, I.; Melgarejo, P.; De Cal, A. Characterization of *Fusarium solani* populations associated with Spanish strawberry crops. *Plant Dis.* **2019**, *103*, 1974–1982. [CrossRef]
9. Manici, L.M.; Caputo, F.; Baruzzi, G. Additional experiences to elucidate the microbial component of soil suppressiveness towards strawberry black root rot complex. *Ann. App. Biol.* **2005**, *146*, 421–431. [CrossRef]
10. Ayoubi, N.; Soleimani, M.J. Morphological and molecular identification of pathogenic *Fusarium* spp. on strawberry in Iran. *Sydowia* **2016**, *68*, 163–171.
11. Ceja-Torres, L.F.; Mora-Aguilera, G.; Téliz, D.; Mora-Aguilera, A.; Sánchez-García, P.; Muñoz-Ruíz, C.; Tlapal-Bolaños, B.; De La Torre-Almaraz, R. Ocurrencia de hongos y etiología de la secadera de la fresa con diferentes sistemas de manejo agronómico. *Agrociencia* **2008**, *42*, 451–461.
12. Zhang, Y.; Yu, H.; Hu, M.; Wu, J.; Zhang, C. Fungal pathogens associated with strawberry crown rot disease in China. *J. Fungi* **2022**, *8*, 1161. [CrossRef]
13. Ons, L.; Bylemans, D.; Thevissen, K.; Cammue, B.P.A. Combining Biocontrol Agents with Chemical Fungicides for Integrated Plant Fungal Disease Control. *Microorganisms* **2020**, *8*, 1930. [CrossRef]
14. Sood, M.; Kapoor, D.; Kumar, V.; Sheteiwy, M.S.; Ramakrishnan, M.; Landi, M.; Araniti, F.; Sharma, A. *Trichoderma*: The "Secrets" of a Multitalented Biocontrol Agent. *Plants* **2020**, *9*, 762. [CrossRef]
15. James, J.E.; Santhanam, J.; Cannon, R.D.; Lamps, E. Voriconazole treatment induces a conserved regulatory network of sterol/pleiotropic drug resistance, including an alternative pathway of ergosterol biosynthesis, in the clinically important FSSC species, *Fusarium keratoplasticum*. *J. Fungi* **2022**, *8*, 1070. [CrossRef]
16. Slaboch, J.; Čechura, L.; Malý, M.; Mach, J. The Shadow Values of Soil Hydrological Properties in the Production Potential of Climatic Regionalization of the Czech Republic. *Agriculture* **2022**, *12*, 2068. [CrossRef]
17. Zhang, C.; Wang, W.; Xue, M.; Liu, Z.; Zhang, Q.; Hou, J.; Xing, M.; Wang, R.; Liu, T. The combination of a biocontrol agent *Trichoderma asperellum* SC012 and hymexazol reduces the effective fungicide dose to control *Fusarium* Wilt in Cowpea. *J. Fungi* **2021**, *7*, 685. [CrossRef] [PubMed]
18. Rabølle, M.; Spliid, N.H.; Kristensen, K.; Kudsk, P. Determination of fungicide residues in field-grown strawberries following different fungicide strategies against gray mold (*Botrytis cinerea*). *J. Agric. Food Chem.* **2006**, *54*, 900–908. [CrossRef] [PubMed]
19. Mayo-Prieto, S.; Squarzoni, A.; Carro-Huerga, G.; Porteous-Álvarez, A.J.; Gutiérrez, S.; Casquero, P.A. Organic and conventional bean pesticides in the development of autochthonous strains of *Trichoderma*. *J. Fungi* **2022**, *8*, 603. [CrossRef]
20. Omar, I.; O'neill, T.M.; Rossall, S. Biological control of *Fusarium* crown and root rot of tomato with antagonistic bacteria and integrated control when combined with the fungicide carbendazim. *Plant Pathol.* **2006**, *55*, 92–99. [CrossRef]
21. Spadaro, D.; Gullino, M.L. State of the art and future prospects of the biological control of postharvest fruit diseases. *Int. J. Food Microbiol.* **2004**, *91*, 185–194. [CrossRef]
22. Wisniewski, M.E.; Wilson, C.L.; Hershherger, W. Characterization of inhibition of *Rhizopus stolonifer* germination and growth by *Enterobacter cloacae*. *Can. J. Bot.* **1989**, *67*, 2317–2323. [CrossRef]
23. Abd-El-Khair, H.; Elshahawy, I.E.; Haggag, H.E. Field application of *Trichoderma* spp. combined with thiophanate-methyl for controlling *Fusarium solani* and *Fusarium oxysporum* in dry bean. *Bull. Natl. Res. Cent.* **2019**, *43*, 1–9. [CrossRef]
24. Terrero-Yépez, P.I.; Peñaherrera-Villafuerte, S.L.; Solís-Hidalgo, Z.K.; Vera-Coello, D.I.; Navarret- Cedeño, J.B.; Herrera-Defaz, M.A. Compatibilidad in vitro de *Trichoderma* spp. con fungicidas de uso común en cacao (*Theobroma cacao* L.). *Investig. Agrar.* **2018**, *20*, 146–151. [CrossRef]
25. Ruano-Rosa, D.; Arjona-Girona, I.; López-Herrera, C.J. Integrated control of avocado white root rot combining low concentrations of fluazinam and *Trichoderma* spp. *Crop Prot.* **2018**, *112*, 363–370. [CrossRef]
26. Wang, H.; Chang, K.F.; Hwang, S.F.; Turnbull, G.D.; Howard, R.J.; Blade, S.F.; Callan, N.W. *Fusarium* root rot of coneflower seedlings and integrated control using *Trichoderma* and fungicides. *BioControl* **2005**, *50*, 317–329. [CrossRef]
27. Zeravakis, G.; Philippoussis, A.; Ioannidou, S.; Diamantopoulou, P. Mycelium growth kinetics and optimal temperature conditions for the cultivation of edible mushroom species on lignocellulosic substrates. *Folia Microbiol.* **2001**, *46*, 231–234. [CrossRef]
28. Azza, R.E.; Hala, M.I.; Saná, A.M. The role of storage in Mancozeb fungicidal formulations and its antifungal activity against *Fusarium oxysporium* and *Rhizoctonia solani*. *Arab. J. Chem.* **2021**, *14*, 103322. [CrossRef]
29. Andrade-Hoyos, P.; Luna-Cruz, L.; Osorio-Hernández, E.; Molina-Gayosso, E.; Landero-Valenzuela, N.; Barrales-Cureño, H.J. Antagonismo de *Trichoderma* spp. vs. Hongos Asociados a la Marchitez de Chile. *Rev. Mex. Cienc. Agric.* **2019**, *10*, 1259–1272. [CrossRef]
30. Bell, D.K.; Wells, H.D.; Markham, C.R. In vitro antagonism of *Trichoderma* species against six fungal plant pathogens. *Phytopathology* **1982**, *72*, 379–382. [CrossRef]
31. Reyes-Figueroa, O.; Ortiz-García, C.F.; Torres de la Cruz, M.; Lagunes-Espinoza, L.D.C.; Valdovinos-Ponce, G. Especies de *Trichoderma* del agroecosistema cacao con potencial de biocontrol sobre *Moniliophthora roreri*. *Chapingo Ser. Cienc. Ambiente* **2016**, *22*, 149–163. [CrossRef]
32. De Mendiburu, F.; Simon, R. Agricolae-Ten years of an open-source statistical tool for experiments in breeding, agriculture and biology. *Peer J. PrePrints* **2015**, *3*, e1404v1. [CrossRef]

33. Yan, K.; Han, G.; Ren, C.; Zhao, S.; Wu, X.; Bian, T. *Fusarium solani* infection depressed photosystem performance by inducing foliage wilting in apple seedlings. *Front. Plant Sci.* **2018**, *9*, 479. [CrossRef]
34. Gu, K.X.; Canción, X.S.; Xiao, X.M.; Duan, X.X.; Wang, J.X.; Duan, Y.B.; Hou, Y.P.; Zhou, M.G. A β2-tubulin dsRNA derived from *Fusarium asiaticum* confers plant resistance to multiple phytopathogens and reduces fungicide resistance. *Pestic. Biochem. Physiol.* **2019**, *153*, 36–46. [CrossRef]
35. Essa, T. Response of some commercial strawberry cultivars to infection by wilt diseases in Egypt and their control with fungicides. *Egypt. J. Pathol.* **2015**, *43*, 113–127. [CrossRef]
36. Robledo-Buriticá, J.; Ángel-García, C.; Castaño-Zapata, J. Microscopía electrónica de barrido ambiental del proceso de infección de *Fusarium solani* f. sp. passiflorae en plántulas de maracuyá (*Passiflora edulis f. flavicarpa*). *Rev. Acad. Cienc. Exactas* **2017**, *41*, 213–221.
37. Morales-Mora, L.A.; Andrade-Hoyos, P.; Valencia-de Ita, M.A.; Romero-Arenas, O.; Silva-Rojas, H.V.; Contreras-Paredes, C.A. Caracterización de hongos asociados a fresa y efecto antagonista in vitro de *Trichoderma harzianum*. *Rev. Mex. Fitopatol.* **2020**, *38*, 434–449. [CrossRef]
38. Miguel-Ferrer, L.; Romero-Arenas, O.; Andrade-Hoyos, P.; Sánchez-Morales, P.; Rivera-Tapia, J.A.; Fernández-Pavía, S.P. Antifungal activity of *Trichoderma harzianum* and *T. koningiopsis* against *Fusarium solani* in seed germination and vigor of Miahuateco chili seedlings. *Rev. Mex. Fitopatol.* **2021**, *39*, 228–247. [CrossRef]
39. Guigón-López, C.; Guerrero-Prieto, V.; Vargas-Albores, F.; Carvajal-Millán, E.; Ávila-Quezada, G.D.; Bravo-Luna, L.; Ruocco, M.; Lanzuise, S.; Woo, S.; Lorito, M. Identificación molecular de cepas nativas de *Trichoderma* spp. su tasa de crecimiento in vitro y antagonismo contra hongos fitopaógenos. *Rev. Mex. Fitopatol.* **2010**, *28*, 87–96.
40. Suárez, C.; Fernández, R.; Valero, N.; Gámez, R.; Páez, A. Antagonismo in vitro de *Trichoderma harzianum* Rifai sobre *Fusarium solani* (Mart.) Sacc., asociado a la marchitezen maracuyá. *Rev. Colomb. Biotecnol.* **2008**, *2*, 35–43. [CrossRef]
41. Michel-Aceves, A.C.; Otero-Sánchez, M.A.; Rebolledo-Domínguez, O.; Lezama-Gutiérrez, R.; Ariza-Flores, R.; Barrios-Ayala, A. Producción y efecto antagónico de quitinasas y glucanasas por *Trichoderma* spp. en la inhibiciónde *Fusarium subglutinans* y *Fusarium oxysporum* in vitro. *Rev. Chapingo Ser. Hortic.* **2005**, *11*, 273–278. [CrossRef]
42. Manzar, N.; Kashyap, A.S.; Goutam, R.S.; Rajawat, M.V.S.; Sharma, P.K.; Sharma, S.K.; Singh, H.V. *Trichoderma*: Advent of Versatile Biocontrol Agent, Its Secrets and Insights into Mechanism of Biocontrol Potential. *Sustainability* **2022**, *14*, 12786. [CrossRef]
43. Peláez-Álvarez, A.; Santos-Villalobos, S.D.L.; Yépez, E.A.; Parra-Cota, F.I.; Reyes-Rodríguez, R.T. Synergistic effect of *Trichoderma asperelleum* T8A and Captan against *Colletotrichum gloeosporioides* (Penz.). *Rev. Mex. Cienc. Agric.* **2016**, *7*, 1401–1412.
44. Ruocco, M.; Lanzuise, S.; Vinale, F.; Marra, R.; Turrà, D.; Woo, S.L.; Lorito, M. Identification of a new biocontrol gene in *Trichoderma atroviride*: The role of anABC transporter membrane pump in the interaction with different plant-pathogenic fungi. *Mol. Plant-Microbe Interact.* **2009**, *22*, 291–301. [CrossRef] [PubMed]
45. Singh, G.; Varma, R.K. Compatibility of fungicides and neem products against *Fusariu solani* f.sp. glycines causing root rot of soybean and *Trichoderma* spp. *J. Mycopathol. Res.* **2005**, *43*, 211–214.
46. Gonzalez, M.F.; Magdama, F.; Galarza, L.; Sosa, D.; Romero, C. Evaluation of the sensitivity and synergistic effect of *Trichoderma reesei* and Mancozeb to inhibit under in vitro conditions the growth of *Fusarium oxysporum*. *Commun. Integr. Biol.* **2020**, *13*, 160–169. [CrossRef]
47. Huilgol, S.N.; Pratibha, M.P.; Hegde, G.M.; Banu, H. Evaluation, and compatibility of new fungicides with *Trichoderma harzianum* for managing the charcoal rot of soybean. *Pharma Innov. J.* **2022**, *11*, 659–664.
48. Maheshwary, N.; Gangadhara-Naik, B.; Amoghavarsha-Chittaragi, M.; Naik, S.K.; Nandish, M. Compatibility of *Trichoderma asperellum* with fungicides. *Pharma Innov. J.* **2020**, *9*, 136–140.
49. Hirpara, D.G.; Gajera, H.P. Molecular heterozygosity and genetic exploitations of *Trichoderma interfusants* enhancing tolerance to fungicides and mycoparasitism against *Sclerotium rolfsii* Sacc. *Infect. Genet. Evol.* **2018**, *66*, 26–36. [CrossRef]
50. Sun, J.; Zhang, T.; Li, Y.; Wang, X.; Chen, J. Functional characterization of the ABC transporter TaPdr2 in the tolerance of biocontrol the fungus *Trichoderma atroviride* T23 to dichlorvos stress. *Biological. Control* **2019**, *129*, 102–108. [CrossRef]
51. Hu, X.; Roberts, D.P.; Xie, L.; Yu, C.; Li, Y.; Qin, L.; Liao, X. Use of formulated *Trichoderma* sp. Tri-1 in combination with reduced rates of chemical pesticide for control of *Sclerotinia sclerotiorum* on oilseed rape. *Crop Prot.* **2016**, *79*, 124–127. [CrossRef]
52. Elshahawy, I.E.; Haggag, K.H.E.; Abd-El-Khaira, H. Compatibility of *Trichoderma* spp. with seven chemical fungicides used in the control of soil borne plant pathogens. *Res. J. Pharm. Biol. Chem. Sci.* **2016**, *7*, 1772–1785.
53. Gangopadhyay, S.; Gopal, R.; Godara, S.L. Effect of Fungicides and antagonists on *Fusarium* wilt of Cumin. *J. Mycol. Plant Pathol.* **2009**, *39*, 331–334.
54. Monadjemi, S.; El Roz, M.; Richard, C.; Ter Halle, A. Photoreduction of Chlorothalonil fungicide in plant leaf models. *Entorno Sci. Technol.* **2011**, *45*, 9582–9589. [CrossRef]
55. National Center for Biotechnology Information. Available online: https://pubchem.ncbi.nlm.nih.gov/compound/Chlorothalonil (accessed on 12 April 2023).

Disclaimer/Publisher's Note: The statements, opinions and data contained in all publications are solely those of the individual author(s) and contributor(s) and not of MDPI and/or the editor(s). MDPI and/or the editor(s) disclaim responsibility for any injury to people or property resulting from any ideas, methods, instructions or products referred to in the content.

Article

Fungicidal Effect of Guava Wood Vinegar against *Colletotrichum coccodes* Causing Black Dot Disease of Potatoes

Mansour M. El-Fawy [1], Kamal A. M. Abo-Elyousr [2,*], Nashwa M. A. Sallam [3], Rafeek M. I. El-Sharkawy [1] and Yasser Eid Ibrahim [4]

[1] Plant Pathology Branch, Agricultural Botany Department, Faculty of Agriculture, Al-Azhar University (Assiut Branch), Assiut 71526, Egypt; mansourhassan.5419@azhar.edu.eg (M.M.E.-F.)
[2] Department Agriculture, Faculty of Environmental Science, King Abdulaziz University, Jeddah 80208, Saudi Arabia
[3] Plant Pathology Department, Faculty of Agriculture, Assiut University, Assiut 71526, Egypt; nashwasallam@aun.edu.eg
[4] Plant Protection, Plant Protection Department, College of Food and Agricultural Sciences, King Saud University, Riyadh 11451, Saudi Arabia
* Correspondence: ka@kau.edu.sa or kaaboelyousr@agr.au.edu.eg

Citation: El-Fawy, M.M.; Abo-Elyousr, K.A.M.; Sallam, N.M.A.; El-Sharkawy, R.M.I.; Ibrahim, Y.E. Fungicidal Effect of Guava Wood Vinegar against *Colletotrichum coccodes* Causing Black Dot Disease of Potatoes. *Horticulturae* **2023**, *9*, 710. https://doi.org/10.3390/horticulturae9060710

Academic Editors: Hillary Righini, Stefania Galletti and Roberta Roberti

Received: 4 April 2023
Revised: 5 June 2023
Accepted: 15 June 2023
Published: 16 June 2023

Copyright: © 2023 by the authors. Licensee MDPI, Basel, Switzerland. This article is an open access article distributed under the terms and conditions of the Creative Commons Attribution (CC BY) license (https://creativecommons.org/licenses/by/4.0/).

Abstract: Wood vinegar (WV) by-product of charcoal production is considered one of the most promising alternatives to synthetic pesticide and fertilizer applications, especially for organic production. Our goal in this study is to evaluate the efficacy of guava (*Psidium guajava*) WV to control *Colletotrichum coccodes*, which causes black dot disease, and how it influences potato plant development and yield. This study tested the efficacy of guava WV against the pathogen both in vitro and under greenhouse conditions. Different guava WV concentrations were tested on pathogen growth development, including 0, 0.25%, 0.50%, 1%, 2%, and 3% (v/v). Data revealed that the pathogen's mycelial growth was significantly inhibited at all the concentrations, and the highest inhibition (100%) was obtained at 3% guava WV. In greenhouse trials conducted for two seasons (2021 and 2022), guava WV applied as a foliar spray at the concentration of 2% and 3% considerably reduced the potato black dot severity evaluated as stem colonization (average of 22.9% for 2021, average of 22.5% for 2022), root covering with sclerotia (average of 21.7% for 2021, average of 18.3% for 2022) and wilted plants percentage (average of 27.8% for 2021, average of 33.3% for 2022). Overall, guava WV also showed a positive effect on plant growth by increasing plant height, stem diameter, and tuber yield per plant of treated potato in both seasons. Gas chromatography-mass spectrometry (GC-MS) analyses revealed the presence in guava WV of phenols, esters, organic acids, antioxidants, and alcohols. In conclusion, guava WV could represent a viable alternative for potato black dot disease management and for plant growth promotion.

Keywords: potato; *Colletotrichum coccodes*; guava wood vinegar components; GC-MS technique; growth inhibition; antifungal activity; plant growth stimulation

1. Introduction

Black dot is one of the common potato (*Solanum tuberosum* L.) diseases caused by *Colletotrichum coccodes* (Wallr.) S. Hughes [1]. *C. coccodes* can infect any part of a potato plant, such as leaves, tubers, stolons, roots, and basal stems [1–6]. Early reports of the disease in potatoes and tomatoes date back to the early 19th century and are described in detail by Dickson [7]. In Egypt, the disease was first reported in Salhiya and Abo Swair areas during the 2009–2010 seasons [8]. During storage, the pathogen causes a reduction in tuber weight and quality [9].

Wood vinegar (WV), also known as pyroligneous acid, is a clear brown liquid formed by the condensation of smoke from the charcoal-making process [10]. More than 200 chemical substances, including organic acids, phenols, acetic acid, dimethyl phenol, trimethyl phenol, esters, and nitrogen pyrimidines, are found in WV [11,12]. It is low-cost [13], with a price that is only one-third that of synthetic fungicides. According to Zulkarami et al. [14], WV contains a number of essential elements. These elements perform important roles in plant life cycles and promote photosynthesis. Moreover, the acids, phenol, and other organic compounds in WV have antifungal properties that inhibit the growth of fungi at high concentrations [15].

WV has also been shown to improve soil physicochemical parameters [16] and microbiome, including plant-growth-promoting rhizobacteria [17]. Additionally, WV is able to protect vegetable and horticultural crops from fungal diseases, especially root rots. Sugars, carboxylic acids, hydroxy aldehydes, hydroxy ketones, and phenolic acids are among the major groups of substances in oak, poplar, pine, pruning litter, and forest waste WV [11,18]. It is a low-cost, all-natural product that has no negative effects on living organisms or the environment [19]. It has been demonstrated to suppress a number of soil-borne plant pathogens [20]. The mycelial growth of *Plasmopara viticola* (Berk. and M.A.Curtis) Berl. and De Toni, *Verticillium dahliae* (Klebahn), *Phytophthora capsici* (Leonian), and *Fusarium graminearum* (Schwabe) have been slowed down by WV made from apricot trees [21]. WV causes complete inhibition of the growth of *Alternaria mali* (Roberts), the causal agent of apple Alternaria blight, when applied at a 1:32 dilution [22].

WV ester compounds can increase chlorophyll, photosynthesis, sugar, and amino acids production and stimulate plants' resistance to diseases and pests [23]. *Cryptomeria japonica* (Linnaeus) WV demonstrated potent antifungal activity against *Pythium splendens* (H.Braun), *Phytopthora capsici,* and *Ralstonica solanacearum* (Smith and Yabuuchi) [24]. In a related study, Velmurugan et al. [25] showed that WV of bamboo significantly reduced the growth of the *Ophiostoma species* that cause wood rot in forest trees. Several studies have conclusively demonstrated that the phenolic compounds in WV are responsible for their antifungal properties [26–28]. Additionally, WV increases the abiotic stress tolerance [29], growth, production, and quality of a wide range of crops [30]. WV, when used as a foliar application, increases yields in cucumber, lettuce, and cole, and jasmine rice [31,32]. The objectives of this study were to evaluate the efficacy of guava wood vinegar against *Colletotrichum coccodes* which causes black dot disease and investigate its effect on potato plant growth as well. Determine and identify the bioactive components that are most prevalent in guava WV using the GC-MS technique.

2. Materials and Methods

2.1. Fungal Isolation and Identification

Colletotrichum coccodes virulent strain was recovered from diseased potato tubers with a black dot collected from potato fields in the Nubaria region (El-Beheira Governorate, Egypt; 30° 91125″ N, 29° 97119″ E). Small sections of the infected tubers were cut up and surface treated for 1 min with 1% sodium hypochlorite (commercial bleach) before being rinsed in sterile water and dried with sterile filter paper. The surface-sterilized samples were then plated onto a PDA medium and incubated at 25 °C in the dark. The developed colonies were then purified using the hyphal tip method and identified according to the cultural features and morphological and microscopical characteristics stated by Sutton [33]. Pure cultures were sub-cultured on PDA slants and stored at 4 °C until use.

2.2. Pathogenicity Test

2.2.1. Plant Material

Cara cv. potato tubers were sterilized with sodium hypochlorite 1% for 2 min and placed in the dark for 3 weeks at 18–25 °C for sprouting [34]. Tubers were cut into pieces and left for 48 h before planting in vitro at room temperature to allow for partial wound healing. One piece of potato tuber with three eyes of almost the same size was planted in each plastic pot of 30 cm diameter (2.4 kg soil), filled with a sterilized mixture of clay and sand soil (4:1 w/w). Sowing was done on 1st February in season 2020. The plants were fertilized with NPK (Dotra Fert: 20/20/20) from the Egyptian Dotra Company at a concentration of 2 g/L water. The pots were kept under careful observation in greenhouse conditions in natural light at the Agricultural Botany Dept., Agric., Al-Azhar University, Assiut Branch, and were irrigated when it was needed.

2.2.2. Inoculum Preparation

Colletotrichum coccodes isolate was grown on a PDA medium in Petri dishes for 7 days at 25 °C in the dark. Conidia were harvested from the agar surface with sterile distilled water using a scraper, then filtered through several layers of sterilized cheesecloth. A hemacytometer was used to adjust the conidia suspension to a final concentration of 2×10^6 conidia/mL [35].

2.2.3. Plants Inoculation

Six weeks after sowing, potato stems were inoculated with the pathogen. Using a sterile scalpel, one wound was made into the base of the stem (above the soil's surface). An Agar disc (6 mm) containing the causal pathogen was placed in each wound. On the wounds of the control plants, only agar discs without pathogens were applied. In addition, each pot received 15 mL of the spore suspension (2×10^6 conidia/mL) by hand sprayer, as El-Marzoky [8] explained; in case of the control plant, it was sprayed with 15 mL sterilized distilled water. Three plants were used for each treatment as replicates, and the experiment was repeated twice. After inoculation, all plants were transferred to a greenhouse. To maintain a high humidity level, inoculated plants were covered with transparent plastic bags for 24 h. The plants were then examined every day to check for symptoms. After three days, every inoculated stem exhibited necrosis, but control stems exhibited no signs [36].

2.2.4. Disease Assessment

The disease severity of the black dot on roots was assessed visually using a 0–3 scale based on the percentage of roots covered with sclerotia as follows: 0 = no sclerotia, 1 = 1–30%, 2 = 31–60%, and 3 = >60% [35]. Additionally, the percentage of wilted plants to the total number of inoculated plants was calculated. The disease severity index (DSI) on the aboveground stem was calculated by multiplying the colonization outcome (0 or 1) by the height aboveground from which the segment was removed. The stem segments (1 cm) were cut off at 2, 6, 10, and 14 cm above ground level. The segments were cultured on a PDA medium after being sterilized and incubated at 25 °C in the dark [36]. The fungal colonization of stems was recorded as a binary outcome with 1 = colonized, and 0 = non-colonized. The disease severity index percentage (DSI%) was calculated according to the following equation [37].

$$\text{DSI (\%)} = 100 \times [2 \times (0 \text{ or } 1) + 6 \times (0 \text{ or } 1) + 10 \times (0 \text{ or } 1) + 14 \times (0 \text{ or } 1)]/32$$

2.3. Antifungal Activity

2.3.1. Guava Wood Vinegar Production

WV is a secondary product produced when smoke from charcoal production is cooled by outside air while passing through a chimney or flue pipe. The cooling effect causes condensation of WV, particularly when the temperature of the smoke produced by carbonization ranges between 80 °C and 180 °C [10]. The guava WV used in this study was derived from the charcoal production of guava trees wood (*Psidium guajava*) in Egypt's El-Kalyobia governorate. The obtained guava WV was stored at room temperature for further studies.

2.3.2. Analysis of the Chemical Composition of Guava Wood Vinegar by GC–MS

A gas chromatography-mass spectrometry (GC-MS) system (GC Trace 1300 Thermo Scientific) was used to analyze the components of guava WV at Assiut University's Faculty of Science Department of Chemical. A single quadruple mass spectrometer (ISQ 7000 Thermo Scientific) was used. The carrier (He, 99.999%) flow was 1 mL min^{-1}, split 10:1, and injected volumes 2 µL. The column temperature was maintained initially at 110 °C and held for 5 min at a rate of 10 °C/min, the temperature increased up to 200 °C and held for 5 min, then the temperature increased up to 250 °C at a rate of 5 °C/min and held for 5 min. The injector temperature was 250 °C, and this temperature was held constant during the analysis. The electron impact energy was 70 eV, and the ion source temperature was set at 250 °C. Electron impact (EI) mass scan (m/s) was recorded in the 40–650 amu range [38]. These components were identified using the following parameters: retention time (RT), molecular weight, molecular formula, and area %.

2.3.3. Plate Assays

The inhibition of *C. coccodes* mycelial radial growth by the guava WV was tested at concentrations of 0.25%, 0.50%, 1.00%, 2.00%, and 3.00%. The guava WV was added directly to the autoclaved (20 min at 121 °C) PDA medium cooled to 45 °C, poured into sterile Petri dishes [39], and allowed to set. The center of each test plate was subsequently inoculated with a 6 mm size plug of 15-day-old *C. coccodes* culture and incubated at 27 °C in the dark. As a control, the fungus was cultivated on a PDA medium without guava WV. For each treatment, three replications were performed, and the experiment was repeated twice. When maximum growth was observed in the control plates, the colony's radial growth was measured, and the percentage of growth inhibition (R) was calculated using the formula.

$$R = [(C - B)/C] \times 100$$

whereas: C = radial growth in control and B = radial growth in treatment.

2.3.4. Greenhouse Experiments

During the growing seasons of 2021 and 2022, greenhouse experiments were conducted at the farm of the Faculty of Agriculture, Al-Azhar University, Assiut Branch, Assiut, Egypt, to study the effect of guava WV on the disease. On 1st February in seasons 2021 and 2022, one piece of potato tubers was planted in pots (30 cm in diameter). The seedlings were kept in a greenhouse with regular irrigation and fertilization. The plants were inoculated with the pathogen as described in the pathogenicity test, as previously stated. Following the appearance of the first symptoms of disease, the plants were sprayed with guava WV at 2% and 3% based on laboratory studies. Two sprays were applied to each treatment (10 days between each one). As a control, the fungicide Amistar (Azoxystrobin 25%) from Syngenta Company, Basel, Switzerland, was applied at a concentration of 0.50 cm^3/L water. Three plants were used for each treatment as replicates, and control plants were only sprayed with distilled water. The plants were then examined every day to check for symptoms. Plant height, stem diameter, and tuber yield of potato per pot were measured.

2.4. Statistical Analysis

The MSTAT-C software version 2.1 was used to conduct a data analysis of variance (ANOVA) [40]. All experiments were carried out in three replicates, and data are presented as mean ± standard deviation (SD). The treatment means were compared using the least significant difference (LSD) ($p < 0.05$), according to Gomez and Gomez [41].

3. Results

3.1. Isolation and Identification of the Pathogen

Colletotrichum coccodes were obtained from tubers of infected potatoes with symptoms of black dot disease (Figure 1A). Grown on PDA medium, *C. coccodes* formed a circular colony at first white, then grey (Figure 1B) with age. Conidiomata (Figure 1C) produced aseptate, hyaline, smooth-walled, and straight conidia (Figure 1D).

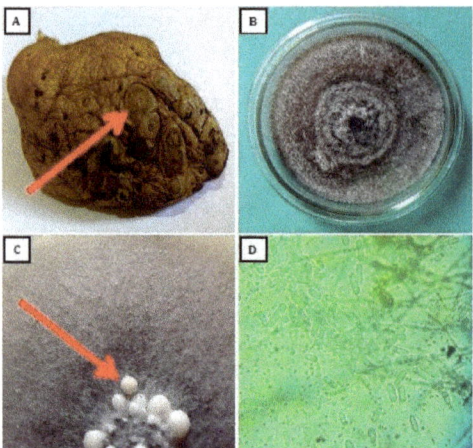

Figure 1. Colletotrichum coccodes symptoms of black dot disease on naturally infected potato tuber (**A**), 15-day-old colony (**B**), conidiomata (**C**), and conidia at 40× magnification (**D**) on Potato Dextrose Agar medium.

3.2. Pathogenicity of C. coccodes Isolate on Potato Plants

The symptoms of the disease were observed on potato plants inoculated with the pathogen. Every injected stem showed necrosis after three days, but control stems showed no symptoms. The black dot disease symptoms on inoculated plants were recorded as stem colonization (SC), root covering with sclerotia (RCS), and wilted plant percentages. Koch's hypotheses were confirmed when the causal pathogen was successfully isolated from inoculated plants.

3.3. Chromatographic Analysis of Guava WV

Data in Figure 2 show that the chromatogram obtained shows the major peaks corresponding to the more abundant compounds present in the sample. The most abundant compound (16.12%) was identified as 2,6-Dimethoxyphenol, which was associated with peak 3, followed by guaiacol (12.82%) associated with peak 1 (Figure 2), and 1,2,4-trimethoxybenzene (6.39%) associated with peak 4. Table 1 describes the overall chemical composition, retention times of each compound, and percentages of each compound present in guava WV. The presence of several bioactive compounds in guava WV, including phenols, esters, alcohols, antioxidants, and organic acids (Table 1 and Figure 3). GC-MS analyses revealed the presence in guava WV of phenols, esters, organic acids, antioxidants, and alcohols.

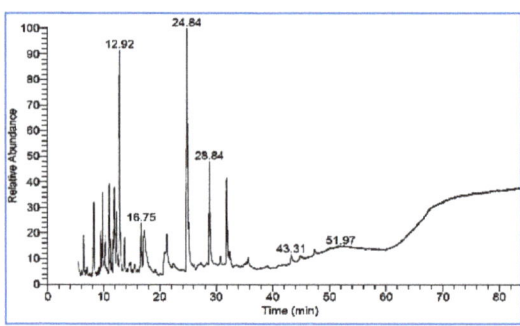

Figure 2. GC-MS chromatogram analysis of guava wood vinegar.

Table 1. The main chemical constituents of the guava WV were analyzed by GC-MS.

Retention Time (min.)	Compound Name	Molecular Formula	Molecular Weight	Area %
6.44	2-Cyclopenten-1-one	C_5H_6O	82	2.39
8.29	Butyrolactone	$C_4H_6O_2$	86	5.74
9.57	2-Cyclopenten-1-one, 3-methyl-	C_6H_8O	96	2.12
9.86	Phenol	C_6H_6O	94	4.32
10.31	1-Cyclopentylethanone	$C_7H_{12}O$	112	2.03
11.05	Cyclotene	$C_6H_8O_2$	112	4.59
12.34	P-Cresol	C_7H_8O	108	3.52
12.91	Guaiacol	$C_7H_8O_2$	124	12.82
16.75	2-Methoxy-5-methylphenol	$C_8H_{10}O_2$	138	3.13
17.28	Catechol	$C_6H_6O_2$	110	3.83
24.83	2,6-Dimethoxyphenol	$C_8H_{10}O_3$	154	16.12
28.84	1,2,4-Trimethoxybenzene	$C_9H_{12}O_3$	168	6.39
31.91	1,2,3-Trimethoxy-5-(methoxymethyl) benzene	$C_{10}H_{14}O_3$	182	5.58
43.31	9, 12, 15-Octadecatrienoic acid, 2-phenyl-1, 3-dioxan-5-yl ester	$C_{28}H_{40}O_4$	440	0.61
51.97	Ethyl iso-allocholate	$C_{26}H_{44}O_5$	436	0.04

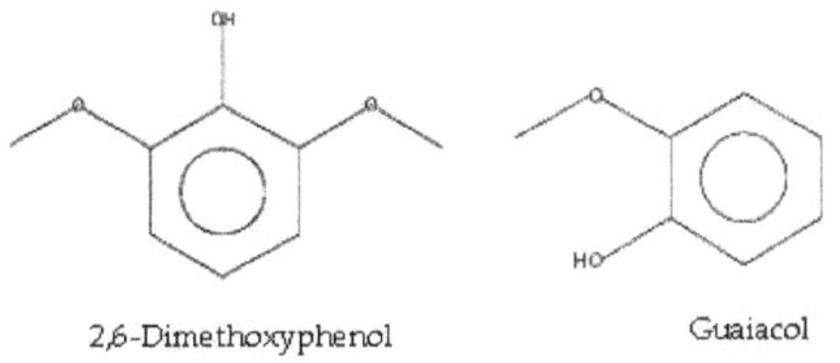

2,6-Dimethoxyphenol   Guaiacol

Figure 3. The chemical structures of the major components of the guava WV.

3.4. Plate Assays

The data in Figures 4 and 5 demonstrate that guava WV treatment significantly ($p \leq 0.05$) reduced *C. coccodes* mycelial growth at all tested concentrations when compared to the control. The percentage of mycelial growth inhibition increased along with increasing concentration. The highest WV concentration (3%) showed a complete inhibition of the pathogen's growth (100%). Furthermore, the data revealed that there were significant differences in the percentage of pathogen growth inhibition among the tested concentrations.

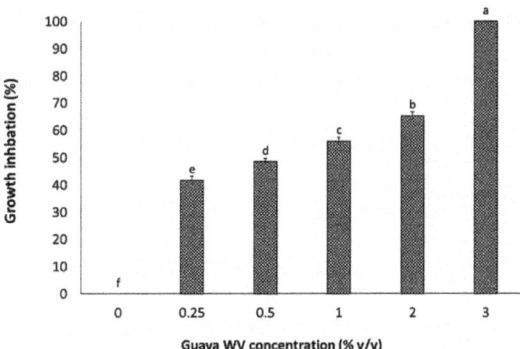

Figure 4. Effect of different guava WV concentrations on in vitro *Colletotrichum coccodes* radial growth inhibition. Values are the mean of three replicates ± SD. In each column, data followed by the same letter do not differ significantly as determined by the LSD test $p = 0.05$.

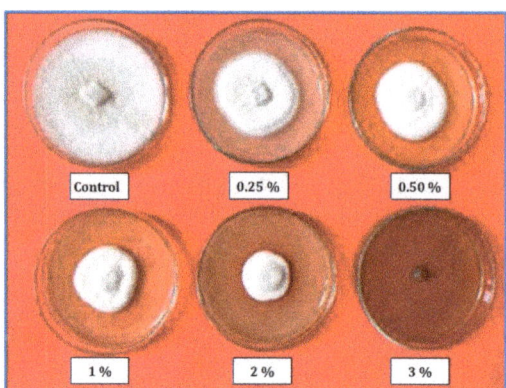

Figure 5. Mycelial growth of Colletotrichum coccodes on Potato Dextrose Agar medium alone (control) or amended with different guava wood vinegar concentrations.

3.5. Greenhouse Experiments

The application of guava WV as a foliar treatment on infected potato plants significantly decreased the disease severity of black dot in all disease parameters, including stem colonization, root covering with sclerotia, and wilt percent percentages (Table 2). Both of the tested concentrations (2% and 3% v/v) had an effect on the disease in greenhouse conditions, but the 3% concentration was more successful in suppressing the disease than the 2% concentration. The fungicide Amistar recorded the best percent of disease inhibition in both seasons of 2021 and 2022. The Amistar 25% treatment showed lower disease severity index values compared to the inoculated and untreated control, indicating its effectiveness in reducing disease severity. The treatment with a concentration of 3% guava WV recorded the greatest reduction in disease severity of all disease parameters, i.e., stem colonization (20.16% and 21.91%), root covering with sclerotia (20 and 23.33), and wilt (22.22% and 33.33%) in seasons 2021 and 2022, respectively. There are significant differences in the effects of all the treatments on the disease, though there are no significant differences in wilting percentage among treatments in the growing season 2022.

Table 2. Disease severity index (%) referred to stem colonization, root covered with sclerotia, and wilt of potato plants inoculated with *Colletotrichum coccodes* and exposed to different treatments under greenhouse conditions during the 2021 and 2022 growing seasons.

Treatments	Disease Severity Index (%)					
	Season 2021			Season 2022		
	SC% *	RCS% **	Wilt%	SC%	RCS%	Wilt%
guava WV 2%	25.63 ± 1.90 [b]	23.33 ± 10.00 [b]	33.33 ± 9.43 [b]	23.14 ± 1.62 [b]	26.67 ± 7.63 [b]	33.33 ± 33.34 [b]
guava WV 3%	20.16 ± 1.70 [c]	20.00 ± 5.77 [c]	22.22 ± 19.05 [c]	21.91 ± 1.80 [c]	23.33 ± 2.89 [b]	33.33 ± 33.34 [b]
Amistar 25%	15.33 ± 1.42 [d]	16.67 ± 2.89 [d]	11.11 ± 19.05 [d]	18.312 ± 1.51 [d]	13.33 ± 5.77 [c]	22.22 ± 19.24 [c]
Inoculate and untreated control	60.41 ± 2.30 [a]	63.33 ± 5.77 [a]	55.56 ± 19.23 [a]	73.15 ± 2.10 [a]	73.33 ± 11.55 [a]	66.67 ± 19.24 [a]
LSD at 0.05	0.65	11.04	5.43	0.52	15.26	6.27

SC% * = Stem colonization%, RCS% ** = Root covering with sclerotia%. Values are the mean of three replicates ± standard deviation (SD). In each column, data followed by the same letter do not differ significantly as determined by the LSD test p = 0.05.

Overall, the results presented in this table suggest that the guava treatments, particularly guava WV 2%, had a positive impact in reducing disease severity (SC, RCS, and Wilt) in comparison to the uninoculated and untreated control.

3.6. Effects of Guava Wood Vinegar on Plant Growth and Tuber Yield of Potato

The data presented in Table 3 show that foliar application of guava WV improved plant growth parameters, such as plant height, stem diameter, and tuber yield per pot of potato, in both seasons. Guava WV at 3% gave the highest values of plant height (38.67 and 38.00 cm), stem diameter (11.00 and 11.33 mm), and tuber yield (274.54 and 281.72 g) in seasons 2021 and 2022, respectively (Figure 6). Furthermore, data also showed that using guava WV increased potato tuber yield per pot when compared to untreated plants.

Table 3. Effect of different treatments on plant height, stem diameter, and tuber yield/pot of potato plants cropped under greenhouse conditions during 2021 and 2022 growing seasons.

Treatments	Growing Season 2021			Growing Season 2022		
	Plant Height (cm)	Stem Diameter (mm)	Tuber Yield/Pot (g)	Plant Height (cm)	Stem Diameter (mm)	Tuber Yield/Pot (g)
Guava WV 2%;	32.00 ± 1.53 [b]	10.33 ± 1.53 [a]	243.52 ± 6.88 [b]	34.00 ± 1.00 [ab]	10.33 ± 2.08 [b]	248.55 ± 2.99 [b]
Guava WV 3%;	38.67 ± 3.06 [a]	11.00 ± 1.00 [a]	274.52 ± 5.45 [a]	38.00 ± 2.65 [a]	11.33 ± 1.15 [a]	281.72 ± 6.85 [a]
Amistar 25%;	30.33 ± 4.04 [b]	10.33 ± 0.58 [a]	208.67 ± 2.31 [cd]	31.67 ± 3.51 [bc]	10.00 ± 1.00 [b]	211.67 ± 2.89 [cd]
Inoculated and untreated control;	21.67 ± 1.53 [c]	7.67 ± 1.16 [b]	195.16 ± 6.01 [d]	22.67 ± 2.08 [d]	7.33 ± 0.58 [c]	198.45 ± 8.06 [d]
Uninoculated and untreated control;	27.00 ± 1.00 [b]	9.67 ± 1.16 [a]	215.29 ± 9.52 [c]	29.00 ± 1.73 [c]	9.33 ± 1.53 [b]	227.38 ± 4.93 [c]
LSD at 0.05	5.24	1.50	17.07	4.50	1.14	20.45

Values are the mean of three replicates ±standard deviation (SD). In each column, data followed by the same letter do not differ significantly as determined by the LSD test $p = 0.05$.

Figure 6. Effect of the guava WV treatment on the vegetative growth of potato plants grown in greenhouses.

The Amistar 25% treatment also showed positive effects on plant height, stem diameter, and tuber yield, although it was generally outperformed by the guava WV 3% treatment. Overall, the results of this table indicate that the guava WV 3% treatment had the most significant positive impact on plant growth and tuber yield among the treatments evaluated in both growing seasons.

4. Discussion

Charcoal production from wood under anaerobic conditions is a very good way to preserve the environment, and at the same time, WV is the result of this process. WV is rich in many phenolic compounds, organic acids, antioxidants [11], and some nutrients necessary for plant growth, such as K, Ca, Fe, P, Zn, and Mo [14]. In this study, the GC-MS analysis of guava WV revealed the presence of several bioactive compounds, including phenols, esters, alcohols, antioxidants, and organic acids. The compound with the highest area percentage of the chromatogram (16.12%) was 2,6-Dimethoxyphenol, followed by guaiacol (12.82%) and 1,2,4-trimethoxybenzene (6.39%). According to the findings of this study, phenols were the most prevalent of the detected compounds in the analysis of the guava WV sample. The findings of our study are fully consistent with those of Li et al., [42] found that 2,6-dimethoxyphenol is the most potent antioxidant in WV. Additionally, 3-methyl-1, 2-cyclopentanedione and 2-methoxyphenol have an important impact on the antioxidant activity of WV. Numerous studies suggest that the phenolic compounds in wood vinegar are what give it its antifungal properties [26–28]. In a related study, Ikergami et al. [43] demonstrated that guaiacol, 4-ethyl, 2, methoxy phenol, 6-2, dimethoxy phenol, and ethyl acetate are the most important phenolic compounds in WV with antifungal properties. Yang et al. [44] found that the strongest antioxidant and antibacterial activity of *Litchi chinensis* WV was due to its highly phenolic composition.

In laboratory tests, guava WV had the ability to inhibit growth and stop the development of the pathogen on the PDA medium at the tested concentrations. It has been shown to inhibit several fungal plant pathogens [20]. Higher concentrations of WV (2% and 3%) effectively inhibited the growth of the pathogen in the medium. The most effective inhibitor, 3% guava WV, completely inhibited *Colletotrichum coccoides* growth on the medium (100%). Our findings from this study are consistent with those of Qiaozhi et al. [21], who noted that the apricot tree WV has inhibited the mycelial growth of *Plasmopara viticola*, *Verticillium dahliae*, *Phytophthora capsici*, and *Fusarium graminearum*. Several studies have found that WV can inhibit the growth of many microbes, including *E. coli* [45], *Bacillus subtilis* [46], *Staphylococcus aureus* [47], and *Listeria monocytogenes* [48]. In a recent study, both *Pestalotiopsis* and *Curvularia* species were suppressed in vitro by high concentrations of WV [15]. Previous studies have reported that bamboo WV reduced the growth of the causal agent of wood rot caused by *Ophiostoma* spp. in forest trees [25]. Furthermore, Chukeatirote and Jenjai [49] revealed that the longan WV had antibacterial activity against all bacterial strains tested. On the other hand, the WV only showed inhibitory activity against the yeast *Candida albicans* (C.P. Robin) Berkhout. In a recent study, Desvita et al. [50] showed that WV made from cocoa pod shells inhibited the diameter growth of *Candida albicans* and *Aspergillus niger*. Oramahi et al. [51] discovered that WV from *Vitex pubescens* Vahl could inhibit the growth of *Fomitopsis palustris* (Berk. and Curtis) Gilb and Ryvarden. According to Kadota and Niimi [27], the antifungal properties of WV depend on its chemical composition and phenolic compound content. In this study, we believe that the inhibitory effect of guava WV against the pathogen is due to the phenolic substances present in it. The phenolic compounds present in WV are toxic to microbes when used in high concentrations [52]. WV produced from cocoa pod shells inhibits the growth of *C. albicans* and *Aspergillus niger* Tiegh. With increasing WV concentration, the diameter of the zone inhibiting microbial growth grows [50]. These findings suggest that WV made from cocoa pod shells has antimicrobial properties.

Applying guava, WV significantly decreased the disease severity of black dot in all disease parameters, including stem colonization, root covering with sclerotia, and wilted plant percentage. Both concentrations (2% and 3% v/v) had an effect on the disease in greenhouse conditions, but the 3% concentration was more successful in suppressing the disease than the 2% concentration when compared to untreated plants. The results of our study are in agreement with those obtained by Chuaboon et al. [23], who demonstrated that WV decreased the occurrence of the diseases brown spot and dirty panicles in rice under greenhouse conditions compared to untreated controls. When applied at a 1:32 dilution, WV completely inhibited *Alternaria mali* growth, the causal agent of apple Alternaria blight [22]. Studies have revealed that WV is capable of effectively inhibiting *Pythium aphanidermatum*, *Penicillium griseofulvum*, *Rhizobium* sp., *Sclerotinia sclerotiorum*, and *Fusarium graminearum* [21,53]. According to Yuan et al. [54], the phenol content in WV is thought to have antifungal properties that work by inhibiting fungi enzymes. Saberi et al. [55] investigated the impact of WV on cucumber damping-off and found that *P. aphanidermatum* and *Phytophthora drechsleri* Tucker mycelial growth significantly decreased. In comparison to untreated control plants, the severity of root and crown rot diseases in greenhouse-cultivated cucumber was significantly decreased at the tested concentrations of WV [29]. The main ways that WV inhibits fungi from growing are by slowing down cell division, rupturing cell membranes, leaking electrolytes, and preventing protein synthesis [56,57].

The foliar application of guava WV improved plant growth characteristics of potato plants, such as plant height, stem diameter, and tuber yield per plant in both seasons (2021 and 2022). In general, the concentration of 3% was the most effective on all studied traits. The results of this study are in agreement with those obtained by Chuaboon et al. [23], who showed that WV treatment improved germination, seedling vigor, shoot height, root length, and fresh weight in rice plants compared to untreated controls. WV increased yield in cucumber, lettuce, and cole when used as a foliar fertilizer [31], and in jasmine rice [32]. Charcoal and WV improve the growth, branching, and survival rate of zinnia when used as a mix in planting materials [27]. In addition to enhancing photosynthesis and increasing chlorophyll, WV esters compounds that may help produce sugar and amino acids [23]. WV has also been demonstrated to enhance the development, yield, and quality of a variety of crops [30]. On the other hand, it can be used as a soil fertilizer in the proper concentration. It has been reported that charcoal and smoke stimulate the soil microbial community [52]. A study on tomatoes found that WV increases various enzyme activities, auxin, gibberellin, while also promoting plant growth and nitrogen uptake [58]. WV applications, according to Wang et al. [59], may simultaneously activate several plant-growth-promoting mechanisms: (1) an accumulation of proteins, adaptation to stress and carbohydrate metabolism; (2) antioxidant enzyme accumulation; and (3) reduced reactive oxygen species (ROS) and the presence of malonaldehydes in the root. At low concentrations, the phenolic compounds in wood vinegar, particularly polyphenols and dihydric phenols, can also significantly promote plant growth [60,61]. In a recent study, Fedeli et al. [62] showed that the application of wood distillate at 0.2% significantly increased the content of soluble sugars, starch, and total carbohydrates in treated potato tubers. Wood distillate (wood vinegar) is an environmentally safe bio-based product stimulating plant growth and yield [63].

In conclusion, the present study, WV gave amazing results, whether in its effect on the disease or on the growth of potato plants. WV is a potent organic agricultural product that is eco-friendly and also stops the spread of fungal infection in plants. Therefore, it is natural content and reasonable price make it a good alternative to pesticides and chemical fertilizers. In future investigations may shed more light on the possibility of separating some components of guava WV and evaluating their effectiveness in controlling some plant diseases. Overall, the use of WV provides a promising approach to control the black dot disease of potato plants, contributing to the development of sustainable and environmentally friendly agricultural practices.

Author Contributions: M.M.E.-F.: Conceptualization, methodology, writing—original draft, writing—review and editing; K.A.M.A.-E.: conceptualization, investigation, supervision, writing—review and editing, resources; N.M.A.S.: data curation, software, visualization, investigation; R.M.I.E.-S.: project administration, writing—review and editing, investigation; Y.E.I.: formal analysis, funding acquisition, resources, validation, data curation. All authors have read and agreed to the published version of the manuscript.

Funding: This research work was funded by Institutional Fund Project under grant no "IFPIP: 256-155-1443". The authors acknowledge the technical and financial support provided by the Ministry of Education and King Abdulaziz University, DSR, Jeddah, Saudi Arabia.

Data Availability Statement: Not applicable.

Acknowledgments: This research work was Funded by Institutional Fund Project under grant no "IFPIP: 256-155-1443". The authors gratefully acknowledge the technical and financial support provided by the Ministry of Education and King Abdulaziz University, DSR, Jeddah, Saudi Arabia.

Conflicts of Interest: The authors declare no conflict of interest.

References

1. Lees, A.K.; Hilton, A.J. Rewiev, black dot (*Colletotrichum coccodes*): An increasingly important disease of potato. *Plant Pathol.* **2003**, *52*, 3–12. [CrossRef]
2. Mohan, S.K.; Davis, J.R.; Sorensen, L.H.; Schneider, A.T. Infection of aerial parts of potato plants by *Colletotrichum coccodes* and its effects on premature vine death and yield. *Am. Potato J.* **1992**, *69*, 547–559. [CrossRef]
3. Johnson, D.A.; Miliczky, E.R. Effects of wounding and wetting duration on infection of potato foliage by *Colletotrichum coccodes*. *Plant Dis.* **1993**, *77*, 13–17. [CrossRef]
4. Johnson, D.A. Effect of foliar infection caused by *Colletotrichum coccodes* on yield of Russett Burbank potato. *Plant Dis.* **1994**, *78*, 1075–1078. [CrossRef]
5. Andrivon, D.; Lucas, J.M.; Guerin, C.; Jouan, B. Colonization of roots, stolons, tubers and stems of various potato (*Solanum tuberosum*) cultivars by the black dot fungus *Colletotrichum coccodes*. *Plant Pathol.* **1998**, *47*, 440–445. [CrossRef]
6. Andrivon, D.; Ramage, K.; Guerin, C.; Lucan, M.; Jouan, B. Distribution and fungicide sensitivity of *Colletotrichum coccodes* in French potato-producing areas. *Plant Pathol.* **1997**, *46*, 722–728. [CrossRef]
7. Dickson, B.T. The black dot disease of potato. *Phytopathology* **1926**, *16*, 23–40.
8. El-Marzoky, H.A. First record of black dot disease caused by *Colletotrichum coccodes* on potato plants in Egypt. *Egypt. J. Phytopathol.* **2013**, *41*, 69–84. [CrossRef]
9. Griffiths, H.M.; Zitter, T.A.; Loeffler, K.; De Jong, W.S.; Menasha, S. First report in North America of atypical symptoms caused by *Colletotrichum coccodes* on field-grown potato tubers during storage. *Plant Health Prog.* **2010**, *11*, 46. [CrossRef]
10. Anonim. Wood vinegar. In *Forest Energy Forum, 9*; FAO of United Nations: Rome, Italy, 2001.
11. Guillen, M.D.; Manzanos, M.J. Study of the volatile composition of an aqueous oak smoke preparation. *Food Chem.* **2002**, *79*, 283–292. [CrossRef]
12. Aguirre, J.L.; Baena, J.; Martin, M.T.; Nozal, L.; Gonzalez, S.; Manjon, J.L.; Peinado, M. Composition, ageing and herbicidal properties of wood vinegar obtained through fast biomass pyrolysis. *Energies* **2020**, *13*, 2418. [CrossRef]
13. Grewal, A.; Abbey, L.; Gunupuru, L.R. Production, prospects and potential application of pyroligneous acid in agriculture. *J. Anal. Appl. Pyrolysis* **2018**, *135*, 152–159. [CrossRef]
14. Zulkarami, B.; Ashrafuzzaman, M.; HusniIsmail, M.R. Effect of pyroligneous acid on growth, yield and quality improvement of rock melon in soilless culture. *Aust. J. Crop. Sci.* **2011**, *5*, 1508–1514.
15. Obeng, J.; Agyei-Dwarko, D.; Teinor, P.; Danso, I.; Lutuf, H.; Lekete-Lawson, E.; Ablormeti, F.K.; Eddy-Doh, M.A. Bioactivity of an organic farming aid with possible fungistatic properties against some oil palm seedling foliar pathogens. *Sci. Rep.* **2023**, *13*, 1280. [CrossRef] [PubMed]
16. Najafi-Ghiri, M.; Boostani, H.R.; Hardie, A.G. Investigation of biochars application on potassium forms and dynamics in a calcareous soil under different moisture conditions. *Arch. Agron. Soil Sci.* **2022**, *68*, 325–339. [CrossRef]
17. Sivaram, A.K.; Panneerselvan, L.; Mukunthan, K.; Megharaj, M. Effect of pyroligneous acid on the microbial community composition and Plant Growth-Promoting Bacteria (PGPB) in soils. *Soil Syst.* **2022**, *6*, 10. [CrossRef]
18. Fengel, D.; Wegener, G. *Wood: Chemistry, Ultrastructure, Reactions*; Walter de Gruyter: Berlin, Germany, 1983.
19. Yatagai, M.; Nishimoto, M.; Hori, K.; Ohira, T.; Shibata, A. Termiticidal activity of wood vinegar, its components and their homologues. *J. Wood Sci.* **2002**, *48*, 338–342. [CrossRef]
20. Yodthong, B.; Jirasak, T.; Nuethip, D.; Yaowalak, S.; Nuanchai, K. Utilization of wood vinegars as sustainable coagulating and antifungal agents in the production of natural rubber sheets. *J. Environ. Sci. Tech.* **2008**, *1*, 157–163. [CrossRef]
21. Mao, Q.; Zhao, Z.; Ma, X. Preparation, toxicity and components analysis of apricot branch wood vinegar. *J. Northwest A F Univ.-Nat. Sci. Ed.* **2009**, *37*, 91–96.
22. Jung, K.H. Growth inhibition effect of pyroligneous acid on pathogenic fungus, *Alternaria mali*, the agent of Alternaria blotch of apple. *Biotechnol. Bioproc. Eng.* **2007**, *12*, 318–322. [CrossRef]
23. Chuaboon, W.; Ponghirantanachoke, N.; Athinuwat, D. Application of wood vinegar for fungal disease controls in paddy rice. *Appl. Environ. Res.* **2016**, *38*, 77–85. [CrossRef]
24. Hwang, Y.; Matsushita, Y.; Sugamoto, K.; Matsui, T. Antimicrobial effect of the wood vinegar from *Crytomeria japonica* sapwood on plant pathogenic microorganisms. *J. Microbiol. Biotechnol.* **2005**, *15*, 1106–1109.
25. Velmurugan, N.; Chun, S.S.; Han, S.S.; Lee, Y.S. Characterization of Chikusaku-eki and Mokusaku-eki and its inhibitory effect on sapstaining fungal growth in laboratory scale. *Int. J. Environ. Sci. Technol.* **2009**, *6*, 13–22. [CrossRef]
26. Bridgewater, A.V. Biomass fast pyrolysis. *Thermal Sci.* **2004**, *8*, 21–50. [CrossRef]
27. Kadota, M.; Niimi, Y. Effects of charcoal with pyroligneous acid and barnyard manure on bedding plants. *Sci. Hortic.* **2004**, *101*, 327–332. [CrossRef]
28. Cowan, M.M. Plant products as antimicrobial agents. *Clin. Microbiol. Rev.* **1999**, *12*, 564–582. [CrossRef]
29. Saberi, M.; Askary, H.; Sarpeleh, A. Integrated effects of wood vinegar and tea compost on root rot and vine decline and charcoal root rot diseases of muskmelon. *Biocontrol Plant Protect.* **2013**, *1*, 91–101. [CrossRef]
30. Nakayama, F.S.; Vinyard, S.H.; Chow, P.; Bajwa, D.S.; Youngquist, J.A.; Muehl, J.H.; Krzysik, A.M. Guayule as a wood preservative. *Ind. Crop. Prod.* **2001**, *14*, 105–111. [CrossRef]
31. Mu, J.; Yu, Z.M.; Wu, W.Q.; Wu, Q.L. Preliminary study of application effect of bamboo vinegar on vegetable growth. *For. Study China* **2006**, *8*, 43–47. [CrossRef]

32. Jothityangkoon, D.; Ruamtakhu, C.; Tipparak, S.; Wanapat, S.; Polthanee, A. Using wood vinegar in increasing rice productivity. In Proceedings of the 2nd International Conference on Rice for the Future, Bangkok, Thailand, 5–9 November 2007; pp. 28–34.
33. Sutton, B.C. The genus Glomerella and its anamorph Colletotrichum. In *Colletotrichum: Biology, Pathology and Control*; Bailey, J.A., Jeger, M.J., Eds.; CAB International: Wallingford, UK, 1992; pp. 1–26.
34. Rehman, F.; Lee, S.K.; Kim, H.S.; Jeon, J.H.; Park, J.; Joung, H. Dormancy breaking and effects on tuber yield of potato subjected to various chemicals and growth regulators under greenhouse conditions. *J. Biol. Sci.* 2001, *1*, 818–820. [CrossRef]
35. Abdel-Hafez, S.I.I.; Abo-Elyousr, K.A.M.; Abdel-Rahim, I.R. Fungicidal activity of extracellular products of cyanobacteria against Alternaria porri. *Eur. J. Phycol.* 2015, *50*, 239–245. [CrossRef]
36. Perez-Mora, J.L.; Cota-Rodriguez, D.A.; Rodriguez-Palafox, E.E.; Garcia-Leon, E.; Beltran-Pena, H.; Lima, N.B.; Tovar-Pedraza, J.M. First confirmed report of *Colletotrichum coccodes* causing black dot on potato in Mexico. *J. Plant Dis. Prot.* 2020, *127*, 269–273. [CrossRef]
37. Nitzan, N.; Evans, M.A.; Cummings, T.F.; Johnson, D.A.; Batchelor, D.L.; Olsen, C.; Haynes, K.G.; Brown, C.R. Field resistance to potato stem colonization by the black dot pathogen *Colletotrichum coccodes*. *Plant Dis.* 2009, *93*, 1116–1122. [CrossRef] [PubMed]
38. Senthilkumar, R.; Manivannan, R.; Balasubramanian, A.; Rajkapoor, B. Antioxidant and hepatoprotective activity of ethanol extract of *Indigofera trita* Linn. On CCl4 induced hepatoxicity in rats. *J. Pharmacol. Toxicol.* 2008, *3*, 344–350. [CrossRef]
39. Zhang, T.; Xu, Y.; Li, J. Inhibitory effect of wood vinegar produced from apricot shell on *Aspergillus fumigatus*. *Agric. Biotechnol.* 2018, *7*, 112–115.
40. MSTAT-C. *A Software Program for the Design, Management and Analysis of Agronomic Research Experiments*; Michigan State University: East Lansing, MI, USA, 1991; p. 400.
41. Gomez, K.A.; Gomez, A.A. Comparison between treatment means. In *Statistical Procedures for Agricultural Research*; Lviley, A., Ed.; Interscience Publication: New York, NY, USA, 1984; pp. 187–680.
42. Li, Z.; Zhang, Z.; Wu, L.; Zhang, H.; Wang, Z. Characterization of five kinds of wood vinegar obtained from agricultural and forestry wastes and identification of major antioxidants in wood vinegar. *Chem. Res. Chin. Univ.* 2019, *35*, 12–20. [CrossRef]
43. Ikergami, F.; Sekin, T.; Fuji, Y. Antidemaptophyte activity of phenolic compounds in Mokusaku-eki. *Yakugaku Zasshi* 1992, *118*, 27–30. [CrossRef]
44. Yang, J.F.; Yang, C.H.; Liang, M.T.; Gao, Z.J.; Wu, Y.W.; Chuang, L.Y. Chemical composition, antioxidant, and antibacterial activity of wood vinegar from *Litchi chinensis*. *Molecules* 2016, *21*, 1150. [CrossRef]
45. Soares, J.M.; da Silva, P.F.; Puton, B.M.S.; Brustolin, A.P.; Cansian, R.L.; Dallago, R.M.; Valduga, E. Antimicrobial and antioxidant activity of liquid smoke and its potential application to bacon. *Innov. Food Sci. Emerg. Technol.* 2016, *38*, 189–197. [CrossRef]
46. Saloko, S.; Darmadji, P.; Setiaji, B.; Pranoto, Y. Antioxidative and antimicrobial activities of liquid smoke nanocapsules using chitosan and maltodextrin and its application on tuna fish preservation. *Food Biosci.* 2014, *7*, 71–79. [CrossRef]
47. Desvita, H.; Faisal, M.; Mahidin, M.; Suhendrayatna, S. Preliminary study on the antibacterial activity of liquid smoke from cacao pod shells (*Theobroma cacao* L.). *IOP Conf. Ser. Mater. Sci. Eng.* 2021, *1098*, 022004. [CrossRef]
48. Morey, A.; Bratcher, C.L.; Singh, M.; dan McKee, S.R. Effect of liquid smoke as an ingredient in frankfurters on *Listeria monocytogenes* and quality attributes. *Poult. Sci. J.* 2012, *91*, 2341–2350. [CrossRef]
49. Chukeatirote, E.; Jenjai, N. Antimicrobial activity of wood vinegar from dimocarpus longan. *Environ. Asia* 2018, *11*, 161–169. [CrossRef]
50. Desvita, H.; Faisal, M.; Mahidin, M.; Suhendrayatna, S. Antimicrobial potential of wood vinegar from cocoa pod shells (*Theobroma cacao* L.) against *Candida albicans* and *Aspergillus niger*. *Mater. Today Proc.* 2022, *63*, S210–S213. [CrossRef]
51. Oramahi, H.A.; Yoshimura, T. Antifungal and antitermitic activities of wood vinegar from *Vitex pubescens* Vahl. *J. Wood Sci.* 2013, *59*, 344–350. [CrossRef]
52. Steiner, C.; Das, K.C.; Garcia, M.; Forster, B.; Zech, W. Charcoal and smoke extract stimulate the soil microbial community in highly weathered xanthic Ferralsol. *Pedobiologia* 2007, *51*, 359–366. [CrossRef]
53. Quan, S. Application of wood vinegar to control diseases. *J. Agric. Sci. Yanbian Univ.* 1994, *2*, 113–116.
54. Yuan, F.; Zhang, C.; Shen, Q.R. Alleviating effect of phenol compounds on cucumber Fusarium wilt and mechanism. *Agric. Sci. China* 2003, *2*, 647–652.
55. Saberi, M.; Sarpeleh, A.; Askary, H. Management of damping-off and increasing of dome growth traits of cucumber in greenhouse culture using citrus wood vinegar. *Appl. Res. Plant Prot.* 2015, *4*, 99–111.
56. Wei, Q.; Ma, X.; Zhao, Z.; Zhang, S.; Liu, S. Antioxidant activities and chemical profiles of pyroligneous acids from walnut shell. *J. Anal. Appl. Pyrolysis* 2010, *88*, 149–154. [CrossRef]
57. Zhang, L.; Wang, L.D.; Gong, W. Chemical constituents and antibacterial activity of Jujube kernel vinegar. *Food Sci.* 2016, *37*, 123–127.
58. Zhu, K.; Gu, S.; Liu, J.; Luo, T.; Khan, Z.; Zhang, K.; Hu, L. Wood vinegar as a complex growth regulator promotes the growth, yield, and quality of rapeseed. *Agronomy* 2021, *11*, 510. [CrossRef]
59. Wang, Y.; Qiu, L.; Song, Q.; Wang, S.; Wang, Y.; Ge, Y. Root proteomics reveals the effects of wood vinegar on wheat growth and subsequent tolerance to drought stress. *Int. J. Mol. Sci.* 2019, *20*, 943. [CrossRef] [PubMed]
60. Lin, K.; Ye, F.; Lin, Y.; Li, Q. Research progress on the mechanism of phenolic substances on soil and plants. *J. Eco-Agric.* 2010, *5*, 1130–1137.

61. Yan, Y.; Lu, X.; Li, L.; Zheng, J.; Pan, G. The composition of straw pyrolysis wood vinegar and its effect on the growth and quality of pepper. *J. Nanjing Agric. Univ.* **2011**, *5*, 58–62.
62. Fedeli, R.; Vannini, A.; Grattacaso, M.; Loppi, S. Wood distillate (pyroligneous acid) boosts nutritional traits of potato tubers. *Ann. Appl. Biol.* **2023**, 1–6. [CrossRef]
63. Fanfarillo, E.; Fedeli, R.; Fiaschi, T.; de Simone, L.; Vannini, A.; Angiolini, C.; Loppi, S.; Maccherini, S. Effects of wood distillate on seedling emergence and first-stage growth in five threatened arable plants. *Diversity* **2022**, *14*, 669. [CrossRef]

Disclaimer/Publisher's Note: The statements, opinions and data contained in all publications are solely those of the individual author(s) and contributor(s) and not of MDPI and/or the editor(s). MDPI and/or the editor(s) disclaim responsibility for any injury to people or property resulting from any ideas, methods, instructions or products referred to in the content.

Article

Natural Products Obtained from Argentinean Native Plants Are Fungicidal against Citrus Postharvest Diseases

Norma Hortensia Alvarez [1,2], María Inés Stegmayer [1], Gisela Marisol Seimandi [1,*], José Francisco Pensiero [1,2], Juan Marcelo Zabala [1,2], María Alejandra Favaro [1,2] and Marcos Gabriel Derita [1,3,*]

[1] ICiAgro LitoraL, Universidad Nacional del Litoral, CONICET, Kreder 2805, Esperanza, Santa Fe 3080HOF, Argentina; nalvarez@fca.unl.edu.ar (N.H.A.); mistegmayer@gmail.com (M.I.S.); jfpensi@fca.unl.edu.ar (J.F.P.); jmzabala@fca.unl.edu.ar (J.M.Z.); mfavaro@fca.unl.edu.ar (M.A.F.)
[2] Facultad de Ciencias Agrarias, Universidad Nacional del Litoral, Kreder 2805, Esperanza, Santa Fe 3080HOF, Argentina
[3] Cátedra de Farmacognosia, Facultad de Ciencias Bioquímicas y Farmacéuticas, Universidad Nacional de Rosario, Suipacha 531, Rosario, Santa Fe S2002LRK, Argentina
* Correspondence: giselaseimandi@hotmail.com.ar (G.M.S.); mderita@fbioyf.unr.edu.ar (M.G.D.); Tel.: +54-9-341-5317769 (M.G.D.)

Abstract: Natural products obtained from plants constitute an alternative to chemically synthesized fungicides, whose improper use might have caused the development of resistant fungal strains. In the present work, 40 products obtained from 20 native Argentinean plant species were tested against three citrus postharvest pathogens: *Penicillium digitatum*, *Penicillium italicum*, and *Geotrichum citri-aurantii*. Natural products were obtained by classical solvent extraction methods and the fungicidal evaluation was carried out by agar diffusion tests using commercial fungicides as negative controls and dimethyl sulfoxide as a positive one. The inhibition percentages were determined 7 and 14 days post inoculation of each fungus. Most of the products tested showed inhibition percentages higher than 50% for *G. citri-aurantii*, but only 20% of them were active against *P. digitatum* and *P. italicum*. The most promising products which inhibited (100%) the growth of at least one of the three phytopathogens were extracted from the following plants: *Orthosia virgata*, *Petiveria alliacea*, *Funastrum clausum*, *Solanum caavurana*, and *Solanum pilcomayense*. These products were tested over inoculated oranges and there were no statistically significant differences between the treatments with a commercial fungicide and the methanolic extract in the control of fruit rot. The products extracted from native plants have fungicide potential, but further studies are required.

Keywords: natural products; post harvest; citrus diseases

Citation: Alvarez, N.H.; Stegmayer, M.I.; Seimandi, G.M.; Pensiero, J.F.; Zabala, J.M.; Favaro, M.A.; Derita, M.G. Natural Products Obtained from Argentinean Native Plants Are Fungicidal against Citrus Postharvest Diseases. *Horticulturae* **2023**, *9*, 562. https://doi.org/10.3390/horticulturae9050562

Academic Editors: Hillary Righini, Roberta Roberti and Stefania Galletti

Received: 31 March 2023
Revised: 28 April 2023
Accepted: 5 May 2023
Published: 9 May 2023

Copyright: © 2023 by the authors. Licensee MDPI, Basel, Switzerland. This article is an open access article distributed under the terms and conditions of the Creative Commons Attribution (CC BY) license (https://creativecommons.org/licenses/by/4.0/).

1. Introduction

Fruits of *Citrus* spp. (Rutaceae) are cultivated in more than one hundred countries and their fruits are widely consumed throughout the world. Postharvest handling tries to achieve the highest quality fruits, increasing their postharvest life and reducing production losses, thus obtaining commercially suitable fruits [1]. According to the latest estimates by Federcitrus (2022), Argentina produces 1,038,168 tons of oranges; 77,000 of them are destined for the export market [2]. Postharvest losses produced by phytopathogenic origin (approximately 30% of the production) are considered of economic importance. Green and blue molds, caused by *Penicillium digitatum* (Pers.: Fr.) Sacc. and *Penicillium italicum* Wehmer, respectively (order Eurotiales, Aspergillaceae family), and sour rot caused by *Geotrichum citri-aurantii* Ferraris (order Saccharomycetales, Dipodascaceae family), are the most economically important postharvest diseases as they affect not only the quantity but also the quality of citrus fruits which, once infected, must be discarded from the production lot [1–5]. *Penicillium digitatum* and *P. italicum* cause citrus diseases by invading wounds in the fruit rind. Wounds occur during the harvest and subsequent handling of fruit in the packinghouse or during commercialization, but some infections can occur before harvest through injuries, cracks, or wounds made by insects in the field. In this case, the fruit infected long before harvest usually drops from the tree, but fruit infected three days before harvest cannot be detected and may be harvested, becoming a source of contamination [2]. A circular area surrounding the infection site (rind wound) appears water-soaked, soft, and decolorized. In the meantime, the fungus grows, and an aerial white mycelium develops in the center of the lesion and expands radially. In the case of green mold, after 7–8 days the central area of the lesion becomes olive green colored and surrounded by a broad band of dense and non-sporulating white mycelium. In contrast, blue mold presents a central sporulating area, blue or bluish green colored [1].

Geotrichum citri-aurantii causes sour rot; this fungus lives in dead plant tissue or orchard soils and it may cause diseases in fruits. Symptoms include maceration and disintegration of the fruit cuticle and pulp. Once the fruits rot, they generate a strong fermented and rancid smell. If high humidity conditions are present, a very thin white mycelium appears on the fruits, rapidly disappearing [3,4].

These three postharvest diseases have been controlled worldwide for many years solely by the application of conventional fungicides after harvest such as (1) Imazalil (IMZ) [(RS)-1-(β-allyloxy-2,4-dichlorophenethyl) imidazole], which constitutes the most common fungicide employed postharvest by the citrus industry; (2) Thiabendazole (TBZ) [2-(1,3-thiazol-4-yl) benzimidazole], which belongs to the benzimidazole group of fungicides and has been used for the control of citrus molds; (3) Sodium ortho-phenylphenate (SOPP) [sodium (1,1′-biphenyl)-2-olate], which is currently used in combinate fungicide formulations; (4) Guazatine and propiconazole, which have been used in Europe, South Africa, and Australia for the control of *G. citri-aurantii*; and finally, pyrimethanil (PYR) (4,6-dimethyl-N-phenyl-2-pyrimidinamine), fludioxonil (FLU) [4-(2,2-difluoro-1,3-benzodioxol-4-yl)-1H-pyrrole-3-carbonitrile, azoxystrobin (AZX) [methyl-(2E)-2-{2-[6-(2-cyanophenoxy) pyrimidin-4-yloxy] phenyl}-3-methoxyacrylate], and Trifloxystrobin (TFX) (methyl-(E)-methoxyimino-{(E)-α-[1-(α,α,α-trifluoro-m-tolyl) ethylideneaminoxy]-o-tolyl}acetate) are classified as "reduced-risk" fungicides [1,5]. IMZ and TBZ are used alone, in mixtures, or separately applied in sequence. These have constituted the most widely used treatments for more than 25 years, which has contributed to the proliferation of resistant fungal strains [6]. Consequently, the excessive or inadequate use of these fungicides represents a risk to human and animal health, due to food contamination and the accumulation of toxic residues in the environment. Due to marketing globalization and climate change, the problem is growing at an accelerated pace [7,8].

In this context, there are some new alternatives for the control of diseases that occur in citrus postharvest, such as biological control, natural products obtained from different sources, and thermotherapy [9]. Biological control is based on the utilization of antagonistic microorganisms to reduce/keep the population of some phytopathogens below the levels that cause economic losses [10]. Another alternative is the use of natural products obtained from plant extracts or essential oils, which have revealed non-toxic action and biodegradability but are potent antibiotics against phytopathogens [8,11,12]. Finally, through cycles of considerable temperature changes on fruits, fungal infections can be reduced [9].

Plants provide unlimited opportunities for evaluating and isolating new antifungal compounds because of their vast chemical diversity; in fact, numerous antifungal compounds currently used against human pathogens have been isolated from them [13,14].

The aim of this study consisted of the discovery of new botanical products obtained from native Argentinean species with potential fungicide properties. For this, we selected 20 native plants from the central zone of Argentina which were evaluated for their fungicide properties against morphologically and molecularly characterized strains of *P. italicum*, *P. digitatum*, and *G. citri-aurantii* (isolated from citrus that presented the correspondent disease symptoms), by using a high-throughput and simple bioassay. Data were collected 7 and 14 days after fungal inoculation and each sample was classified according to its fungal growth inhibition percentage. Statistical analyses were performed based on the part of each plant used and the type of extract in which the most active compounds might be present. The methanolic extract obtained from *S. pilcomayense* was evaluated over postharvest oranges artificially inoculated with *P. digitatum*. Finally, a discussion on the most active species, their ethnopharmacological uses, and phytochemistry, was accomplished.

2. Materials and Methods

2.1. Plant Material

The plants were collected from farms and at the side of roads in areas surrounding the Litoral region of Argentina, between the years 2019 and 2020. Each plant sample was identified, and a *Voucher Specimen* was deposited in the Herbarium of the FCA-UNL "Arturo Ragonese" (Herbario SF), Kreder 2805-(3080HOF)-Esperanza, Argentina (Table 1). After harvesting, the plants were dried in a suitable environment (a dark room with low relative humidity), while their different parts (leaves, flowers, fruits, or the whole plant) were separated according to the type of extract that had to be prepared. Three species (*Orthosia virgata*, *Solanum argentinum*, and *Solanum pilcomayense*) were collected in different zones to detect differences in their bioactivities which may be associated with phytochemical changes determined by the environment.

Table 1. Plant species used for extracts preparation to be tested against the orange pathogens *Penicillium digitatum*, *P. italicum*, and *Geotrichum citri-aurantii*. Botanic family, *Voucher specimen*, collection data, and place of collection or acquisition are also reported.

Family	Plant Scientific Name	V. specimen	Collection Data and Place of Acquisition
Apocynaceae	*Araujia brachystephana* (Griseb.) Fontella & Goyder	Pensiero et Zabala 13800	13/08/2019, San Javier, Santa Fe, 30°03′23.2″ S 59°51′34.8″ W
	Forsteronia glabrescens Müll. Arg.	Pensiero et Zabala 13804	13/08/2019, Gral. Obligado, Santa Fe, 28°42′54.8″ S 59°23′13.7″ W
	Orthosia virgata (Poir.) E. Fourn.	Pensiero et Zabala 13802	13/08/2019, San Javier, Santa Fe, 30°03′23.2″ S 59°51′34.8″ W
	Orthosia virgata (Poir.) E. Fourn.	Pensiero et Zabala 13805	13/08/2019, Gral. Obligado, Santa Fe, 28°42′54.8″ S 59°23′13.7″ W
	Funastrum clausum (Jacq.) Schltr.	Pensiero et Zabala 13797	13/08/2019, San Javier, Santa Fe, 30°03′23.2″ S 59°51′34.8″ W
Bromeliaceae	*Tillandsia tricholepis* Baker	Pensiero et Zabala 13798	13/08/2019, San Javier, Santa Fe, 30°03′23.2″ S 59°51′34.8″ W
Fabaceae	*Albizia inundata* (Mart.) Barneby & J.W. Grimes	Pensiero et Zabala 13816	14/08/2019, Gral. Obligado, Santa Fe, 28°12′53.9″ S 59°08′58.3″ W
	Vachellia caven (Molina) Seigler & Ebinger	Derita M. 39	17/03/2019, Las Colonias, Santa Fe, 31°26′30.3″ S 60°56′24.7″ W
Phytolaccaceae	*Petiveria alliacea* L.	Pensiero et Zabala 13975	13/08/2019, San Javier, Santa Fe, 30°03′23.2″ S 59°51′34.8″ W
Polygonaceae	*Polygonum stelligerum* Cham.	Derita M. 55	16/03/2019, San Pedro, Bs As, 33°40′11.6″ S 59°39′38.4″ W
	Polygonum lapathifolium L.	Derita M. 54	16/03/2019, San Pedro, Bs As, 33°40′11.6″ S 59°39′38.4″ W
Ranunculaceae	*Clematis montevidensis* Spreng.	Pensiero et Zabala 13824	17/08/2019, Vera, Santa Fe, 29°28′17.6″ S 60°05′27.7″ W
Nictaginaceae	*Pisonia zapallo* Griseb	Pensiero et Zabala 13808	13/08/2019, Gral. Obligado, Santa Fe, 28°42′54.8″ S 59°23′13.7″ W
Solanaceae	*Solanum argentinum* Bitter & Lillo	Pensiero et Zabala 13820	16/08/2019, Tapenagá, Chaco, 27°53′17.9″ S 59°51′39.4″ W
	Solanum argentinum Bitter & Lillo	Pensiero et Zabala 13817	16/08/2019, Lib. San Martín, Chaco, 26°07′13.9″ S 59°56′25.1″ W
	Solanum granulosum-leprosum Dunal	Pensiero et Zabala 13806	13/08/2019, Gral. Obligado, Santa Fe, 28°42′54.8″ S 59°23′13.7″ W
	Solanum pilcomayense Morong	Pensiero et Zabala 13815	14/08/2019, Gral. Obligado, Santa Fe, 28°13′42.2″ S 59°06′52.6″ W
	Solanum pilcomayense Morong	Pensiero et Zabala 13801	13/08/2019, San Javier, Santa Fe, 30°03′23.2″ S 59°51′34.8″ W
	Solanum caavurana Vell.	Pensiero et Zabala 13813	14/08/2019, Gral. Obligado, Santa Fe, 28°13′42.2″ S 59°06′52.6″ W
Verbenaceae	*Lantana megapotamica* (Spreng.) Tronc.	Pensiero et Zabala 13796	13/08/2019, San Javier, Santa Fe, 30°03′23.2″ S 59°51′34.8″ W

2.2. Extraction Process

For the preparation of the extracts, air-dried aerial parts of each species (100 g) were pulverized and successively macerated with 250 mL of dichloromethane (DCM) and 250 mL of methanol (MeOH) under mechanical stirring (3 × 24 h each) to obtain the corresponding extracts after filtration and evaporation. These solvents were selected based on their differential extraction capability of less polar metabolites (DCM) and more polar metabolites (MeOH). In this way, most of the compounds present in the plant species under study were extracted.

2.3. Fungal Strains

Monosporic strains of *Geotrichum* and *Penicillium* spp. called NA1, NA2, and NA3 were obtained from orange fruits that presented typical symptoms of sour rot, green mold rot, and blue mold rot, and cultivated in Potato-Dextrose-Agar medium (PDA) [15–18]. The strains were morphologically characterized and conserved in the Mycology Reference Center (CEREMIC-UNR, Rosario, Argentina) under the codes CCC-5-2019 (NA1), CCC-102 (NA2), and CCC-101 (NA3). Isolates were also conserved at $-20\ °C$ on dried filter paper in the mycological collection of ICiAgro Litoral, UNL, CONICET, FCA, Argentina.

2.4. Molecular Characterization

The identity of the isolates was confirmed through molecular characterization. For that, fungal genomic DNA was extracted from 7-day-old cultures grown on PDA at 20–25 °C as described in Gupta et al. [15] and was used as the template for PCR amplification of a segment of the ITS (Internal Transcribed Spacer) region of ribosomal nuclear DNA (rDNA) using the primers ITS4 (TCCTCCGCTTATTGATATGC) and ITS5 (GGAAGTAAAAGTCG-TAACAAGG) [16]. PCR reactions were performed on a Techne TC-312 thermal cycler (Techne, Cambridge, UK) in 20 μL reaction mixtures containing 1× PCR buffer, 2.5 mM $MgCl_2$, 0.4 μM each primer, 0.2 mM dNTPs, 1 U of Taq DNA polymerase (PB-L, Productos Bio-Lógicos®, Rosario, Argentina), and 100 ng of genomic DNA. Amplifications were programmed to carry out an initial denaturation at 94 °C for 5 min, followed by 36 cycles, each consisting of a denaturation step at 94 °C for 30 s, an annealing step, and an extension step at 57 °C for 30 s and at 72 °C for 30 s. The final extension was carried out at 72 °C for 7 min. PCR products were visualized under ultraviolet light on 1.5% (w/v) agarose gel in 1× TAE buffer stained with GelRed (Biotium Inc., Fremont, CA, USA). A UST-30M-8E, Biostep transilluminator (Biostep, Jahnsdorf, Germany) was used. The amplified products were purified and sequenced with the same primers in Macrogen (Seoul, Republic of Korea). The identification was performed by comparing the sequences with all fungal sequences of the GenBank Nucleotide Database hosted by the National Center for Biotechnology Information (NCBI, https://www.ncbi.nlm.nih.gov/ accessed on 15 December 2022) using BLASTn (Nucleotide Basic Local Alignment Search Tool) [17]. All the sequences generated in this study were deposited in GenBank.

The inoculum for bioassays was obtained according to the Clinical & Laboratory Standards Institute reported procedures and adjusted to 1×10^4 Colony Forming Units (CFU) mL^{-1} [18].

2.5. In Vitro Susceptibility Test

Diffusion tests were carried out using 9 cm diameter sterile Petri dishes provided with four divisions so that each sample was tested in quadruplicate. Plant extract solutions were prepared at a concentration of 50 mg mL^{-1} in DMSO and once dissolved, 400 μL of this stock solution was diluted in 20 mL of molten PDA culture medium. After vigorous shaking and before the mixture was solidified, 5 mL was poured into each of the four compartments of the Petri dishes and cooled down. A conidia concentration between 10^4 and 10^5 CFU mL^{-1} was inoculated inside a well located in the center of each compartment once the medium containing 1000 ppm of each plant extract was solidified, according to our previous methodology developed and published [7,13]. A negative control was prepared using the commercial fungicide Imazalil® and the solvent DMSO without plant extract served as a positive (growth) control. Once the mycelium of the control plates completely covered the surface of the medium (approximately 7 days), the mycelium diameter of each plant-treated plate was carried out by scanning the plates with ImageJ® software [19]. The average inhibition percentages with their standard deviations were determined at 7 and 14 days after inoculation.

2.6. Fungicidal Assays on Wounded Oranges Using S. pilcomayense Methanolic Extract

Oranges (cv. "Salustiana") were harvested from the Experimental Field of Intensive and Forestry Crops (*Facultad de Ciencias Agrarias, Universidad Nacional del Litoral*) at the mature stage and sorted based on size and absence of physical injuries or disease infection. They were artificially inoculated with *P. digitatum* following the methodology described by Di Liberto et al. [20]. Three groups of 10 oranges were used: (1) control with sterile water, (2) treatment with 3000 ppm aqueous solution of *S. pilcomayense* methanolic extract, and (3) treatment with 3000 ppm aqueous solution of Imazalil®. The treatments were carried out after 2 h of inoculation, by 5 s immersion of each fruit into a beaker with the corresponding solutions mentioned above. For the in vivo test, the 3000 ppm concentration of the extract was chosen because we did not have a good response when we treated the fruits with a 1000 ppm concentration. So, we decided to hardly increase the fungicidal concentration resulting from the in vitro assays.

The fruits were stored at 25 °C and 85% RH for 14 days. After storage, the degree of *P. digitatum* sporulation on the surface of decayed fruits was evaluated on a 0 to 4 scale (sporulation index). With this scale, the severity of the disease was visually quantified assuming the following values: 0 (negligible sporulation); 1 (fruit with lesions on up to 10% of its surface); 2 (fruit with injuries on between 10 and 30% of its surface); 3 (fruit with lesions on between 30 and 50% of its surface); and 4 (refers to dense fungal sporulation over the entire fruit that infected more than 50% of its surface) [20]. The sporulation index for each fruit was treated as a replicate, and each treatment mean was subjected to statistical analysis.

2.7. Statistical Analysis

For the in vitro test, the differences in the mean percentage of fungal growth in the presence of each extract were compared with positive and negative controls by statistical analysis with a 95% Confidence Interval (CI) according to our published results [13].

For the in vivo test, experimental data were analyzed statistically by one-way ANOVA followed by Tukey's multiple comparison tests ($\alpha = 0.05$) using the GraphPad Prism 7.0 software and according to Di Liberto et al. [20].

3. Results

3.1. Plant Extract Yields

The part of each plant used and the corresponding DCM and MeOH extract yields (g/100 g of dried plant material) are provided in Table 2. For the cases of duplicated species under study, *O. virgata* showed similar crude extract percentages (2.4 and 2.3 for DCM extracts; 4.1 and 5.0 for MeOH extracts), but *S. argentinum* and *S. pilcomayense* presented differences in DCM yields and both (DCM and MeOH) yields, respectively.

Table 2. Part of each plant used and the corresponding DCM and MeOH extract yields (g/100 g of dried plant material).

Plant Scientific Name	Part Used	DCM Extract Yield (%)	MeOH Extract Yield (%)
A. brachystephana	Aerial parts	2.4	10.7
F. glabrescens	Aerial parts	1.9	7.3
O. virgata	Aerial parts	2.4	4.1
O. virgata	Aerial parts	2.3	5.0
F. clausum	Aerial parts	3.7	7.2
T. tricholepis	Whole plant	3.4	4.4
A. inundata	Aerial parts	1.1	3.4
V. caven	Flowers	1.3	7.3
P. alliacea	Aerial parts	0.4	3.0
P. stelligerum	Leaves	0.7	2.8
P. lapathifolium	Leaves	0.6	3.0
C. montevidensis	Aerial parts	1.0	10.2
P. zapallo	Aerial parts	1.3	7.7
S. argentinum	Aerial parts	0.7	7.0
S. argentinum	Aerial parts	3.5	7.7
S. granulosum-leprosum	Aerial parts	3.2	5.0
S. pilcomayense	Aerial parts	1.3	7.6
S. pilcomayense	Aerial parts	0.5	3.2
S. caavurana	Aerial parts	1.4	10.3
L. megapotamica	Aerial parts	0.4	4.5

3.2. Fungicidal Evaluation of the Selected Plant Extracts against P. digitatum, P. italicum, and G. citri-aurantii

Fungal isolates NA1, NA2, and NA3 were morphologically identified by CEREMIC-UNR as *G. citri-aurantii* (CCC-5-2019), *P. digitatum* (CCC-102), and *P. italicum* (CCC-101), respectively. Figure 1 shows the typical symptoms and signs caused by each fungus on oranges, the morphological characteristics of the colonies developed after isolation, and the microscopic features of their conidia. Typical green, blue, and white sporulation were observed above the rots and in the plates.

To confirm morphological identification, DNA was extracted from the three fungal isolates, and the ITS region was amplified and sequenced. BLASTn searches revealed that the fungal isolate sequences NA1, NA2, and NA3 presented 99.4%, 100%, and 100% of identity with previously characterized strains of *G. citri-aurantii* (EU131181, [21]), *P. digitatum* (MT448740, [22]), and *P. italicum* (MK736929, [23]), respectively. The nucleotide sequences were deposited in GenBank (accession numbers OQ132875, OQ132876, and OQ132877 for NA1, NA2, and NA3 isolates, respectively).

Table 3 shows the fungicidal activities (percentages of inhibition in quadruplicate using the agar diffusion test) of 40 extracts obtained from different parts of 20 plant species evaluated against the orange phytopathogens *P. digitatum*, *P. italicum*, and *G. citri-aurantii* determined 7 days post inoculation. Table 4 shows the fungicidal activities of the plant extracts under study, but now determined 14 days post inoculation.

From Tables 3 and 4, many comments should be highlighted: Among all the DCM extracts obtained from the 20 species evaluated against *P. digitatum*, none of them presented 100% of inhibition, but the most active ones after 7 days of incubation were obtained from *T. tricholepis*, *A. inundata*, and *P. stelligerum* (80.0, 79.3, and 75.2% of fungal growth inhibition). These percentages decreased after 14 days of incubation to 76.8, 67.2%, and 50.2%, respectively. *Penicillium italicum* inhibition by DCM extracts was lower than for *P. digitatum* since no extract of this type exceeded 70% growth inhibition at 7 days post inoculation. In contrast, many DCM extracts showed high inhibition against *G. citri-aurantii* (*O. virgata*, *T. tricholepis*, *A. inundata*, *P. lapathifolium*, and some species of *Solanum* diminished the fungal growth by more than 70%) and remarkably, *F. clausum* and *P. alliacea* completely interrupted the growth of *G. citri-aurantii* even 14 days after inoculation, representing potent fungicide products.

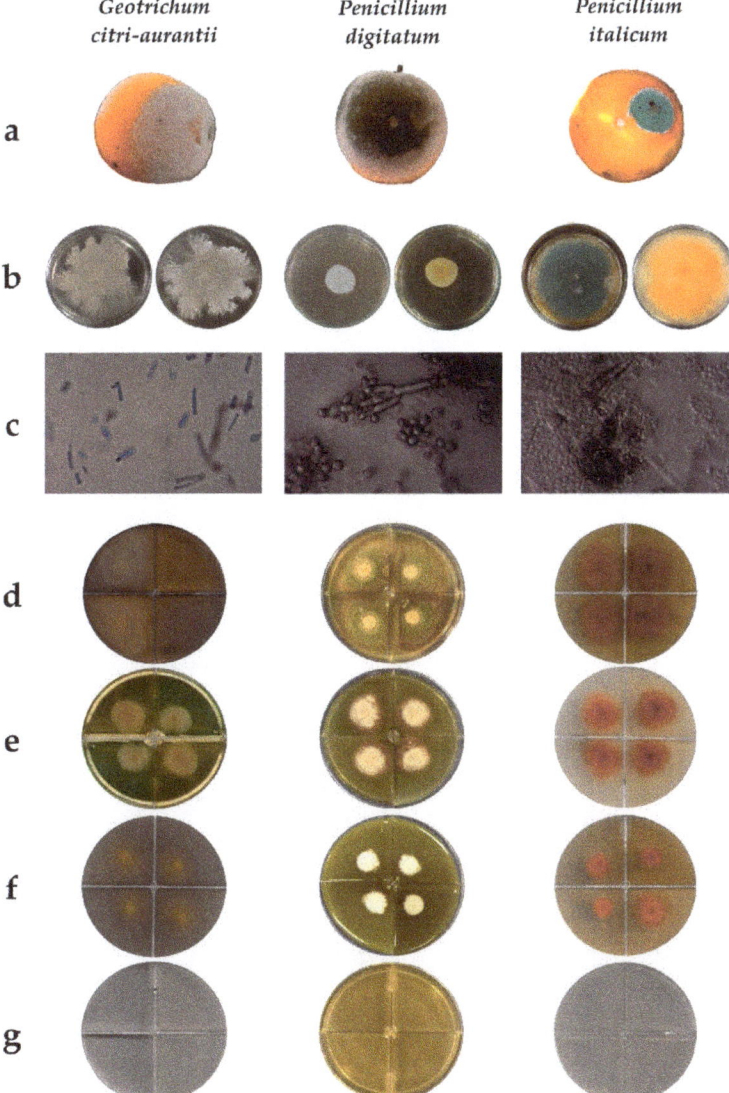

Figure 1. (**a**) Symptoms and signs caused by *Geotrichum citri aurantii*, *Penicillium digitatum*, and *Penicillium italicum* in orange; (**b**) morphological characteristics of morphologically and molecularly characterized colonies isolates on PDA media; (**c**) typical microscopic conidia features observed at magnification 40×; (**d**) tested plates with 0% inhibition; (**e**) tested plates with 50% inhibition; (**f**) tested plates with 80% inhibition; and (**g**) tested pates with 100% inhibition. Note: in some cases, the coloration of the medium may be due to the plant extract dissolved in it.

Table 3. Fungicidal activity % of inhibition (calculated as the average of the inhibition percentage of the four sectors in which the Petri dish was divided) of 40 extracts obtained from different parts of 20 plant species evaluated against the orange phytopathogens *Penicillium digitatum*, *P. italicum*, and *Geotrichum citri-aurantii* determined 7 days post inoculation. Different letters mean significant statistical differences.

Plant Scientific Name	P. digitatum		P. italicum		G. citri-aurantii		Controls	
	DCM	MeOH	DCM	MeOH	DCM	MeOH	+(DMSO)	−(IMZ)
A. brachystephana	61.0 ± 5.2 [c]	44.3 ± 6.2 [a]	30.9 ± 6.5 [a]	30.5 ± 5.3 [a]	65.4 ± 5.0 [c]	66.6 ± 1.8 [c]	0.0 ± 0.0 [a]	100.0 ± 0.0 [b]
F. glabrescens	52.9 ± 7.1 [c]	75.1 ± 1.5 [c]	45.8 ± 6.9 [a]	54.1 ± 5.4 [c]	71.0 ± 0.4 [c]	70.1 ± 2.8 [c]	0.0 ± 0.0 [a]	100.0 ± 0.0 [b]
O. virgata	55.3 ± 2.9 [c]	50.9 ± 5.9 [c]	33.2 ± 6.5 [a]	48.4 ± 6.5 [a]	88.0 ± 4.9 [b]	50.8 ± 1.9 [c]	0.0 ± 0.0 [a]	100.0 ± 0.0 [b]
O. virgata	70.6 ± 2.0 [c]	86.3 ± 3.8 [b]	31.6 ± 4.9 [a]	48.5 ± 4.7 [a]	99.2 ± 0.7 [b]	100.0 ± 0.0 [b]	0.0 ± 0.0 [a]	100.0 ± 0.0 [b]
F. clausum	62.4 ± 3.8 [c]	81.5 ± 1.5 [b]	40.8 ± 4.4 [a]	43.1 ± 6.9 [a]	100.0 ± 0.0 [b]	33.0 ± 5.2 [a]	0.0 ± 0.0 [a]	100.0 ± 0.0 [b]
T. tricholepis	80.0 ± 2.2 [bc]	85.9 ± 0.7 [b]	60.8 ± 7.0 [c]	67.1 ± 3.6 [c]	87.8 ± 1.5 [b]	54.9 ± 1.0 [c]	0.0 ± 0.0 [a]	100.0 ± 0.0 [b]
A. inundata	79.3 ± 2.1 [bc]	100.0 ± 0.0 [b]	35.7 ± 5.2 [a]	100.0 ± 0.0 [b]	79.3 ± 7.0 [bc]	97.3 ± 0.5 [b]	0.0 ± 0.0 [a]	100.0 ± 0.0 [b]
V. caven	59.5 ± 4.0 [c]	76.8 ± 3.5 [c]	36.2 ± 3.8 [a]	58.1 ± 6.5 [c]	56.5 ± 5.1 [c]	32.3 ± 4.2 [a]	0.0 ± 0.0 [a]	100.0 ± 0.0 [b]
P. alliacea	72.5 ± 1.7 [c]	83.3 ± 3.8 [b]	53.2 ± 2.8 [c]	81.3 ± 2.2 [b]	100.0 ± 0.0 [b]	62.9 ± 3.9 [c]	0.0 ± 0.0 [a]	100.0 ± 0.0 [b]
P. stelligerum	75.2 ± 2.3 [c]	80.1 ± 2.8 [b]	32.8 ± 6.5 [a]	57.0 ± 5.3 [c]	76.0 ± 2.1 [c]	38.4 ± 4.1 [a]	0.0 ± 0.0 [a]	100.0 ± 0.0 [b]
P. lapathifolium	64.2 ± 6.4 [c]	80.7 ± 0.9 [bc]	33.0 ± 3.2 [a]	43.8 ± 5.2 [a]	82.1 ± 3.5 [b]	40.2 ± 3.9 [a]	0.0 ± 0.0 [a]	100.0 ± 0.0 [b]
C. montevidensis	49.4 ± 5.2 [a]	56.4 ± 5.1 [c]	24.6 ± 3.2 [a]	45.8 ± 2.7 [a]	65.5 ± 2.8 [c]	52.7 ± 4.6 [c]	0.0 ± 0.0 [a]	100.0 ± 0.0 [b]
P. zapallo	73.2 ± 4.0 [c]	81.7 ± 2.2 [b]	37.5 ± 4.7 [a]	51.8 ± 4.7 [c]	90.6 ± 7.9 [b]	70.0 ± 4.2 [c]	0.0 ± 0.0 [a]	100.0 ± 0.0 [b]
S. argentinum	45.2 ± 6.3 [a]	62.3 ± 5.3 [c]	29.0 ± 3.3 [a]	62.7 ± 5.1 [c]	63.8 ± 0.9 [c]	73.3 ± 1.9 [c]	0.0 ± 0.0 [a]	100.0 ± 0.0 [b]
S. argentinum	73.3 ± 4.0 [c]	70.0 ± 5.6 [c]	27.9 ± 4.0 [a]	51.6 ± 4.8 [c]	75.0 ± 6.8 [c]	48.6 ± 4.9 [a]	0.0 ± 0.0 [a]	100.0 ± 0.0 [b]
S. granulosum-leprosum	54.7 ± 6.0 [c]	49.5 ± 4.5 [a]	44.6 ± 4.7 [a]	40.0 ± 4.3 [a]	41.3 ± 5.4 [a]	61.5 ± 5.1 [c]	0.0 ± 0.0 [a]	100.0 ± 0.0 [b]
S. pilcomayense	44.5 ± 4.9 [a]	100.0 ± 0.0 [b]	15.3 ± 3.4 [a]	100.0 ± 0.0 [b]	74.4 ± 2.4 [c]	100.0 ± 0.0 [b]	0.0 ± 0.0 [a]	100.0 ± 0.0 [b]
S. pilcomayense	46.4 ± 4.7 [a]	100.0 ± 0.0 [b]	41.9 ± 5.3 [a]	100.0 ± 0.0 [b]	71.0 ± 1.9 [c]	100.0 ± 0.0 [b]	0.0 ± 0.0 [a]	100.0 ± 0.0 [b]
S. caavurana	60.0 ± 5.6 [c]	94.6 ± 1.5 [b]	35.5 ± 3.5 [a]	87.8 ± 1.3 [b]	79.8 ± 4.5 [c]	100.0 ± 0.0 [b]	0.0 ± 0.0 [a]	100.0 ± 0.0 [b]
L. megapotamica	75.0 ± 2.2 [c]	56.3 ± 4.4 [c]	34.9 ± 4.8 [a]	41.6 ± 3.4 [a]	70.3 ± 1.2 [c]	62.9 ± 5.1 [c]	0.0 ± 0.0 [a]	100.0 ± 0.0 [b]

Table 4. Fungicidal activity % of inhibition (calculated as the average of the inhibition percentage of the four sectors in which the Petri dish was divided) of 40 extracts obtained from different parts of 20 plant species evaluated against the orange phytopathogens *Penicillium digitatum*, *P. italicum*, and *Geotrichum citri-aurantii* determined 14 days post inoculation. Different letters mean significant statistical differences.

Plant Scientific Name	P. digitatum		P. italicum		G. citri-aurantii		Controls	
	DCM	MeOH	DCM	MeOH	DCM	MeOH	+(DMSO)	−(IMZ)
A. brachystephana	40.8 ± 2.5 [a]	18.4 ± 3.4 [a]	0.0 ± 0.0 [a]	7.2 ± 2.3 [a]	13.7 ± 3.2 [a]	39.5 ± 4.3 [a]	0.0 ± 0.0 [a]	100.0 ± 0.0 [b]
F. glabrescens	38.0 ± 3.8 [a]	38.8 ± 4.3 [a]	30.6 ± 4.6 [a]	10.3 ± 2.1 [a]	26.1 ± 3.7 [a]	37.2 ± 4.9 [a]	0.0 ± 0.0 [a]	100.0 ± 0.0 [b]
O. virgata	37.9 ± 5.2 [a]	33.7 ± 4.3 [a]	7.7 ± 2.1 [a]	33.8 ± 4.6 [a]	80.0 ± 4.4 [bc]	29.7 ± 3.2 [a]	0.0 ± 0.0 [a]	100.0 ± 0.0 [b]
O. virgata	45.9 ± 3.9 [a]	57.0 ± 3.6 [c]	0.0 ± 0.0 [a]	16.9 ± 3.2 [a]	90.4 ± 6.0 [b]	100.0 ± 0.0 [b]	0.0 ± 0.0 [a]	100.0 ± 0.0 [b]
F. clausum	50.3 ± 4.5 [ac]	53.6 ± 3.7 [c]	15.1 ± 3.6 [a]	31.2 ± 4.6 [a]	100.0 ± 0.0 [b]	16.0 ± 1.2 [a]	0.0 ± 0.0 [a]	100.0 ± 0.0 [b]
T. tricholepis	76.8 ± 2.7 [c]	69.1 ± 1.8 [c]	55.0 ± 2.7 [c]	39.1 ± 6.5 [a]	26.7 ± 1.7 [a]	46.9 ± 2.8 [a]	0.0 ± 0.0 [a]	100.0 ± 0.0 [b]
A. inundata	60.3 ± 2.7 [c]	100.0 ± 0.0 [b]	34.3 ± 6.2 [a]	100.0 ± 0.0 [b]	71.5 ± 5.3 [c]	92.5 ± 0.2 [b]	0.0 ± 0.0 [a]	100.0 ± 0.0 [b]
V. caven	42.3 ± 3.7 [a]	33.0 ± 6.1 [a]	30.0 ± 5.8 [a]	11.0 ± 3.0 [a]	0.0 ± 0.0 [a]	22.0 ± 5.1 [a]	0.0 ± 0.0 [a]	100.0 ± 0.0 [b]
P. alliacea	67.2 ± 2.9 [c]	58.8 ± 6.7 [c]	41.0 ± 2.0 [a]	57.8 ± 3.3 [c]	100.0 ± 0.0 [b]	34.4 ± 5.8 [a]	0.0 ± 0.0 [a]	100.0 ± 0.0 [b]
P. stelligerum	50.2 ± 4.3 [ac]	42.3 ± 4.3 [a]	0.0 ± 0.0 [a]	0.0 ± 0.0 [a]	59.1 ± 4.1 [c]	21.5 ± 1.6 [a]	0.0 ± 0.0 [a]	100.0 ± 0.0 [b]
P. lapathifolium	37.8 ± 6.5 [a]	33.0 ± 5.3 [a]	0.0 ± 0.0 [a]	0.0 ± 0.0 [a]	22.1 ± 3.6 [a]	20.5 ± 1.2 [a]	0.0 ± 0.0 [a]	100.0 ± 0.0 [b]
C. montevidensis	41.4 ± 6.2 [a]	10.4 ± 2.1 [a]	0.0 ± 0.0 [a]	11.6 ± 4.5 [a]	0.0 ± 0.0 [a]	31.6 ± 4.3 [a]	0.0 ± 0.0 [a]	100.0 ± 0.0 [b]
P. zapallo	52.9 ± 5.9 [c]	39.5 ± 4.4 [a]	18.3 ± 4.4 [a]	0.0 ± 0.0 [a]	10.3 ± 2.7 [a]	44.6 ± 5.3 [a]	0.0 ± 0.0 [a]	100.0 ± 0.0 [b]
S. argentinum	29.5 ± 4.3 [a]	40.2 ± 6.7 [a]	0.0 ± 0.0 [a]	13.8 ± 2.3 [a]	0.0 ± 0.0 [a]	37.5 ± 5.8 [a]	0.0 ± 0.0 [a]	100.0 ± 0.0 [b]
S. argentinum	50.1 ± 6.3 [ac]	50.1 ± 5.2 [ac]	0.0 ± 0.0 [a]	27.8 ± 3.5 [a]	39.3 ± 4.5 [a]	33.1 ± 2.8 [a]	0.0 ± 0.0 [a]	100.0 ± 0.0 [b]
S. granulosum-leprosum	38.4 ± 4.3 [a]	10.4 ± 3.7 [a]	29.9 ± 4.7 [a]	0.0 ± 0.0 [a]	0.0 ± 0.0 [a]	37.4 ± 3.6 [a]	0.0 ± 0.0 [a]	100.0 ± 0.0 [b]
S. pilcomayense	32.4 ± 4.7 [a]	100.0 ± 0.0 [b]	0.0 ± 0.0 [a]	100.0 ± 0.0 [b]	17.8 ± 4.7 [a]	100.0 ± 0.0 [b]	0.0 ± 0.0 [a]	100.0 ± 0.0 [b]
S. pilcomayense	32.8 ± 4.8 [a]	100.0 ± 0.0 [b]	27.4 ± 3.7 [a]	100.0 ± 0.0 [b]	0.0 ± 0.0 [a]	100.0 ± 0.0 [b]	0.0 ± 0.0 [a]	100.0 ± 0.0 [b]
S. caavurana	50.4 ± 5.5 [ac]	88.3 ± 2.2 [b]	18.2 ± 3.2 [a]	55.7 ± 4.3 [c]	71.0 ± 3.9 [c]	100.0 ± 0.0 [b]	0.0 ± 0.0 [a]	100.0 ± 0.0 [b]
L. megapotamica	62.9 ± 5.4 [c]	23.6 ± 5.3 [a]	34.6 ± 4.2 [a]	22.5 ± 4.4 [a]	0.0 ± 0.0 [a]	34.6 ± 3.4 [a]	0.0 ± 0.0 [a]	100.0 ± 0.0 [b]

Regarding MeOH extracts, it can be observed that they were more active than DCM extracts. *Orthosia virgata*, *F. clausum*, *T. tricholepis*, *P. alliacea*, and *S. caavurana* methanolic extracts inhibited the growth of *P. digitatum* by more than 80%. Moreover, *A. inundata* and the two collections of *S. pilcomayense* presented 100% *P. digitatum* inhibition even more than 14 days after inoculation. *P. italicum* inhibition by methanolic extracts was not as high as *P. digitatum* inhibition: *P. alliacea* and *S. caavurana* extracts inhibited 81.3 and 87.8%, respectively, but this activity diminished after 14 days. Remarkably, *A. inundata* and both collections of *S. pilcomayense* maintained their *P. italicum* 100% inhibition activity for more than 14 days. Concerning *G. citri-aurantii* inhibition by methanolic extracts, it was 100% inhibited for more than 14 days by *O. virgata* and both collections of *S. pilcomayense* and *S. caavurana*, with the activity of *A. inundata* also being remarkable, which showed more than 90% inhibition even at day 14 post inoculation.

An interesting topic to remark on is the case of *O. virgata* and *S. pilcomayense* which were collected from two different locations on the same date. The high fungicidal activity of *O. virgata* was only observed for one of the collections (Gral. Obligado, Santa Fe, 28°42′54.8″ S 59°23′13.7″ W); meanwhile, the potent activity of *S. pilcomayense* was shown for both collections (Gral. Obligado, Santa Fe, 28°13′42.2″ S 59°06′52.6″ W and San Javier, Santa Fe, 30°03′23.2″ S 59°51′34.8″ W).

3.3. Analysis of the Results by Categories of Fungicidal Action

The results were divided into the following categories for percentage analysis: inactive extract (fungal growth inhibition less than 50%), moderately active extract (fungal growth inhibition between 50 and 80%), and active extract (fungal growth inhibition more than 80%).

None of the DCM extracts tested against *P. digitatum* were rated as active according to the above classification, but 80% of them were moderately active and 20% appeared to be inactive after 7 days of inoculation. The percentage of moderately active DCM extracts decreased from 80% to 45% and the percentage of inactive ones increased from 20% to 55%. These results may suggest that some DCM extracts could be fungistatic for a short period but not fungicidal. For *P. italicum*, most of the DCM extracts were inactive (90%), with only 10% of them moderately active after 7 days. This fungus represents the most resistant to DCM extracts among the three phytopathogenic strains under study. In contrast, *G. citri-aurantii* was the most susceptible phytopathogen to DCM extracts, since 35% of them were active at 7 days but decreased to 20% at 14 days after inoculation, 60% were moderately active after 7 days and decreased to 15% after 14 days, and the percentage of inactive DCM extracts against *G. citri-aurantii* was 5% after 7 days, increasing to 65% after 14 days (Figure 2).

Regarding MeOH extracts (Figure 3), 55% were active, 35% were moderately active, and 10% were inactive against *P. digitatum* 7 days post inoculation, and after 14 days 20% were active and 25% of them were moderately active. Additionally, 25, 35, and 40% of MeOH extracts were active, moderately active, and inactive, respectively, against *P. italicum* 7 days after inoculation; meanwhile, the percentages of active and moderately active extracts decreased after 14 days to 15 and 10%, respectively. The percentages of active MeOH extracts against *G. citri-aurantii* were the same 7 and 14 days post inoculation (25%), but the moderately active extracts after 7 days (50%) decreased to 0% after 14 days, increasing the percentages of inactive extracts. These results may suggest that the MeOH extracts that displayed strong activities against *G. citri-aurantii* have potential as fungicides due to their fungicidal activity being maintained for more than 14 days.

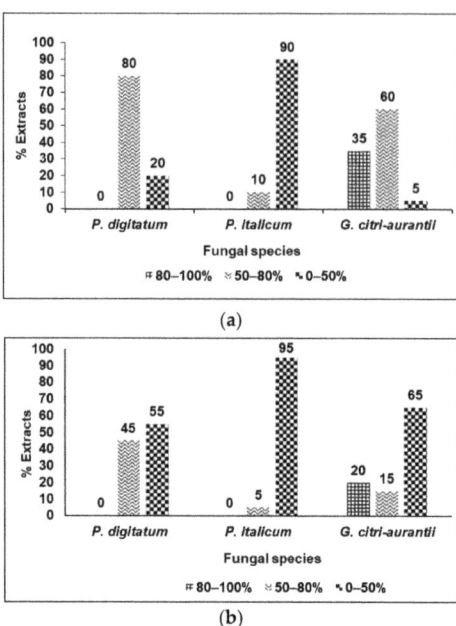

Figure 2. Percentage of active, moderately active, and inactive DCM extracts evaluated (**a**) 7 days and (**b**) 14 days after inoculation.

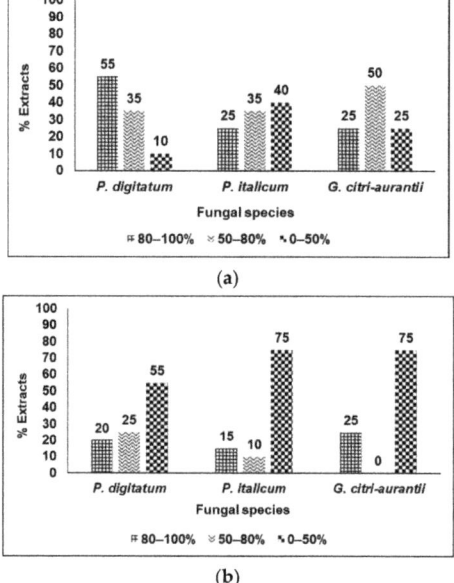

Figure 3. Percentage of active, moderately active, and inactive MeOH extracts evaluated (**a**) 7 days and (**b**) 14 days after inoculation.

3.4. Curative Effect of S. pilcomayense Methanolic Extract over Oranges Infected with P. digitatum

The effect of the most active extract from *S. pilcomayense* was examined on postharvest fresh oranges cv. "Salustiana" infected with an inoculum of *P. digitatum* (Figure 4), where the wound-inoculated fruits were treated with the bioactive extract by dipping as described in the Section 2. After 14 days of the treatment applications, no significant differences in the disease severity were observed for both treatments (3000 ppm aqueous solution of *S. pilcomayense* methanolic extract, and 3000 ppm aqueous solution of Imazalil®). The control fruits (those which had been only inoculated with the pathogen, but not treated with any fungicide) showed marked symptoms of the disease caused by the pathogen (see the sporulation index in each fruit of the control set). Contrarily, both treatments (*S. pilcomayense* methanolic extract and commercial imazalil) reduced the green mold sporulation index significantly ($p = 0.03$) with respect to the control oranges, showing no significant differences between them.

(a) (b) (c)

Figure 4. Sporulation index (0 to 4 scale, in which the value 0 was assigned to negligible sporulation and 4 referred to dense fungal sporulation over the entire fruit) after 14 days of inoculation of *P. digitatum* on wound-inoculated oranges: (**a**) control fruits, (**b**) fruits treated with 3000 ppm aqueous solution of *S. pilcomayense* methanolic extract, and (**c**) fruits treated with 3000 ppm aqueous solution of commercial imazalil.

4. Discussion

Although a considerable number of native Argentinean plant species were evaluated in this work for their fungicidal properties against three economically important citrus pathogens, only six of them have demonstrated high fungal inhibition. Additionally, the place of collection played an important role in the bioactivity of *O. virgata*, indicating the importance of environmental conditions on the production of bioactive secondary metabolites [24]. MeOH extracts obtained from all the species produced higher yields than DCM extracts, and this fact may be due to the number and type of molecules that can be extracted by the solvent methanol [24]. Moreover, MeOH extracts were more active than DCM extracts, especially against both species of *Penicillium*, the two phytopathogens which were more difficult to control in this work. Regarding *Geotrichum*, both types of extracts (DCM and MeOH) displayed more than 50% inhibition.

A literature review of the six most active species lets us state that they have been hardly evaluated against other microbiological activities, and in some cases a few references were found. Remarking on the importance of our native sources' knowledge, a discussion about their biological activities and phytochemical studies found in the literature is presented below:

Orthosia virgata (Figure 5a) belongs to the Apocynaceae family and it is commonly known as "*Liana de leche*" [25]. To the best of our knowledge, there are no reports about the bioactivities and phytochemistry of this species and even its genus, which turns *O. virgata* into an interesting source to go on studying. The Apocynaceae family has been reported widely for indole-containing compounds which are under clinical use including vinblastine and vincristine (anticancer), atevirdine (anti-HIV), yohimbine (erectile dysfunction), reserpine (antihypertension), ajmalicine (vascular disorders), ajmaline (anti-arrhythmic), and vincamine (vasodilator). The main genera of the Apocynaceae family are *Alstonia*, *Rauvolfia*, *Kopsia*, *Ervatamia*, and *Tabernaemontana*. These genera consist of 400 members which represent 20% of Apocynaceae species, but only 30 (7.5%) species were investigated, whereas the rest are promising to be investigated. *Orthosia* does not belong to the most studied genus of this family, a fact that makes it even more attractive. Monoterpene Indole Alkaloids (MIAs) present in this family, deserve the curiosity and attention of researchers due to their chemical diversity and biological activities. These compounds were considered an impending source of drug lead [26].

Figure 5. Pictures of the most active plant species acting as fungicides in the present study: (**a**) *Orthosia virgata*, (**b**) *Funastrum clausum*, (**c**) *Albizia inundata*, (**d**) *Petiveria alliacea*, (**e**) *Solanum pilcomayense*, and (**f**) *Solanum caavurana*. Source: database Irupé [20].

Funastrum clausum (Figure 5b) also belongs to the Apocynaceae family and it is commonly known as "*Bejuco sapo*" [27]. These plants exude sticky white latex containing the endopeptidases funastrain and funastrain c II which exhibit cysteine proteolytic activity [28]. This fact may be the reason why the products obtained from this species provided potent fungicidal action during the assays developed in this work. This was the only report found in the literature for this species and its genus, so it constitutes one of the promising species belonging to the Apocynaceae family mentioned above [26].

Albizia inundata (Figure 5c) belongs to the Fabaceae family and it is commonly known as "*Pacará*", "*Timbó blanco*" or "*Palo flojo*" [29]. The *Albizia* genus is phytochemically characterized by the presence of lignanoids, flavonoids, and phenolic glycosides [30]. Moreover, olean-type triterpene saponins isolated from *A. inundata* exhibited cytotoxicity against melanoma cells and human head and neck squamous cell carcinoma [31,32]. Also, these compounds showed high anti-plasmodial and anticandidal activities probably due to triterpene saponins' capacity for disrupting cellular membranes [33]. So, it is possible that saponins from *A. inundata* may be responsible for the anti-phytopathogenic activities found in this work.

Petiveria alliacea (Figure 5d) belongs to the Phytolaccaceae family and it is commonly known as *"Guiné"*, *"Pipi"* or *"Mucuracaá"* [34,35]. Pharmacological studies reported that extracts obtained from different parts of *P. alliacea* showed diverse effects on animal behavior affecting the central nervous system [35]. Rosado-Aguilar et al. [36] demonstrated that methanolic extracts obtained from the stem and leaves presented 100% larvae and adult mortality of *Rhipicephalus microplus*. This result proposed that *P. alliacea* may be a promising acaricide against resistant strains of *R. microplus* [36]. The antimicrobial activities of *P. alliacea* have been widely reported: (1) the hexane extracts inhibited *Staphylococcus aureus* with a Minimum Inhibitory Concentration (MIC) of 240 µg/mL; (2) the methanolic extract showed activity against *Enterococcus faecalis* (MIC = 240 µg/mL); (3) the hydroalcoholic extract was active against *Candida parapsilosis* (MIC = 250 µg/mL), *C. kefyr* and *C. albicans* (MIC = 760 µg/mL); (4) The tearful principle (Z)-thiobenzaldehyde-S-oxide obtained from *P. alliacea* demonstrated antimicrobial activity against *C. albicans*, *Klebsiella pneumoniae*, *Escherichia coli*, *S. aureus* and *S. agalactiae* [37]. Moreover, thiosulfonates isolated from this plant, and their degradation products inhibited at low concentrations (MIC values $\leq$ 64 µg/mL) the following microorganisms: *Bacillus cereus*, *Mycobacterium smegmatis*, *Micrococcus luteus*, *S. agalactiae*, *S. aureus*, *E. coli*, *Stenotrophomonas maltophila*, *K. pneumoniae*, and the fungus *Aspergillus flavus*, *Mucor racemosus*, *Pseudallescheria boydii*, *C. albicans*, *C. tropicalis* and *Issatchenkia orientalis* [37,38]. A compound named DTS, which is a polysulfide often identified in *P. alliacea*, led to a reversible disassembly of microtubules through the decrease of the total expression of tubulin (0.1 µM) and caused a decrease in phosphorylation of erk1/erk2 protein kinases (0.5 µM) in SH-SY5Y cell line, what may be the mode of action for its biocidal activities [37]. Some studies affirm that the bioactivity of *P. alliacea* leaf extracts could be attributed to the main triterpene metabolite named isoarborinol [39].

Solanum pilcomayense (Figure 5e) belongs to the Solanaceae family and it is commonly known as *"Tomatillo del monte"* [40]. Few reports have been found in the literature about its components or bioactivities; the most highlighting study was performed by Muelas-Serrano et al. [41] who demonstrated some antiprotozoal activity of aerial part extracts against *Trypanosoma cruzi* and *Trichomonas vaginalis* [41]. More recently, a study examined the antioxidant compositions and their capacities in three New Zeeland tamarillo (*Solanum* spp.) cultivars. It showed that tamarillo peels possessed higher amounts of phenolic compounds, total phenolic content, and antioxidant activity than the pulps. Pulps had higher anthocyanins concentration than peels. The antioxidant capacity of tamarillos exhibited relatively high values and was strongly correlated with high total phenolic content. The presence of these bioactive compounds highlights the potential of tamarillo for further utilization in the food and pharmaceutical industries. Tamarillo remains underutilized despite its bioactive components [42].

Finally, *S. caavurana* (Figure 5f) also belongs to the Solanaceae family and it is commonly known as *"Palo aguí"*, *"Laranjinha do mato"*, or *"Jurubeba-branca"* [40]. The ripe fruits of *S. caavurana* contain a wide variety of steroidal alkaloids such as 4-tomatiden-3-one, 5α-tomatidan-3-one, and caavuranamide as well as glycoalkaloids. Regarding the nutritional components of the fruit, it contains proteins (4.2%), lipids (1.5%), and carbohydrates (56.7%). The bioactive caavuranamide (steroid alkaloid) showed significant antibacterial activity (MIC = 135 µg/mL) against *Rhodococcus equi*, a similar value to the commercial chloramphenicol (MIC = 124 µg/mL) [43]. Solanaceae fruits constitute a source of food in many countries. In a recent study, the proximate composition of two Solanaceae fruits from Brazilian *Cerrado* was evaluated. The results showed that the pulp had a high moisture content (74.6–85.4 g/100 g) and soluble fiber (1.3–2.0 g/100 g) content, and low fat, protein, and ash content. Potassium is the main mineral found in both fruits. The major components revealed 24 phenolic compounds, most being hydroxycinnamic acids derivatives and chlorogenic acid. The antioxidant capacity of the fruits ranged from 1.3 to 11.5 μmol TE/100 mL of extract [44]. These results indicated that the *Solanum* genus can be interesting for the Brazilian fruit market and that it has the potential to be exploited for agroindustry diversification of fruit products. In addition, our work demonstrated the fungicidal capacity of a related species of *Solanum*.

5. Conclusions

Many efforts have been made to promote reducing the use of chemical fungicides on postharvest fruits worldwide. In this sense, the use of microorganisms as biological controllers has been highly developed in recent decades, but their safety has sometimes been questioned. The high mutation capacity may turn a benefic bacterium into a pathogenic one which could cause plant or human diseases. On the other hand, plant species contain a complex mixture of molecules that should be explored, not only to treat human diseases but also to prevent and control fruit or plant diseases. So, this type of study should be encouraged to reduce the use of agrochemicals, especially in small or familiar farms. In this work, we isolated and characterized (by morphological and molecular means) three strains of the most important citric phytopathogens in our region. Through a high-throughput screening method, we demonstrated that six Argentinean native plants (*O. virgata, P. alliacea, F. clausum, A. inundata, S. caavurana*, and *S. pilcomayense*) inhibited 100% of the growth of at least one of the three orange phytopathogens tested (*P. digitatum, P. italicum*, and *G. citri-auranti*). Additionally, an aqueous solution of 3000 ppm of *S. pilcomayense* crude methanolic extract concentration was highly fungicidal on postharvest oranges inoculated with *P. digitatum*. However, more studies are needed, such as on the chemical standardization of the extracts, their evaluation against other phytopathogens to detect spectra and mechanisms of action, human cytotoxicity, and technological formulations to determine the correct dose for the control of each disease on fruits.

Author Contributions: Conceptualization, N.H.A., G.M.S. and M.G.D.; methodology, N.H.A., M.A.F. and M.G.D.; software, G.M.S., J.M.Z. and M.I.S.; validation, J.F.P., J.M.Z. and M.G.D.; formal analysis, N.H.A., G.M.S., M.A.F. and M.G.D.; investigation, N.H.A., G.M.S. and M.I.S.; resources, J.F.P., M.A.F. and M.G.D.; data curation, J.M.Z., M.A.F. and M.G.D.; writing—original draft preparation, N.H.A. and M.G.D.; writing—review and editing, J.F.P., M.A.F. and M.G.D.; visualization, M.I.S. and G.M.S.; supervision, M.G.D.; project administration, J.F.P., M.A.F. and M.G.D.; funding acquisition, J.F.P., M.A.F. and M.G.D. All authors have read and agreed to the published version of the manuscript.

Funding: This work was supported by the ANPCyT under grant PICT-2020-SERIEA-02504 and PICT-2021-CAT-II-00097, and by CONICET under grant PIP 11220210100388CO.

Data Availability Statement: Not applicable.

Acknowledgments: N.H.A., M.I.S. and G.M.S. thank CONICET for their scholarships. J.M.Z. and J.F.P. acknowledge the ProDoCoVa program.

Conflicts of Interest: The authors report there are no competing interests to declare.

References

1. Palou, L. Penicillium digitatum, *Penicillium italicum* (green mold, blue mold). In *Postharvest Decay, Control Strategies*; Bautista-Baños, S., Ed.; Academic Press: Valencia, Spain, 2014; pp. 45–102. [CrossRef]
2. La Actividad Citrícola. Available online: https://www.federcitrus.org/wp-content/uploads/2022/07/La-Actividad-Citricola-2022.pdf (accessed on 12 March 2023).
3. Eckert, J.W.; Eaks, I.L. Postharvest disorders and diseases of citrus fruits. In *The Citrus Industry: Division of Agriculture and Natural Resources*; Reuter, W., Calavan, E.C., Carman, G.E., Eds.; California Press: Los Angeles, CA, USA, 1989; pp. 179–260.
4. Ogawa, J.M.; Zehr, E.; Bird, G.W.; Ritchie, D.F.; Uriu, K.; Uyemoto, J.K. *Compendium of Stone Fruit Diseases*; American Phytopathological Society: Saint Paul, MN, USA, 1995; pp. 1–168.
5. Smilanick, J.L.; Mansour, M.F.; Gabler, F.M.; Sorenson, D. Control of citrus postharvest green mold and sour rot by potassium sorbate combined with heat and fungicides. *Postharvest Biol. Technol.* **2008**, *47*, 226–238. [CrossRef]
6. Ismail, M.; Zhang, J. Post-harvest citrus diseases and their control. *Outlooks Pest Manag.* **2004**, *15*, 29–35. [CrossRef]

7. Stegmayer, M.I.; Álvarez, N.H.; Favaro, M.A.; Fernandez, L.N.; Carrizo, M.E.; Reutemann, A.G.; Derita, M.G. Argentinian wild plants as controllers of fruits phytopathogenic fungi: Trends and perspectives. In *Wild Plants: The Treasure of Natural Healers*; Rai, M., Bhattarai, S., Feitosa, C., Eds.; CRC Press: Boca Raton, FL, USA, 2020; pp. 121–137.
8. Di Liberto, M.G.; Stegmayer, M.I.; Svetaz, L.A.; Derita, M.G. Evaluation of Argentinean medicinal plants and isolation of their bioactive compounds as an alternative for the control of postharvest fruits phytopathogenic fungi. *Rev. Bras. Farmacogn.* **2019**, *29*, 686–688. [CrossRef]
9. Moura, V.S.; Moretto, R.K.; Machado, B.I.; Kupper, K.C. Alternativas de controle de doenças de pós-colheita em citros. *Citrus Res. Technol.* **2019**, *40*, e1044. [CrossRef]
10. Carmona-Hernandez, S.; Reyes-Pérez, J.J.; Chiquito Contreras, R.G.; Rincon-Enriquez, G.; Cerdan-Cabrera, C.R.; Hernandez-Montiel, L.G. Biocontrol of postharvest fruit fungal diseases by bacterial antagonists: A review. *Agronomy* **2019**, *9*, 121. [CrossRef]
11. Stegmayer, M.I.; Fernández, L.; Álvarez, N.; Olivella, L.; Gutiérrez, H.; Favaro, M.A.; Derita, M. Aceites esenciales provenientes de plantas nativas para el control de hongos fitopatógenos que afectan a frutales. *Rev. FAVE Cienc. Agrar.* **2021**, *20*, 317–329. [CrossRef]
12. Rodriguez, A.; Derita, M.; Borkosky, S.; Socolsky, C.; Bardón, A.; Hernández, M. Bioactive farina of *Notholaena sulphurea* (Pteridaceae): Morphology and histochemistry of glandular trichomes. *Flora* **2018**, *240*, 144–151. [CrossRef]
13. Stegmayer, M.I.; Fernández, L.N.; Álvarez, N.H.; Seimandi, G.M.; Reutemann, A.G.; Derita, M.G. In vitro antifungal screening of Argentine native or naturalized plants against the phytopathogen *Monilinia fructicola*. *Comb. Chem. High Throughput Screen.* **2022**, *25*, 1158–1166. [CrossRef]
14. Svetaz, L.; Zuljan, F.; Derita, M.; Petenatti, E.; Tamayo, G.; Cáceres, A.; Cechinel Filho, V.; Giménez, A.; Pinzón, R.; Zacchino, S.A.; et al. Value of the ethnomedical information for the discovery of plants with antifungal properties. A survey among seven Latin American countries. *J. Ethnopharmacol.* **2010**, *127*, 137–158. [CrossRef]
15. Gupta, V.K.; Tuohy, M.G.; Gaur, R. Methods for high-quality DNA extraction from fungi. In *Laboratory Protocols in Fungal Biology*; Springer: New York, NY, USA, 2012; pp. 403–406.
16. White, T.J.; Bruns, T.D.; Lee, S.B.; Taylor, J.W. Amplification and direct sequencing of fungal ribosomal RNA Genes for phylogenetics. In *PCR Protocols: A Guide to Methods and Applications*; Academic Press: Cambridge, MA, USA, 1990; pp. 315–322.
17. Boratyn, G.M.; Camacho, C.; Cooper, P.S.; Coulouris, G.; Fong, A.; Ma, N.; Zaretskaya, I. BLAST: A more efficient report with usability improvements. *Nucleic Acids Res.* **2013**, *41*, 29–33. [CrossRef]
18. CLSI (Clinical and Laboratory Standards Institute). *Reference Method for Broth Dilution Antifungal Susceptibility Testing of Filamentous Fungi*, 3rd ed.; CLSI Standard M38; CLSI: Berwyn, PA, USA, 2017; pp. 1–35.
19. Rasband, W.S. Available online: https://imagej.nih.gov/ij/ (accessed on 7 February 2020).
20. Di Liberto, M.G.; Seimandi, G.M.; Fernández, L.N.; Ruiz, V.E.; Svetaz, L.A.; Derita, M.G. Botanical control of citrus green mold and peach brown rot on fruits assays using a *Persicaria acuminata* phytochemically characterized extract. *Plants* **2021**, *10*, 425. [CrossRef] [PubMed]
21. Hernández-Montiel, L.G.; Holguín-Peña, R.J.; Latisnere-Barragan, H. First report of sour rot caused by *Geotrichum citri-aurantii* on Key Lime (*Citrus aurantifolia*) in Colima State, Mexico. *Plant Dis.* **2010**, *94*, 488. [CrossRef] [PubMed]
22. Deng, J.; Kong, S.; Wang, F.; Liu, Y.; Liao, I.; Lu, Y.; Zhang, F.; Wu, J.; Wang, L.; Li, X. Identification of a new *Bacillus sonorensis* strain KLBC GS-3 as a biocontrol agent for postharvest green mold in grapefruit. *Biol. Control* **2020**, *151*, 104393. [CrossRef]
23. Khan, I.H.; Javaid, A. Molecular characterization of *Penicillium italicum* causing blue mold on lemon in Pakistan. *J. Plant Pathol.* **2022**, *104*, 845–846. [CrossRef]
24. Derita, M.; Leiva, M.; Zacchino, S. Influence of plant part, season of collection and content of the main active constituent, on the antifungal properties of *Polygonum acuminatum* Kunth. *J. Ethnopharmacol.* **2009**, *124*, 377–383. [CrossRef]
25. Paye, I.A.; Pensiero, J.F.; Grenón, D.A.; Exner, E. Irupé: Sitio web de imágenes de la flora nativa de Argentina asociado al herbário SF. *Rev. FAVE Cienc. Agrar.* **2021**, *20*, 19–31. [CrossRef]
26. Mohammed, A.E.; Abdul-Hameed, Z.H.; Alotaibi, M.O.; Bawakid, N.O.; Sobahi, T.R.; Abdel-Lateff, A.; Alarif, W.M. Chemical diversity and bioactivities of monoterpene indole alkaloids (MIAs) from six Apocynaceae genera. *Molecules* **2021**, *26*, 488. [CrossRef] [PubMed]
27. Rodríguez, I.L.; Guadarrama Olivera, M.A.; Ortiz Guadarrama, M.; Labastida Astudillo, M.; Jiménez Pérez, N.D.C. *Funastrum clausum* (Jacq.) Schltr.: El Bejuco De Leche. *Kuxulkab* **2021**, *27*, 59–61. [CrossRef]
28. Morcelle, S.R.; Barberis, S.; Priolo, N.; Caffini, N.O.; Clapés, P. Comparative behavior of proteinases from the latex of *Carica papaya* and *Funastrum clausum* as catalysts for the synthesis of Z-Ala-Phe-OMe. *J. Mol. Catal. B Enzym.* **2006**, *41*, 117–124. [CrossRef]
29. Zabala, J.M.; Exner, E.; Cerino, C.; Buyatti, M.; Cuffia, C.; Marinoni, L.; Kern, V.; Pensiero, J.F. Recursos fitogenéticos forestales, forrajeros, de interés apícola y paisajístico nativos de la provincia de Santa Fe (Argentina). *Rev. FAVE Cienc. Agrar.* **2021**, *20*, 99–131. [CrossRef]
30. He, Y.; Wang, Q.; Ye, Y.; Liu, Z.; Sun, H. The ethnopharmacology, phytochemistry, pharmacology and toxicology of genus *Albizia*: A review. *J. Ethnopharmacol.* **2020**, *257*, 112677. [CrossRef] [PubMed]
31. Hussain, M.M. A short review on phytoconstituents from genus *Albizzia* and *Erythrina*. *Bangladesh Pharmacol. J.* **2018**, *21*, 160–172. [CrossRef]
32. Zhang, H.P.; Samadi, A.K.; Rao, K.V.; Cohen, M.S.; Timmermann, B.N. Cytotoxic oleanane-type saponins from *Albizia inundata*. *J. Nat. Prod.* **2011**, *74*, 477–482. [CrossRef] [PubMed]

33. Shetu, H.J. Phytochemical and Biological Evaluation of *Albizia richadiana* (Benth. Fabaceae Family). Bachelor's Thesis, BRAC University, Dhaka, Bangladesh, 2015.
34. Luz, D.A.; Pinheiro, A.M.; Silva, M.L.; Monteiro, M.C.; Prediger, R.D.; Maia, C.S.F.; Fontes-Junior, E.A. Ethnobotany, phytochemistry and neuropharmacological effects of *Petiveria alliacea* L. (Phytolaccaceae): A review. *J. Ethnopharmacol.* **2016**, *185*, 182–201. [CrossRef]
35. Blainski, A.; Piccolo, V.K.; Mello, J.C.P.; de Oliveira, R.M. Dual effects of crude extracts obtained from *Petiveria alliacea* L. (Phytolaccaceae) on experimental anxiety in mice. *J. Ethnopharmacol.* **2010**, *128*, 541–544. [CrossRef]
36. Rosado-Aguilar, J.A.; Aguilar-Caballero, A.; Rodriguez-Vivas, R.I.; Borges-Argaez, R.; Garcia-Vazquez, Z.; Mendez-Gonzalez, M. Acaricidal activity of extracts from *Petiveria alliacea* (Phytolaccaceae) against the cattle tick, Rhipicephalus (Boophilus) microplus (Acari: Ixodidae). *Vet. Parasitol.* **2010**, *168*, 299–303. [CrossRef]
37. Silva do Nascimento, S.; Okabe, D.; Pinto, A.; de Oliveira, F.; da Paixão, T.; Souza Siqueira, M.L.; Baetas, A.C.; de Andrade, M. Antimicrobial and anticancer potential of *Petiveria alliacea* L. (Herb to "Tame the Master"): A review. *Pharmacogn. Rev.* **2018**, *12*, 85–93. [CrossRef]
38. Ramos, M.A.; Machado, L.P. Potencial antifúngico de tipi (*Petiveria alliacea* L.) em fungos de *Aspergillus flavus*. *Rev. Cient. FAESA* **2020**, *16*, 32–41.
39. Zavala-Ocampo, L.M.; Aguirre-Hernández, E.; Pérez-Hernández, N.; Rivera, G.; Marchat, L.A.; Ramírez-Moreno, E. Antiamoebic activity of *Petiveria alliacea* leaves and their main component, isoarborinol. *J. Microbiol. Biotechnol.* **2017**, *27*, 1401–1408. [CrossRef]
40. Palchetti, M.V.; Cantero, J.J.; Barboza, G.E. Solanaceae diversity in South America and its distribution in Argentina. *Anais Acad. Brasil. Cienc.* **2020**, *92*, 1–17. [CrossRef]
41. Muelas-Serrano, S.; Nogal, J.J.; Martınez-Dıaz, R.A.; Escario, J.A.; Martınez-Fernández, A.R.; Gómez-Barrio, A. In vitro screening of American plant extracts on *Trypanosoma cruzi* and *Trichomonas vaginalis*. *J. Ethnopharmacol.* **2000**, *71*, 101–107. [CrossRef] [PubMed]
42. Diep, T.; Pook, C.; Yoo, M. Phenolic and anthocyanin compounds and antioxidant activity of tamarillo (*Solanum betaceum* Cav.). *Antioxidants* **2020**, *9*, 169. [CrossRef] [PubMed]
43. Vaz, N.P.; Costa, E.V.; Santos, E.L.; Mikich, S.B.; Marques, F.A.; Braga, R.M.; Delarmelina, C.; Duarte, M.; Ruiz, A.N.; Souza, V.H.; et al. Caavuranamide, a novel steroidal alkaloid from the ripe fruits of *Solanum caavurana* Vell. (Solanaceae). *J. Braz. Chem. Soc.* **2012**, *23*, 361–366. Available online: https://www.scielo.br/pdf/jbchs/v23n2/a25v23n2.pd (accessed on 26 May 2022). [CrossRef]
44. Aparecida Pereira, A.P.; Figueiredo Angolini, C.F.; Paulino, B.N.; Chatagnier Lauretti, L.B.; Orlando, E.A.; Siqueira Silva, J.G.; Neri-Numa, I.A.; Pimentel Souza, J.D.; Lima Pallone, J.A.; Nogueira Eberlin, M.; et al. A comprehensive characterization of *Solanum lycocarpum* St. Hill and *Solanum oocarpum* Sendtn: Chemical composition and antioxidant properties. *Food Res. Int.* **2019**, *124*, 61–69. [CrossRef] [PubMed]

Disclaimer/Publisher's Note: The statements, opinions and data contained in all publications are solely those of the individual author(s) and contributor(s) and not of MDPI and/or the editor(s). MDPI and/or the editor(s) disclaim responsibility for any injury to people or property resulting from any ideas, methods, instructions or products referred to in the content.

Article

Resistance of Tunisian Melon Landraces to *Podosphaera xanthii*

Hela Chikh-Rouhou [1,*], Ana Garcés-Claver [2,3], Lydia Kienbaum [4,5], Abdelmonem Ben Belgacem [1] and Maria Luisa Gómez-Guillamón [6]

[1] Regional Research Centre on Horticulture and Organic Agriculture (CRRHAB) BP 57, Chott-Mariem LR21AGR03-Production and Protection for a Sustainable Horticulture, University of Sousse, Sousse 4042, Tunisia
[2] Department of Plant Science, Agrifood Research and Technology Centre of Aragon (CITA) Avda. Montañana 930, 50059 Zaragoza, Spain
[3] AgriFood Institute of Aragon—IA2 (CITA-University of Zaragoza), 50059 Zaragoza, Spain
[4] Institute of Plant Breeding, Seed Science and Population Genetics, University of Hohenheim, 70599 Stuttgart, Germany
[5] Selecta One (Klemm & Sohn GmbH & Co. KG), Hanfäcker 10, 70378 Stuttgart, Germany
[6] Instituto de Hortofruticultura Subtropical y Mediterranea-La Mayora, CSIC-UMA, Avda. Dr. Wienberg s/n., Algarrobo, 29750 Malaga, Spain
* Correspondence: hela.chikh.rouhou@gmail.com

Abstract: Powdery mildew caused by *Podosphaera xanthii* is among the most threatening fungal diseases affecting melons on the Mediterranean coast. Although the use of genetic resistance is a highly recommended alternative to control this pathogen, many races of this fungus have been described and, therefore, resistance is usually overcome; thus, breeding for resistance to this pathogen is a challenge. Several melon genotypes carrying resistance to powdery mildew have been described but their agronomical and fruit characters are usually far away from the required melon types in many commercial markets. Taking this into consideration, looking for novel sources of resistance in Tunisian landraces is a very convenient step to obtain new resistant melon varieties/hybrids suitable for Mediterranean markets. Several Tunisian melon landraces have been tested against three common races in Mediterranean regions (Race 2, Race 3.5, and Race 5), using phenotypic approaches in two independent experiments (artificial inoculations in a growth chamber and natural conditions of infection in a greenhouse). The results of the artificial inoculations showed that all the tested landraces were susceptible to Race 3.5 and Race 5 and several landraces were resistant to Race 2. Under natural conditions of infection, Race 2 of *P. xanthii* was the race prevalent in the plot and the resistance of TUN-16, TUN-19, and TUN-25 was confirmed. The found resistances were race-specific and underlie a high genetic influence reflected in the high value of the estimated heritability of 0.86. These resistant landraces should be considered as a potential source of resistance in breeding programs of melons belonging to inodorus and reticulatus groups, but further research is necessary to elucidate the genetic control of the found resistances and to provide useful molecular markers linked to *P. xanthii* Race 2 resistance.

Keywords: powdery mildew; phenotypic evaluation; heritability

1. Introduction

Powdery mildew is a major problem in melon production worldwide since it occurs all year round regardless of the growing system in cultivated fields or greenhouses [1]. The disease is caused by two pathogens: *Podosphaera xanthii* (Castagne) U. Braun and Shishkoff (synonym *Sphaerotheca xanthii* (Castagne) L. Junell) and *Golovinomyces orontii* (DC) V.P. Heluta (synonyms *Erysiphe cichoracearum* DC, *G. cichoracearum* DC). The most prevalent species is *P. xanthii* but in some countries, *G. orontii* can also be found [2,3]. *P. xanthii* occurs more frequently in temperate subtropical and tropical areas, while *G. orontii* occurs more frequently in temperate areas under field conditions [2]. The infection is evident by the development of white mycelia on leaves and stems and, in severe cases, also affects fruits and floral structures. Severely infected leaves may become chlorotic, or even necrotic and brittle. As a consequence, the fungus causes a reduction in plant growth, premature desiccation of the leaves, and consequent reduction in the quality and marketability of the fruits [3]. No previous research has been published on the incidence and pathogenic variability of powdery mildew in melon crops in Tunisia. However, it has been observed during surveys that the severity of the attacks is devasting if not treated.

Spraying chemicals is the most extended method used to control this disease but the fungus is able to develop resistance, leading to a lack of efficiency in chemical control in many cases [4,5]. So, the genetic control of resistance against powdery mildew fungi represents a serious challenge to researchers and breeders. Breeding of melons for resistance to powdery mildew is hindered partially due to the diverse genetic pool from which many sources of genetic resistance have been identified [1,6].

Many physiological races of *P. xanthii* have been identified according to their reactions on different melon lines [1,3]. The predominant race in powdery mildew populations changes depending on the melon cultivar, the growing season, and the geographical area [7]. It has been suggested that a large fraction of the reported races are not relevant to the majority of *P. xanthii* resistance breeding, which may have to be performed on a regional basis for subsets of races. Only a limited number of races highly harmful to cucurbit crops are distributed over a large area [1]. Races 1, 2, 3.5, and 5 are reported to be the most extended in Southern European and Mediterranean regions [8–10]. Resistance to Race 1 could be controlled by one recessive gene in the Indian line 'PI313970' [6], while resistance to Race 2 is more complex, and both partial and total resistances have been described [10]. Resistance to Race 5 has been described under monogenic and dominant control in the Indian melon line 'PI124112' [11], and Yuste-Lisbona et al. [10] described a dominant–recessive epitasis controlling resistance to Race 5 as well as to Races 1 and 2 in the Zimbabwean melon line 'TGR-1551'. So far, no resistance against Race 3.5 has been published/studied, although in current experiments carried out by the International Seed Federation Disease Resistance Terminology Working Group (ISF DRT WG), melon lines PI313970, PI124112, Arum, and SV1105, among others, have shown complete or intermediate resistance to several isolates of this race [12]. Most of the sources of resistance to *P. xanthii* originate from India (e.g., PI124112, PI414723, PI134198, PI313970) and several genes have been described controlling mainly Races 1, 2, and 5 [10,13]. Although most of them are race-specific genes, there are melon genotypes carrying genes controlling resistance to more than one race (e.g., PI124112, PI414723, PI313970, TGR-1551). To date, 19 powdery mildew resistance genes and 12 major quantitative trait loci (QTLs) associated with this trait have been reported [6,11,14,15].

Breeding programs require sources of resistance and a few have been reported in melon, mostly in *momordica* and *acidulus* horticultural groups [10,16]. Vegetable seed companies are using these sources of resistance to develop commercial melon varieties with resistance to *P. xanthii* [12]. However, the existence of many races reducing the durability of the resistance [1,3] makes it necessary to find new resistant genotypes with different genetic backgrounds. Local landraces represent a valuable genetic resource for breeding in a changing environment. They exhibit fine adaptation to the specific environment in which they have evolved under years of domestication. The agronomic traits and fruit characteristics of Tunisian landraces are similar to the commercial types demanded by Mediterranean markets, and they also carry adaptation to the environment and cultivation methods [17,18]. Thus, Tunisian melon germplasm, unexplored to date for powdery mildew resistance, could be of great potential and should be exploited. This study is aimed at identifying novel sources of resistance in Tunisian local germplasm for powdery mildew resistance, in order to obtain new resistant melon varieties/hybrids adapted to their cultivation in Mediterranean areas.

2. Materials and Methods

2.1. Plant Material and Fungal Isolates

Tunisian melon landraces belonging to different botanical groups, collected from local farmers of the Centre-East region of the country and maintained at the Regional Research Centre on Horticulture and Organic Agriculture (CRRHAB, Tunisia), were used in the experiments (Table 1). Most of these landraces have been characterized for their morpho-agronomic traits [18], fruit quality [19], and also evaluated for their resistance to Fusarium wilt and *A. gossypii* [17,20].

Table 1. Tunisian melon landraces used and their main characteristics regarding some biotic stress resistances.

Codes	Accessions	Horticultural Group	Resistance to Fusarium wilt [17]	Resistance to Aphid [20]
TUN-2	Maazoun Menzel Chaker	*inodorus*	+	−
TUN-3	Maazoun Mehdia (MM2009)	*inodorus*	+	−
TUN-4	Maazoun Fethi	*inodorus*	−	−
TUN-5	Fakous (FL)	*flexuosus*	+	−
TUN-7	Trabelsi	*inodorus*	−	−
TUN-8	Galaoui	*reticulatus*	−	−
TUN-9	Dziri (DZ P5 2011)	*inodorus*	+	−
TUN-13	Arbi1	*inodorus*	−	−
TUN-16	Sarachika	*inodorus*	+	−
TUN-18	Rupa	*cantalupensis*	+	−
TUN-19	Chamem (Ananas type)	*reticulatus*	+	+
TUN-24	Maazoun (Kairouan)	*inodorus*	−	−
TUN-25	Asli	*inodorus*	−	−
TUN-26	Stambouli	*inodorus*	+	−

(+) and (−) indicate the presence and absence of resistance, respectively.

A set of differential melon lines were used to confirm and identify the race in both assays (artificial and natural conditions of infection): 'PMR-45', resistant to Races 0 and 1, and susceptible to Races 2, 3, 3.5, 4, and 5 of *P. xanthii*; 'Edisto-47', resistant to Races 0, 1, 2, 3, and 4 and susceptible to Races 3.5 and 5; 'PMR-5', resistant to Races 0, 1, 2, 3, 4, and 5 and susceptible to Race 3.5; and 'PI414723', with resistance to all races. The Spanish cultivar 'Bola de Oro', susceptible to all *P. xanthii* races, was used as a susceptible control. The identification of the species causing powdery mildew in the greenhouse was conducted by examination of conidia for the presence of fibrosin bodies and the production of forked germ tubes [3,21].

Three races of *P. xanthii* were used in the artificial inoculations: Race 2 (isolate SF30), Race 3.5 (isolate M23), and Race 5 (isolate C8), which are the most frequent and widespread races in melon production in Mediterranean areas. These isolates came from melon crops from different areas of Spain (Almeria, Murcia, and Malaga, respectively) and were maintained in axenic conditions on cotyledons of the zucchini cultivar 'Black Beauty' and the Spanish susceptible melon cultivar 'Amarillo' as described by Yuste-Lisbona et al. [10].

2.2. Phenotypic Evaluation by Artificial Inoculations

Seeds of the melon accessions were sown into sterilized sand and grown in a growth chamber under 24 °C day/16 °C night conditions with a 16:8 h (light:dark) photoperiod. Fourteen accessions and twelve plants per accession were inoculated. For artificial inoculations of Races 2 and 5, inoculation was carried out by depositing a small amount of conidia, taken from two–three infected cotyledons maintained in axenic conditions, at two spots (at each side of, and equidistant from, the main leaf vein) on the second true leaf of each plant [10] (Figure 1A). Once inoculated, plants were maintained in a controlled growth chamber under 32 °C day/22 °C night conditions with a 16:8 h (light:dark) photoperiod.

Figure 1. (**A**) Plants inoculated artificially by depositing a small amount of conidia at two spots on the second true leaf of melon plants. (**B**) Leaf disk inoculation method.

Inoculations with Race 3.5 were carried out on leaf discs (Figure 1B). For the leaf disc inoculation method, one leaf disc of 18 mm in diameter was taken from the second true leaf of each plant at the 3-4 leaf stage; discs were placed adaxially upwards on moistened filter paper in clear polystyrene boxes. Each box was placed on the base of a settling tower of plastic (124 × 580 × 580 mm) and a cover with a centered hole of 150 mm of diameter. Inoculation was performed by blowing, through the hole, the conidia present on the surface of two–three infected cotyledons, obtained in axenic conditions. Inoculated discs were incubated in growth chambers under 25 °C/18 °C day/night conditions and a 12 h photoperiod as suggested by [22,23].

Inoculations were repeated two times for each race. For each accession and *P. xanthii* race, symptoms were evaluated 12–15 days post-inoculation. Plants were scored according to the level of fungus sporulation, using a scale of 0 to 3 [10] as follows: 0 = no visible sporulation; 1 = low level of sporulation; 2 = moderate level of sporulation; and 3 = profuse sporulation. Melon accessions with a mean disease score between 0 and 1 were considered resistant (R) (the infection did not progress), whereas those >1, where the infection progressed, were considered susceptible (S) [10].

2.3. Phenotypic Evaluation under Natural Infection

Accessions were evaluated under normal daylight conditions in a greenhouse at the Experimental Station of Sahline, Tunisia (N 35°45′02″, E 10°42′44″). Tun-9 and Tun-18 were not included in the assay due to the poor number of available seeds. The experiment was carried out, in a randomized complete block design with three replications of 6 plants each, and was repeated two times. The evaluation was performed when all plants of the *P. xanthii*-susceptible control showed severe symptoms of infection. The percentage of affected plants was scored for each melon accession.

Powdery mildew on each infected leaf was evaluated based on the following rating scale: 0 = no disease; 1 = less than 5% of the leaf area infected; 2 = 5% to 20% of the leaf area with symptoms; 3 = 21% to 50% of the leaf area infected; and 4 = more than 50% of the leaf area infected. For each melon accession, the mean and standard error of the disease score were calculated. The disease incidence was also calculated for both methods (artificial inoculation and natural infection) as follows:

$$D(\%) = \frac{\sum(n_i\, X_i)}{(4 * N)} \times 100$$

where:

X_i: the rating scale;
n_i: the number of plants with the same rating scale;
N: total number of plants.

2.4. Statistical Analyses

To analyze the responses of the accessions and the influence of the methodology used (artificial inoculation in the growth chamber or natural conditions of infection) on the response to the disease caused by *P. xanthii* Race 2, an ANOVA using Type III error was performed. All analyses were carried out using R 4.1.2 [24] with the statistical packages emmeans [25], lme4, and car. For plotting the results, the R library ggpplot2 [26] was used. Model selection by Akaike's Information Criterion (AIC) resulted in the set-up of a final model:

$$y_{ijk} = \mu + m_i + mb_{ij} + a_k + ma_{ik} + e_{ijk}$$

where:

μ = general effect;
m_i = effect of the i-th method;
mb_{ij} = nested effect of the i-th method within the j-th block;
a_k = effect of the k-th accession;
ma_{ik} = interaction effect between the i-th method and the k-th accession;
e_{ijk} = residual error of observation y_{ijk}.

Cullis broad-sense heritability [27] was calculated from the model estimations to analyze the ratio of genetic variance compared to the total phenotypic variance observed. The least squares means were obtained for the significant interaction effect between accessions and methods and were compared in the form of a letter display using the Tukey test ($p < 0.05$). BLUP (best linear unbiased prediction) estimates were calculated across both methods for the estimation of the pure genetic effects of the accessions regarding their response to *P. xanthii* Race 2. Additionally, the estimated genetic values were calculated and correlation analyzed to estimate the relationship between the performance of accessions under the two methodologies (artificial inoculation and natural conditions of infection).

3. Results

3.1. Phenotypic Evaluation by Artificial Inoculations

All plants of the susceptible control 'Bola de Oro' showed symptoms when inoculated with Races 2, 5, and 3.5; five Tunisian accessions, TUN-9, TUN-16, TUN-18, TUN-19, and TUN-25, were found to be resistant to Race 2 since they did not show any powdery mildew symptoms or the disease did not progress (≤ 1) (Figure 2; Table 2); however, no resistance to Race 5 nor to Race 3.5 was found as all of the landraces showed highly susceptible responses (> to 2.5).

Figure 2. (**A**) Symptoms in susceptible control 'Bola de Oro' in comparison to the resistant accessions, (**B**) Tun-19 (Chamem) and (**C**) Tun-25 (Asli). All of them were inoculated with Race 2 of *P. xanthii*, with subsequent symptom evaluation 15 days post-inoculation in a growth chamber. Plants were scored according to the level of fungus sporulation using a scale of 0 to 3.

Table 2. Phenotypic evaluation of the resistance to *P. xanthii* Races 2, 5, and 3.5, by artificial inoculation in growth chamber conditions. Values are mean (n = 12) ± SE.

	Disease Score			Reaction Classification *		
	Race 2	Race 5	Race 3.5	Race 2	Race 5	Race 3.5
TUN-2	3.00 ± 0.00	2.85 ± 0.12	2.79 ± 0.39	S	S	S
TUN-3	1.30 ± 0.30	2.65 ± 0.18	3.00 ± 0.00	S	S	S
TUN-4	1.50 ± 0.25	3.00 ± 0.00	2.75 ± 0.39	S	S	S
TUN-5	1.10 ± 0.20	3.00 ± 0.00	-	S	S	-°
TUN-7	3.00 ± 0.00	2.85 ± 0.14	2.95 ± 0.14	S	S	S
TUN-8	2.70 ± 0.20	3.00 ± 0.00	2.83 ± 0.40	S	S	S
TUN-9	1.00 ± 0.50	3.00 ± 0.00	-	R	S	-
TUN-13	1.40 ± 0.40	2.60 ± 0.20	3.00 ± 0.00	S	S	S
TUN-16	0.00 ± 0.00	2.50 ± 0.24	3.00 ± 0.00	R	S	S
TUN-18	0.00 ± 0.00	2.65 ± 0.20	2.93 ± 0.17	R	S	S
TUN-19	0.40 ± 0.25	2.00 ± 0.35	-	R	S	-
TUN-24	2.00 ± 0.54	3.00 ± 0.00	3.00 ± 0.00	S	S	S
TUN-25	0.20 ± 0.20	2.50 ± 0.20	2.95 ± 0.15	R	S	S
TUN-26	3.00 ± 0.00	2.85 ± 0.14	2.75 ± 0.39	S	S	S
Differential set						
Bola Oro	2.00 ± 0.10	2.85 ± 0.14	3.00 ± 0.00	S	S	S
PMR45	2.00 ± 0.00	2.50 ± 0.20	3.00 ± 0.00	S	S	S
Edisto47	0.00 ± 0.00	2.85 ± 0.20	3.00 ± 0.00	R	S	S
PMR5	0.00 ± 0.00	0.50 ± 0.10	2.00 ± 0.20	R	R	S
PI414723	0.00 ± 0.00	0.00 ± 0.00	0.37 ± 0.25	R	R	R

* Reaction classification as resistant (R; disease score ≤ 1) and susceptible (S; disease score > 1), ° Due to their poor germination ability, TUN-5, TUN-9, and TUN-19 were not tested against Race 3.5.

3.2. Phenotypic Evaluation under Natural Infection and Identification of The Prevalent Race

The disease of powdery mildew was recognized by the presence of a visual white powdery mass, mainly composed of mycelia and conidia, on leaf surfaces (Figure 3A,B), petioles, and young stems. The microscopic observation allowed for the identification of *P. xanthii* as the causal agent and the race of *P. xanthii* prevalent in the greenhouse plot was determined as Race 2 based on the resistance or susceptibility of the differential standard host cultivars used. Thus, 'Bola de Oro' and 'PMR-45' were rated as susceptible (disease score = 4) while 'Edisto-47', 'PMR-5', and 'PI414723' were rated as resistant with score means of 0.33, 0.16, and 0.33, respectively (Table S1).

Figure 3. Assessment of Powdery mildew symptoms under natural infection. (**A**) 'PMR-45' and (**B**) 'Bola de Oro' (both accessions were susceptible to Race 2), (**C**) Resistant accession Tun-25 (Asli). Symptom evaluation was made when all plants of the susceptible accession showed severe symptoms of infection.

Most of the powdery mildew-infected plants exhibited white colonies that covered the leaves, except plants of TUN-16, TUN-19, and TUN-25 (Figure 3C), which showed mild symptoms and the disease incidence was not higher than 10% (Figure 4), confirming the results of the artificial inoculations. All remaining landraces were susceptible, showing symptoms exceeding 30% in the susceptible lines 'Bola de Oro' and 'PMR-45' (Figures 3 and 4).

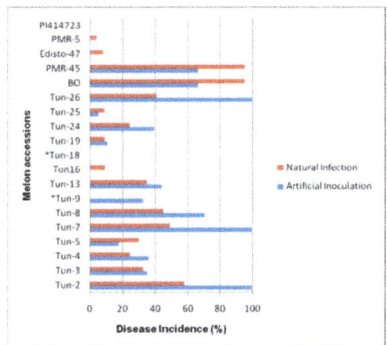

Figure 4. Disease incidence (%) of melon landraces and control genotypes evaluated to *P. xanthii* Race 2, under natural infection and under artificial inoculation. 'Bola de Oro' (BO) and 'PMR-45' were the susceptible controls; 'Edisto-47', 'PMR-5', and 'PI414723' were the resistant controls. *Tun-9 and *Tun-18 were not evaluated under natural infection due to poor numbers of available seeds.

The behavior of the different accessions against Race 2 was evaluated under artificial inoculation in the growth chamber and under natural conditions of infection. The disease score data were analyzed using a two-factorial analysis of variance. For both evaluations, the effects of the accession (G) and method of evaluation (M) were highly significant, and the interaction GxM was also observed (Table 3). However, replication effects were not significant (Table 3), indicating that there were no measurable systematic environmental effects between the blocks.

Table 3. Analysis of variance for effects of the melon accession (G), method of evaluation (M), and their interaction (GxM) on the *P. xanthii* Race 2 disease score.

Sources	df	Sum of Square	F Value	P (>F)
Block:Method	2	0.831	0.796	0.395
Accession (G)	15	222.89	28.673	<2.2 $\times$ 10^{-16} ***
Method (M)	1	11.26	21.730	3.276 $\times$ 10^{-7} ***
Interaction G $\times$ M	15	32.76	4.214	1.175 $\times$ 10^{-5} ***
Residuals	158	81.88		

df: Degrees of freedom, *** highly significant at $p < 0.001$.

The regression line of the *P. xanthii* Race 2 responses between the methods of evaluation and the confidence interval was determined, showing that accessions outside the grey area are deviating, highlighting differences in their reaction among the methods (Figures 5 and S1). These differences were specially observed in the susceptible accessions, where their susceptible responses were more or less pronounced depending on the method of evaluation. For example, the accession Tun-26 showed profuse sporulation by artificial inoculation, but it showed only 5% to 20% infection of the total leaf area under natural infection conditions.

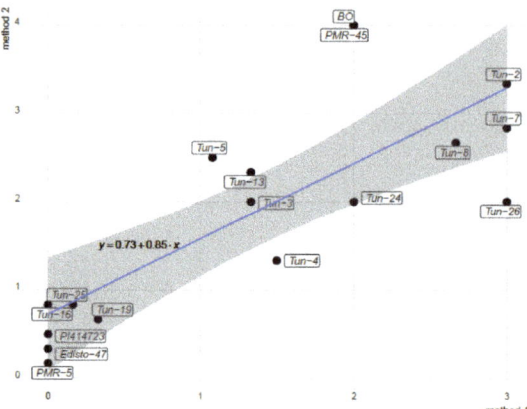

Figure 5. Display of the differences in the *P. xanthii* Race 2 responses between the methods of evaluation (Method 1: artificial inoculation in growth chamber; Method 2: natural conditions of infection under greenhouse). Accessions lying near the regression line show the same reaction whatever the method used. The confidence interval is marked with a grey area. All accessions outside the grey area are deviating, showing differences in their reaction among the methods. The straight regression line and the highly significant correlation (0.79***) of genetic estimations between Methods 1 and 2 generally indicate a high correspondence of landrace responses between the two methods. Tun-9 and Tun-18 were evaluated in Method 1 but not in Method 2, so they are not presented in this figure.

The extent of variance components influencing the response to *P. xanthii* Race 2 showed that 57% of the total variance could be explained by the accession, i.e., genetic variance, with a high value of 0.86 for the estimated heritability (Figure 6). The interaction between accession and method of evaluation explained 11% of the total variance and the method used contributed only 6% to the total phenotypic variance, suggesting a minor influence. Residual variance captured a quarter of the total variance, highlighting that only a small part of the total variance could not be explained by any of the applied factors.

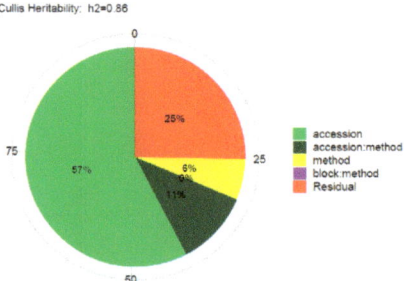

Figure 6. Display the extent of variance components influencing the responses to *P. xanthii* Race 2 analyzed over both methods. The accession covers with 57% most of the variance, followed by the residual variance with 25%. The interaction between accession and method explains 11% and the method effect 6%. 0% of the total variance is captured by the design effect block:method.

When analyzing variance components separately for each method (Figure S2), it becomes clear that artificial inoculation in the growth chamber had a larger genetic variance (81%) than natural infection in the greenhouse (67%). The residual component, which might refer to other environmental effects, was higher under natural infection (33% vs. 19%).

Estimated BLUPs (the pure genetic estimation values) of the accessions regarding the response of the accessions to Race 2 of *P. xanthii* were calculated (Table 4), where the genetic variance was separated from other influencing factors (method of evaluation, accession x method, and residual effects). The breeding values that the breeder would use span a range of 0.267 (resistant) to 2.987 (susceptible) and showed a consistent phenotyping ranking. In addition, the identified resistances in TUN-16, TUN-19, and TUN-25 were validated by low BLUPs values which ranged between 0.561 and 0.634.

Table 4. Estimated BLUPs of the accessions regarding their response to *P. xanthii* Race 2. All other investigated effects (block, method, method:accession, residuals) are separated from these effects, leading to pure genetic estimations for resistance to Race 2.

Accession	BLUP for Disease Scoring
PMR-5	0.267
Edisto-47	0.340
PI414723	0.414
TUN-16	0.561
TUN-19	0.634
TUN-25	0.634
TUN-4	1.443
TUN-3	1.663
TUN-5	1.774
TUN-13	1.811
TUN-24	1.958
TUN-26	2.399
TUN-8	2.546
TUN-7	2.766
BO	2.840
PMR-45	2.840
TUN-2	2.987

4. Discussion

The existence of many races of *P. xanthii* reduces the durability of the resistance, which is usually overcome; this makes it necessary to find new resistant genotypes with different genetic backgrounds. Although several genotypes carrying resistance to powdery mildew have been described and used to develop commercial melon varieties with resistance to *P. xanthii*, their agronomical and fruit characteristics are usually far away from the required melon types in certain markets. Thus, looking for novel sources of resistance in unexplored landrace collections, such as those held in the CRRHAB (Tunisia), is a very convenient step to test the possibility of obtaining new resistant melon varieties/hybrids.

The predominant pathogens of powdery mildew populations change depending on the melon cultivar, the growing season, and the geographical area [7]. In Tunisia, no previous research has been carried out leading to the identification of powdery mildew species and races. Under the greenhouse conditions observed during this study, *P. xanthii* Race 2 was identified as the pathogen responsible for powdery mildew. Determination of the predominant races of *P. xanthii* in Tunisian melon-growing regions is of substantial significance both for combating the disease and for breeding purposes. Race composition can vary over time and screening tests are required every year.

The ANOVA analysis showed that the response to *P. xanthii* Race 2 infection was significantly different among accessions. Additionally, significant differences were observed between the employed methods: both methodologies were useful and accurate in identifying resistance, but accessions were more susceptible when artificial inoculation in the growth chamber was used. Significant effects of the interaction accession x method have been also observed. This interaction was observed only in some susceptible accessions; thus, the susceptible response shown by accessions TUN-4 and TUN-26 was higher when artificial inoculation in the growth chamber was used; however, TUN-8, TUN-13, BO, and

PMR-45, showed higher susceptibility in natural conditions of infection. Genetic effects explained 57% of the total phenotypic variation observed, followed by the interaction effect between the method of evaluation and the accession (11%) and, thereafter, the method effect (6%).

Correlation analysis revealed a highly significant relationship ($r^2 = 0.791$***, $p < 0.05$) in the response of the accessions between the artificial inoculation in the growth chamber and the natural conditions of infection, which confirms that the use of a controlled growth chamber and artificial inoculations could speed up the evaluation of the response to *P. xanthii* Race 2 in breeding programs and also confirms that the evaluation is not dependent on the crop cycle and can be carried out at any time. Obviously, greenhouse or field trials are needed to complement and confirm the results of artificial inoculation in growth chambers.

In our study, a very strong genetic influence on the response to *P. xanthii* Race 2 was found. This was reflected in the high value of the estimated heritability (0.86), showing that selection regarding resistance in this broad gene pool and the successive development of resistant varieties are very promising. A complete genetic characterization of the found resistances is necessary to identify the number of loci/alleles controlling this resistance that seems to be race-specific.

Most of the resistances found in *P. xanthii* are race-specific. Many commercial varieties with specific resistances have been developed but they become susceptible to infection shortly after they are exploited, because of frequent changes in pathogen populations [1]. Therefore, the pyramiding of resistance genes to specific races in a single genotype, or the use of resistance that is not race-specific, should give more comprehensive and durable protection [1]. Consequently, the search for and utilization of new genes that confer resistance to powdery mildew are still primary objectives in melon breeding. The results obtained after artificial inoculation showed that all the melon landraces were susceptible to Races 3.5 and 5 and five landraces were resistant to Race 2. Under natural conditions of infection, TUN-16, TUN-19, and TUN-25 showed low symptoms and the disease incidence was not higher than 10%, confirming the results of the artificial inoculations. Only a few sources of resistance to powdery mildew have been reported in melon, mostly belonging to *flexuosus*, *conomon*, *momordica*, and *acidulus* horticultural groups [10,16]. Resistance to different races of *P. xanthii* has been generally found in accessions from India [28]. Most of these sources are quite different from Tunisian commercial types, which makes it difficult to eliminate undesirable melon characteristics introgressed into the commercial melon lines together with the resistance gene/s. The resistant landraces identified in this study belong to *inodorus* (Dziri: TUN-9, Sarachika: TUN-16, and Asli: TUN-25), *cantalupensis* (Rupa: TUN-18), and *reticulatus* (Chamem: TUN-19) horticultural groups, which are similar to commercial types of melons regarding their agronomic traits and fruit characteristics. Special attention should be given to the resistant landrace TUN-19 Chamem, of Ananas type, since it also carries the *Vat* gene of resistance to *Aphis gossypii* [20] and the *Fom-1* gene of resistance to Races 0 and 2 of *Fusarium oxysporum* f.sp *melonis* [17]. TUN-19 is a potential landrace with high value as a donor of different diseases and pest resistances for melon breeding programs to develop commercial melons of the Ananas type, which are highly appreciated not only in Tunisia but also in other Mediterranean countries. Furthermore, these Tunisian landraces are productive [18], are adapted to local cultivation conditions, and do not have non-desirable traits [29]. Thus, the use of these new resistance sources will promote breeding programs to obtain melon varieties/hybrids.

Supplementary Materials: The following supporting information can be downloaded at: https://www.mdpi.com/article/10.3390/horticulturae8121172/s1, Table S1: Phenotypic evaluation of the resistance to *P. xanthii* Race 2, under natural infection, Figure S1: Least squares mean for the significant interaction effect between method and accession. Figure S2. Display of the extent of variance influencing the *P. xanthii* Race 2 response in each method used.

Author Contributions: Conceptualization and methodology, H.C.-R. and M.L.G.-G.; formal analysis, L.K. and H.C.-R.; writing—original draft preparation, H.C-R. and A.G.-C.; writing—review and editing, H.C.-R., A.G.-C., L.K. and M.L.G.-G. Regarding A.B.B., he collaborated in the artificial inoculation with Races 2 and 5. All authors have read and agreed to the published version of the manuscript.

Funding: This research was funded, in part, by the Spanish Ministry Economy and Competitivity co-funded with FEDER (AGL2017-85563-C2-2-R), the PID2020-116055RB-C22 I+D+I project funded by MCIN/AEI/10.13039/501100011003, and the A11-20R project funded by the Aragon Government.

Institutional Review Board Statement: Not applicable.

Informed Consent Statement: Not applicable.

Data Availability Statement: The data presented in this study are available on request from the corresponding author.

Acknowledgments: The Ministry of Higher Education and Scientific Research of Tunisia (MESRST) through the funding allocated to the Research Laboratory LR21AGR03. The authors thank Rafika Sta-Baba for providing seeds of some landraces and Mohamed Ferjani (Sahline Station) for his help in the greenhouse experiment.

Conflicts of Interest: The authors declare no conflict of interest.

References

1. Mccreight, J. Melon-Powdery Mildew Interactions Reveal Variation in Melon Cultigens and *Podosphaera xanthii* Races 1 and 2. *J. Am. Soc. Hort. Sci.* **2006**, *131*, 59–65. [CrossRef]
2. Křístková, E.; Lebeda, A.; Sedláková, B. Species Spectra, Distribution and Host Range of Cucurbit Powdery Mildews in the Czech Republic, and in Some Other European and Middle Eastern Countries. *Phytoparasitica* **2009**, *37*, 337–350. [CrossRef]
3. Lebeda, A.; Křístková, E.; Sedláková, B.; McCreight, J.D.; Coffey, M.D. Cucurbit Powdery Mildews: Methodology for Objective Determination and Denomination of Races. *Eur. J. Plant Pathol.* **2016**, *144*, 399–410. [CrossRef]
4. Hollomon, D.W.; Wheeler, I.E. Controlling Powdery Mildews with Chemistry. In *The Powdery Mildews: A Comprehensive Treatise*; American Phytopathological Society (APS Press): St. Paul, MN, USA, 2002; pp. 249–255.
5. Pérez-García, A.; Romero, D.; Fernández-Ortuño, D.; López-Ruiz, F.; De Vicente, A.; Torés, J.A. The Powdery Mildew Fungus *Podosphaera fusca* (Synonym *Podosphaera xanthii*), a Constant Threat to Cucurbits. *Mol. Plant Pathol.* **2009**, *10*, 153–160. [CrossRef]
6. McCreight, J.D. Genes for Resistance to Powdery Mildew Races 1 and 2U.S. in Melon PI 313970. *HortScience* **2003**, *38*, 591–594. [CrossRef]
7. Hosoya, K.; Kuzuya, M.; Murakami, T.; Kato, K.; Narisawa, K.; Ezura, H. Impact of Resistant Melon Cultivars on *Sphaerotheca fuliginea*. *Plant Breed.* **2000**, *119*, 286–288. [CrossRef]
8. Bardin, M.; Nicot, P.C.; Normand, P.; Lemaire, J.M. Virulence Variation and DNA Polymorphism in *Sphaerotheca fuliginea*, Causal Agent of Powdery Mildew of Cucurbits. *Eur. J. Plant Pathol.* **1997**, *103*, 545–554. [CrossRef]
9. Del Pino, D.; Olalla, L.; Pérez-García, A.; Rivera, M.; García, S.; Moreno, R.; Vicente, A.; Tores, J. Occurrence of Races and Pathotypes of Cucurbit Powdery Mildew in Southeastern Spain. *Phytoparasitica* **2002**, *30*, 459–466. [CrossRef]
10. Yuste-Lisbona, F.J.; López-Sesé, A.I.; Gómez-Guillamón, M.L. Inheritance of Resistance to Races 1, 2 and 5 of Powdery Mildew in the Melon TGR-1551. *Plant Breed.* **2010**, *129*, 72–75. [CrossRef]
11. Perchepied, L.; Bardin, M.; Dogimont, C.; Pitrat, M. Relationship between Loci Conferring Downy Mildew and Powdery Mildew Resistance in Melon Assessed by Quantitative Trait Loci Mapping. *Phytopathology* **2005**, *95*, 556–565. [CrossRef]
12. Grimault, V.; Houdault, S.; Himmel, P. Development of Differential Hosts to Identify Commercially Relevant Races of Melon *Podosphaera xanthii* against Which Vegetable Seed Companies Make Claims of Resistance. *Cucurbit Genet. Coop. Rpt.* **2020**, *43*, 1–3.
13. Pitrat, M. Melon Genetic Resources: Phenotypic Diversity and Horticultural Taxonomy. In *Genetics and Genomics of Cucurbitaceae*; Grumet, R., Katzir, N., Garcia-Mas, J., Eds.; Springer International Publishing: Cham, Switzerland, 2017; pp. 25–60. [CrossRef]
14. Yuste-Lisbona, F.J.; Capel, C.; Sarria, E.; Torreblanca, R.; Gómez-Guillamón, M.L.; Capel, J.; Lozano, R.; López-Sesé, A. Genetic Linkage Map of Melon (*Cucumis melo* L.) and Localization of a Major QTL for Powdery Mildew Resistance. *Mol. Breed.* **2011**, *27*, 181–192. [CrossRef]
15. Howlader, J.; Hong, Y.; Natarajan, S.; Sumi, K.R.; Kim, H.-T.; Park, J.-I.; Nou, I.-S. Development of Powdery Mildew Race 5-Specific SNP Markers in *Cucumis melo* L. Using Whole-Genome Resequencing. *Hortic. Environ. Biotechnol.* **2020**, *61*, 347–357. [CrossRef]
16. Nunes, E.W.; Esteras, C.; Ricarte, A.O.; Martínez-Perez, E.; Gómez-Guillamón, M.L.; Nunes, G.H.; Picó, M.B. Brazilian Melon Landraces Resistant to *Podosphaera xanthii* Are Unique Germplasm Resources. *Ann. Appl. Biol.* **2017**, *171*, 214–228. [CrossRef]
17. Chikh-Rouhou, H.; Gómez-Guillamón, M.L.; González, V.; Sta-Baba, R.; Garcés-Claver, A. *Cucumis melo* L. Germplasm in Tunisia: Unexploited Sources of Resistance to Fusarium Wilt. *Horticulturae* **2021**, *7*, 208. [CrossRef]

18. Chikh-Rouhou, H.; Mezghani, N.; Mnasri, S.; Mezghani, N.; Garcés-Claver, A. Assessing the Genetic Diversity and Population Structure of a Tunisian Melon (*Cucumis melo* L.) Collection Using Phenotypic Traits and SSR Molecular Markers. *Agronomy* **2021**, *11*, 1121. [CrossRef]
19. Chikh-Rouhou, H.; Tlili, I.; Ilahy, R.; R'Him, T.; Sta-Baba, R. Fruit Quality Assessment and Characterization of Melon Genotypes. *Int. J. Veg. Sci.* **2021**, *27*, 3–19. [CrossRef]
20. Chikh-Rouhou, H.; Ben Belgacem, A.M.; Sta-Baba, R.; Tarchoun, N.; Gómez-Guillamón, M.L. New Source of Resistance to Aphis Gossypii in Tunisian Melon Accessions Using Phenotypic and Molecular Marker Approaches. *Phytoparasitica* **2019**, *47*, 405–413. [CrossRef]
21. Ballantyne, B. Powdery Mildew on Cucurbitaceae: Identity, Distribution, Host Range and Sources of Resistance. *Proc. Linn. Soc. New South Wales* **1975**, *99*, 100–120.
22. Lebeda, A.; Sedláková, B. Screening for Resistance to Cucurbit Powdery Mildew (*Golovinomyces cichoracearum, Podosphaera xanthii*). In *Mass Screening Techniques for Selecting Crops Resistant to Diseases*; IAEA: Vienna, Austria, 2010; Volume 19, pp. 295–307. ISBN 978-92-0-105110-3. Available online: https://www.iaea.org/publications/7758/mass-screening-techniques-for-selecting-crops-resistant-to-disease (accessed on 1 August 2021).
23. McGrath, M.T. Heterothallism in *Sphaerotheca fuliginea*: Mycologia: Volume 86, No 4. Available online: https://www.tandfonline.com/doi/abs/10.1080/00275514.1994.12026445 (accessed on 1 August 2021).
24. R Core Team. *R: A Language and Environment for Statistical Computing*; R Core Team: Vienna, Austria, 2013.
25. Lenth, R.V. R Package Emmeans: Estimated Marginal Means. R. 3 August 2022. Available online: https://github.com/rvlenth/emmeans (accessed on 1 August 2022).
26. Wickham, H. Ggplot2 Book. Perl. 11 August 2022. Available online: https://github.com/hadley/ggplot2-book (accessed on 1 August 2022).
27. Cullis, B.R.; Smith, A.B.; Coombes, N.E. On the Design of Early Generation Variety Trials with Correlated Data. *J. Agric. Biol. Environ. Stat.* **2006**, *11*, 381. [CrossRef]
28. Dhillon, N.; Monforte, A.; Pitrat, M.; Pandey, S.; Singh, P.; Reitsma, K.; Garcia-Mas, J.; Sharma, A.; Mccreight, J. Melon Landraces of India: Contributions and Importance. *Plant Breed Rev.* **2011**, *35*, 85–150. [CrossRef]
29. Chikh-Rouhou, H.; Gómez-Guillamón, M.L.; Garcés-Claver, A. Melon Germplasm from Tunisia with Immense Breeding Value. *CGC Rep.* **2021**, *44*, 7–11.

Article

A New Promising Plant Defense Stimulator Derived from a By-Product of Agar Extraction from *Gelidium sesquipedale*

Rromir Koçi [1,2], Fabrice Dupuy [1], Salim Lebbar [2], Vincent Gloaguen [1] and Céline Faugeron Girard [1,*]

[1] Laboratoire E2Lim (UR24133), Faculté des Sciences et Techniques, Université de Limoges, 123, Avenue Albert Thomas, 87060 Limoges, France
[2] SETEXAM, Km 7, Rte de Tanger, BP 210, Cité Assam, Kenitra 14000, Morocco
* Correspondence: celine.girard@unilim.fr; Tel.: +33-5-55-45-74-76

Abstract: Stimulation of plant defenses by elicitors is an alternative strategy to reduce pesticide use. In this study, we examined the elicitor properties of a by-product of the industrial extraction of agar from the red alga *Gelidium sesquipedale*. Agar extraction process leads to the formation of an alkaline residue which is poorly valorized. This by-product has been analyzed for its chemical composition. It contains 44% minerals and, among the organic compounds, sugars are the most represented and encompass 12.5% of the dry matter. When sprayed on tomato plants, this by-product enhanced the levels of defense markers such as peroxidase or phenylalanine ammonia lyase activities. Furthermore, this treatment increased the expression levels of the pathogenesis-related gene, *PR9* encoding peroxidase. A field trial conducted on grapevine revealed that spraying treatment with this by-product resulted in a reduction of the macroscopic disease symptoms induced by *Plasmospora viticola*, with 40 to 60% efficacy. These results indicate that this agar extraction by-product could be used as a plant defense stimulator.

Keywords: tomato; grapevine; elicitor; plant defense; by-product; *Gelidium sesquipedale*

1. Introduction

Plant pathogens are responsible for significant reductions in crop yields. Large quantities of pesticides are being used to reduce their impact. However, an international study published in 2017 revealed that pesticides are responsible for the death of 5000–20,000 and the poisoning of 500,000 to a million people [1]. Despite their controversial side effects, 2.6 million tons of pesticides have been used worldwide in 2020 [2]. In addition, these chemicals might alter the quality of water and soil and could be harmful to farmers. Furthermore, pesticide residues are frequently found in commercialized plants. In order to reduce the use of pesticides, an alternative strategy can be adopted consisting in improving the natural defenses of plants against their pathogens. Indeed, plants naturally defend themselves from pathogen attacks by reinforcing their natural barriers or by developing direct attacks against the pathogens. Plants benefit from passive as well as active defense mechanisms. Passive mechanisms are permanent and consist of a series of physical barriers like cuticles or cell walls [3], in addition to constitutively produced molecules like phenolic compounds, known for their antimicrobial properties [4]. However, pathogens are able to produce cell wall-degrading enzymes (cutinases, polygalacturonases, etc.) that weaken this barrier of the plant cell [5]. In case of failure of the passive defenses, plants build a series of active defenses in response to pathogen aggression. Active defenses include a hypersensitive response (HR), leading to an auto-destruction of the damaged cells, and a local acquired resistance (LAR), consisting in isolating the cells close to the damage site. LAR can later become a SAR (systemic acquired resistance), a resistance that involves the whole plant, thanks to the production of PR proteins (pathogenesis related proteins), which are targeted towards specific pathogens [3]. The responses to the pathogen occur in

the following order: recognition of the elicitor (plant defense stimulator, PDS), signaling to other cells, and defensive reactions. The initial event is, therefore, the recognition of a signal of pathogen attack, for example the recognition of pathogen molecules, called microbe-associated molecular patterns (MAMPs), or also of molecules released from the damaged host cells, called the damage-associated molecular patters (DAMPs). The recognition process directly implies specific receptors called PRRs (patterns recognition receptors). These are usually receptor-like kinases (RLKs) or receptor-like proteins (RLPs) which harbor an extracellular domain recognizing MAMPs/DAMPs. Among MAMPs, bacterial flagellin, peptidoglycans, lipopolysaccharides, fungal chitin, oomycete β-glucans or toxins and degradation enzymes like endopolyglucanases can be mentioned. On the other hand, DAMPs are endogenous molecules, for example oligogalacturonides, which are released from the walls of the damaged cells. The recognition of MAMPs/DAMPs gives place to the aforementioned plant defense mechanisms like the production of Pathogenesis-Related (PR) proteins and phytoalexins, HR activation and cell wall thickening. For example, deposition of callose and lignification render the plant cell wall more resistant to pathogen attack [5]. These plant defenses might be induced by elicitors defined as compounds which mimic either MAMPs or DAMPs. Therefore, elicitors could be used to protect the plant from pathogen attacks [6].

An elicitor is usually a molecule produced by a pathogen, but hormonal and abiotic elicitors (some metals for example) have also been described [7]. Moreover, some environmental stresses, like excess light, hyperosmolarity, drought, thermal variations, etc., can induce defense responses [7]. Among elicitor compounds, various types have been characterized, including hormonal source elicitors, carbohydrate-containing polymers, like (glyco)peptides and (glyco)proteins, and algae-derived elicitors. Hormonal elicitors are phytohormones that are implicated in the signaling of defense responses, for example salicylic acid [8,9]. Carbohydrate elicitors are secreted by microorganisms or derived from the cell walls of fungi, bacteria or from the host plants themselves [8,9]. Also, many carbohydrate-based polymers whose structures are close to those present in the pathogens or the plant have been found to act as elicitors. Among them xyloglucans, oligogalacturonides, chitin or chitosan oligomers and ramified β-(1,3)-(1,6)-glucans, exhibit elicitor activities in different plant species and trigger pathogen defense responses. Seaweed polysaccharides such as fucans, laminarin, or carrageenans have also been described as plant elicitors [10]. We have shown in a previous work that an extract from the red alga *Gelidium sesquipedale* contained some original molecules, such as glycerol-galactosides, endowed with elicitor activities [11,12]. However, this extract was obtained at a laboratory scale. As *G. sesquipedale* is used at the industrial level for the production of agar, large volumes of by-products are generated which might potentially contain molecules exhibiting biological activities. We describe in the present work, the chemical characterization and the determination of the principal components of a by-product of the industrial agar extraction from *G. sesquipedale*. Then, the biological activities of this extract were assessed, in controlled conditions, on tomato plantlets by measuring some defense responses such as peroxidase or phenylalanine ammonia lyase (PAL) activities. Transcription levels of some PR protein genes were also determined. Furthermore, the capacity of this by-product to protect plants from pathogens was evaluated in a field trial conducted on grapevine infected by *Plasmopara viticola* (downy mildew grapevine disease).

2. Materials and Methods

2.1. By-Product Origin, Production and Preparation

The agar by-product was obtained from the industrial treatment of *G. sesquipedale*. Algae harvested from the Moroccan seashore during summer 2019 were processed by the SETEXAM company (Kenitra, Morocco). The extraction process comprises an alkaline treatment of the dried algae with a 5% NaOH solution during 1 h. Algae are then separated from the alkaline solution. The recovered solution is analyzed and its NaOH content is readjusted, if necessary, in order to be used for the treatment of a new batch of alga. This step is repeated 35 times (i.e., 35 batches of alga). The final alkaline solution, recovered *in loco*, neutralized to pH = 7 with 1% sulfuric acid, concentrated and finally freeze-dried, constitutes the lyophilizate, called SL35, which was used in the present study.

2.2. Elemental Analyses

Samples of SL35 were calcined at 800 °C for 8 h and around 250 mg batches of the resulting residue were weighed and digested as described by Nguyen et al. [13]. Elements, except P, C, H, N and S, were analyzed by MP-AES as in [13]. C, H, N and S were quantified by CHNS as described by Bascle et al. [14]. The digested samples were also quantified for their phosphorus content using the LCK349™ and LCK350™ kits (Hach©, Düsseldorf, Germany).

2.3. Organic Compounds Assays

2.3.1. Carbohydrate Assays

The method of Dubois et al. [15] was used for the quantification of neutral sugars. Galactose and galacturonic acid were used for the calibrating curve with respective ranges of 0–100 and 0–200 µg mL^{-1}. Uronic acids were quantified by the method of Blumenkrantz and Asboe-Hansen [16] using meta-hydroxydiphenyl as reagent. The calibrating curve was made with galactose and galacturonate as standards with respective ranges of 0–500 and 0–100 µg mL^{-1}.

2.3.2. Protein Assay

Protein concentration was determined as described by Bradford et al. [17] using bovine serum albumin as a standard.

2.3.3. Phenolic Compounds Assay

To quantify the total amount of phenolic compounds the method from Piló-Veloso et al. [18] was used with a calibration curve of gallic acid from 0 to 100 µg mL^{-1}.

2.4. Greenhouse Experiments

The experiments were conducted as described by Faugeron-Girard et al. [19] except that the present treatment was done with the lyophilized agar extraction by-product, SL35.

Briefly, tomato plants (*Marmande* variety), at least 1 month old, cultured in greenhouse conditions were treated three times at days 1, 3 and 5 by foliar spraying. SL35 was applied in solution with a surfactant (Tween®80 0.05%, Thermo Fisher 76870 Kandel, Germany). The concentration range of SL35 was from 10 to 1000 mg L^{-1} and 0.05% Tween®80 was applied as a control. Plants were inoculated on the 8th day by infiltration of leaves with conidial suspensions of *Botrytis cinerea* (UBOC 117017). The leaves above the point of inoculation were collected and subjected to biochemical analyses.

2.5. Plant Defense Activities

Peroxidase activity was determined as described by Faugeron-Girard et al. [19]. Briefly, foliar samples were ground in presence of 100 mM phosphate buffer (pH = 7). Then, extracts were assayed in a 250 mM acetate buffer (pH = 4.4) containing 500 µM 2,2′-azino-bis (3-ethylbenzothiazoline-6-sulfonic acid and 250 µM H_2O_2. The reaction was followed spectrophotometrically at λ = 412 nm.

Phenylalanine ammonia lyase (PAL) activity was determined as described by Francini et al. [20]. Extraction was done in 100 mM borate buffer (pH = 8.8) containing 14 mM β-mercaptoethanol. Assay medium consisted of 1 mL of the borate buffer mixed with 200 µL of 100 mM phenylalanine and the formation of *trans*-cinnamic acid at room temperature was deduced from absorbance of the medium at λ = 290 nm.

Callose content of tomato leaves was determined as described by Hirano et al. [21]. Samples from 15 different plants treated the same way were dried and ground together and 3 replicas were created by weighing 30 mg of the mixed powder obtained from the 15 plants. Phenolic compounds were removed by addition of polyvinylpyrrolidone followed with ethanol washings until no visible green color could be seen. The remaining pellet was resuspended in 1 M NaOH to extract callose. A calibration curve was prepared with a 15 µg mL^{-1} β-1,3-glucan (Megazyme/Libios) solution dissolved in 1 M NaOH (concentration range: 0–15 µg mL^{-1}). Callose was measured using aniline blue chromophore in 1 M glycine buffer (pH = 9.5).

2.6. Quantification of Defense Gene Expression by Real Time q-PCR

The experiment was carried out as described by Grassot et al. [22] with slight modifications. Here, samples consisted of 3 tomato foliar disks of 1 cm diameter. A NanoDrop One (ThermoFisher Scientific™, F67403 Illkirch, France) was used to measure RNA amounts and quality was assessed with RNA easy kit (Qiagen). Reverse transcription was effectuated with a Thermo Scientific TM RevertAid First Strand cDNA Synthesis Kit using 1 µg of RNA. Finally, 50 ng of cDNA were used with 10 nM of primers in each well of the PCR 96-well plates. The analyzed genes were *PR2*, *PR8*, *PR9*, *PR14*, *PR15* and *PAL*, with *β-actin* as housekeeping reference gene. The sequences of the primers used are presented in Table 1.

Table 1. List of primers used in real time q-PCR experiments on tomato plants (*Solanum lycopersicum*). F: Forward; R: Reverse.

Gene	References	Primers in the Corresponding Gene
PR2b	-	F: CCGTTGGAAACGAAGTTGAT R: TCATCAGCATGGCCAAAATA
PR8	[23]	F: TGC AGG AAC ATT CAC TGG AG R: TAA CGT TGT GGC ATG ATG GT
PR9	[24] [25]	F: GCTTTGTCAGGGGTTGTGAT R: TGCATCT CTAGCAACCAACG
PR14	[26]	F: CTCCATGCCTCCCTTATCTTC R: CATGCTGTCTTTCGATCCG
PR15	[27]	F: GGGCTAAATCCACCTCA R: GGCACCACGAACATCTC
PAL	[28]	F: CTTTGATGCAGAAGCTGAGACA R: TCGTCCTCGAAAGCTACAATCT
β-actin	[23]	F: AGG CAC ACA GGT GTT ATG GT R: AGC AAC TCG AAG CTC ATT GT

DNA quantification was possible though the SYBR GREEN method. To calculate the expression of the genes of interest we chose to set their expression relative to the reference gene based on threshold values (Ct) using the ΔΔCt method (fold gene expression = $2^{-\Delta\Delta Ct}$). The formula ($-1/2^{-\Delta\Delta Ct}$) was applied to values between 0 and 1, in order to optimize their graphical representation. For ΔΔCt relative quantification, only Ct(s) below 35 were taken into consideration.

2.7. Field Trial on Grapevine against P. viticola Artificially Inoculated

The tested plants belong to *Vitis vinifera* species (Merlot variety) cultivated at Sainte-Livrade sur Lot (47), France. Plants were watered twice a day. The trial was organized in 25 plots (with 7 plants per elementary plot), randomly disposed, and including 4 replicas of each treatment. The treatments consisted in spraying the products over the leaves at 7-to-10-day intervals. The treatment dates are mentioned in Table 2. The different treatments were: not-treated control (NTC), a positive control as IODUS 2 CULTURES SPECIALISEES® with laminarin as the active molecule applied at 90 g active ingredient per ha, and the by-product SL35 applied at different doses: 5 g SL35 per ha (SL35x1), 35 g SL35 per ha (SL35x7), and 70 g SL35 per ha (SL35x14). All SL35 solutions were done in 0.05% (w/v) Tween 80. Artificial contamination was performed on May 28 at the beginning of the flowering (BBCH 61). An inoculum was prepared with frozen grapevine leaves infected by *P. viticola*, harvested in 2019. It was applied by spraying on ten leaves belonging to a grapevine plant positioned at the end of each plot. This plant was further excluded from the rating of the symptoms. To evaluate the progression of the disease, 100 leaves from the same canopy level (similar age) per plot or 50 bunches per plot were used for each rating. The severity was defined as the average surface of leaves or bunches showing symptoms. The incidence was determined as the average percentage of leaves or bunches showing symptoms.

Table 2. Treatment schedule: dates of application and vegetative development stage of the grapevine plants at each time of application.

Dates	Developmental Stage
5 May	First leaf unfolded (BBCH 11)
12 May	Inflorescences clearly visible (BBCH 53)
20 May	Inflorescences fully developed; flowers separating (BBCH 57)
27 May	Beginning of flowering: 10% of flowerhoods fallen (BBCH 61)
8 June	Fruit set: young fruits beginning to swell, remains of flowers lost (BBCH71)
15 June	Pea-sized berries, hanging bunches (BBCH 75)
24 June	Berries beginning to touch (BBCH 77)

2.8. Statistical Analyses

For the biochemical experiments 15 samples per treatment were used. Callose quantification was realized on 3 analytical replicates, each one using the material collected on 15 plants. Real time qPCR was independently realized on 5 biological replicas for each treatment. Statistical analyses were carried out with the Past free software (version 2.17c). The one-way ANOVA or the Kruskal Wallis tests were used (noted on each figure). p values under 0.05 were considered as significant.

For the field trials, the statistical significances of differences among treatments were estimated using one-way ANOVA followed by the Newman and Keuls test ($p < 0.05$) and the Kurtosis analysis.

3. Results

3.1. SL35 Composition

Chemical analyses were done to determine the contents of the major compounds of the SL35 by-product. Mineral compounds represent on the average 44% (± 1.7) of the dry mass (Table 3). Na and K are the most important in quantity with respectively 140 (± 9) and 40 (± 0.7) mg g^{-1} DW. Elemental analysis revealed that carbon is quite abundant with 170 (± 7.4) mg g^{-1} DW. Among organic compounds, sugars were the most represented since total sugar content was estimated to 125 (± 0.2) mg g^{-1} DW. Comparatively, proteins and phenolic compounds were less present as their contents were respectively 7 to 16 times lower compared to the sugar content. Also, the pH value of the SL35 solution measured at a final concentration of 25 mg DW mL^{-1} was 7.0.

3.2. Biological Activity of SL35: Effect of the Concentration on the Stimulation of Tomato Plant Defenses

To evaluate the effect of the treatment on a model plant (tomato), enzymatic activities related to defense mechanisms, peroxidase and PAL specific activity, and callose content were measured. Peroxidase is involved in hydrogen peroxide detoxification and in the synthesis of lignin (*PR9*), PAL transforms phenylalanine into *trans*-cinnamic acid, a precursor of phenolic compounds and, therefore, contributes to the cell wall thickening. Phenolic compounds also possess antimicrobial activity, while callose is one of the main contributors to cell wall thickening.

When tomato plantlets were pre-treated with SL35, a significant increase in peroxidase specific activity was observed for SL35 concentrations of 50 and 1000 mg L^{-1} as compared with control (Figure 1). A tendency to increase was also observed with 200 mg L^{-1} despite the fact that this increase is not significant. A concentration of 10 mg L^{-1} seems not to be sufficient to stimulate this activity.

Table 3. Chemical composition of SL35 (average ± standard deviation, n = 3). CHNS: elemental analysis; DW: Dry Weight; MP-AES: Microwave Plasma Atomic Emission Spectroscopy.

Types of Components	Analysis Method	CONTENTS
Mineral content	calcination	440 ± 17 mg g−1 DW
Na	MP-AES	140 ± 9 mg g−1 DW
K	MP-AES	40.5 ± 0.7 mg g−1 DW
Si	MP-AES	0.48 ± 0.0 mg g−1 DW
Ca	MP-AES	0.43 ± 0.1 mg g−1 DW
Elemental analysis		
C	CHNS	170 ± 7.4 mg g−1 DW
H	CHNS	27.6 ± 1.5 mg g−1 DW
N	CHNS	24 ± 1 mg g−1 DW
S	CHNS	86 ± 6.4 mg g−1 DW
P	LCK349TM and LCK350TM kits	4.3 ± 0.1 mg g−1 DW
Organic compounds		
Total sugars	Colorimetry	125.4 ± 0.2 mg g−1 DW
Uronic acids *	Colorimetry	9.3 ± 0.2%
Proteins	Colorimetry	16.9 ± 2 mg g−1 DW
Phenolic compounds	Colorimetry	7.7 ± mg g−1 DW

* Expressed as % of total sugars.

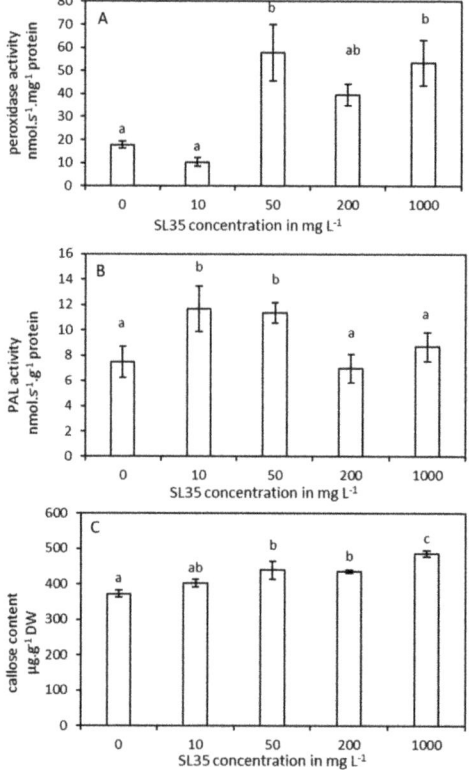

Figure 1. Peroxidase (**A**) and PAL (**B**) specific activities and callose content (**C**) of tomato leaves after treatment with increasing concentration of SL35 diluted in 0.05% Tween 80 (days 1, 3 and 5) and inoculation with *B. cinerea* at day 8. All samples were collected at day 15 (mean ± SD; n = 15 for peroxidase and PAL activities and n = 3 for callose contents). No letters in common between treatments indicate significant differences ($p < 0.05$; one-way ANOVA).

PAL specific activities were also significantly more important in plants treated with 10 or 50 mg SL35 L^{-1} as compared with control. An increase in SL35 concentration of sprayed solution at 200 and 1000 mg L^{-1} seemed not to increase PAL specific activities which were close to control values. Data suggest that concentrations over 200 mg L^{-1} are supra-optimal for the PAL stimulation.

Treatments of tomato plants with SL35 concentrations of 50, 200 and 1000 mg L^{-1} significantly increased callose contents above control.

Finally, SL35 concentration of 50 mg L^{-1} seemed to be the optimal concentration to induce an increase of all the three defense markers: peroxidase and PAL specific activities, together with callose contents.

3.3. Biological Activities of SL35 in Presence or Not of Pathogen: Comparison with Commercial Standard

The elicitor activity of SL35 was compared with that one of laminarin (Vacciplant®; Goëmar laboratories, Saint Malo 35400, France) which is a known PDS, containing a polysaccharide composed of (1–3)-β-D-glucan with β-(1–6) branching [29]. It was applied by spraying a solution at a concentration of 300 mg L^{-1}. The SL35 concentration (50 mg L^{-1}) was chosen as the lowest concentration leading to a significant difference with control (see Figure 1). Then, a part of the tomato plants was inoculated with *B. cinerea* while another part was not inoculated.

As seen in Figure 2, peroxidase activity was not significantly increased in presence of laminarin, whether the plants were inoculated or not. However, a small increase of peroxidase activities was obtained in presence of laminarin as compared with control. Even if the SL35 treatment induced a lower peroxidase activity in non-inoculated plants in comparison with laminarin-treated plants, this activity was significantly higher in the inoculated specimens as compared with control and plants treated with laminarin. The profile of PAL activities was quite different. In absence of inoculation, the laminarin treatment seemed to significantly reduce the PAL activity. The peak value of activity probably occurred before the 7th day after the last laminarin treatment. According to previous works on suspended tobacco cells treated with laminarin, the peak of activity was reached 4 h after treatment [30]. Laminarin treatment of the inoculated plants induced a significant increase of PAL activity in comparison with inoculated plants devoid of laminarin treatment. Besides, SL35 treatment significantly stimulated PAL activity in non-inoculated plants in comparison with their control (non-inoculated and SL35-free plants). Finally, inoculation by itself did not lead to any difference with non-inoculated control specimens.

For the callose contents, the laminarin treatment induced a significant difference with the control in the case of non-inoculated plants: 284 ± 5 µg g^{-1} compared with 207 ± 2 µg g^{-1} respectively. Nevertheless, no statistical difference was observed in the inoculated counterpart, 315 ± 17 µg g^{-1} for plants treated with laminarin vs. 298 ± 16 µg g^{-1} for the control ones. However, a difference appeared between the inoculated plants and the non-inoculated control ones sprayed with water. The same goes for SL35, where there was a statistical difference in non-inoculated samples (SL35-treated plants compared with water-treated control) with a higher value at 318 ± 14 µg g^{-1}. However, inoculated samples did not present any notable difference with the control, since only 286 ± 21µg g^{-1} of callose was found in the former samples. The statistical difference between inoculated SL35 and non-inoculated water control persists, as does the difference between inoculated and non-inoculated controls. These data give the information that pathogen, elicitor, or a combination of both, stimulate callose formation at the same level since all of these treatments led to statistical differences with the non-inoculated water control but not between themselves. It might be possible that callose accumulation reach a maximum value which cannot be increased even if other stimuli are perceived.

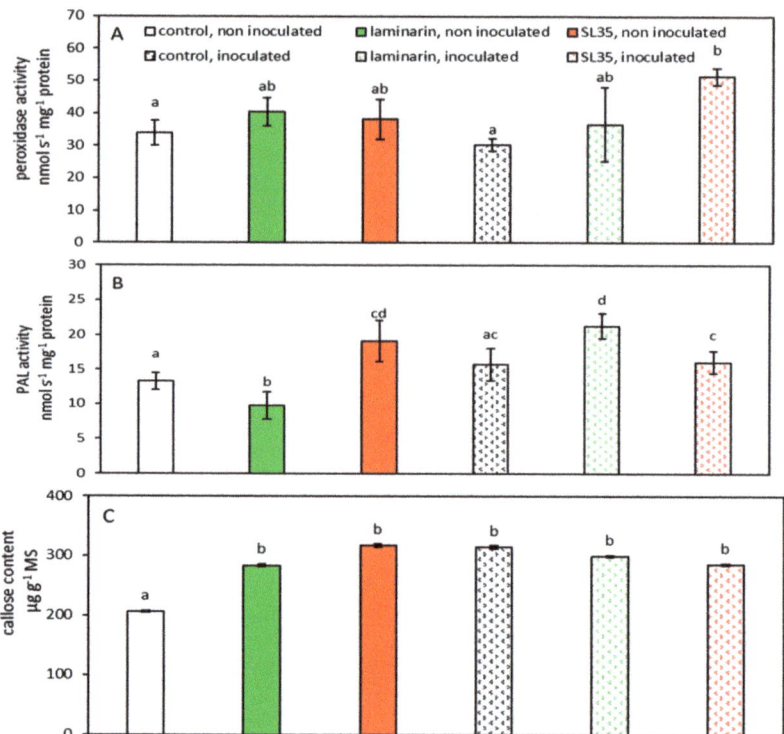

Figure 2. Peroxidase (**A**) and PAL (**B**) specific activities and callose content (**C**) determined in tomato leaves after treatment with SL35 (50 mg L^{-1}) or laminarin (days 1, 3 and 5). At day 8, inoculation by *B. cinerea* was applied or not (non-inoculated). All samples were collected at day 15. Peroxidase values are means of 15 samples ± SE and Kruskal Wallis test was performed. PAL values are means of 15 samples ± SD and one-way ANOVA was performed. Callose values are mean of 3 samples ± SD and one-way ANOVA was performed. Different letters indicate significant differences ($p < 0.05$).

3.4. Evaluation of Defense Responses by Quantification of Gene Expression

Gene expression in relation to some PR proteins and PAL was quantified in tomato plantlet leaves one day following inoculation (day 9) after pre-treatment with laminarin (300 mg L^{-1}) or SL35 (50 mg L^{-1}) comparatively to a control treated with water. The same experiment was done on plantlets without inoculation (harvested on day 9). Relative expressions refer to the appropriate water controls (non-inoculated water-treated plants for the inoculation-free treatments and inoculated water-treated plantlets for the treatments with inoculation).

For non-inoculated plants, no induction was recorded for all of the tested genes in comparison with the positive control (laminarin treatment). Moreover, a significant 5-fold repression of the *PR9* gene was observed relative to the control (Figure 3). For the inoculated samples, laminarin significantly induced increased expression of *PR2*, *PR8* and *PR15*, respectively 7.7, 3.4 and 5.6-fold. A slight increase for the *PR9* gene expression was visible (2.1-fold), but the difference was not sufficient to generate a statistical significance. Expression of the other *PR* or *PAL* genes did not show any significant difference.

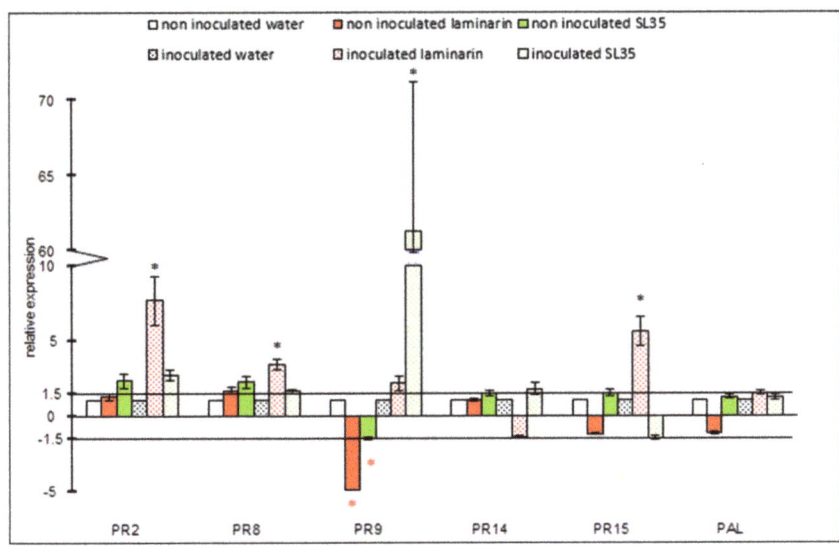

Figure 3. Relative expression levels of some genes encoding PR proteins and PAL in tomato leaves subjected to SL35 or laminarin treatment by foliar spraying (days 1, 3 and 5). Plants were infiltrated with *B. cinerea* conidia at day 8 (dotted histograms). All samples were harvested at day 9. Gene expressions were calculated relative to the appropriate water controls (non-inoculated water for non-inoculated laminarin and non-inoculated SL35, inoculated water for inoculated laminarin and inoculated SL35). Values are means of at least 5 independent samples ± SE. One-way ANOVA was performed to evaluate statistical differences. Only relative expression values out of the interval (+1.5, −1.5), were taken in consideration. * Means significant statistical differences in comparison with controls ($p < 0.05$).

When tomato plantlets were treated with SL35 without inoculation, *PR2* and *PR8* genes were overexpressed 3.1 and 2.2-fold respectively, but no significant difference was observed. *PR14*, *PR15* and *PAL* genes showed almost no increased expression. *PR9* gene expression seemed to be significantly reduced (by 1.5-fold) by the SL35 treatment compared with control. The same tendency was observed for laminarin treatment as described above. For this gene, inoculation seemed to increase the level of expression as compared with the plantlets treated with either laminarin or SL35. Indeed, the induction by SL35 was quite high, reaching 61.3-fold over control. On the other hand, for *PR2*, *PR8* and *PR14* genes, the expression showed a tendency to increase (respectively 2.7, 1.6 and 1.8-fold as compared with control). Again, *PR15* and PAL did not show any significant expression change, whatever the plant treatment.

When SL35 was compared with laminarin, known as an inducer of defense genes, the expression profiles followed the same tendency for the *PR2*, *PR8* and *PR9* genes, i.e., an increase in expression, *PR9* being the only significantly overexpressed of these three genes in SL35-treated plants. Opposite results were obtained for *PR9* in non-inoculated plants as compared with inoculated ones. The only stronger effect of laminarin was observed with *PR15* which presented a significant induction although no significant effect was observed with SL35 treatment.

3.5. Field Trial on Grapevine against P. viticola Artificially Inoculated

The preliminary work performed in greenhouse on tomato as model plant, allowed to assess the elicitor activity and the optimal concentration of SL35. The field trial was conducted on the pathosystem grapevine-*P. viticola* instead of tomato-*B. cinerea* because of the economic importance of grapevine in France. The long distance of the experimental field from the laboratory did not allow to carry out the analysis on gene expression.

The field trial was localized in Sainte-Livrade sur Lot (France, 47), during the year 2020. The SL35 capacity to protect grapevine from *P. viticola* was assessed in comparison with a commercial laminarin (IODUS 2 CEREALES®, Goëmar laboratories, Saint Malo 35400, France). The control condition corresponded to treatment with water. The concentrations of the sprayed solutions of SL35 were 50 mg SL35 L^{-1} named SL35x1, 350 mg SL35 L^{-1} named SL35x7 and 700 mg SL35 L^{-1} named SL35x14. These choices were based on the previous greenhouse experiments in which the lowest concentration leading to a significant increase of tomato defense markers was 50 mg L^{-1} (Figure 1). While the two other concentrations, 350 mg L^{-1} and 700 mg L^{-1}, were chosen because field treatments are known to generate losses in applied product (drift during spraying, or rainwater leaching, depending on weather conditions). The treatments consisted in spraying the products over the leaves at 7-to-10-day intervals. An artificial contamination with spores of *P. viticola* was done on 28 May, one day after the 4th application of the treatments since the disease did not occur naturally.

Figure 4 shows the results recorded on 4 June, 7 days after the artificial inoculation (for grapes), on 18 June, 21 days after artificial inoculation (for leaf incidence) and on 30 June, 33 days after inoculation (for leaf severity). For leaf incidence, even if a decrease in incidence was observed on plants treated with laminarin and with SL35 (six treatments done), no significant difference was observed whatever the concentration (Figure 4A).

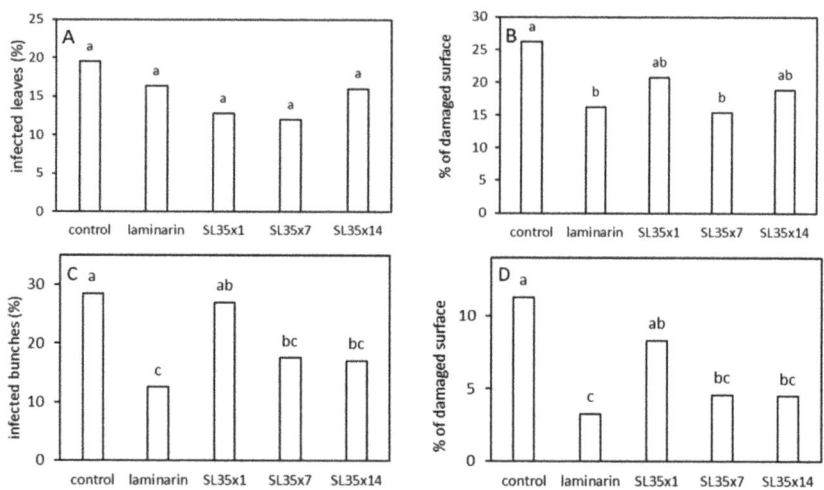

Figure 4. Effects of SL35 treatments on incidence (**A**) of downy mildew on grapevine (field trial, notation 18 June 2020) expressed as % of infected leaves and on severity (**B**) expressed as % of damaged surface (notation 30 June 2020). The same parameters were measured on bunches: incidence (**C**) and severity (**D**) (notation 4 June 2020). SL35 was applied at different concentrations (50 mg L^{-1} for SL35x1, 350 mg L^{-1} for SL35x7, 700 mg L^{-1} for SL35x14) beginning from May 5 and repeated at 7-12-day intervals (see Table 2 for details). Control corresponds to water treatment. Laminarin corresponds to application of IODUS 2 CEREALES®. Means followed by the same letter do not significantly differ ($p < 0.05$, one-way ANOVA followed by Newman and Keuls test and Kurtosis analysis).

However, the control showed an advanced stage of the disease in terms of percentage of infected leaves (19.5% on June 18). On 30 June (seven treatments done and 33 days after the artificial inoculation), a significant reduction in severity was recorded with both laminarin and SL35x7 which showed efficacies of 38% and 41%, respectively (Figure 4B;

see also Table S1). Treatments with SL35x1 and SL35x14 could possibly decrease the severity of the disease but these results were not significantly different from the control which showed a severity of 26.2%, compared with 20.7% and 18.8% for SL35x1 and SL35x14, respectively. Taken together, the results obtained during this field trial showed that 350 mg L^{-1} was an optimal concentration of SL35, able to reduce the symptoms of mildew on grapevine and that this concentration was statistically as efficient as the reference used in this study (laminarin, applied as IODUS 2 CEREALES).

Downy mildew symptoms were also observed on sampled bunches of grapes. They were significantly less important with SL35x7 and SL35x14 treatments as compared with control, showing an efficacy to reduce the disease symptoms around 38% (Figure 4C). This activity was equivalent to the laminarin treatment. The same profile was observed for damaged surface of bunches and again SL35x1 seemed to be insufficient to reduce the disease symptoms, even if a slight, but not significant, reduction was observed as compared with control.

4. Discussion

Agar extraction from red algae such as *G. sesquipedale* is a source of large amounts of by-products which are poorly valorized. We have previously shown that alkaline treatment of this seaweed could extract some original components such as low molecular weight carbohydrates. In this study, the alkaline solution obtained from the industrial process was recovered, neutralized and dried. Afterwards, it was analyzed for its chemical composition. It is characterized by a quite high content of minerals representing 44% of DW, Na being the main element (140 mg g^{-1} DW) which could be explained by the NaOH solution used for alkaline treatment. K content reached 40 mg g^{-1} DW probably due to the marine provenience of the alga, NaCl and KCl being very present salts of seawater. The high level of the S element (86 ± 6.4 mg g^{-1} DW) might be due to the neutralization of the by-product by sulfuric acid before lyophilization. Among organic compounds, carbohydrates are the most represented, their content reached 125.4 mg g^{-1} DW. They could be extracted from algal cell wall during the alkaline treatment; the cell wall is known to be rich in polysaccharides such as agar (a galactan) and cellulose (in lower quantity) [31]. Among them, uronic acids represent only 9.3%. This value might also be overestimated because of the important concentration in salts of the by-product which interfere with the colorimetric assay. Nevertheless, previous studies showed the presence of uronic acids (a little more than 20%) in the sugar composition of polysaccharides extracted from red algae of the *Gelidium* genus [32,33]. Protein and phenolic compounds are minor components, representing 16.9 and 7.7 mg g^{-1} DW, respectively.

The ability of SL35 to induce defense responses in plant was firstly studied in controlled conditions on tomato plantlets. A significant three-fold increase in peroxidase specific activity was observed for concentrations of 50 and 1000 mg SL35 L^{-1} (Figure 1) as compared with the control. This biomarker is often used to evaluate the plant response and 2-3-fold increases in specific activity have previously been described [19,34,35]. Indeed, peroxidase is involved in hydrogen peroxide detoxification and in the synthesis of lignin (*PR9*). PAL-specific activities also increased, by about a factor of two, when tomato plants were treated with 10 or 50 mg SL35 L^{-1} as compared with the control. The same trend had already been observed, for example, in soybean treated with algino-oligosaccharides [36]. As PAL transforms phenylalanine into *trans*-cinnamic acid, a precursor of phenolic compounds, this enzyme contributes to the cell wall thickening. The phenolic compounds also possess antimicrobial activity. Concerning callose, the minimum concentration of SL35 needed to significantly induce an increase in callose content is 50 mg L^{-1}. Induction of callose deposition has also been described as a plant defense against pathogens [37] since it can reinforce the plant cell wall structure and so diminish the ability of some pathogens to attack plant cells [38]. From data presented in Figure 1, it can be concluded that SL35 can induce the plant defense responses, notably the three biomarkers chosen in this study (peroxidase and PAL activities and callose content) at an optimal concentration of 50 mg L^{-1}.

In the literature, elicitors are able to induce plant defense responses even if the plant is not subjected to a pathogen attack [34,36]. Laminarin, used here as a positive control, did not induce any significant increase in PAL and peroxidase activities in absence of pathogen; only callose formation was stimulated in comparison with the control. PAL activity significantly increased after pathogen inoculation in plants treated with laminarin, exemplifying the priming process already described for this elicitor [39]. Unfortunately, to our knowledge, data on peroxidase activity in plants treated with laminarin are not available so far in the literature. In the case of SL35, the increased response of defense biomarkers was confirmed with PAL and callose levels, even if the plants were not first inoculated; concerning peroxidase, even if a slight increase was observed, it was not significant in comparison with control. Concerning this activity, the presence of the pathogen seems to be necessary to fully express the potential of SL35. Therefore, SL35 treatment seemed to be efficient, also as a preventive tool designed to help the plant to resist the pathogen attack.

When biological activities are considered at the molecular level, i.e., the regulation of expression of defenses genes (some PR proteins and PAL-encoding genes), a strong increase in *PR2* gene expression was observed after laminarin treatment, but only after inoculation. This result is equivalent to previously reported data [30,40]. Since this gene encodes a β-1,3-glucanase, this activation could be implied in the destruction of the fungus cell wall. SL35 treatment was also able to induce an activation of the *PR2* gene after inoculation, even if our results are not statistically significant.

PR8 encodes a type III chitinase, which is responsible for the hydrolysis of chitin, a polysaccharide of fungal cell walls. Concerning *PR8* gene expression in laminarin-treated plants, the profile was equivalent to what was observed with *PR2* gene. Indeed, Xin et al. [41] have reported an increase of total chitinase activities in green tea leaves two days after laminarin spraying. And, more recently, Borba et al. [40] signaled a significant increase in wheat chitinase activity after laminarin treatment. SL35 was also able to induce an increased expression of the *PR8* gene, whether the plants were inoculated or not.

The *PR9* group peroxidases are specific enzymes that only intervene in cell wall thickening [42,43]. The expression profile for this PR group is similar in laminarin- and SL35-treated tomato plants, i.e., both treatments led to a statistically significant repression in absence of pathogen inoculation, while presenting an induction when plants were first inoculated. This induction is strongly marked and statistically different from control in presence of SL35. Despite of that, many studies did report increases in H_2O_2 content consecutive to laminarin treatment [44,45] but, to our knowledge, the peroxidase (H_2O_2 detoxifier) gene expression has not been recorded yet after treatment with laminarin. However, a closely comparable result has been reported with poplar cells treated with laminarin [46]. The authors observed a peak of expression of the peroxidase gene 10 h after treatment, as revealed by Northern blot analysis. Since our assay was done 24 h after inoculation (day 9), the maximal expression might not be revealed. The marked expression induced by SL35 in presence of pathogen may explain the increase observed in peroxidase activity at day 15, which was 10 days after the last treatment (Figure 2).

PR14 encodes lipid transfer proteins implicated in the reinforcement of cutin and might also have a direct effect on the pathogen by inducing its membrane permeabilization [47]. The weak induction of *PR14* by SL35 treatment of tomato plants after inoculation with the pathogen needs to be confirmed.

PR15 encodes oxalate oxidase-type proteins. Expression of this gene has been found to increase in tomatoes infected with *B. cinerea* strains [27]. The only treatment which induced a significant increase of the *PR15* gene expression is laminarin after inoculation with the pathogen. Data concerning this *PR* do not allow us to deduce any effect of SL35 on the expression of its gene, but we cannot exclude that oxidative stress might be a defense response of the plant. A quantification of the H_2O_2 content of leaf tissues might give an answer to this question.

The PAL gene expression after laminarin treatment presents a profile suggesting a priming process according to the enzymatic activity shown in Figure 2, despite an absence of significant results about the expression of this gene. SL35 treatments do not induce any change in PAL gene expression. Keeping in consideration the enzymatic activities assayed at day 15, maybe the timelapse for the characterization of the gene expression, i.e., day 9, was not optimal.

To assess the efficiency of SL35 to stimulate the defense mechanisms of plant cells in order to make the plant more resistant to pathogen attack, a field trial conducted on grapevine artificially inoculated with *P. viticola* was designed. Laminarin treatment was efficient to reduce the severity of the symptoms on leaves and on bunches with an efficacy of 38 and 71%, respectively. The SL35x7 treatment showed a similar activity (59% on bunches and 41% on leaves). These effects could be explained by the results showing the activation of some markers of plant defense implied either in cell wall reinforcement (callose, peroxidase, PAL), or in the direct attack against the pathogenic fungus (*PR2* and *PR8*). The optimal spray concentration of SL35 estimated in field conditions was 350 mg L^{-1}. This is seven-fold higher than during the experiment in controlled conditions, but the field conditions might reduce the biological activity, due to partial drift during spraying, or leaching by rainwater. This optimal concentration led to an equivalent dose of 88 g of SL35 per ha which is in line with agricultural practices. Since downy mildew symptoms have been not fully eliminated by SL35 treatment, the use of synthetic pesticide could be useful under strong pathogen pressure. However, the integration of SL35 with synthetic products at low doses could be considered in integrated pest management strategies. In order to progress in the comprehension of the biological activity of SL35, a more detailed chemical composition has to be established to ascertain which component(s) is (are) responsible for the stimulation of the plant defenses.

5. Conclusions

The results of this study show that a raw alkaline extract of the red alga *G. sesquipedale* might be considered as a new promising plant defense stimulator. Further studies will be focused on the evaluation of the properties of some components of this crude extract, in particular carbohydrates, since oligosaccharides have been shown to elicit plant defense responses. These data would be of interest in the comprehension of the mode of action of this seaweed-originated by-product and could provide enhanced value to the economic sector of agarophytes.

Supplementary Materials: The following supporting information can be downloaded at: https://www.mdpi.com/article/10.3390/horticulturae8100958/s1, Table S1: Photographs of vine plants taken on 23 June 2020 showing the symptoms of infection by Plasmopara viticola on grapes and leaves.

Author Contributions: Conceptualization, C.F.G., V.G., S.L., F.D. and R.K.; methodology, C.F.G., V.G., S.L., F.D. and R.K.; resources, S.L. and C.F.G.; investigation, C.F.G., S.L., F.D. and R.K.; writing—original draft preparation, R.K. and C.F.G.; writing—review and editing: C.F.G., V.G., S.L., F.D. and R.K.; funding acquisition, S.L., C.F.G. and V.G. All authors have read and agreed to the published version of the manuscript.

Funding: This research was funded by SETEXAM (Morocco).

Institutional Review Board Statement: Not applicable.

Informed Consent Statement: Not applicable.

Data Availability Statement: Not applicable.

Acknowledgments: The authors acknowledge M. Guilloton for his help in manuscript editing and ANADIAG S.A.S. company which have carried out the field trial on grapevine.

Conflicts of Interest: SETEXAM collected the seaweed, supplied the industrial by-product and approved the publication of the results presented herein.

References

1. Yadav, I.C.; Devi, N.L. Pesticides Classification and Its Impact on Human and Environment. In *Environmental Science and Engineering*; Chapter 7; Studium Press LLC: Houston, TX, USA, 2017; Volume 6, pp. 140–158.
2. Food and Agriculture Organization of the United Nations. Available online: https://www.fao.org/faostat/en/#data/RP/visualize (accessed on 17 August 2022).
3. Hammerschmidt, R. Introduction: Definitions and some history. In *Induced Resistance for Plant Defence*; Walters, D., Newton, A., Lyon, G., Eds.; Blackwell Publishing: Hoboken, NJ, USA, 2007; pp. 1–8.
4. Proestos, C.; Chorianopoulos, N.; Nychas, G.J.; Komaitis, M. RP-HPLC analysis of the phenolic compounds of plant extracts. Investigation of their antioxidant capacity and antimicrobial activity. *J. Agric. Food Chem.* **2005**, *53*, 1190–1195. [CrossRef]
5. Bellincampi, D.; Cervone, F.; Lionetti, V. Plant cell wall dynamics and wall-related susceptibility in plant–pathogen interactions. *Front. Plant Sci.* **2014**, *5*, 228. [CrossRef] [PubMed]
6. Héloir, M.-C.; Adrian, M.; Brulé, D.; Claverie, J.; Cordelier, S.; Daire, X.; Dorey, S.; Gauthier, A.; Lemaître-Guillier, C.; Negrel, J.; et al. Recognition of Elicitors in Grapevine: From MAMP and DAMP Perception to Induced Resistance. *Front. Plant Sci.* **2019**, *10*, 1117. [CrossRef] [PubMed]
7. Poornananda, M.; Jameel, K. Impact of abiotic elicitors on in vitro production of plant secondary metabolites: A review. *J. Adv. Res. Biotech.* **2016**, *1*, 7.
8. Trouvelot, S.; Héloir, M.-C.; Poinssot, B.; Gauthier, A.; Paris, F.; Guillier, C.; Combier, M.; Trdá, L.; Daire, X.; Adrian, M. Carbohydrates in plant immunity and plant protection: Roles and potential application as foliar sprays. *Front. Plant Sci.* **2014**, *5*, 592. [CrossRef] [PubMed]
9. Abdul Malik, N.A.; Kumar, I.S.; Nadarajah, K. Elicitor and Receptor Molecules: Orchestrators of Plant Defense and Immunity. *Int. J. Mol. Sci.* **2020**, *21*, 963. [CrossRef] [PubMed]
10. Vera, J.; Castro, J.; Gonzalez, A.; Moenne, A. Seaweed polysaccharides and derived oligosaccharides stimulate defense responses and protection against pathogens in plants. *Mar. Drugs* **2011**, *9*, 2514–2525.
11. Lebbar, M.S.; Faugeron-Girard, C.; Gloaguen, V. Use of An Extract or An Extract Fraction of Agarophyte Red Algae as A Plant Defense Elicitor/Stimulator and Application of Said Extract or Said Extract Fraction. French Patent FR 1870542, 7 May 2018.
12. Lebbar, S.; Fanuel, M.; Le Gall, S.; Falourd, X.; Ropartz, D.; Bressollier, P.; Gloaguen, V.; Faugeron-Girard, C. Agar extraction by-products from *Gelidium sesquipedale* as a source of glycerol-galactosides. *Molecules* **2018**, *23*, 3364. [CrossRef]
13. Nguyen, D.N.; Grybos, M.; Rabiet, M.; Deluchat, V. How do colloid separation and sediment storage methods affect water-mobilizable colloids and phosphorus? An insight into dam reservoir sediment. *Colloids Surf.* **2020**, *606*, 125505. [CrossRef]
14. Bascle, S.; Bourven, I.; Baudu, M. Nature and accessibility of organic matter in lacustrine sediment. *J. Soils Sediments* **2021**, *21*, 1–19. [CrossRef]
15. Dubois, M.; Gilles, K.A.; Hamilton, J.K.; Rebers, P.A.; Smith, F. Colorimetric Method for Determination of Sugars and Related Substances. *Anal. Chem.* **1956**, *28*, 350–356. [CrossRef]
16. Blumenkrantz, N.; Asboe-Hansen, G. New method for quantitative determination of uronic acids. *Anal. Biochem.* **1973**, *54*, 484–489. [CrossRef]
17. Bradford, M.M. A rapid and sensitive method for the quantification of microgram quantities of protein utilizing the principle of protein-dye binding. *Anal. Biochem.* **1976**, *72*, 248–254. [CrossRef]
18. Piló-Veloso, D.; Nascimento, E.A.; Morais, S.A.L. Isolamento e análise estrutural de ligninas. *Química Nova* **1993**, *16*, 435–448.
19. Faugeron-Girard, C.; Gloaguen, V.; Koçi, R.; Célérier, J.; Raynaud, A.; Moine, C. Use of a *Pleurotus ostreatus* complex cell wall extract as elicitor of plant defenses: From greenhouse to field trial. *Molecules* **2020**, *25*, 1094. [CrossRef]
20. Francini, A.; Nali, C.; Picchi, V.; Lorenzini, G. Metabolic changes in white clover clones exposed to ozone. *Environ. Exp. Bot.* **2007**, *60*, 11–19. [CrossRef]
21. Hirano, Y.; Pannatier, E.G.; Zimmermann, S.; Brunner, I. Induction of callose in roots of Norway spruce seedlings after short-term exposure to aluminium. *Tree Physiol.* **2004**, *24*, 1279–1283. [CrossRef]
22. Grassot, V.; Da Silva, A.; Saliba, J.; Maftah, A.; Dupuy, F.; Petit, J.M. Highlights of glycosylation and adhesion related genes involved in myogenesis. *BMC Genom.* **2014**, *15*, 621. [CrossRef]
23. Chun, S.C.; Chandrasekaran, M. Chitosan and chitosan nanoparticles induced expression of pathogenesis-related proteins genes enhances biotic stress tolerance in tomato. *Int. J. Biol. Macromol.* **2019**, *125*, 948–954. [CrossRef]
24. Babu, A.N.; Jogaiah, S.; Ito, S.I.; Nagaraj, A.K.; Tran, L.S.P. Improvement of growth, fruit weight and early blight disease protection of tomato plants by rhizosphere bacteria is correlated with their beneficial traits and induced biosynthesis of antioxidant peroxidase and polyphenol oxidase. *Plant Sci.* **2015**, *231*, 62–73. [CrossRef]
25. Jogaiah, S.; Abdelrahman, M.; Tran, L.S.P.; Shin-Ichi, I. Characterization of rhizosphere fungi that mediate resistance in tomato against bacterial wilt disease. *J. Exp. Bot.* **2013**, *64*, 3829–3842. [CrossRef] [PubMed]
26. D'Agostino, N.; Buonanno, M.; Ayoub, J.; Barone, A.; Monti, S.M.; Rigano, M.M. Identification of non-specific Lipid Transfer Protein gene family members in Solanum lycopersicum and insights into the features of Sola l 3 protein. *Sci. Rep.* **2019**, *9*, 1–16. [CrossRef] [PubMed]
27. Sun, G.; Feng, C.; Zhang, A.; Zhang, Y.; Chang, D.; Wang, Y.; Ma, Q. The dual role of oxalic acid on the resistance of tomato against *Botrytis cinerea*. *World J. Microbiol. Biotechnol.* **2019**, *35*, 1–7. [CrossRef] [PubMed]

28. Wang, F.; Feng, G.; Chen, K. Defense responses of harvested tomato fruit to burdock fructooligosaccharide, a novel potential elicitor. *Postharvest Biol. Technol.* **2009**, *52*, 110–116. [CrossRef]
29. Becker, S.; Scheffel, A.; Polz, M.F.; Hehemann, J.H. Accurate quantification of laminarin in marine organic matter with enzymes from marine microbes. *Appl. Environ. Microbiol.* **2017**, *83*, e03389-16. [CrossRef] [PubMed]
30. Klarzynski, O.; Plesse, B.; Joubert, J.M.; Yvin, J.C.; Kopp, M.; Kloareg, B.; Fritig, B. Linear β-1, 3 glucans are elicitors of defense responses in tobacco. *Plant Physiol.* **2000**, *124*, 1027–1038. [CrossRef]
31. Usov, A.I. Polysaccharides of the red algae. *Adv. Carbohydr. Chem. Biochem.* **2011**, *65*, 115–217.
32. Izumi, K. Chemical heterogeneity of the agar from *Gelidium amansii*. *Carbohydr. Res.* **1971**, *17*, 227–230. [CrossRef]
33. Cui, M.; Wu, J.; Wang, S.; Shu, H.; Zhang, M.; Liu, K.; Liu, K. Characterization and anti-inflammatory effects of sulfated polysaccharide from the red seaweed *Gelidium pacificum* Okamura. *Int. J. Biol. Macromol.* **2019**, *129*, 377–385. [CrossRef]
34. Zhang, P.Y.; Wang, J.C.; Liu, S.H.; Chen, K.S. A novel burdock fructooligosaccharide induces changes in the production of salicylates activates defence enzymes and induces systemic acquired resistance to Colletotrichum orbiculare in cucumberseedlings. *J. Phytopathol.* **2009**, *157*, 201–207. [CrossRef]
35. Abouraïcha, E.; El Alaoui-Talibi, Z.; El Boutachfaiti, R.; Petit, E.; Courtois, B.; Courtois, J.; El Modafar, C. Induction of natural defense and protection against Penicillium expansum and Botrytis cinerea in apple fruit in response to bioelicitors isolated from green algae. *Sci. Hortic.* **2015**, *181*, 121–128. [CrossRef]
36. An, Q.D.; Zhang, G.L.; Wu, H.T.; Zhang, Z.C.; Zheng, G.S.; Luan, L.; Murata, Y.; Li, X. Alginate-deriving oligosaccharide production by alginase from newly isolated Flavobacterium sp. LXA and its potential application in protection against pathogens. *J. Appl. Microbiol.* **2009**, *106*, 161–170. [CrossRef]
37. Feng, H.; Xia, W.; Shan, C.; Zhou, T.; Cai, W.; Zhang, W. Quaternized chitosan oligomers as novel elicitors inducing protection against B. cinerea in Arabidopsis. *Int. J. Biol. Macromol.* **2015**, *72*, 364–369. [CrossRef]
38. Pogorelko, G.; Lionetti, V.; Bellincampi, D.; Zabotina, O. Cell wall integrity. *Plant Signal. Behav.* **2013**, *8*, 9. [CrossRef]
39. Gauthier, A.; Trouvelot, S.; Kelloniemi, J.; Frettinger, P.; Wendehenne, D.; Daire, X.; Joubert, J.M.; Ferrarini, A.; Delledonne, M.; Flors, V.; et al. The sulfated laminarin triggers a stress transcriptome before priming the SA- and ROS-dependent defenses during grapevine's induced resistance against Plasmopara viticola. *PLoS ONE* **2014**, *6*, e88145. [CrossRef]
40. De Borba, M.C.; Velho, A.C.; de Freitas, M.B.; Holvoet, M.; Maia-Grondard, A.; Baltenweck, R.; Stadnik, M.J. A Laminarin-Based Formulation Protects Wheat Against Zymoseptoria tritici via Direct Antifungal Activity and Elicitation of Host Defense-Related Genes. *Plant Dis.* **2022**, *106*, 1408–1418. [CrossRef]
41. Xin, Z.; Cai, X.; Chen, S.; Luo, Z.; Bian, L.; Li, Z.; Ge, L.; Chen, Z. A disease resistance elicitor laminarin enhances tea defense against a piercing herbivore Empoasca (Matsumurasca) onukii Matsuda. *Sci. Rep.* **2019**, *9*, 814. [CrossRef]
42. Sinha, M.; Singh, R.P.; Kushwaha, G.S.; Iqbal, N.; Singh, A.; Kaushik, S.; Kaur, P.; Sharma, S.; Singh, T.P. Current overview of allergens of plant pathogenesis related protein families. *Sci. World J.* **2014**, *2014*, 543195. [CrossRef]
43. Wanderley-Nogueira, A.C.; Belarmino, L.C.; Soares-Cavalcanti, N.D.M.; Bezerra-Neto, J.P.; Kido, E.A.; Pandolfi, V.; Abdelnoor, R.V.; Binneck, E.; Carazzole, M.F.; Benko-Iseppon, A.M. An overall evaluation of the resistance (R) and pathogenesis-related (PR) superfamilies in soybean, as compared with Medicago and Arabidopsis. *Genet. Mol. Biol.* **2012**, *35*, 260–271. [CrossRef]
44. Aziz, A.; Poinssot, B.; Daire, X.; Adrian, M.; Bézier, A.; Lambert, B.; Pugin, A. Laminarin elicits defense responses in grapevine and induces protection against Botrytis cinerea and Plasmopara viticola. *Mol. Plant Microbe Interact.* **2003**, *16*, 1118–1128. [CrossRef]
45. Ménard, R.; de Ruffray, P.; Fritig, B.; Yvin, J.C.; Kauffmann, S. Defense and resistance-inducing activities in tobacco of the sulfated β-1, 3 glucan PS3 and its synergistic activities with the unsulfated molecule. *Plant Cell Physiol.* **2005**, *46*, 1964–1972. [CrossRef]
46. Bae, E.K.; Lee, H.; Lee, J.S.; Noh, E.W.; Jo, J. Molecular cloning of a peroxidase gene from poplar and its expression in response to stress. *Tree Physiol.* **2006**, *26*, 1405–1412. [CrossRef]
47. Sels, J.; Mathys, J.; de Coninck, B.M.A.; Cammue, B.P.A.; de Bolle, M.F.C. Plant pathogenesis-related (PR) proteins: A focus on PR peptides. *Plant Physiol. Biochem.* **2008**, *46*, 941–950.

Article

Jania adhaerens Primes Tomato Seed against Soil-Borne Pathogens

Hillary Righini [1], Roberta Roberti [1,*], Silvia Cetrullo [2], Flavio Flamigni [2], Antera Martel Quintana [3], Ornella Francioso [1], Veronica Panichi [2], Stefano Cianchetta [4] and Stefania Galletti [4]

[1] Department of Agricultural and Food Sciences, Alma Mater Studiorum, University of Bologna, Viale G. Fanin 40, 40127 Bologna, Italy
[2] Department of Biomedical and Neuromotor Sciences, Alma Mater Studiorum, University of Bologna, Via Irnerio 48, 40126 Bologna, Italy
[3] Banco Español de Algas, Instituto de Oceanografía y Cambio Global, IOCAG, Universidad de Las Palmas de Gran Canaria, 35214 Telde, Spain
[4] Council for Agricultural Research and Economics, Research Centre for Agriculture and Environment, 40128 Bologna, Italy
* Correspondence: roberta.roberti@unibo.it

Abstract: Managing soil-borne pathogens is complex due to the restriction of the most effective synthetic fungicides for soil treatment. In this study, we showed that seed priming with *Jania adhaerens* water-soluble polysaccharides (JA WSPs) was successful in protecting tomato plants from the soil-borne pathogens *Rhizoctonia solani*, *Pythium ultimum*, and *Fusarium oxysporum* under greenhouse conditions. WSPs were extracted from dry thallus by autoclave-assisted method, and the main functional groups were characterized by using FT-IR spectroscopy. WSPs were applied by seed treatment at 0.3, 0.6 and 1.2 mg/mL doses, and each pathogen was inoculated singly in a growing substrate before seeding/transplant. Overall, WSPs increased seedling emergence, reduced disease severity and increased plant development depending on the dose. Transcriptional expression of genes related to phenylpropanoid, chlorogenic acid, SAR and ISR pathways, and chitinase and β-1,3 glucanase activities were investigated. Among the studied genes, HQT, HCT, and PR1 were significantly upregulated depending on the dose, while all doses increased PAL and PR2 expression as well as β-1,3 glucanase activity. These results demonstrated that, besides their plant growth promotion activity, JA WSPs may play a protective role in triggering plant defense responses potentially correlated to disease control against soil-borne pathogens.

Keywords: biological control; *Jania adhaerens*; water-soluble polysaccharides; seed priming; soil-borne pathogens; plant-induced resistance; tomato; FT-IR

1. Introduction

Jania adhaerens (J.V. Lamour) (JA) is a red calcareous macroalga belonging to the Corallinaceae family of the Corallinales order, Planta kingdom. The genus Jania is widely present on the Mediterranean and Atlantic coasts [1], and in the Pacific, the Caribbean, and Gulf of Mexico coasts of North and Central America [2]. The typical red color is due to phycoerythrin, the major pigment in red algae [3]. The cell wall of JA is particularly rich in carbohydrates similar to other marine macroalgae [4]. Sulfated galactans are the main polysaccharides constituting the cell walls of most red seaweeds and are also produced by the Corallinales order including JA [1,5]. It has been reported that polysaccharides are rich in functional groups able to bind to some microelements and metal ions having an important role in plant growth [6]. Furthermore, polysaccharides are well-known potent elicitors of plant resistance to fungi, bacteria and viruses, and abiotic stresses [4,7]. The elicitor-induced host response involves the activation of complex signaling cascades followed by the synthesis of defense signaling molecules such as salicylic acid, jasmonic acid, and ethylene that activate the SAR and ISR metabolic pathways. Accordingly, several antimicrobial compounds such as phytoalexins, and pathogenesis-related proteins accumulate in elicited plants [8,9].

Soil-borne organisms are characterized by a subterranean part of their life cycle [10]. This definition encompasses many types of organisms such as bacteria, oomycetes, fungi and nematodes, amongst which some species are plant pathogens. Among the most important soil-borne pathogens that cause economic losses in horticulture, there are the oomycete *Pythium ultimum* and the fungi *Rhizoctonia solani* and *Fusarium oxysporum*. *Pythium ultimum* is a common inhabitant of fields, ponds, streams, and decomposing vegetation worldwide. This oomycete forms hyphae (without septa), oospores, and sporangia which are capable of long-term survival [11]. Mycelia and oospores in the soil can initiate infections of seeds or roots, leading to wilting, reduced yield, root rot, and mortality on >300 diverse hosts [12]. *Rhizoctonia solani* (multinucleate) is a collective species consisting of several unrelated strains, divided into anastomosis groups (AGs), differing in their host range and pathogenicity [13]. Species do not produce spores but are composed of hyphae and sclerotia, living in soil or on organic debris. Symptoms vary according to the host and the plant part affected: damping-off of seedlings is probably the most common disease, but also root and stem rots, stem cankers and fruit rot can occur [14]. *Fusarium oxysporum* is a species complex that comprises a multitude of saprophytic strains that grow and survive on organic matter in soil and the rhizosphere of many plant species. Some species are pathogenic to plants: spores germinate, and the hyphae penetrate the roots causing either root rot or vascular wilt diseases. Pathogenic species show a high level of host specificity, so they are classified into more than 120 *formae speciales* and races [15,16]. Symptoms include vein clearing and leaf epinasty, followed by stunting, yellowing of the lower leaves, progressive wilting, defoliation, and, finally, death of the plant. In addition, the vascular tissue turns brown due to fungal colonization [17].

Managing the control of soil-borne pathogens is problematic, because of the prohibition or restriction on the use of the most effective synthetic fungicides for soil treatment, according to Reg. (EC) No. 1107/2009. This regulation establishes the principles of integrated pest management and gives priority to eco-friendly alternatives wherever possible to ensure a high level of protection of human and animal health and the environment. Therefore, soil-borne disease management is now directed towards sustainable strategies such as the use of antagonistic microorganisms and opportune agronomic techniques. Other control means that have shown some efficacy against several plant pathogens include plant extracts and plant essential oils. Indeed, many plant secondary metabolites, such as phenols, are bioactive substances with a defensive role against biotic stresses [18]. Several plant extracts showed some activity against tomato soil-borne pathogens following different kinds of applications. For example, extracts from *Calotropis procera Thymus vulgaris*, *Eugenia caryophyllata*, *Syzygium aromaticum*, *Stevia rebaudiana*, and *Allium tuncelianum* applied by root dipping or in soil/growing substrate effectively reduced *F. oxysporum* wilt disease in tomato and enhanced plant growth parameters [19–23]. Still, in tomato, *Rhizoctonia solani* root rot was reduced by a clove extract and by extracts from *Monsonia burkeana* and *Moringa oleifera* applied by soil treatment [24,25].

Concerning seed treatment, the application of plant extracts might be an important tool for priming seeds against soil-borne pathogens, as well as triggering the plant defense and improving the overall performance of the plant [26]. Moreover, seed treatments require lower amounts of products than those required for soil irrigation or root soaking. Very few scientific papers report results with plant extracts against tomato soil-borne pathogens in in vivo experiments. An extract of *Ocimum basilicum* is an example of seed treatment effective against *F. oxysporum* in tomato [27].

As far as we know, studies regarding the use of JA extract for the control of soil-borne pathogens are very limited. This species has been primarily studied for its priming effects on tomato seeds against the soil-borne pathogen *R. solani* [26]. Among algal compounds, polysaccharides have been reported to be effective in stimulating biological activities which play an important role as inducers of plant resistance. Indeed, they increased the activity of various defense-related enzymes such as chitinase, β-1,3-glucanase, peroxidase, polyphenol oxidase, phenylalanine ammonia-lyase, and lipoxygenase [28–31]. Usually, the methods in use for polysaccharides extraction are hot water alone or combined with autoclaving, microwaving, and ultrasonication [32–35], alkaline or acidic solutions, and enzymatic treatments [36,37]. These methods appear to influence the characteristics of the extracted polysaccharides [38]. In a study conducted with an aqueous extract of JA at 50 °C under overnight agitation, a considerable reduction in disease severity was detected, as well as an increased seedling dry weight, stem caliber, plant chitinase activity, and deposition of lignin in root tissues [26].

To the best of our knowledge, there are no studies on water-soluble polysaccharides (WSPs) extracted from JA by the autoclave-assisted procedure. It was hypothesized that these polysaccharides possess enhanced bioactivity potentially useful in controlling the soil-borne pathogens *R. solani*, *P. ultimum*, and *F. oxysporum*. For this purpose, we used the tomato as a model plant, since it is an important economic crop. For the first time, the test was performed by seed treatment with JA WSPs for pathogen control and the elicitation of plant defense responses was also evaluated. The purpose of this paper meets the Sustainable Development Goals (SDGs) of the 2030 Agenda, which is based on positive and immediate benefits for both agriculture and the environment.

2. Materials and Methods

2.1. Alga and Preparation of the Water-Soluble Polysaccharides

The alga JA was provided by the Spanish Bank of Algae (BEA), University of Las Palmas de Gran Canaria, Spain. The alga was harvested at Bocabarranco beach, Las Palmas, East coast of Gran Canaria, washed in fresh water, and dried using an air heating drying system (B. Master, Tauro Essiccatori Srl, Camisano Vicentino, VI, Italy) at 65 °C for 24 h. The dry thallus was then ground to a fine powder with a mortar and pestle. A total of 10 mg of dry thallus was suspended in 10 mL of sterile distilled water, autoclaved for 20 min at 100 °C, 1 bar, and then centrifuged 3 times at 5000 rpm for 20 min (Beckman Coulter Allegra 21R centrifuge, Inc., Krefeld, Germany). The supernatant was filtered through a 0.45 µm syringe filter, frozen at −80 °C, and then lyophilized until experiments. The yield related to the WSPs was calculated according to [39] with modifications: (WSPs g/dry thallus g) × 100.

In all experiments, the JA WSPs were used at three concentrations, 0.3, 0.6, and 1.2 mg/mL, prepared by serial dilution (1:2) with sterile distilled water. A commercial product containing laminarin (45 g/L) was used at the recommended dose of 2 mL/L (0.09 mg laminarin/mL) as a reference treatment (RT).

2.2. FT-IR Analyses of the Extract

The FT-IR spectra of WSPs from JA and of the commercial product based on laminarin were performed by using a Bruker Tensor FT-IR instrument (Bruker Optics, Ettlingen, Germany) equipped with an accessory for analysis in micro-Attenuated Total Reflection (ATR). The sampling device contained a microdiamond crystal, a single reflection with an angle of incidence of 45° (Specac Quest ATR, Specac Ltd., Orpington, Kent, UK). Spectra were carried out from 4000 to 400 cm^{-1}, with a spectral resolution of 4 cm^{-1} and 64 scans. Background spectra were also taken against air under the same conditions before each sample. Spectra were handled with the Grams/386 spectroscopic software (version 6.00, Galactic Industries Corporation, Salem, NH, USA).

2.3. Plant Material and Pathogens

The tomato cv. Marmande (L'Ortolano, Savini Vivai, Italy) was used as a model plant because it is among the crops that are most responsive to priming [40] and has already been tested for priming with a JA water extract [26].

The fungi *R. solani* DAFS3001 (RS) and an isolate belonging to *F. oxysporum* species complex DISTAL2019 (FO) of the collection of the Department of Agricultural and Food Sciences, University of Bologna, and the oomycete *Pythium ultimum* 22 (PU) belonging to the collection of the Council for Agricultural Research and Economics (CREA), Research Centre for Agriculture and Environment of Bologna, Italy, were used in this study. All the pathogen isolates were maintained on Potato Dextrose Agar (PDA, 3.9%) medium in tubes at 4 °C until use.

2.4. Seed Treatment and Growing Substrate

The seeds were treated by immersion in a 1000 µL aliquot of JA WSPs at 0.3, 0.6, and 1.2 mg/mL and of the RT (2 mL/L) overnight at room temperature in the dark [41] (with modifications). Sterile distilled water was used as a control. After the treatment, seeds were firstly washed with sterile distilled water to remove any treatment residues and then were left to dry on sterile filter paper in a laminar air flow hood for 10 min and then sown in a sterile peat/sand mix (7/3; weight/weight) used as a growth substrate. The different concentrations of JA WSPs were first tested for their effect on seed germination and seedling emergence by blotting paper and greenhouse assays, respectively. For the blotting paper assay, seeds were firstly surface sterilized with NaOCl water solution (2.5%) for 2 min, rinsed three times, and treated as above described with JA WSPs at 0.3, 0.6, and 1.2 mg/mL, RT at 2 mL/L, and water (control). Then, 10 seeds were sown on wet sterile paper inside sterile glass tubes (20 cm height, 2.7 cm diam.) and vertically incubated at 24–25 °C for 48 h in a growth chamber with a 12 h/12 h day/night photoperiod. Tubes were sealed with a cotton plug to avoid contamination and periodically irrigated with sterile distilled water. Three tubes were considered for each treatment and the control. After 15 days of incubation, the germinated seeds were assessed. The experiment was repeated three times ($n = 3$). For the greenhouse assay, seeds were treated as above described. A total of 50 seeds were sown in the growing substrate in a plastic pot (13.5 × 11.5 × 7.5 cm), with 3 pots per treatment and the control, and then incubated under greenhouse conditions at 24–26 °C (day), 20–22 °C (night), with a 12 h/12 h day/night photoperiod, 70% relative humidity. The seedling emergence was assessed every 3 days from the 6th day after sowing (DAS) until 15 DAS.

2.5. Systemic-Induced Resistance Bioassays in Greenhouse Pot Experiments
2.5.1. *Rhizoctonia solani* (RS) and *Pythium ultimum* (PU)

A bioassay was set up to test the effectiveness of the seed treatment with the JA WSPs in inducing defense response in the tomato plants against *R. solani* and *P. ultimum*. The substrate was inoculated with each pathogen singly before the seeding. For inoculation, 10-day-old colonies of RS or PU grew on a PDA medium were homogenized in sterile distilled water with a kitchen blender and then mixed with the substrate (2% weight/weight, pathogen/substrate). The inoculated substrate was covered with a black plastic film and incubated at 25 °C for 2 days in the greenhouse and then distributed in plastic pots (13.5 × 11.5 × 7.5 cm). A total of 50 seeds treated, as described above, with 0.3, 0.6, and 1.2 mg/mL of JA WSPs or with 2 mL/L of RT were sown in each pot inoculated with RS or PU. Seeds treated with water and sown in substrate infected with each pathogen were used as a positive control (C+), while seeds treated with water and sown in growing substrate not infected were used as a negative control (C-). Three pots per treatment and the controls were considered. Plants were grown under greenhouse conditions at 24–26 °C (day), 20–22 °C (night), with a 12 h/12 h day/night photoperiod, and 70% relative humidity.

The effect of treatment against RS and PU was assessed by recording seedling emergence over time from 6 days after sowing (DAS) until 10 DAS. To determine disease symptoms, and plant and root length, 20 DAS seedlings were carefully removed.

For RS, disease severity was visually evaluated as root necrosis symptoms by using a five-point scale modified from [26], where: 0, absence of necrosis (0% of symptoms); 1, very slight root necrosis (up to 5% of root with symptoms); 2, slight necrosis (6–20% of root with symptoms); 3, moderate root necrosis (20–50% of root with symptoms); 4, severe root necrosis (51–70% of root with symptoms); 5, severe root and crown necrosis (>70% of root with symptoms).

For PU, disease severity was measured on the whole plant using a visual disease assessment scale based on [42] with modifications, as follows: 0 = no visible disease symptoms; $1 \leq 20\%$ moderate level of general decay; 2 = extensive general decay, and with an obvious reduction in overall plant development (but <50% of root system missing); 3 = very severe levels of general decay associated with an extensive reduction in overall roots (>50% root system missing); and 4 = dead plant.

Each experiment was repeated three times (n = 3).

2.5.2. *Fusarium oxysporum* (FO)

The bioassay with JA WSPs against FO was carried out by transplanting tomato plantlets into a substrate previously inoculated with the pathogen. Firstly, JA-WSPs-treated seeds, the RT-treated seeds, and the water-treated control were sown in alveolate trays filled with non-inoculated substrate and grown under the same greenhouse conditions cited above until the second true leaf stage. After 20 days of incubation, plantlets were carefully removed and transplanted (3 plantlets/pot, 16 cm diameter) into the inoculated substrate (2% weight/weight, pathogen/substrate) prepared as reported for RS and PU. Plantlets from water-treated seeds transplanted in substrate infected with FO were used as a positive control (C+), while plantlets from water-treated seeds transplanted in the not infected substrate were used as a negative control (C-). Three pots per treatment and the controls were considered. Plants were grown under the same greenhouse conditions.

Disease severity was evaluated 20 days after the transplant (DAT) by using a 6-point scale based on [43], where: 0, plant without disease symptoms; 1, very slight wilt (mild chlorosis on lowest leaves only); 2, lower leaves dead and some upper leaves wilt slight chlorosis; 3, lower leaves dead and some upper leaves wilted; 4, lower leaves dead and severe wilt of upper leaves; 5, dead plant. The experiment was repeated three times (n = 3).

2.6. Expression of PR Protein and Polyphenol Pathway Genes and Enzymatic Activities

Seeds treated with 0.3, 0.6, and 1.2 mg/mL of JA WSPs, 2 mL/L of RT, or water (control) as above described were sowed in the not-infected substrate in plastic pots (13.5 × 11.5 × 7.5 cm), and incubated under greenhouse conditions at 24–26 °C (day), 20–22 °C (night), with a 12 h/12 h day/night photoperiod, and 70% relative humidity. A total of 50 seeds were sown in each pot, with 6 pots per treatment, and the control. The 6 pots per treatment and the control were randomly divided into two groups of 3 pots each: one group (G1) was used for gene expression analysis, and the other group (G2) was used for enzymatic activity assays; 20-day-old plantlets randomly chosen from each group pot were gently removed from the substrate, washed with tap water, and dried on filter paper. The plantlets from pots within each treatment were pooled together and snap-frozen in liquid nitrogen. Frozen tissues were then finely ground by using a pre-chilled mortar and pestle.

2.6.1. Expression of PR Protein and Polyphenol Pathway Genes

Cellular RNA was extracted from G1 ground frozen tissues with RNAeasy Plant Mini kit (Qiagen, Hilden, Germany), according to the manufacturer's instructions. The RNA pellets were quantified by using a spectrophotometer (Nanovue, GE Healthcare Life Sciences, Buckinghamshire, UK), and the same amount of total RNA (250 ng) was reverse-transcribed by using PrimeScript™ RT Reagent Kit with gDNA Eraser (Takara, Kusatsu, Japan). The cDNA mixture (2 µL) was used in real-time PCR analysis in a LightCycler instrument (Roche Molecular Biochemicals, Basel, Swiss) using the TB Green Premix Ex Taq II (Takara). The following PCR conditions were used: activation of HotStart Taq DNA polymerase at 95 °C for 10 s, amplification (40 cycles: 95 °C for 5 s followed by appropriate annealing temperature for each target, as detailed below, for 20 s). Gene expression levels were calculated by the ΔΔ cycle threshold (Ct) method. The amount of

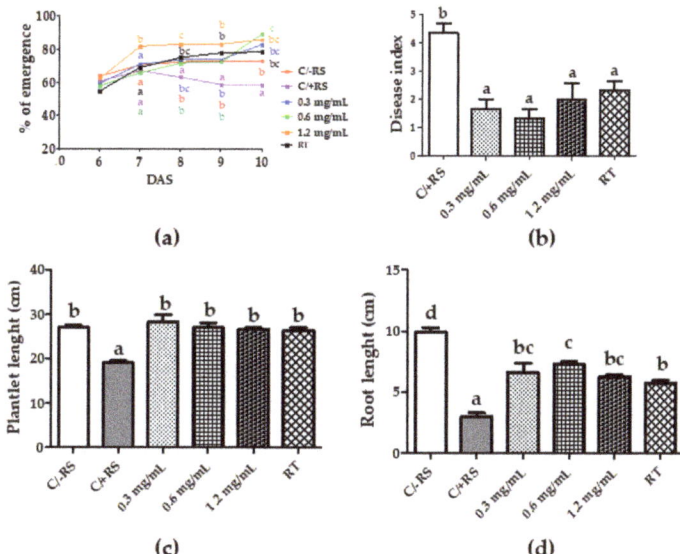

Figure 3. Effect of tomato seed treatment with different concentrations (0.3, 0.6, and 1.2 mg/mL) of *Jania adhaerens* water-soluble polysaccharides on the emergence percentage of seedlings (**a**), disease severity (**b**), plant length (**c**), and root length (**d**) in a growing substrate infected with Rhizoctonia solani (RS). C/-RS = non infected control; C/+RS = infected control; RT = reference treatment (2 mL/L); DAS = days after sowing. In panel (**a**), each value is the mean of 3 independent experiments ($n = 3$); different letters indicate significant differences among treatments and controls within each DAS, according to the LSD test ($p < 0.05$). In panels (**b–d**), columns are mean values of 3 independent experiments ($n = 3$) ± SD. Different letters indicate significant differences among treatments and controls according to the LSD test ($p < 0.05$).

Figure 4. Effect of tomato seed treatment with different concentrations (0.3, 0.6, and 1.2 mg/mL) of *Jania adhaerens* water-soluble polysaccharides on the disease severity caused by *Rhizoctonia solani* and plant growth. RT = reference treatment (2 mL/L); C/+RS = infected control.

3.2. Effect of Seed Treatment on Seed Germination and Seedling Emergence

In the blotting paper assay (Figure 2a), the JA WSPs and RT have significantly increased the seed germination percentage as compared with the water-treated control. No differences among the JA WSPs concentrations were observed. In the greenhouse assay (Figure 2b), all JA concentrations and RT treatment significantly improved the seedling emergence over time with respect to the water treatment (control). At 15DAS, the JA WSPs concentration of 1.2 mg/mL showed the highest percentage emergence (95.0%).

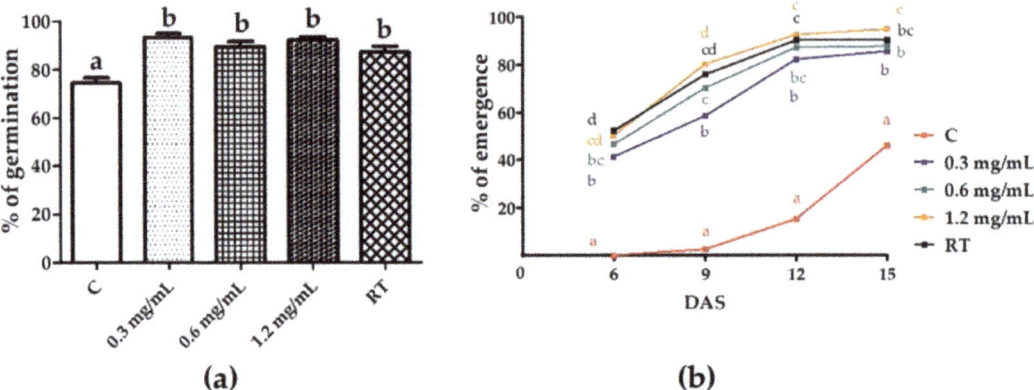

Figure 2. Preliminary assays on the effect of tomato seed treatment with different concentrations (0.3, 0.6, and 1.2 mg/mL) of *Jania adhaerens* water-soluble polysaccharides on the seed germination percentage in blotting paper test (**a**) and on the percentage of seedling emergence in the experiment under greenhouse conditions (**b**). C = water control; RT = reference treatment (2 mL/L); DAS = days after sowing. In panel (**a**), columns are mean values of 3 independent experiments ($n = 3$) ± SD. Different letters indicate significant differences among treatments and control according to the LSD test ($p < 0.05$). In panel (**b**), each value is the mean of 3 independent experiments ($n = 3$); different letters indicate significant differences among treatments and the control within each DAS, according to the LSD test ($p < 0.05$).

As the treatments had no adverse effect on either seed germination or seedling emergence, they were further investigated in greenhouse experiments.

3.3. Systemic-Induced Resistance Bioassays in Greenhouse Pot Experiments

The treatment with JA WSPs at all doses increased the seedling emergence percentage (Figure 3a) in the growing substrate inoculated with RS with respect to the inoculated control (C/+RS). This was very noticeable and significant from 8DAS until 10DAS. In particular, at 10DAS 0.3, 0.6, and 1.2 mg/mL doses significantly increased the emergence percentage with respect to the inoculated control (C/+RS) by 41.9, 52.1, and 46.5%, respectively. No statistical difference was observed between the three doses of JA WSPs and RT, which increased the emergence by 34.0%.

All JA concentrations significantly reduced similarly the disease severity with respect to the C/+RS by 57.6% on average. The disease reduction obtained with JA WSPs was not statistically different from RT (Figures 3b and 4). The WSPs significantly increased seedling length by 43.8% on average without differences among concentrations with respect to C/+RS and similarly to C/-RS and RT (increase by 38.4% vs. the C/+RS) (Figure 3c). The root lengths of seedlings from seeds treated with JA WSPs were significantly greater than that of C/+RS, even though they were smaller than C/-RS. The concentrations of 0.6 mg/mL significantly increased root length by 25.8% compared to RT treatment (Figure 3d).

2.7. Statistical Analysis

All experiments were arranged according to a complete randomized design. All data were analyzed by ANOVA, and, if the p-value was less than 0.05, the means were separated by the LSD test ($p < 0.05$). All analyses were performed with GraphPad Prism software, San Diego, CA, USA, version 5.01.

3. Results

The content of crude WSPs extracted from JA was 6% of JA dry thallus.

3.1. FT-IR Analyses

The spectroscopic profiles of JA WSPs and RT samples are shown in Figure 1.

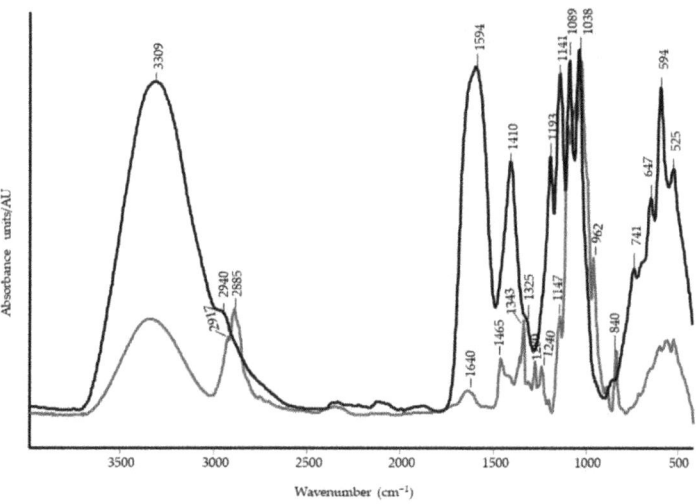

Figure 1. FT-IR spectra of *Jania adhaerens* water-soluble polysaccharide (black line) and RT a commercial product based on laminarin (green line).

Both samples showed, in the range from 4000 cm^{-1} to 2000 cm^{-1}, a broad intense band at 3309 cm^{-1} assigned to OH stretching, a shoulder at 2940 cm^{-1} in JA polysaccharide, and 2917 cm^{-1} in RT assigned to CH$_2$ asymmetric stretching in methylene chain, respectively. Conversely, RT exhibited a strong band at 2885 cm^{-1} due to CH$_3$ symmetric stretching vibration very close to the oxygen atom [47]. In the range from 1800 cm^{-1} to 1220 cm^{-1}, both spectra differed in their functional groups: JA WSPs displayed two strong bands at 1594 cm^{-1} and 1410 cm^{-1} assigned to asymmetric and symmetric stretching motions in carboxylate of alduronic acids such as guluronic acid or mannuronic acid (Sterner et al. 2016); by contrast, in RT a small peak appeared at 1640 cm^{-1}, attributed to the bending vibration of water. This was also observed in the laminarin standard FT-IR spectrum [48]. As for the CH deformation vibrations, these were in the region between 1465 and 1343 cm^{-1} of both spectra. Only in JA WSPs, however, did they seem like weak shoulders. The appearance of the bands in the region 1193–1141 cm^{-1} is typical of the glycosidic linkage formation in polysaccharides [49]. Additionally, the bands from 900 to 800 cm^{-1} region (anomeric region), are used to distinguish the anomeric carbon in α and β configurations [49]. However, in the RT sample, the band at 840 cm^{-1} may be related to the bands at 1200–1280 cm^{-1}, and 1038 cm^{-1} assigned to S–O and C–O–S stretching in sulfonyl groups [49]. The bands below 800 cm^{-1} are related to the carbohydrate skeletal vibrations [50].

mRNA was normalized to Actin-7 expression as a housekeeping gene in each sample. The sequences of primers (from Merck) are shown in Table 1. Finally, melting curves were evaluated to check amplicon specificity.

Table 1. Primer sequences used in PCR amplification.

Gene Names	Forward	Reverse	Annealing T (°C)
Actin-7	GGGATGGAGAAGTTTGGTGGTGG	CTTCGACCAAGGGATGGTGTAGC	61
PAL5	CACTGTAAGCCAAGTAGCCAAA	CTGCAGGGGTCATCAGCATA	59
HCT	CGGACGTTACCATCACTGGA	AAGGAGGACTCAGTAGCTTTG	59
HQT	GGTGTTTTGTTTGTTGAGGCTG	GACTCCGCCACACTTGAAAC	59
FLS	GATTTGGCCTCCTCCTGCTA	TCCAAACCAAGCCCAAGTGA	59
PR1a	AGGATGCAACACTCTGGTGG	GCACAACCAAGACGTACCGA	60
PR2	GGTGGATCCAATTCGCAAGC	ACCTGAGAACCCACCAGACT	59
PR3	AGAGTTCCAGGGTACGGTGT	CCAATTCGACTTTCCGCTGC	59
PR4	GATGCTGACAAGCCTCTGGA	CCCTCAAGCATCTACCGCAT	59

2.6.2. Chitinase and β-1,3-Glucanase Activities

Ground frozen tissues from the G2 group (§ 2.6) were used for protein extraction with 20 mM sodium acetate buffer pH 5.2 (1 mL/g of fresh weight) added with polyvinylpolypyrrolidone (1%) from Sigma-Aldrich (St Louis, MO, USA) under continuous gentle stirring at 4 °C for 90 min [44]. The protein crude extract was centrifuged twice at 12,000 rpm for 20 min at 4 °C, and then the supernatant was filtered using a GV Millex® Syringe Filter Unit (pore size 0.22 µm, Millipore Corporation, Burlington, MA, USA). Protein concentration was determined at the spectrophotometer (Tecan NanoQuant, Infinite M200PRO, Tecan Trading AG, Männedorf, Swiss) by the protein–dye-binding Bradford method [45] in a 96-well microplate (Greiner CELLSTAR®, Merck KGaA, Darmstadt, Germany), by using bovine serum albumin (Bio-Rad Laboratories, Inc., Segrate, Italy) as the standard.

For both chitinase and glucanase activities, the plate assay with modification was used [46]. For chitinase activity, each sample was assayed in triplicate in agarose gel (1.5%) containing 0.01% glycol chitin in a glass Petri plate (14 cm diam.). A total of 200 µL (20 µg of proteins) of each replicate along with chitinase standard (Streptomyces griseus (Sigma-Aldrich St Louis, MO, USA), were added to 7 mm-diameter wells cut in the agar. After incubation at 37 °C for 24 h, 50 mL of 500 mM Tris–HCl (pH 8.9) containing 0.01% fluorescent brightener was added to each plate for 2 h. The plates were rinsed three times with distilled water, flooded with distilled water overnight in the dark, and then observed under a 302 nm UV light source to visualize non-fluorescent lytic zones corresponding to the enzyme activity on a fluorescent background. Images of gels were taken with a digital camera, then the light intensity/mm² chitinase activity lytic zone was calculated with the Quantity One 1-D analysis software v. 4.6.6 (Bio-Rad Laboratories Inc., Hercules, CA, USA). The specific chitinase activity was expressed as units. One unit corresponds to the mg of N-acetyl-D-glucosamine released/h/mg of protein in comparison to the standard.

For β-1,3-glucanase activity, 200 µL (20 µg of proteins) for each sample, in replicates of 3, was added to 7 mm-diameter wells in a 1.5% agarose gel containing 0.5 mg/mL laminarin (from Laminaria digitata, Sigma Aldrich St Louis, MO, USA) in a 14 cm-diameter glass Petri plate. As a standard, β-1,3-D-glucanase (from *Helix pomatia*, Sigma Aldrich) was used. After incubation at 37 °C for 24 h, 50 mL of 2,3,5-triphenyl tetrazolium chloride (0.15% in NaOH 1M) was added to each plate. The pink lytic zones on a white background corresponding to the glucanase activity were visible after further incubation at 37 °C for 30 min. Images of gels were taken and processed by the Quantity One 1-D analysis software as above described for chitinase activity. Glucanase activity was expressed as units. One unit corresponds to the release of 1 µmol of glucose from laminarin/min in comparison to the lytic zone of the standard.

Each assay was repeated three times ($n = 3$).

As regards the effect of JA WSPs treatment against PU, Figure 5a shows that all concentrations significantly increased the seedling emergence at all DAS with respect to the inoculated control (C/+PU). Notably, RT treatment was not significantly different from C/+PU at all DAS. Overall, 0.3 and 0.6 mg/mL gave the highest emergence percentages from 8 to 10DAS (in average 86.7–90.3%). As regards the effect of the treatments against the disease, all JA WSPs concentrations and RT significantly reduced the disease index with respect to the C/+PU (Figure 5b). The highest disease reduction (52.6%) was obtained with 0.6 mg/mL. This concentration was 28% more effective than RT. Plant length was similar with the three concentrations (30.7, 31.7, and 31.3 cm for 0.3, 0.6, and 1.2 mg/mL, respectively) and higher than C/+PU (16.0 cm), C/-PU (25.3 cm), and RT (23.3 cm) (Figure 5c). The root length following seed treatment with all doses of JA WSPs was significantly higher than that of C/+PU and lower than C/-PU. The concentration of 0.6 mg/mL increased root length more effectively than RT by 40.6%, while the 0.3 and 1.2 mg/mL doses increased root length as well as RT treatment (Figure 5d).

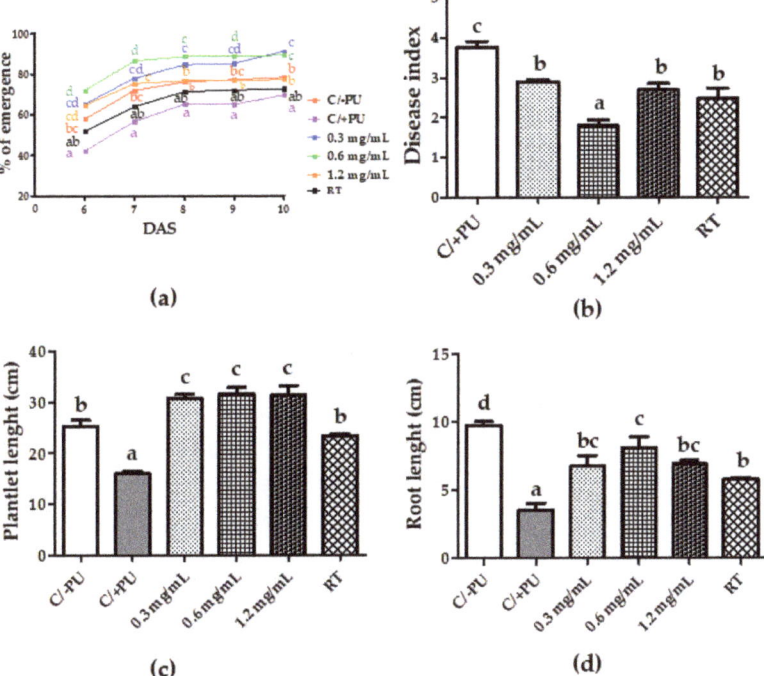

Figure 5. Effect of tomato seed treatment with different concentrations (0.3, 0.6, and 1.2 mg/mL) of *Jania adhaerens* water-soluble polysaccharides on the emergence of seedlings (**a**), plant length (**b**) and root length (**c**) in a growing substrate infected with *Pythium ultimum* (PU). C/-PU = non-infected control; C/+PU = infected control; RT = reference treatment (2 mL/L); DAS = days after sowing. In panel (**a**), each value is the mean of 3 independent experiments ($n = 3$); different letters indicate significant differences among treatments and controls within each DAS, according to the LSD test ($p < 0.05$). In panels (**b**–**d**), columns are mean values of 3 independent experiments ($n = 3$) ± SD. Different letters indicate significant differences among treatments and controls according to the LSD test ($p < 0.05$).

Figure 6 shows the effect of seed treatment with different JA WSPs concentrations in the case of substrate inoculated with FO on disease severity and plant growth. The concentrations of 0.6 and 1.2 mg/mL displayed a significant disease index reduction with respect to C/+FO by 14.3 and 34.7%, respectively (Figure 6a). The highest statistical reduction was obtained with the 1.2 mg/mL concentration which was also 28.9% more effective than RT. Plant length was also increased significantly by 0.6 and 1.2 mg/mL (Figure 6b). The concentration of 1.2 mg/mL showed the highest plant length (54.7 cm) with regards to the C/+FO (25.7 cm). This value was even similar to that of C/-FO (63.0 cm). The three WSPs doses significantly increased root length with respect to the C/+FO (Figure 6c). The highest increase (228.3% vs. C/+FO) was obtained with the 1.2 mg/mL dose. Plant dry weight showed the highest significant values at 0.6 mg/mL (0.751 g) and 1.2 mg/mL (0.872 g) (Figure 6d). In particular, the plant dry weight with 1.2 mg/mL dose was significantly similar to that of the C/-FO control.

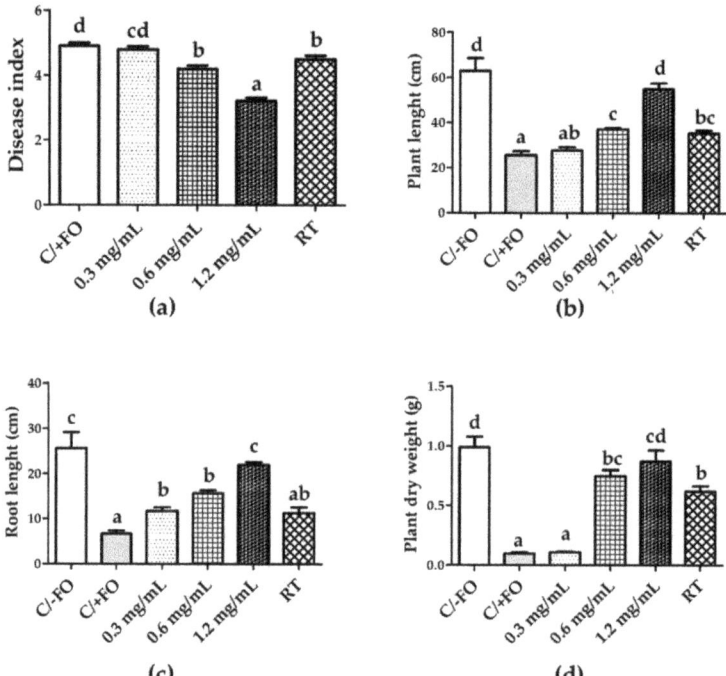

Figure 6. Effect of tomato seed treatment with different concentrations (0.3, 0.6, and 1.2 mg/mL) of *Jania adhaerens* water-soluble polysaccharides on the disease index (**a**), plant and root length (**b**,**c**) and plant dry weight (**d**) in a growth substrate infected with *Fusarium oxysporum* (FO). C/-FO = non infected control; C/+FO = infected control; RT = reference treatment (2 mL/L). Columns are mean values of 3 independent experiments ($n = 3$) ± SD. Different letters indicate significant differences among treatments and controls according to the LSD test ($p < 0.05$).

3.4. Expression of PR Protein and Polyphenol Pathway Genes

To shed some light on the molecular mechanisms associated with the induced resistance elicited by JA WSPs, the mRNA expression of pathogenesis-related (PR) proteins and key enzymes of polyphenol pathways was investigated in plantlets 20 days after seed treatment with JA WSPs or RT. Figure 7a shows that the transcript levels of phenylalanine ammonia-lyase (PAL5), a key enzyme in the phenylpropanoid biosynthetic pathway, were increased by JA WSPs at any concentration employed and similarly to RT. The expression pattern of two enzymes that are located downstream PAL, i.e., hydroxycinnamoyl CoA quinate hydroxycinnamoyl transferase (HQT) and hydroxycinnamoyl CoA shikimate hydroxycinnamoyl transferase (HCT) was also investigated and illustrated in Figure 7b. Seed treatment with JA WSPs or RT enhanced the expression level of HQT, while HCT was less affected. Even the expression of PR proteins involved in the systemic acquired resistance

(SAR), i.e., PR1A and PR2, was enhanced by JA WSPs and RT (Figure 7c). On the contrary, we failed to observe an increase in the transcript level of the induced systemic resistance (ISR) protein, PR3 (Figure 7d), which actually resulted lower after JA WSPs compared to the control. However, the expression level of PR4, another protein involved in ISR, was significantly, but slightly increased by JA WSPs 1.2 mg/mL and, more markedly, by RT.

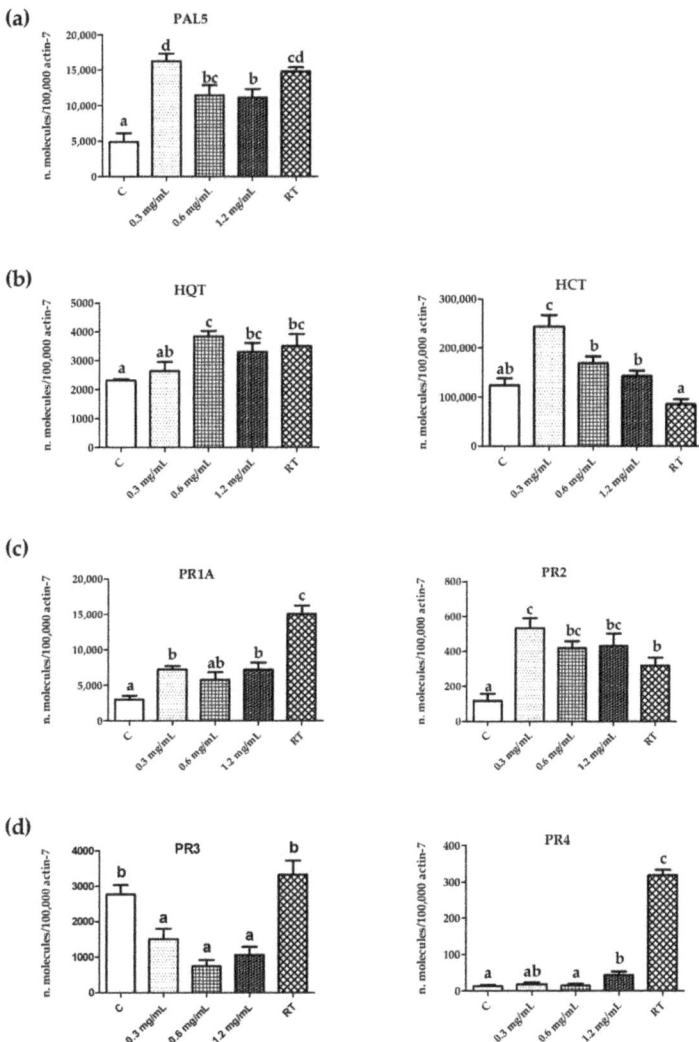

Figure 7. Transcriptional expression levels of phenylpropanoid (**a**), chlorogenic acid (**b**), systemic acquired resistance (**c**), and induced systemic resistance (**d**) biosynthetic pathways genes in tomato plants as a response to seed treatment with different concentrations (0.3, 0.6, and 1.2 mg/mL) of *Jania adhaerens* water-soluble polysaccharides. C = control; RT = reference treatment (2 mL/L); PAL5, phenylalanine ammonia-lyase; HQT, hydroxycinnamoyl CoA quinate hydroxycinnamoyl transferase; HCT, hydroxycinnamoyl CoA shikimate hydroxycinnamoyl transferase; PR1A, PR2, PR3 and PR4, pathogenesis-related genes. Columns are mean values of 3 independent experiments (n = 3) ± SD. Different letters indicate significant differences among treatments and the control according to the LSD test ($p < 0.05$).

3.5. Chitinase and β-1,3-D-Glucanase Activities

Seed treatment with JA WSPs at different concentrations significantly enhanced the β-1,3-D-glucanase activity of seedlings with respect to the control by 25.3% on average (Figure 8a). No differences were observed among JA WSPs doses and RT. Chitinase activity was increased only at the higher dose, 1.2 mg/mL (44.6%), which produced a result similar to the RT treatment (Figure 8b).

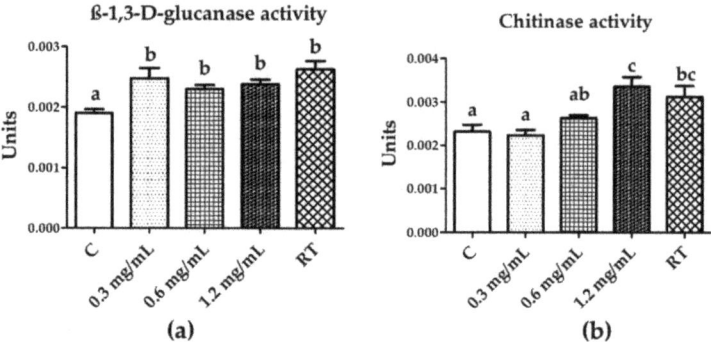

Figure 8. β-1,3-D-glucanase (**a**) and chitinase (**b**) activities determined in protein extract of tomato seedlings following seed treatment with different concentrations (0.3, 0.6, and 1.2 mg/mL) of *Jania adhaerens* water-soluble polysaccharides. For β-1,3-D-glucanase activity, one unit corresponds to the release of 1 μmol of glucose from laminarin/min. in comparison to the standard; for chitinase activity, one unit corresponds to the mg of N-acetyl-D-glucosamine released/h/mg of protein in comparison to standard. C = control; RT = reference treatment (2 mL/L). Columns are mean values of 3 independent experiments (n = 3) ± SD. Different letters indicate significant differences among treatments and the control according to the LSD test ($p < 0.05$).

4. Discussion

At present, plant pest control solutions for environmentally sustainable agriculture are highly recommended by European standards (Reg. EC 1107/2008; Dir. 128/2009). In particular, soil-borne disease management requires innovative strategies because the most effective pesticides for soil treatment underwent prohibition or restriction on their use.

Seaweed extracts are a source of several bioactive compounds that have a broad range of biological activity with a potential role in reducing the need for pesticides in agriculture [4,28,51–55]. Seaweeds are already widely used in agriculture due to their content of mineral substances, amino acids, vitamins, and plant growth regulators that ameliorate plant growth and enhance crop productivity [56,57]. Generally, the algal extracts are obtained with organic solvents such as methanol, ethanol, and acetone that on the one hand can optimize the extraction of specific components [55,58], but, on the other hand, represent a hazard to human health. On the contrary, our previous research showed that seaweed aqueous extracts that were obtained at 50 °C could be a sustainable and effective alternative to organic solvents to be used in plant disease management [26,59]. In particular, the study by [26] highlighted the potential of JA applied by seed priming technique for the control of *R. solani* on tomato plants and has led us to further research on this issue.

Seed priming is an alternative, low-cost, and feasible technique that involves seed imbibition with a low amount of water and various priming agents, such as osmoregulators, salts, plant growth regulators, beneficial microorganisms, magnetic waves, nanoparticles, and macro- and micronutrients, that allow the advancement of metabolic processes helping plants to overcome biotic and abiotic stress [60,61].

With the present study, we have demonstrated that JA WSPs could be a new potential tool for the control of the soil-borne pathogens *R. solani*, *P. ultimum*, and *F. oxysporum* in tomato by seed treatment. Moreover, we have highlighted that JA WSPs seed treatment may elicit plant defense responses likely involved in pathogen control. Although crude polysaccharide extracts from microalgae have previously been reported to induce immunity in plants [62], seed priming with JA WSPs is the first case study. FT-IR analysis demonstrated that JA WSPs differed in functional groups from laminarin used as reference treatment (RT). JA WSPs were characterized by typical carboxylate bands (1594 cm^{-1} and 1410 cm^{-1}), most likely related to uronic acids. Our results are in agreement with those by Hentati et al. [1], who found uronic acids in the amount of about 6% in *J. adhaerens*. Moreover, ref. [62] found uronic acids in the red alga *Porphyridium* sp. Furthermore, based on their composition in functional groups, a structure–bioactivity relationship can be speculated. As a consequence, we can infer that carboxylate groups contributed to the anionic properties of JA WSPs, probably promoting interactions with other molecules capable of triggering several biochemical processes in treated tomato seeds. Indeed, polysaccharides are perceived in the plant as 'danger' molecules similar to what happens during the first interaction between a pathogen and the plant at the level of the cell wall and the plasma membrane [63]. To highlight the potential induced resistance effect by JA WSPs, treated seeds were water-washed after soaking, to remove any extract residues that could have a direct antifungal effect. Polysaccharides such as laminarin were reported to exert an antigerminative effect against the conidia of some fungi [64,65]. The polysaccharides extracted from JA might explain the elicitation of plant physiological responses against pathogens with the improvement of seedling emergence and plant growth, which may have concurred to the disease reduction. It is well known that the extracellular matrix of red seaweeds contains sulfated galactans as major components [66–70] which are elicitors of plant defense responses, as in the case of *Phytophthora nicotianae* on tobacco plants [7].

The seed treatment with JA WSPs showed a biostimulant-like effect in the preliminary assays since it increased seed germination and seedling emergence. Similarly, increases in germination were observed for tomato seeds treated with a different aqueous extract from the same red alga without pathogen challenge [26]. A biostimulant effect on plants was obtained with red alga extract applied by spraying and soil treatments as in the case of *J. rubens* on chickpea and maize [71,72] and by a foliar application with the red alga *Kappaphycus alvarezii* on wheat [73,74].

In the present study, the increase in the emergence rate was also obtained in response to the *R. solani* and *P. ultimum* challenge; indeed, JA WSPs at all doses increased the seedling emergence from 7DAS and 6DAS up to the last assessment, respectively. The plant growth promotion by the extract in the substrate inoculated with *R. solani*, *P. ultimum* and *F. oxysporum* was also observed in terms of plant and root length.

It is likely that the disease control is correlated with the plant growth promotion obtained with WSPs in substrates inoculated with each pathogen. All the concentrations reduced disease severity and increased both seedling and root length in a similar way, without showing a dose-dependent response. However, the 1.2 mg/mL concentration was the most effective in reducing *F. oxysporum* disease severity. We hypothesize that the disease control exerted by JA WSPs is due to the induction of plant defense responses, as shown by the increase in plant chitinase and glucanase activity observed and by the transcript levels of defense genes. This is in line with [75], who found a correlation between *F. oxysporum* disease control and the induction of plant defense responses by applying an extract of the brown alga *Ascophyllum nodosum* to cucumber plants. Chitinases and β-1,3-glucanase are enzymes constitutively produced by plants during their growth [76,77]. Some chitinases accumulate during their developmental program, in seeds of several species, while others, called defense enzymes, are produced as a response to a microbial attack or wounding [76]. During seed germination, both chitinases and β-1,3-glucanase are particularly important because seeds are exposed to soil-borne pathogens and the embryo becomes vulnerable to pathogen attack when the radicle crosses the endosperm tissue. Our findings showed that seed treatment with JA WSPs may have contributed to increasing the synthesis of chitinases and β-1,3-glucanase making the plant protected against pathogen attack. Ghule et al. [78] demonstrated that the activity of both these defense enzymes was increased in fenugreek plants derived from seeds treated with the marine polysaccharide chitosan. Following this treatment, plants were protected against root rot disease by *Fusarium solani*.

Two forms of induced resistance are described in response to stimuli that enhance plant defensive capacity versus pathogens: systemic acquired resistance (SAR), which involves the accumulation of PR1 and PR2 proteins and induced systemic resistance (ISR), which involves PR3 and PR4 proteins. According to the results shown in the present study (Figure 7), SAR response was elicited by JA polysaccharide seed treatment as judged by the increase in PR1A and PR2 expression. Instead, JA WSPs caused a decrease in PR3 gene expression, used as a marker of ISR response. It has been reported that crude polysaccharides extracted from *Acanthophora spicifera* (another red alga) enhanced rubber tree defenses against *Phytophthora palmivora* infection [79]. In accordance with our results, JA WSPs also induced PR1 and PR2 gene expression but suppressed ISR-related gene expression. Moreover, we have found an increase in the expression of PAL, known to be a key enzyme involved in plant development and defense responses to pathogens. PAL catalyzes the first and committed step in the biosynthesis of phenylpropanoids, which can give rise to a wide range of secondary metabolites, such as flavonoids and lignin. Among phenylpropanoid-derived phenolics and flavonoids, chlorogenic acid is known to activate the SAR pathway [80]. In this regard, we have also found increases in the expression of HQT and HCT, which are key enzymes of the biosynthetic pathway of chlorogenic acid.

5. Conclusions

Algae are considered an important resource of many biologically active compounds, and among them are polysaccharides. The chemical characteristics and potential biological properties make polysaccharides promising for ecological disease management in agriculture. In particular, seed priming with polysaccharides may be an attractive alternative to conventional fungal disease treatments. The present study actually shows that tomato seed priming with water-soluble polysaccharides extracted from the red alga *Jania adhaerens* increased seedling emergence and plant development, while reduced disease severity caused by three soil-borne pathogens, i.e., *Rhizoctonia solani*, *Pythium ultimum* and *Fusarium oxysporum* under greenhouse conditions. Furthermore, the protective effects were associated with the increased activity of defense enzymes and increased expression of genes involved in induced resistance and related pathways. In conclusion, we have described a red algal-based seed treatment that promotes plant growth and triggers plant defense responses against soil-borne pathogens.

This study may contribute to environmentally sustainable agriculture. However, there are still some aspects that need further investigation such as standardized protocols for extraction procedures and the isolation and purification of polysaccharides. These will help in determining the chemical structure and biological activity of polysaccharides. Currently, these procedures are limited to lab research and have not yet been applied on an industrial scale.

Author Contributions: Conceptualization, H.R., S.G., R.R. and S.C. (Silvia Cetrullo); methodology, S.C. (Stefano Cianchetta) and V.P.; formal analysis, H.R. and R.R.; investigation, H.R., S.G., R.R., O.F., S.C. (Silvia Cetrullo), V.P. and A.M.Q.; resources, R.R.; data curation, H.R., R.R., O.F., F.F. and S.C. (Silvia Cetrullo); writing—original draft preparation, H.R., R.R., S.G., S.C. (Silvia Cetrullo), O.F. and A.M.Q.; writing—review and editing, H.R., R.R., O.F. and F.F.; supervision, F.F.; funding acquisition, S.G. and A.M.Q. All authors have read and agreed to the published version of the manuscript.

Funding: This research was funded by the Italian Ministry of Agricultural, Food, Forestry and Tourism Policies (MiPAAFT) under the DIBIO-BIOPRIME project, (DM N. 3400, 20 December 2018) and by Interreg MAC Program 2014–2020, grant number MAC2/1.1b/269: REBECA-CCT.

Institutional Review Board Statement: Not applicable.

Informed Consent Statement: Not applicable.

Data Availability Statement: The data presented in this study are available upon request from the corresponding author.

Conflicts of Interest: The authors declare no conflict of interest. The funders had no role in the design of the study; in the collection, analyses, or interpretation of data; in the writing of the manuscript; or in the decision to publish the results.

References

1. Hentati, F.; Barkallah, M.; Ben Atitallah, A.; Dammak, M.; Louati, I.; Pierre, G.; Fendri, I.; Attia, H.; Michaud, P.; Abdelkafi, S. Quality characteristics and functional and antioxidant capacities of algae-fortified fish burgers prepared from common barbel (*Barbus barbus*). *Biomed. Res. Int.* **2019**, *2019*, 2907542. [CrossRef] [PubMed]
2. Kousha, M.; Daneshvar, E.; Esmaeli, A.R.; Jokar, M.; Khataee, A.R. Optimization of Acid Blue 25 removal from aqueous solutions by raw, esterified and protonated *Jania adhaerens* biomass. *Int. Biodeter. Biodegr.* **2012**, *69*, 97–105. [CrossRef]
3. Ramu Ganesan, A.; Kannan, M.; Karthick Rajan, D.; Pillay, A.A.; Shanmugam, M.; Sathishkumar, P.; Johansen, J.; Tiwari, B.K. Phycoerythrin: A pink pigment from red sources (rhodophyta) for a greener biorefining approach to food applications. *Crit. Rev. Food Sci. Nutr.* **2022**, 1–19. [CrossRef] [PubMed]
4. Hamed, S.M.; Abd El-Rhman, A.A.; Abdel-Raouf, N.; Ibraheem, I.B. Role of marine macroalgae in plant protection & improvement for sustainable agriculture technology. *Beni-Suef Univ. J. Basic. Appl. Sci.* **2018**, *7*, 104–110.
5. Navarro, D.A.; Stortz, C.A. The system of xylogalactans from the red seaweed *Jania rubens* (Corallinales, Rhodophyta). *Carbohyd. Res.* **2008**, *343*, 2613–2622. [CrossRef] [PubMed]
6. Kaplan, D.; Christiaen, D.; Arad, S.M. Chelating properties of extracellular polysaccharides from *Chlorella* spp. *Appl. Environ. Microbial.* **1987**, *53*, 2953–2956. [CrossRef] [PubMed]
7. Mercier, L.; Lafitte, C.; Borderies, G.; Briand, X.; Esquerré-Tugayé, M.T.; Fournier, J. The algal polysaccharide carrageenans can act as an elicitor of plant defence. *New Phytol.* **2001**, *149*, 43–51. [CrossRef]
8. Lefevere, H.; Bauters, L.; Gheysen, G. Salicylic acid biosynthesis in plants. *Front. Plant Sci.* **2020**, *11*, 338. [CrossRef]
9. Seyfferth, C.; Tsuda, K. Salicylic acid signal transduction: The initiation of biosynthesis, perception and transcriptional reprogramming. *Front. Plant Sci.* **2014**, *5*, 697. [CrossRef]
10. Bruehl, G.W. *Soilborne Plant Pathogens*; Macmillan Publishing Co.: New York, NY, USA, 1987; ISBN 0029491304.
11. Martin, F.N.; Loper, J.E. Soilborne plant diseases caused by *Pythium* spp.: Ecology, epidemiology, and prospects for biological control. *Crit. Rev. Plant Sci.* **1999**, *18*, 111–181. [CrossRef]
12. Kamoun, S.; Furzer, O.; Jones, J.D.; Judelson, H.S.; Ali, G.S.; Dalio, R.J.; Roy, G.S.; Schena, L.; Zambounis, A.; Panabières, F.; et al. The Top 10 oomycete pathogens in molecular plant pathology. *Mol. Plant. Pathol.* **2015**, *16*, 413–434. [CrossRef]
13. Lübeck, M. Molecular characterization of *Rhizoctonia solani*. In *Applied Mycology and Biotechnology*, 1st ed.; Khachatourians, G.G., Arora, D.K., Eds.; Volume 4: Fungal genomics; Elsevier Science B.V.: Amsterdam, The Netherlands, 2004; pp. 205–224, ISBN 0444516425.
14. Marcou, S.; Wikström, M.; Ragnarsson, S.; Persson, L.; Höfte, M. Occurrence and anastomosis grouping of *Rhizoctonia* spp. inducing black scurf and greyish-white felt-like mycelium on carrot in Sweden. *J. Fungi* **2021**, *7*, 396. [CrossRef] [PubMed]
15. Gordon, T.R. *Fusarium oxysporum* and the Fusarium wilt syndrome. *Annu. Rev. Phytopathol.* **2017**, *55*, 23–39. [CrossRef] [PubMed]
16. Fravel, D.; Olivain, C.; Alabouvette, C. *Fusarium oxysporum* and its biocontrol. *New Phytol.* **2003**, *157*, 493–502. [CrossRef] [PubMed]
17. Michielse, C.B.; Rep, M. Pathogen profile update: *Fusarium oxysporum*. *Mol. Plant. Pathol.* **2009**, *10*, 311. [CrossRef] [PubMed]
18. Galletti, S.; Cianchetta, S.; Righini, H.; Roberti, R. A Lignin-rich extract of giant reed (*Arundo donax* L.) as a possible tool to manage soilborne pathogens in horticulture: A preliminary study on a model pathosystem. *Horticulturae* **2022**, *8*, 589. [CrossRef]
19. Abo-Elyousr, K.A.; Ali, E.F.; Sallam, N.M. Alternative control of tomato wilt using the aqueous extract of *Calotropis procera*. *Horticulturae* **2022**, *8*, 197. [CrossRef]
20. La Torre, A.; Caradonia, F.; Matere, A.; Battaglia, V. Using plant essential oils to control Fusarium wilt in tomato plants. *Eur. J. Plant Pathol.* **2016**, *144*, 487–496. [CrossRef]
21. Sharma, A.; Rajendran, S.; Srivastava, A.; Sharma, S.; Kundu, B. Antifungal activities of selected essential oils against *Fusarium oxysporum* f. sp. lycopersici 1322, with emphasis on Syzygium aromaticum essential oil. *J. Biosci. Bioeng.* **2017**, *123*, 308–313.
22. Ramírez, P.G.; Ramírez, D.G.; Mejía, E.Z.; Ocampo, S.A.; Díaz, C.N.; Martínez, R.I.R. Extracts of *Stevia rebaudiana* against *Fusarium oxysporum* associated with tomato cultivation. *Sci. Hortic.* **2020**, *259*, 108683. [CrossRef]
23. Ozkaya, H.O.; Ergun, T. The effects of *Allium tuncelianum* extract on some important pathogens and total phenolic compounds in tomato and pepper. *Pak. J. Bot.* **2017**, *49*, 2483–2490.
24. Sanaullah, N.A.R.; Atiq, M.; Rehman, A.; Khan, S.; Bashir, M.; Hameed, A.; Khan, B.; Shakir Baloch, M.; Kachelo, G.A. Antifungal potency of three plant extracts against *Rhizoctonia solani* damping-off disease in tomato. *Int. J. Biosci.* **2018**, *13*, 309–316.
25. Hlokwe, M.T.P.; Kena, M.A.; Mamphiswana, D.N. Application of plant extracts and *Trichoderma harzianum* for the management of tomato seedling damping-off caused by *Rhizoctonia solani*. *S. Afr. J. Sci.* **2020**, *116*, 1–5. [CrossRef]
26. Righini, H.; Francioso, O.; Di Foggia, M.; Prodi, A.; Quintana, A.M.; Roberti, R. Tomato seed biopriming with water extracts from *Anabaena minutissima*, *Ecklonia maxima* and *Jania adhaerens* as a new agro-ecological option against *Rhizoctonia solani*. *Sci. Hortic.* **2021**, *281*, 109921. [CrossRef]

27. Akladious, S.A.; Isaac, G.S.; Abu-Tahon, M.A. Induction and resistance against Fusarium wilt disease of tomato by using sweet basil (*Ocimum basilicum* L.) extract. *Can. J. Plant. Sci.* **2015**, *95*, 689–701. [CrossRef]
28. Righini, H.; Roberti, R.; Baraldi, E. Use of algae in strawberry management. *J. Appl. Phycol.* **2018**, *30*, 3551–3564. [CrossRef]
29. Vera, J.; Castro, J.; Gonzalez, A.; Moenne, A. Seaweed polysaccharides and derived oligosaccharides stimulate defense responses and protection against pathogens in plants. *Mar. Drugs* **2011**, *9*, 2514–2525. [CrossRef]
30. Vicente, T.F.L.; Lemos, M.F.L.; Félix, R.; Valentão, P.; Félix, C. Marine macroalgae, a source of natural inhibitors of fungal phytopathogens. *J. Fungi* **2021**, *7*, 1006. [CrossRef]
31. Ben Salah, I.; Aghrouss, S.; Douira, A.; Aissam, S.; El Alaoui-Talibi, Z.; Filali-Maltouf, A.; El Modafar, C. Seaweed polysaccharides as bio-elicitors of natural defenses in olive trees against verticillium wilt of olive. *J. Plant Interact.* **2018**, *13*, 248–255. [CrossRef]
32. Ansari, F.A.; Shriwastav, A.; Gupta, S.K.; Rawat, I.; Guldhe, A.; Bux, F. Lipid extracted algae as a source for protein and reduced sugar: A step closer to the biorefinery. *Bioresour. Technol.* **2015**, *179*, 559–564. [CrossRef]
33. Chen, Y.Y.; Xue, Y.T. Optimization of microwave assisted extraction, chemical characterization and antitumor activities of polysaccharides from *Porphyra haitanensis*. *Carbohydr. Polym.* **2019**, *206*, 179–186. [CrossRef] [PubMed]
34. Garcia-Vaquero, M.; O'Doherty, J.V.; Tiwari, B.K.; Sweeney, T.; Rajauria, G. Enhancing the extraction of polysaccharides and antioxidants from macroalgae using sequential hydrothermal-assisted extraction followed by ultrasound and thermal technologies. *Mar. Drugs* **2019**, *17*, 457. [CrossRef] [PubMed]
35. Elarrouss, H.; Elmernissi, N.; Benhima, R.; El Kadmiri, I.M.; Bendaou, N.; Smouni, A.; Wahbya, I. Microalgae polysaccharides a promising plant growth biostimulant. *J. Algal Biomass Util.* **2016**, *7*, 55–63.
36. Ozawa, T.; Yamamoto, J.; Yamagishi, T.; Yamazaki, N.; Nishizawa, M. Two fucoidans in the holdfast of cultivated *Laminaria japonica*. *J. Nat. Med.* **2006**, *60*, 236–239. [CrossRef] [PubMed]
37. Sinurat, E.; Rosmawaty, P.; Saepudin, E. Characterization of fucoidan extracted from Binuangeun's Brown Seaweeds. *Int. J. Chem. Environ. Biol. Sci.* **2015**, *3*, 329–332.
38. Shen, S.; Cheng, H.; Li, X.; Li, T.; Yuan, M.; Zhou, Y.; Ding, C. Effects of extraction methods on antioxidant activities of polysaccharides from camellia seed cake. *Eur. Food Res. Technol.* **2014**, *238*, 1015–1021. [CrossRef]
39. Álvarez-Gómez, F.; Korbee, N.; Figueroa, F.L. Analysis of antioxidant capacity and bioactive compounds in marine macroalgal and lichenic extracts using different solvents and evaluation methods. *Cienc. Mar.* **2016**, *42*, 271–288. [CrossRef]
40. Nawaz, A.; Amjad, M.; Khan, S.M.; Afzal, I.; Ahmed, T.; Iqbal, Q.; Iqbal, J. Tomato seed invigoration with cytokinins. *J. Anim. Plant. Sci.* **2013**, *23*, 121–128.
41. Burnett, F.J.; Boor, T.; Dussart, F.; Smith, J. Developing Sustainable Management Methods for Clubroot. Project Report No. 608. Agriculture and Horticulture Development Board. 2019. Available online: https://pure.sruc.ac.uk/ws/portalfiles/potal/18991123/pr608_final_project_report.pdf (accessed on 12 July 2022).
42. Li, Q.; Yu, P.; Chen, X.; Li, G.; Zhou, D.; Zheng, W. Facilitative and inhibitory effect of litter on seedling emergence and early growth of six herbaceous species in an early successional old field ecosystem. *Sci. World J.* **2014**, *2014*, 101860. [CrossRef]
43. De Cal, A.; Pascual, S.; Melgarejo, P. Infectivity of chlamydospores vs microconidia of *Fusarium oxysporum* f. sp. lycopersici on tomato. *J. Phytopathol* **1997**, *145*, 231–233. [CrossRef]
44. Roberti, R.; Galletti, S.; Burzi, P.L.; Righini, H.; Cetrullo, S.; Perez, C. Induction of defence responses in zucchini (*Cucurbita pepo*) by *Anabaena* sp. water extract. *Biol. Control* **2015**, *82*, 61–68. [CrossRef]
45. Bradford, M.M. A rapid and sensitive method for the quantitation of microgram quantities of protein utilizing the principle of protein-dye binding. *Anal. Biochem.* **1976**, *72*, 248–254. [CrossRef]
46. Bargabus, R.L.; Zidack, N.K.; Sherwood, J.E.; Jacobsen, B.J. Screening for the identification of potential biological control agents that induce systemic acquired resistance in sugar beet. *Biol. Control* **2004**, *30*, 342–350. [CrossRef]
47. Rao, C.N.R. *Chemical Applications of Infrared Spectroscopy*; Academic Press, Inc.: New York, NY, USA; London, UK, 1963; p. 681.
48. Xu, R.B.; Yang, X.; Wang, J.; Zhao, H.T.; Lu, W.H.; Cui, J.; Cheng, C.L.; Zou, P.; Huang, W.W.; Wang, P.; et al. Chemical composition and antioxidant activities of three polysaccharide fractions from pine cones. *Int. J. Mol. Sci.* **2012**, *13*, 14262–14277. [CrossRef]
49. Hong, T.; Yin, J.Y.; Nie, S.P.; Xie, M.Y. Applications of infrared spectroscopy in polysaccharide structural analysis: Progress, challenge and perspective. *Food Chem. X* **2021**, *12*, 100168. [CrossRef]
50. Wiercigroch, E.; Szafraniec, E.; Czamara, K.; Pacia, M.Z.; Majzner, K.; Kochan, K.; Malek, K. Raman and infrared spectroscopy of carbohydrates: A review. *Spectrochim. Acta A* **2017**, *185*, 317–335. [CrossRef]
51. Shukla, P.S.; Mantin, E.G.; Adil, M.; Bajpai, S.; Critchley, A.T.; Prithiviraj, B. *Ascophyllum nodosum*-based biostimulants: Sustainable applications in agriculture for the stimulation of plant growth, stress tolerance, and disease management. *Front. Plant Sci.* **2019**, *10*, 655. [CrossRef]
52. Arioli, T.; Mattner, S.W.; Winberg, P.C. Applications of seaweed extracts in Australian agriculture: Past, present and future. *J. Appl. Phycol.* **2015**, *27*, 2007–2015. [CrossRef]
53. Khan, W.; Rayirath, U.P.; Subramanian, S.; Jithesh, M.N.; Rayorath, P.; Hodges, D.M.; Critchley, A.T.; Craigie, J.S.; Norrie, J.; Prithiviraj, B. Seaweed extracts as biostimulants of plant growth and development. *J. Plant Growth Regul.* **2009**, *28*, 386–399. [CrossRef]
54. O' Keeffe, E.; Hughes, H.; McLoughlin, P.; Tan, S.P.; McCarthy, N. Methods of analysis for the in vitro and in vivo determination of the fungicidal activity of seaweeds: A mini review. *J. Appl. Phycol.* **2019**, *31*, 3759–3776. [CrossRef]

55. Sharma, H.S.; Fleming, C.; Selby, C.; Rao, J.R.; Martin, T. Plant biostimulants: A review on the processing of macroalgae and use of extracts for crop management to reduce abiotic and biotic stresses. *J. Appl. Phycol.* **2014**, *26*, 465–490. [CrossRef]
56. Abdel-Raouf, N.; Al-Homaidan, A.A.; Ibrahim, I.B.M. Agricultural importance of algae. *Afr. J. Biotechnol.* **2012**, *1*, 11648–11658. [CrossRef]
57. Calvo, P.; Nelson, L.; Kloepper, J.W. Agricultural uses of plant biostimulants. *Plant Soil* **2014**, *383*, 3–41. [CrossRef]
58. Arunkumar, K.; Sivakumar, S.R.; Rengasamy, R. Review on bioactive potential in seaweeds (marine macroalgae): A special emphasis on bioactivity of seaweeds against plant pathogens. *Asian J. Plant Sci.* **2010**, *9*, 227–240. [CrossRef]
59. Righini, H.; Somma, A.; Cetrullo, S.; D'Adamo, S.; Flamigni, F.; Martel Quintana, A.; Roberti, R. Inhibitory activity of aqueous extracts from *Anabaena minutissima*, *Ecklonia maxima* and *Jania adhaerens* on the cucumber powdery mildew pathogen in vitro and in vivo. *J. Appl. Phycol.* **2020**, *32*, 3363–3375. [CrossRef]
60. Arun, M.N.; Hebbar, S.S.; Senthivel, T.; Nair, A.K.; Padmavathi, G.; Pandey, P.; Singh, A. Seed priming: The way forward to mitigate abiotic stress in crops. In *Plant Stress Physiology-Perspectives in Agriculture*; Hasanuzzaman, M., Nahar, K., Eds.; IntechOpen: London, UK, 2022; Volume 11, p. 173.
61. El-Mougy, N.S.; Abdel-Kader, M.M. Long term activity of bio-priming seed treatment for biological control of faba bean root rot pathogens. *Aust. Plant Pathol.* **2008**, *37*, 464–471. [CrossRef]
62. Rachidi, F.; Benhima, R.; Kasmi, Y.; Sbabou, L.; Arroussi, H.E. Evaluation of microalgae polysaccharides as biostimulants of tomato plant defense using metabolomics and biochemical approaches. *Sci. Rep.* **2021**, *11*, 101860. [CrossRef]
63. Chaliha, C.; Rugen, M.D.; Field, R.A.; Kalita, E. Glycans as modulators of plant defense against filamentous pathogens. *Front. Plant Sci.* **2018**, *9*, 928. [CrossRef]
64. Lagogianni, C.S.; Tsitsigiannis, D.I. Effective biopesticides and biostimulants to reduce aflatoxins in maize fields. *Front. Microbiol.* **2019**, *10*, 2645. [CrossRef]
65. de Borba, M.C.; Velho, A.C.; de Freitas, M.B.; Holvoet, M.; Maia-Grondard, A.; Baltenweck, R.; Magnin-Robert, M.; Randoux, B.; Hilbert, J.L.; Reignault, P.; et al. A laminarin-based formulation protects wheat against *Zymoseptoria tritici* via direct antifungal activity and elicitation of host defense-related genes. *Plant Dis.* **2022**, *106*, 1408–1418. [CrossRef]
66. Aruna, P.; Mansuya, P.; Sridhar, S.; Kumar, J.S.; Babu, S. Pharmacognostical and antifungal activity of selected seaweeds from Gulf of Mannar region. *Recent Res. Sci. Technol.* **2010**, *2*, 115–119.
67. Damonte, E.B.; Matulewicz, M.C.; Cerezo, A.S. Sulfated seaweed polysaccharides as antiviral agents. *Curr. Med. Chem.* **2004**, *11*, 2399–2419. [CrossRef] [PubMed]
68. Matsuhiro, B.; Conte, A.F.; Damonte, E.B.; Kolender, A.A.; Matulewicz, M.C.; Mejías, E.G.; Pujol, C.A.; Zúñiga, E.A. Structural analysis and antiviral activity of a sulfated galactan from the red seaweed *Schizymenia binderi* (Gigartinales, Rhodophyta). *Carbohyd. Res.* **2005**, *340*, 2392–2402. [CrossRef] [PubMed]
69. Pujol, C.A.; Scolaro, L.A.; Ciancia, M.; Matulewicz, M.C.; Cerezo, A.S.; Damonte, E.B. Antiviral activity of a carrageenan from *Gigartina skottsbergii* against intraperitoneal murine *Herpes simplex* virus infection. *Planta Med.* **2006**, *72*, 121–125. [CrossRef]
70. Souza, B.W.S.; Cerqueira, M.A.; Bourbon, A.I.; Pinheiro, A.C.; Martins, J.T.; Teixeira, J.A.; Coimbra, M.A.; Vicente, A.A. Chemical characterization and antioxidant activity of sulfated polysaccharide from the red seaweed *Gracilaria birdiae*. *Food Hydrocoll.* **2012**, *27*, 287–292. [CrossRef]
71. Abdel Latef, A.A.H.; Srivastava, A.K.; Saber, H.; Alwaleed, E.A.; Tran, L.S.P. *Sargassum muticum* and *Jania rubens* regulate amino acid metabolism to improve growth and alleviate salinity in chickpea. *Sci. Rep.* **2017**, *7*, 10537. [CrossRef]
72. Safinaz, F.; Ragaa, A.H. Effect of some red marine algae as biofertilizers on growth of maize (*Zea mays* L.) plants. *Int. Food Res. J.* **2013**, *20*, 1629–1632.
73. Zodape, S.T.; Mukherjee, S.; Reddy, M.P.; Chaudhary, D.R. Effect of *Kappaphycus alvarezii* (Doty) doty ex silva. extract on grain quality, yield and some yield components of wheat (*Triticum aestivum* L.). *Int. J. Plant Prod.* **2009**, *3*, 97–101.
74. Zodape, S.T.; Gupta, A.; Bhandari, S.C.; Rawat, U.S.; Chaudhary, D.R.; Eswaran, K.; Chikara, J. Foliar application of seaweed sap as biostimulant for enhancement of yield and quality of tomato (*Lycopersicon esculentum* Mill.). *J. Sci. Ind. Res.* **2011**, *70*, 215–219.
75. Jayaraman, J.; Norrie, J.; Punja, Z.K. Commercial extract from the brown seaweed *Ascophyllum nodosum* reduces fungal diseases in greenhouse cucumber. *J. Appl. Phycol.* **2011**, *23*, 353–361. [CrossRef]
76. Witmer, X.; Nonogaki, H.; Beers, E.P.; Bradford, K.J.; Welbaum, G.E. Characterization of chitinase activity and gene expression in muskmelon seeds. *Seed Sci. Res.* **2003**, *13*, 167–178. [CrossRef]
77. Morohashi, Y.; Matsushima, H. Development of β-1, 3-glucanase activity in germinated tomato seeds. *J. Exp. Bot.* **2000**, *51*, 1381–1387. [CrossRef]
78. Ghule, M.R.; Ramteke, P.K.; Ramteke, S.D.; Kodre, P.S.; Langote, A.; Gaikwad, A.V.; Holkar, S.K.; Jambhekar, H. Impact of chitosan seed treatment of fenugreek for management of root rot disease caused by *Fusarium solani* under in vitro and in vivo conditions. *3 Biotech* **2021**, *11*, 290. [CrossRef] [PubMed]
79. Pettongkhao, S.; Bilanglod, A.; Khompatara, K.; Churngchow, N. Sulphated polysaccharide from *Acanthophora spicifera* induced *Hevea brasiliensis* defense responses against *Phytophthora palmivora* infection. *Plants* **2019**, *8*, 73. [CrossRef]
80. Xu, X.; Chen, Y.; Li, B.; Zhang, Z.; Qin, G.; Chen, T.; Tian, S. Molecular mechanisms underlying multi-level defense responses of horticultural crops to fungal pathogens. *Hortic. Res.* **2022**, *9*, uhac066. [CrossRef] [PubMed]

Article

In Search of Antifungals from the Plant World: The Potential of Saponins and Brassica Species against *Verticillium dahliae* Kleb.

Caterina Morcia [1], Isabella Piazza [1], Roberta Ghizzoni [1], Stefano Delbono [1], Barbara Felici [2], Simona Baima [2], Federico Scossa [2], Elisa Biazzi [3], Aldo Tava [3], Valeria Terzi [1,*] and Franca Finocchiaro [1]

[1] Council for Agricultural Research and Economics, Research Centre for Genomics and Bioinformatics, Via San Protaso 302, I-29017 Fiorenzuola d'Arda, Italy
[2] Council for Agricultural Research and Economics, Research Centre for Genomics and Bioinformatics, Via Ardeatina 546, I-00178 Roma, Italy
[3] Council for Agricultural Research and Economics, Centre for Animal Production and Aquaculture, Viale Piacenza 29, I-26900 Lodi, Italy
* Correspondence: valeria.terzi@crea.gov.it; Tel.: +39-0523-983758

Abstract: Control methods alternative to synthetic pesticides are among the priorities for both organic and conventional farming systems. Plants are potential sources of compounds with antimicrobial properties. In this study, the antifungal potentialities of saponins derived from Medicago species and oat grains and of brassica sprouts have been explored for the control of *Verticillium dahliae*, a widely distributed fungal pathogen that causes vascular wilt disease on over 200 plant species. All the tested plant extracts showed antifungal properties. Such compounds, able to reduce mycelium growth and conidia formation, deserve deeper in vivo evaluation, even in combination with a delivery system.

Keywords: antifungal compounds; saponins; glucosinolates; polyphenols; *Verticillium*; conidia

1. Introduction

The priority objectives that fall under the European Green Deal (2019) [1], more precisely in the From Farm to Fork Strategy (F2F; 2020) and the European Union (UN) Biodiversity Strategy for 2030 (2020), include the development of organic farming in the EU and the reduction of pesticides, antimicrobials, and synthetic fertilizers used in agriculture and animal husbandry. By 2030, member countries will undertake to halve the use of chemical pesticides, further reducing those most harmful to the environment and human health and to allocate 25% of UAA (Utilized Agricultural Area) to organic farming.

In recent years, biocontrol products and control measures alternative to synthetic pesticides, have attracted considerable interest by farmers. Biocontrol is generally defined as a method for insect, weed, and disease management, using natural enemies and natural products [2]. Biocontrol tools therefore include live organisms (generally microbes), chemicals of semisynthetic origin and natural substances extracted from plant, animal, or mineral sources. Among plant-natural products, various classes of secondary metabolites have been shown to possess potential as biocontrol agents, including saponins, polyphenols, and glucosinolates.

Saponins are a large family of secondary metabolites found in a wide range of plant species: their presence has been reported in more than 100 plant families and in some marine sources, such as starfish and sea cucumbers [3]. They are glycosidic substances consisting of a steroidal (C 27) or triterpenic backbone (C 30), known as aglycone or sapogenin, and a variable number of monosaccharide units, both pentose and hexose, joined by glycosidic bonds which give them an amphiphilic character [4–7] (see Figures 1 and 2). They are synthesized from the cytosolic mevalonate pathway, and they derive from the triterpenoid or steroid cyclization products of 2,3-oxidosqualene [8]. Steroidal saponins, which are generally found in monocotyledonous angiosperms (but not exclusively), consist of a steroidal aglycone, a C 27 spirostane skeleton. In some cases, the hydroxyl group at position 26 is engaged in a glycosidic bond, and therefore the aglycone structure remains pentacyclic [9] (Figure 2A). Triterpene saponins, which are instead commonly found in dicotyledonous angiosperms, consist of a triterpenoid aglycone (a C 30 skeleton), comprising a pentacyclic structure (Figures 1 and 2B).

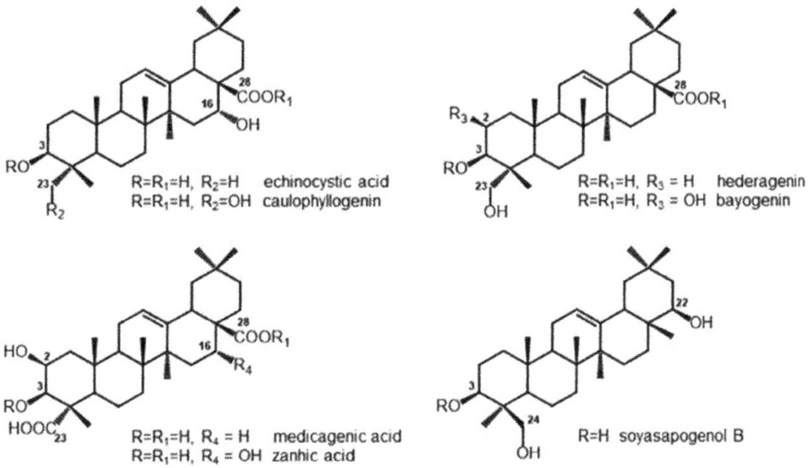

Figure 1. Chemical structures of the most abundant sapogenins detected in the Medicago spp. plant extracts. Saponins: R = sugar or sugar chain, R_1 = H: monodesmosides. R = R_1 = sugar or sugar chain: bidesmosides.

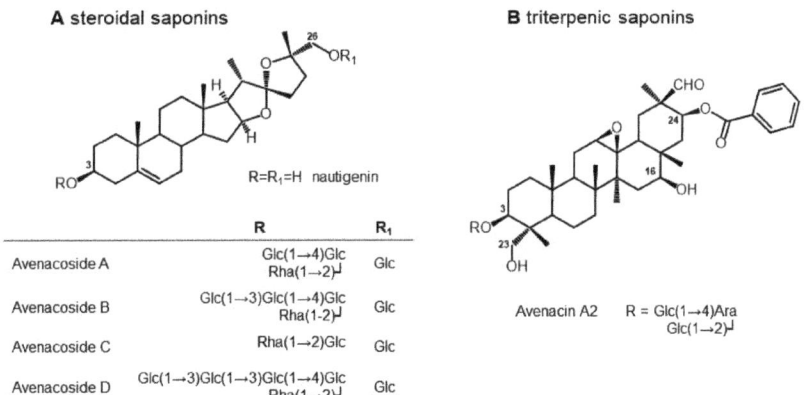

Figure 2. Chemical structures of the most abundant saponins identified in the *Avena sativa* seed extract, i.e., steroidal saponins (**A**) and triterpeninc saponins (**B**).

Cereals are generally known to lack saponins, with the exception of oats, which accumulate both triterpenoid (avenacins) and steroidal (avenacosides) saponins. Their distribution is mutually exclusive, avenacosides have been reported to accumulate in the leaves while avenacins accumulate in the roots [8].

In several plant species, saponin production is induced in response to biotic (attack by herbivores and pathogens) and abiotic (humidity, nutrient deficiency, light, temperature) stresses.

Saponins showed a multitude of biochemical properties, such as being used as drugs and medicines, precursors for hormone synthesis, cholesterol-lowering agents, adjuvants, foaming agents, sweeteners, taste modifiers, and cosmetics.

Saponins have also been intensively studied as antimicrobial and biocontrol agents against human and plant pathogenic microorganisms and harmful insects [4]. Saponin extracts have also been tested against numerous Gram-positive and Gram-negative bacteria, yeasts, and molds [7]. Although the results reported so far are difficult to generalize, due to the high structural and biological diversity of plant saponins, antifungal activities were generally found to be stronger with respect to their antibacterial properties [6].

Another class of plant derived products with interesting potential as biocontrol agents are glucosinolates (GLSs). This class of secondary metabolites is produced almost exclusively by plants belonging to the *Brassicaceae* family and is composed of β-thioglucoside *N*-hydroxysulfates, consisting of a D-thioglucose group linked to a sulfonated aldoxime group and a variable side chain derived from amino acids [10]. Based on the structure of different amino acids precursors, GLSs have been divided into three classes: aliphatic (derived from methionine, isoleucine, leucine, or valine), aromatic (derived from phenylalanine or tyrosine) and indolic (derived from tryptophan). Out of more than 120 different GLSs identified, only some show a high abundance in *Brassicaceae*. Cultivar, developmental stage, organ, agronomic, and environmental conditions are known to significantly affect GLSs content and profile [11–13].

The involvement of GLSs in plant defense response mechanisms is well known as they are induced after wounding, pathogen infection, or insect and herbivore attack. The biological activity of GSLs depends on their enzymatic hydrolysis catalyzed by degradative enzymes known as myrosinases and by specific proteins acting as cofactors that release various toxic products (isothiocyanates, nitriles, epithioalkanes, and thiocyanates). As GLSs and myrosinase are stored in different cellular compartments, the hydrolysis products are released only after cells are mechanically damaged. Interestingly, the glucosinolate–myrosinase system, also known as the mustard-oil bomb, provides a plant defense response not only against herbivores and insect pests but also against soil borne pathogens and pests, such as nematodes, fungi, and some weeds. Due to the presence of such antimicrobial, antifungal, and biocidal compounds, *Brassicaceae* are used in agriculture for biofumigation through the preparation of commercial fumigants or growth as green manure or rotation crops.

Elicitation, i.e., application of biotic and abiotic stress factors during growth, such as extreme light or temperatures, saline or osmotic stress, elicitors or hormones, has been shown to further increase the content of bioactive molecules, including GLSs, in *Brassicaceae* [11,14,15]. Among others, treatment with sucrose has been reported to elicit the accumulation of GLSs and to induce the synthesis of anthocyanins in broccoli sprouts [15–19]. In previous works, Ferruzza et al. [20] obtained two aqueous juices from dark grown and sucrose-treated broccoli sprouts, showing different biological activities on a human intestinal cell line. Juices from broccoli sprouts were also shown to be protective in a cellular model of Alzheimer's disease and in Spontaneously Hypertensive Stroke Prone rats [21–23]. The composition analyses of these juices revealed in the sucrose-treated sprouts a marked increase of anthocyanins and higher levels of 14 phenolic acids, including flavonoids [20].

Verticillium dahliae Kleb. is a widely distributed fungal pathogen that causes vascular wilt disease on over 200 plant species [24], including economically important crops and ornamental plants, native species, weeds, including both woody and herbaceous plants. The main economic hosts of *V. dahliae* include artichoke, eggplant, pepper, cotton, hops, lettuce, mint (*Mentha* spp.), rapeseed, olive, potato, strawberry, and tomato. It has also been isolated from root and crown tissues of cereals including barley, ryegrass, and winter wheat [25]. In fact, the fungus infects the roots and invades the xylem tissue, causing obstruction of the vascular tissue and the typical symptoms of vascular discoloration and wilting.

In addition, *V. dahliae* has the ability to survive for many years in the soil in the form of microsclerotia, small rigid survival structures capable of withstanding extreme temperatures and dehydration [25]. The wide host range of this pathogen and the lack of host resistance make this disease particularly difficult to manage [24].

In the present study, we tested the antifungal activity against *V. dahliae* of a panel of saponins extracts obtained from *Medicago* spp. leaves and roots and *Avena sativa* (common oat) seeds. Homogenates from *Brassica oleracea* sprouts, derived from either control or elicited conditions, were also tested to evaluate their potential activities to inhibit or delay the growth of *V. dahliae*. Both saponin-enriched extracts and Brassica homogenates were shown to inhibit, with different magnitudes, the mycelium growth in vitro. The potential practical applicability of the compounds for horticultural crop protection has been confirmed by the lack of phytotoxicity at the antifungal concentrations.

2. Materials and Methods

2.1. Plant Sources

We tested nine different extracts (including two homogenates from sprouts of *Brassica oleracea*) for their antifungal activity against *V. dahliae*. The plant sources from which we obtained the extracts are listed in Table 1.

Table 1. The tested compounds and their sources are reported.

Plant Species	Plant Tissue	Main Compound	Code
Medicago arborea	leaves	saponins	Sap1
Medicago polymorpha 22507	leaves		Sap2
Medicago polymorpha 155004	leaves		Sap3
Medicago sativa	leaves		Sap4
Medicago sativa	leaves	prosapogenins	Pros5
Medicago sativa	roots	saponins	Sap6
Avena sativa	seeds	saponins	Sap70
Brassica oleracea	etiolated sprouts	Brassica homogenates	A
Brassica oleracea	purple sprouts		B

2.2. Extraction, Purification, and Characterization of Saponins from Medicago spp.

Medicago plants used in this study were grown at the Research Centre for Animal Production and Aquaculture (CREA-ZA, Lodi, Italy). Tops from *Medicago arborea* L., *M. polymorpha* 25570, *M. polymorpha* 15504, and *M. sativa* L. were collected at plant anthesis, while *M. sativa* roots were collected at the end of the growing season. Saponins were extracted and purified following general procedures previously reported [26,27]. In addition, saponins from *M. sativa* were subjected to basic hydrolysis [28] to extract the related prosapogenins, which were also evaluated in this study.

The purified mixtures of saponins were obtained as whitish powders in high pure grade (85–90% purity) and characterized for their qualitative and quantitative aglycone composition by gas chromatography (GC) and gas chromatography-mass spectrometry (GC-MS) analyses of derivative sapogenins obtained after acid hydrolysis, as already reported [4,29]. To obtain information on saponin composition (e.g., chemical structure and monodesmoside/bidesmoside compounds), the saponin mixtures were then analyzed by HPLC-PDA and LC-MS and the results compared with available data [26–28].

2.3. Extraction and Characterization of Saponins from Avena Sativa Seeds

Oat grains from (*Avena sativa* cv. Novella Antonia) were milled with a Cyclotec Sample Mill (Foss Italia S.p.A., Padova, Italy) equipped with a 0.5 mm screen.

The flour was defatted in a Soxhlet apparatus with chloroform for two days. The extraction of the compounds of interest has been carried out in Soxhlet device using 100% methanol.

The methanol extract has been diluted to 20% with water, filtered, and ultracentrifuged with Beckman ultracentrifuge Model J2-21, set at temperature of 0 °C, for 25 min at 14,000 rpm. The supernatant was purified by open column chromatography filled with a C18 stationary phase, and four different fractions were collected by sequential elution with Methanol 20%, 50%, 70%, and 100%. All subsequent analyses were conducted on the fraction eluted with 70% Methanol (named Sap70 hereafter, see Table 1), given that most of the oat saponins were exclusively recovered in this fraction.

The saponin-enriched extract Sap70 was evaporated to dryness in a rotary evaporator set at 40 °C, yielding a crude saponin mixture powder. The presence of saponins in the extract was confirmed through a TLC.

The enriched oat saponin fraction Sap70 was subsequently characterized by LC–ESI-MS, with a C18 reversed-phase column (150 × 2.1 mm, 5 µm, KinetexR Core Shell, Phenomenex, Bologna, Italy). The mobile phase contained: (A) 5% acetonitrile and 0.1% formic acid solution and (B) acetonitrile 0.1% formic acid. The gradient elution program was as follows: 50% B (0–15 min); 50% B (15–24 min); 90% B (24–28 min); 90% B (28–30 min); 13% B (30–32 min); and 13% B (32–35 min). The flow rate was 0.2 mL/min and the injection volume was 10 µL. MS analysis was performed on the LTQ-XL ion trap (ThermoFisher, Monza, Italy) with an electrospray ionization (ESI) source in the negative ion mode. The mass scan range was set to 100–2000 m/z, with sheath gas 45 arb, auxiliary gas 20, capillary temperature set to 275 °C, and a spray voltage of 3.6 kV. MS^2 data were obtained from a data-dependent approach, acquiring MS^2 spectra on three most intense ions from the initial full scan event. Metabolites annotation was assigned on the basis of a combination of authentic standards, MS^2 data and confirmation of the presence of the metabolites from phytochemical data already available in *Avena* spp.

2.4. Brassica Sprouts Juices Preparation

Homogenates from sprouts of *Brassica oleracea* convar. *botrytis* var. *cimosa* were obtained as described by Ferruzza et al. [20]. Briefly, after sterilization, seeds were transferred to the Vitaseed germinator (SUBA & UNICO, Longiano, Italy). Sprouts were grown for 5 days at 21 °C and 70% humidity in a dedicated climatic chamber (Weiss Gallenkamp, Loughborough, UK) in the dark (type A sprouts) or with 16 h of lighting and 8 h of darkness (type B sprouts). After the first 3 days of growth, type B sprouts were treated for 48 h with a 176 mM sucrose solution. After 5 days of growth, the seedlings were weighted and cold-pressed with an Angel Juicer 8500S (Living Juice Ltd., Lecco, Italy) for juice production. The juices obtained were centrifuged at 4000 g for 30 min at 4 °C. The supernatant was immediately frozen in liquid nitrogen and stored at −80 °C. At the time of use, the extracts were filtered sequentially first through 1.2 um, then 0.45, and finally 0.22 um filters in order to eliminate any bacterial and fungal loads that may be present.

2.5. Antifungal Activity In Vitro Test

The antifungal activity of saponin-enriched extracts and Brassica homogenates were evaluated against a *V. dahliae* strain isolated from tomato (*S. lycopersicum*) and stored at the Fungal Repository of the Università Cattolica del Sacro Cuore, Piacenza, Italy.

A 10% saponin stock was prepared with 100 mg of saponin extracts dissolved in 1 mL of sterile H_2O + DMSO (900 µL of H_2O + 100 µL of DMSO for saponin 1, 3, 4 and 6; 800 µL of H_2O + 200 µL of DMSO for prosapogenin 5; 600 µL of H_2O + 400 µL of DMSO for saponin 2) and different volumes were added to the Potato Dextrose Agar (PDA) medium before it solidified (at a temperature of about 60 °C) to obtain final concentrations of 0.25%, 0.5%, 1%, and 1.5%. For brassica sprouts juices, different volumes were added to the PDA medium before it solidified (at a temperature about 60 °C) to obtain final concentrations of 6%, 3%, and 1%. The medium thus prepared was distributed in 60 mm Petri dishes (about 5 mL of PDA / plate), which were subsequently inoculated with a rod of *Verticillium dahliae* from a holding plate. The plates were incubated at 20 °C with 12 h light and 12 h dark photoperiod. The diameter of the fungal mycelium was evaluated 3 to 12 days after inoculation. As controls, plates with PDA as such and added with DMSO (solution in which the saponins are dissolved) were used. The results were expressed as growth inhibition (I) calculated with the formula I = [(C−T)/C] × 100, where C = control mean and T = treatment mean. Control samples have 0% growth inhibition values. The experimental design consisted of six replicates per thesis. Moreover, mycelium morphology in control and treated samples was observed with the microscope Olympus DP50 (Olympus, Milan, Italy) and with the stereomicroscope Zeiss Discovery V8 at increasing magnifications.

2.6. Phytotoxicity Tests

The phytotoxicity of the best saponins in term of antifungal activity was evaluated in tomato seedlings. A panel of tomato cultivar (Cuore di Bue, Sailor, Mariner, Wilson and Rossoro) were used. A total of 20 seeds/variety were inserted into 2 mL centrifuge tubes containing 1mL of treatment solution. The tubes were kept at room temperature and gently stirring for 2 h. At the end of the incubation, the seeds were drained and dried on absorbent paper for 5 min. They were then transferred into Petri dishes (6 cm in diameter), containing 2 sterile paper filters wetted with 1ml of H_2O. In each plate, 20 seeds were placed and the experiment was conducted in duplicate. As controls, seeds treated with DMSO or simply soaked with H_2O were used.

After 10 days, the seedlings were measured (shoot and main radicle) and the percentage of germination was evaluated. The straightened roots and shoots were measured with a Vernier manual caliper, with a precision of 0.02 mm. Furthermore, for the different varieties and treatments, the vigor of the seed was calculated in accordance with Abdul-Baki and Anderson [30] by applying the formula:

$$\text{vigor} = \% \text{ viability} \times (\text{mm coleoptile length} + \text{mm main root length})$$

To analyze the potential phytotoxicity of the brassica sprouts juices, their effects on cereal seed germination was evaluated. Seeds of *Zea mays* (class 300) were washed for 6h in tap water, sterilized with NaClO 5% for 5′ and then drained and dried on absorbent paper for 5 min. They were then transferred into Petri dishes containing three layers of filter paper wetted with distilled water (control samples) or with the two types of brassica sprouts juices (treatment) at five different concentrations (0.1, 1, 10, 1000, and 5000 ppm). Each vessel contained 10 kernels and the experiment was repeated 3 times. The seeds were kept in the dark at 25 °C, 65% humidity, for 72 h to soak, and the percentage of seed germination was calculated after 72 h according to the UNICHIM 1651:2003 method.

2.7. Statistical Analysis

Fungal data and tomato vigor germination data were analyzed with *t*-test (R version 4.1.2 and R-studio 2021.09.01 build 372) with $p < 0.05$. Seed germination data from maize phytotoxicity tests were analyzed with a Kruskal–Wallis test, and post hoc comparison was done using the Mann–Whitney test. Statistical significance was established at $p < 0.08$.

3. Results

3.1. Composition of Saponins and Related Sapogenins from Medicago spp.

The chemical composition of *Medicago* saponin extracts differed according to the plant species. The composition of the most abundant sapogenins in the saponin mixtures is reported in Table 2. Based on the relative content of the dominant sapogenins after acid hydrolysis of the corresponding glycosides, saponins from *M. arborea* leaves (Sap1) were characterized by a higher amount of medicagenic and zanhic acids (27.5% and 45.8%, respectively, Figure 1). Echinocystic acid (Figure 1) was the dominant sapogenin in *M. polymorpha* 25570 leaves (Sap2), representing 76.5% of the total aglycones, while hederagenin (85.9%) (Figure 1) was the dominant sapogenin detected in *M. polymorpha* 15504 (Sap3). Medicagenic acid and zanhic acid were also the most abundant aglycones from saponins of *M. sativa* leaves (Sap4), accounting for 43.4% and 44.7%, respectively. *M. sativa* root saponins (Sap6) were instead characterized by a higher amount of medicagenic acid and hederagenin (Figure 1, 64.7% and 19.3%, respectively). Based on the HPLC-PDA/LC-MS comparison with authentic saponin standards [26–28], all the saponin mixtures here evaluated were found to be mainly constituted by bidesmosidic type saponins (70–80%). The *M. sativa* prosapogenins (Pros5), obtained after basic hydrolysis of the corresponding saponins, were instead entirely made up by monodesmosides (Figure 1).

Table 2. Composition of the most abundant sapogenins in the saponin mixtures from *Medicago* spp., expressed as percentage (%) of the total sapogenins.

		Echinocystic Acid	Caulophyllogenin	Hederagenin	Bayogenin	Medicagenic Acid	Zanhic Acid	Soyasapogenol B
M. arborea leaves	Sap1	-	-	0.1	3.5	27.5	45.8	10.1
M. polymorpha 25570 leaves	Sap2	76.5	4.6	9.7	1.5	-	-	2.0
M. polymorpha 15504 leaves	Sap3	0.3	0.1	85.9	1.2	-	-	2.3
M. sativa leaves	Sap4	-	-	0.2	1.4	43.4	30.2	13.1
M. sativa roots	Sap6	-	-	19.3	2.3	64.7	2.9	3.1

3.2. Composition of Saponins from Oat Seeds

In order to simplify the composition of the extract from oat grains and to obtain a saponin-enriched fraction, we fractionated the raw extract through RP-18 open column chromatography, with sequential elution steps with methanol washes at increasing concentrations. Of the 4 eluted fractions (obtained from sequential elution at 20, 50, 70, and 100% Methanol), TLC, HPLC, and LC-MS analysis showed that the 70% methanol fraction contained most of the saponins present in the initial raw extract. The 70% fraction was therefore analyzed by LC-ESI- MS^2 analysis (Figure S1) to allow elucidation of the chemical structure of avenacins and avenacosides, which was achieved based on their chromatographic behavior, MS^2 fragmentation spectra, and available literature data. When authentic standards were available, the annotation of putative saponin peaks was confirmed through co-elution of the corresponding pure compound. This enriched saponin fraction was shown to contain both avenacins and avenacosides (Table 3), in addition to over 70 authentic molecular ion signals for which a tentative annotation could not be provided.

Saponins were tentatively identified based on the molecular ion [M-H]$^-$, on key fragment ions and other MS observations. In general, the loss of 146 m/z was indicative of deoxyhexose (e.g., rhamnose) and the loss of 162 m/z was indicative of hexose (e.g., glucose). The most abundant tentatively identified saponins are reported in Figure 2. Altogether, avenacins and avenacosides represent 35–40% of the total fraction, and their relative percentage composition was evaluated by LC-MS analyses and reported in Table 3.

The most abundant saponins in the oat 70% fraction were avenacoside B (compound **5**, 35.3% of the total detected saponins) and avenacoside A (compound **6**, 28.9%) followed by compound **1** (avenacin A2, 16.5%) and compound **4** (one of the isomer of avenacoside D, eluting at RT11.12, 7.1%). Avenacin C (both isomers 1 and 2, compounds **7** and **8**) were also detected in lesser amount (see Table 3).

Table 3. LC-ESI-MS2 analysis of the *Avena sativa* enriched saponin extract. "A" grade refers to a metabolite annotation confirmed by co-elution of the authentic standard; "B" grade refers to a metabolite annotation confirmed by MS2 analysis and previous reports from the literature; "C" grade refers to a metabolite annotation confirmed by MS2 analysis and previous reports from the literature.

n	t_R	Molecular Formula	Monoisotopic Mass	$[M-H]^-$ (m/z)	Metabolite Annotation	Grade	MS2 (m/z)	% of Total Detected Saponins	Reference
1	10.09	$C_{54}H_{80}O_{21}$	1064.51918	1063.7	Avenacin A2	C		16.5	Crombie et al., 1984 [31] (oat roots); Hu and Sang (2020) (sprouted oat bran) [32]
2	10.29	$C_{63}H_{102}O_{33}$	1386.6303	1385.8	Avenacoside D (isomer 1)	B	1223.3 $[M-H-Hexose]^-$; 1061.0 $[M-H-2Hexose]^-$	1.4	Yang et al., 2016 [33] (oat bran)
3	10.52	$C_{57}H_{92}O_{28}$	1224.57748	1223.7	Isomer to Avenacoside B	B	1077.4 $[M-H-Rha]^-$; 1061.5 $[M-H-Hex]^-$; 1043.3 $[M-H-Hex-H_2O]^-$; 915.3 $[M-H-Hex-Rha]^-$; 899.4 $[M-H-2Hex]^-$	2.7	Yang et al., 2016 [33] (oat bran)
4	11.12	$C_{63}H_{102}O_{33}$	1386.6303	1385.8	Avenacoside D (isomer 2)	B	1223.3 $[M-H-Hex]^-$; 1061.0 $[M-H-2Hex]^-$	7.1	Yang et al., 2016 [33] (oat bran)
5	11.34	$C_{57}H_{92}O_{28}$	1224.57748	1223.7	Avenacoside B	C		35.3	Yang et al., 2016 [33] (oat bran)
6	11.57	$C_{51}H_{82}O_{23}$	1062.52466	1061.7	Avenacoside A	A	915.3 $[M-H-Rha]^-$; 899.3 $[M-H-Hex]^-$; 753.3 $[M-H-Rha-Hex]^-$; 736.3 $[M-H-2Hex]^-$	28.9	Tschesche et al. Chem Ber 1969 [34] (seeds and leaves)
7	11.98	$C_{45}H_{72}O_{18}$	900.47184	899.6	Avenacoside C	B	753.0 $[M-H-Rha]^-$; 737.3 $[M-H-Hex]^-$	6.4	Pecio et al., 2013 [35] (seeds)
8	12.52	$C_{45}H_{72}O_{18}$	900.47184	899.6	Isomer to Avenacoside C	C		1.9	Pecio et al., 2013 [35] (seeds)

3.3. Medicago Saponins Antifungal Activity

All tested saponins had an inhibitory effect against *Verticillium dahliae* at concentrations greater than 0.5%. The most effective was the prosapogenin extract from *Medicago sativa* leaves, with a reduction in fungal growth of 50.8–52.5% at the lowest concentrations tested and up to an inhibition of 54.1–59% at the highest concentrations (12 days after inoculation). The saponin 6 extract from *Medicago sativa* roots reduced the growth of the pathogen by 40–46% at the lowest concentrations, up to an inhibition of 47.3–49.3% at the highest concentrations (12 days after inoculation). Saponins extracts 3 and 4 (from the leaves of *Medicago polymorpha* and *Medicago sativa*) were less effective in reducing the fungal growth, which at the highest doses reduced the development of the pathogen by 35.3–42.7%, respectively (at 12 days after inoculation) (Figure 3).

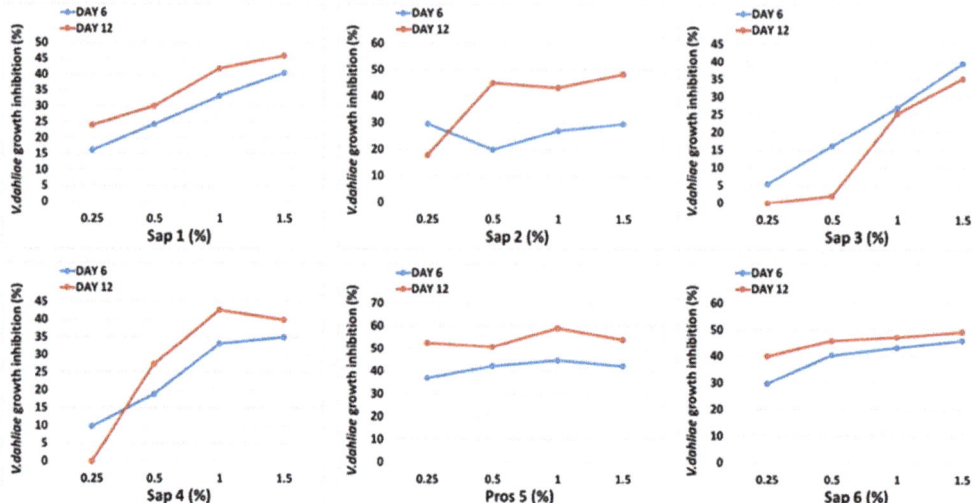

Figure 3. Effect on the in vitro growth of *Verticillium dahliae* mycelium by saponins added to the growth medium after 6 and 12 days from inoculation. Sap1) Saponins extracted from leaves of *M. arborea*, Sap2) and Sap3) Saponins extracted from leaves of *M. polymorpha*, Sap4) Saponins extracted from leaves of *M. sativa*, Pros5) Prosapogenins extracted from leaves of *M. sativa*, Sap6) Saponins extracted from roots of *M. sativa*. Values are reported as inhibition percentages as compared to the control ± SD. The concentration percentages are intended as volume/volume. All the treatments with saponins and prosapogenins concentrations equal or higher than 0.5% are significant (*t*-test, $p \leq 0.05$) in mycelium growth inhibition in comparison with not treated controls.

3.4. Oat Saponins Antifungal Activity

The saponin-enriched seed oat extract Sap70 significantly inhibited *Verticillium dahliae* growth at concentrations greater than 0.5% (Figure 4). Sap70 impact on mycelium growth is significant at 6 days after inoculation, with reductions ranging from 26% to 61% at increasing concentrations. The reduction in fungal growth ranged from 45% at 0.5% Sap70 concentration to 66% at the highest concentrations 12 days after inoculation.

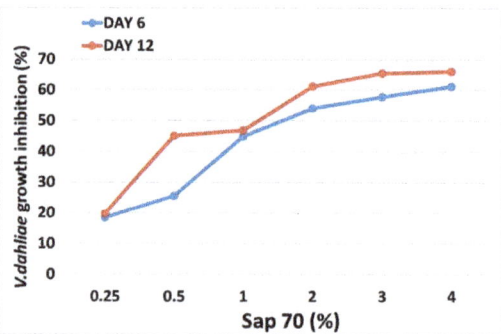

Figure 4. Effect on the in vitro growth of *Verticillium dahliae* mycelium by the saponin-enriched seed oat extract Sap70 added to the growth medium after 6 and 12 days from inoculation. Values are reported as inhibition percentages as compared to the control ± SD. The concentration percentages are intended as volume/volume. The mycelium growth reduction is significant (*t*-test, $p \leq 0.05$) in the presence of Sap70 concentrations greater than 0.5% after 6 days and equal or greater than 0.5% after 12 days of treatment.

In addition to seeing a strong effect on the growth of the fungus, the morphology of the mycelium is shaped by the saponins presence (Figure 5), with the transition from a cottony mycelium to a very compact one. This effect is particularly evident at the higher concentrations of the extract. It is noteworthy that Sap70 at the highest inhibited even the fungal conidia production.

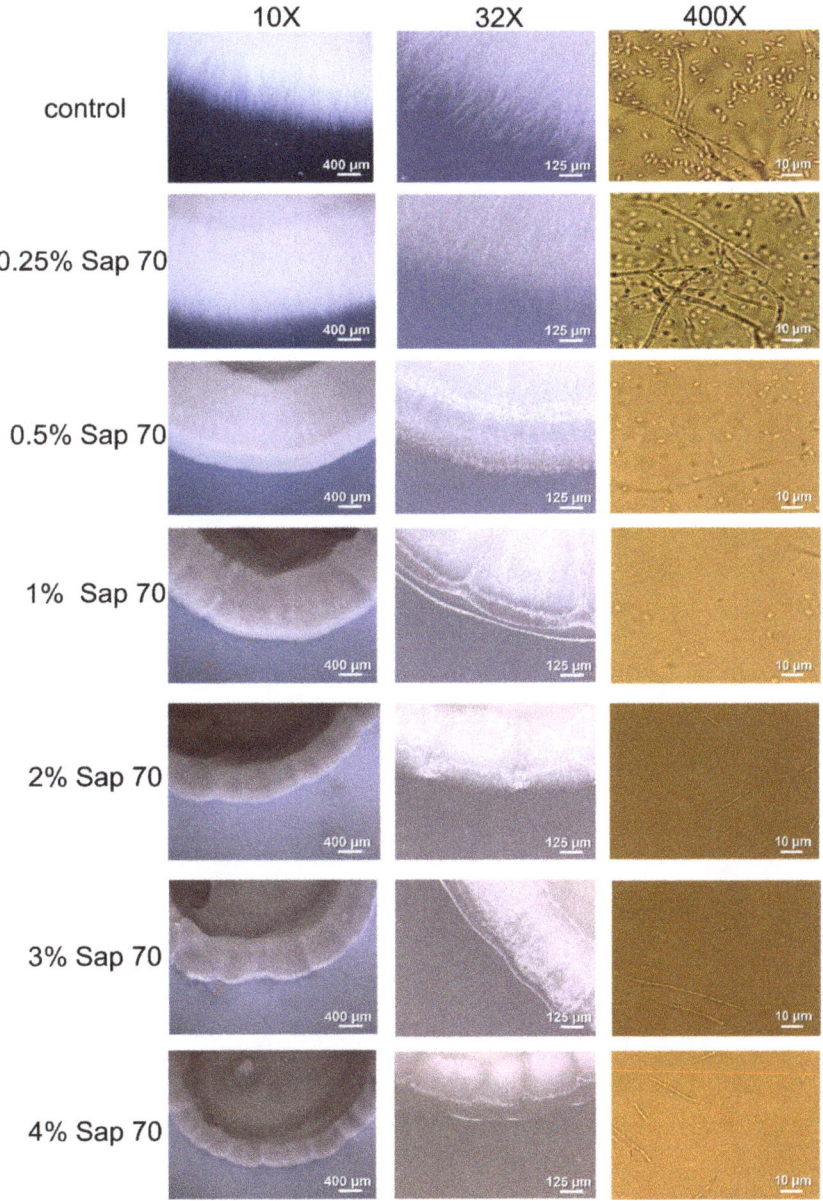

Figure 5. *Verticillium dahliae* mycelium images taken with the microscope and stereomicroscope using 3 magnifications (10×, 32×, 400×) after 10 days of growth in the absence (control) or in the presence of increasing Sap 70 concentrations.

3.5. Brassica Sprouts Juices Antifungal Activity

The two brassica sprouts juices, A and B, were in vitro tested to evaluate the fungicidal-fungistatic effect on the pathogen *Verticillium dahliae* (Figure 6). The two juices significantly affect the mycelium growth at all the tested concentrations. The B juice was found more effective in limiting fungal growth, managing to reduce, after 12 days from inoculation, the diameter of the mycelium by 49.4%, when used at the highest concentration tested and by 32.2% at the lowest. The A juice reduced the radial growth of the pathogen by 18.8–23.9%, without significant differences between the three quantities tested.

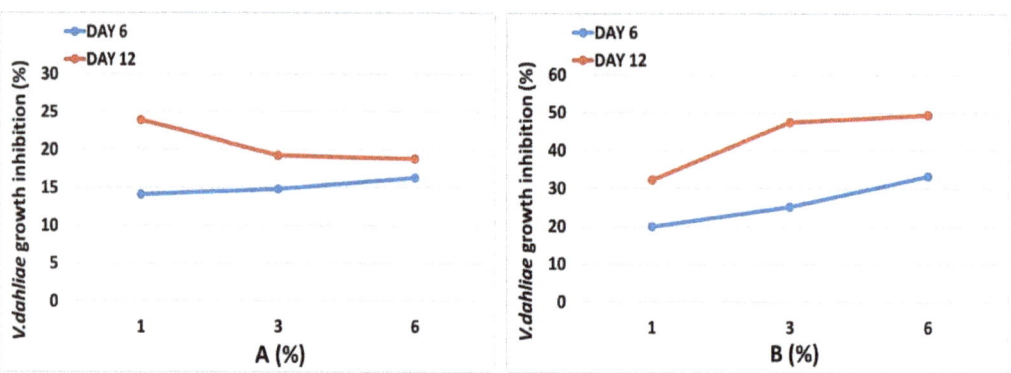

Figure 6. Effect of A and B brassica sprouts juices on the in vitro growth of *Verticillium dahliae* mycelium at 6 and 12 days after inoculation. Three concentrations (volume/volume) have been tested. Values are reported as inhibition percentages as compared to the control ± SD. All the treatments reduced significantly the mycelium growth in comparison with control (*t*-test, $p \leq 0.05$).

3.6. Saponins Phytotoxicity

Phytotoxicity evaluations were carried out on five tomato varieties using the test described in the Materials and Methods, which allows for the measuring of the germinative vigor of seeds treated with saponin extracts in comparison with untreated control seeds. The tests were conducted using the two saponin extracts that have shown the higher antifungal activity, i.e., Sap 6 and Sap 70, at the highest concentrations used to inhibit the mycelial growth of *V. dahliae*.

No significant differences in germination vigor were found between the control seeds and the treated seeds of the five varieties (Figure S2). In some varieties (i.e., Wilson, Mariner and Rossoro) a trend of greater vigor was observed in the treated seeds, although not statistically significant (Figure S2).

3.7. Brassica Sprouts Juices Phytotoxicity

There were no statistically significant differences in germination between control seeds and seeds treated with different concentrations of both types of brassica sprouts juices (Figure 7). Therefore the germination tests did not show any toxic effects of the analyzed extracts at concentrations up to 5000 ppm.

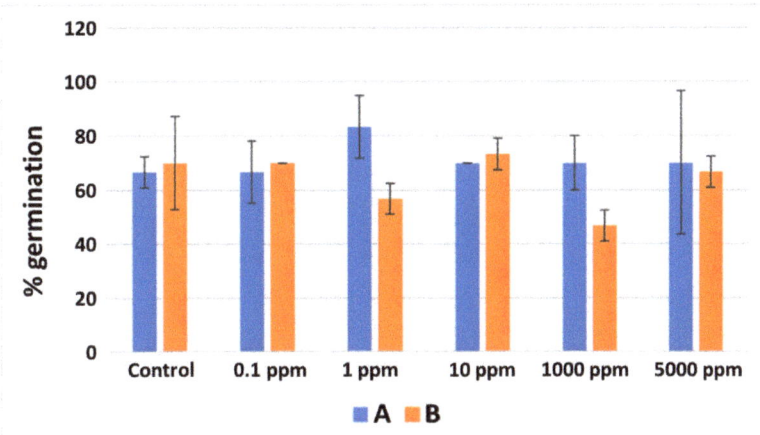

Figure 7. Effect of A and B brassica sprouts juices at the concentrations reported on x axis on the maize seed germination percentages (y axis). No statistically significant differences were detected between control and treatments at all the concentrations tested (Kruskal–Wallis test, $p \leq 0.08$).

4. Discussion

The antimicrobial activity of plant saponins has been long known, having been reported since the initial purifications reported for this class of compounds [6]. Saponins are primarily implicated in the plant defense response to pathogen infections. Their presence in plant roots and leaves provides a chemical barrier to intruding soil-borne and phyllosphere microbes infecting plant tissue [36]. The pathogenic attack in some plants causes the hydrolysis of saponins to derivatives with strong antibiotic activity.

Saponins, as components of exogenous plant defense treatments, have been mainly proposed to control insects affecting crop production and bacterial and fungal pathogens of relevance for human and animal health. Less knowledge is available on the saponins potential as antifungal agents. This study contributed to the increase of such knowledge, finding that saponins extracted from different Medicago species and from oat seeds are effective antifungal agents against *V. dahliae*.

Regarding saponins present in Medicago spp., some studies have found that sapogenins (i.e., saponin aglycones) generally have higher antifungal activity with respect to intact saponins (which are extensively glycosylated), suggesting that the sugar moiety is not important for antimicrobial efficacy. In particular, the biological activities of *M. sativa* and *M. arborea* extracts were found to be related to the content of medicagenic acid [37]. Moreover, the antifungal properties of saponin mixtures from Medicago tops and roots, the corresponding mixtures of prosapogenins from tops, and purified saponins and sapogenins have been successfully evaluated against the fungus *Pyricularia oryzae*. The in vitro trials clearly demonstrated the antifungal effects of prosapogenin mixture from alfalfa tops [38].

Oats, the only saponin-accumulating cereal, store such molecules in leaves and roots (steroidal avenacosides and triterpene avenacins), where they have phytoprotectant activity, as demonstrated by the increased susceptibility to pathogens of avenacin-free mutant genotypes. Oat grains and husks have been reported as being rich in the inactive biological forms of the steroidal avenacosides A and B, which can then be converted to their active forms, 26-desglucoavenacoside A and B, by the action of glycosidases resulting from tissue damage or pathogen attack [39].

Despite the avenacins biosynthetic pathway elucidated in *Avena strigosa* root tissue [40,41], it is still unknown how the expression of genes involved in the pathway impacts on the seed avenacins content.

A pattern of molecular mechanisms has been proposed for saponins antimicrobial activities, including membrane lipid re-arrangement, pores formation, and cell lysis [42]. Recently, several studies suggested that many saponins do not solubilize the lipid layers of biological membranes but fluidize them, causing an increase of their permeability [43].

It is noteworthy that saponins extracted from *Medicago* and from oat seeds impact on fungus life cycle—inhibiting, at least at higher saponin concentrations, the production of fungal spores (Figures 5 and 8). Such peculiarity of saponin treatments can be of particular interest to slow or to stop the vascular colonization of the plant, which occurs when conidia are sucked into the plant into the so-called trapping sites, where they germinate and invade adjacent vessel elements to continue the plant colonization.

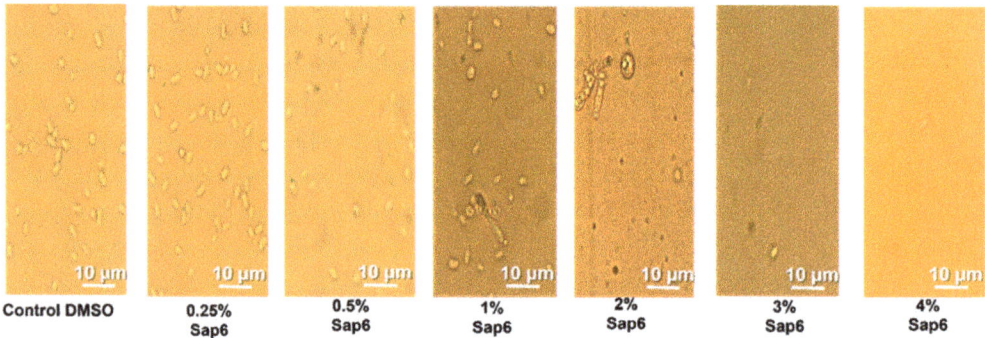

Figure 8. *Verticillium dahliae* mycelium images taken at 400× magnification after 10 days of growth in absence (control) or in presence of increasing Sap6 concentrations.

Brassica juice from purple sprouts has been found significantly more efficient in inhibiting *V. dahliae* mycelium growth in comparison with brassica juice from etiolated sprouts. The juice from purple sprouts is enriched not only with GLSs but also with polyphenols and in particular with anthocyanins, all compounds which may also have antimicrobial and antifungal properties [44]. In common with other flavonoids, certain anthocyanins have demonstrable antiviral, antibacterial, and fungicidal activities. They have the potential, therefore, to protect plants from infections by pathogenic microorganisms. In general, however, the antimicrobial activities of anthocyanins are appreciably less effective than those of other phenolic compounds, such as key flavanols and hydroxycinnamic acids. On the other hand, anthocyanins have not been found to be toxic to any higher animal species. Aphid survival rates, for example, are unaffected by anthocyanins direct chemical defense is unlikely to be a major function of these pigments in plants [45].

Interestingly, the presence of increased levels of anthocyanins and other polyphenols in addition to GLSs in sucrose treated brassica sprouts juices can potentially be one of the reason for its higher protective effect.

5. Conclusions

The overall aim of this work has been to identify plant extracts and mixtures characterized by antifungal activity, low toxicity level and low preparation costs, as a potential application in fungal control in horticultural crops. Brassica sprouts, Medicago tissues, and oat seeds are all low-cost sources of natural antifungals. All tested extracts and mixtures were found effective in inhibiting the growth of the widespread pathogen *V. dahliae* at concentrations that are not toxic for the plant. To minimize the losses caused by the wilt, an integrated approach, i.e., a combination of cultural, chemical, biological, and genetic actions is now proposed [46]. Mazzotta et al. [47] recently proposed to deliver olive leaf extracts to tomato plants using chitosan nanoparticles as a carrier. In this frame, saponins able to reduce mycelium growth but first of all to block conidia formation deserve deeper in vivo evaluation, even in combination with delivery system, e.g., chitosan nanoparticles. Their applicability in open field and greenhouse environments can be proposed in the frame of organic productions.

Supplementary Materials: The following supporting information can be downloaded at: https://www.mdpi.com/article/10.3390/horticulturae8080729/s1, Figure S1. LC-MS analysis of the Avena sativa saponin-enriched extract. (a.) Base peak chromatogram of the saponin fraction eluted from the 70% MeOH wash; (b.) extracted ion chromatogram of m/z 1064.4, annotated as Avenacin A2; (c.) extracted ion chromatogram of m/z 1385.8, yielding two peaks annotated as the putative isomers of Avenacoside D; (d.) extracted ion chromatogram of m/z 1223.7, yielding two peaks annotated as the putative isomers of Avenacoside B; (e.) extracted ion chromatogram of m/z 1061.7, annotated as Avenacoside A by coelution of the corresponding authentic standard; (f.) extracted ion chromatogram of m/z 899.6, yielding two peaks annotated as the putative isomers of Avenacoside C. For details of MS identifications see Table 3; Figure S2. Effect of Sap70 and Sap 6 extracts at antifungal concentrations on tomato (five cultivars) seed germinative vigor. The germination vigor is reported on Y axis and was calculated as reported in Materials and Method section.

Author Contributions: Conceptualization, C.M., A.T., S.B., F.F. and V.T.; methodology, C.M., I.P., R.G., S.D., E.B. and B.F.; resources, V.T., A.T., S.B. and F.F.; data curation, C.M., S.D., S.B., F.S., A.T. and F.F.; writing—original draft preparation, V.T., C.M., S.B. and A.T.; writing—review and editing, F.S. and F.F.; funding acquisition, V.T. All authors have read and agreed to the published version of the manuscript.

Funding: This research was funded by the Italian Ministry of Agriculture and Forestry in the DiBio–BIOPRIME project (Prot. 76381, MiPAAF PQAI I).

Institutional Review Board Statement: Not applicable.

Informed Consent Statement: Not applicable.

Data Availability Statement: Not applicable.

Conflicts of Interest: The authors declare no conflict of interest.

References

1. The EU Green Deal–A Roadmap to Sustainable Economies 2019. Available online: https://www.switchtogreen.eu/the-eu-green-deal-promoting-a-green-notable-circular-economy/ (accessed on 7 July 2022).
2. Alabouvette, C.; Olivain, C.; Steinberg, C. Biological control of plant diseases: The European situation. *Eur. J. Plant Pathol.* **2006**, *114*, 329–341. [CrossRef]
3. Kregiel, D.; Berlowska, J.; Witonska, I.; Antolak, H.; Proestos, C.; Babic, M.; Babic, L.; Zhang, B. Saponin-based, biological-active surfactants from plants. In *Application and Characterization of Surfactants*; Najjar, R., Ed.; InTech: Rijeka, Croatia, 2017; pp. 184–205.
4. Tava, A.; Pecetti, L. Chemical Investigation of Saponins from Twelve Annual Medicago Species and their Bioassay with the Brine Shrimp Artemia salina. *Nat. Prod. Commun.* **2012**, *7*, 837–840. [CrossRef] [PubMed]
5. Bonilla, H.; Carbajal, Y.; Gonzales, M.; Vásquez, V.; López, A. Determinación de la actividad insecticida de la saponina de la quinua (*Chenopodium quinoa*) en larvas de *Drosophila melanogaster*. *Sci. Agropecu.* **2019**, *10*, 39–45. [CrossRef]
6. Zaynab, M.; Sharif, Y.; Abbas, S.; Afzal, M.Z.; Qasim, M.; Khalofah, A.; Ansari, M.J.; Khan, K.A.; Tao, L.; Li, S. Saponin toxicity as key player in plant defense against pathogens. *Toxicon* **2021**, *193*, 21–27. [CrossRef] [PubMed]
7. Rai, S.; Acharya-Siwakoti, E.; Kafle, A.; Devkota, H.P.; Bhattarai, A. Plant-Derived Saponins: A Review of Their Surfactant Properties and Applications. *Sci* **2021**, *3*, 44. [CrossRef]
8. Osbourn, A.E. Saponins in cereals. *Phytochemistry* **2003**, *62*, 1–4. [CrossRef]
9. Sparg, S.G.; Luce, M.E.; Van Staden, J. Biological activities and distribution of plant saponins. *J. Ethnopharmacol.* **2004**, *94*, 219–243. [CrossRef]
10. Blažević, I.; Montaut, S.; Burčul, F.; Olsen, C.E.; Burow, M.; Rollin, P.; Agerbirk, N. Glucosinolate structural diversity, identification, chemical synthesis and metabolism in plants. *Phytochemistry* **2020**, *169*, 112100. [CrossRef]
11. Björkman, M.; Klingen, I.; Birch, A.N.E.; Bones, A.M.; Bruce, T.J.A.; Johansen, T.J.; Meadow, R.; Mølmann, J.; Seljåsen, R.; Smart, L.E.; et al. Phytochemicals of Brassicaceae in plant protection and human health—Influences of climate, environment and agronomic practice. *Phytochemistry* **2011**, *72*, 538–556. [CrossRef]
12. Possenti, M.; Baima, S.; Raffo, A.; Durazzo, A.; Giusti, A.M.; Natella, F. Glucosinolates in food. In *Glucosinolates, Reference Series in Phytochemistry*; Merillon, J.M., Ramawat, K.G., Eds.; Springer International Publishing: Cham, Switzerland, 2017; pp. 87–132.
13. Ilahy, R.; Tlili, I.; Pék, Z.; Montefusco, A.; Siddiqui, M.W.; Homa, F.; Hdider, C.; R'Him, T.; Lajos, H.; Lenucci, M.S. Pre- and post-harvest factors affecting glucosinolate content in broccoli. *Front. Nutr.* **2020**, *7*, 147. [CrossRef]

14. Jahangir, M.; Abdel-Farid, I.B.; Kim, H.K.; Choi, Y.H.; Verpoorte, R. Healthy and unhealthy plants: The effect of stress on the metabolism of *Brassicaceae*. *Environ. Exp. Bot.* **2009**, *67*, 23–33. [CrossRef]
15. Baenas, N.; Garcia-Viguera, C.; Moreno, D.A. Elicitation: A tool for enriching the bioactive composition of foods. *Molecules* **2014**, *19*, 13541–13563. [CrossRef] [PubMed]
16. Fahey, J.W.; Zhang, Y.; Talalay, P. Broccoli sprouts: An exceptionally rich source of inducers of enzymes that protect against chemical carcinogens. *Proc. Natl. Acad. Sci. USA* **1997**, *94*, 10367–10372. [CrossRef]
17. Guo, R.; Yuan, G.; Wang, Q. Sucrose enhances the accumulation of anthocyanins and glucosinolates in broccoli sprouts. *Food Chem.* **2011**, *129*, 1080–1087. [CrossRef]
18. Guo, R.; Yuan, G.; Wang, Q. Effect of sucrose and mannitol on the accumulation of health-promoting compounds and the activity of metabolic enzymes in broccoli sprouts. *Sci. Hortic.* **2011**, *128*, 159–165. [CrossRef]
19. Natella, F.; Maldini, M.; Nardini, M.; Azzini, E.; Foddai, M.S.; Giusti, A.M.; Baima, S.; Morelli, G.; Scaccini, C. Improvement of the nutraceutical quality of broccoli sprouts by elicitation. *Food Chem.* **2016**, *201*, 101–109. [CrossRef]
20. Ferruzza, S.; Natella, F.; Ranaldi, G.; Murgia, C.; Rossi, C.; Trošt, K.; Mattivi, F.; Nardini, M.; Maldini, M.; Giusti, A.M.; et al. Nutraceutical improvement increases the protective activity of broccoli sprout juice in a human intestinal cell model of gut inflammation. *Pharmaceuticals* **2016**, *9*, 48. [CrossRef] [PubMed]
21. Masci, A.; Mattioli, R.; Costantino, P.; Baima, S.; Morelli, G.; Punzi, P.; Giordano, C.; Pinto, A.; Donini, L.M.; d'Erme, M.; et al. Neuroprotective effect of *Brassica oleracea* sprouts crude juice in a cellular model of alzheimer's disease. *Oxid. Med. Cell. Longev.* **2015**, *2015*, 781938. [CrossRef]
22. Rubattu, S.; Di Castro, S.; Cotugno, M.; Bianchi, F.; Mattioli, R.; Baima, S.; Stanzione, R.; Madonna, M.; Bozzao, C.; Marchitti, S.; et al. Protective effects of *Brassica oleracea* sprouts extract toward renal damage in high-salt-fed SHRSP: Role of AMPK/PPARα/UCP2 axis. *J. Hypertens.* **2015**, *33*, 1465–1479. [CrossRef]
23. Rubattu, S.; Stanzione, R.; Bianchi, F.; Cotugno, M.; Forte, M.; Della Ragione, F.; Fioriniello, S.; D'Esposito, M.; Marchitti, S.; Madonna, M.; et al. Reduced brain UCP2 expression mediated by microRNA-503 contributes to increased stroke susceptibility in the high-salt fed stroke-prone spontaneously hypertensive rat. *Cell Death Dis.* **2017**, *8*, 2891. [CrossRef]
24. Hu, X.; Puri, K.D.; Gurung, S.; Klosterman, S.J.; Wallis, C.M.; Britton, M.; Durbin-Johnson, B.; Phinney, B.; Salemi, M.; Short, D.P.G.; et al. Proteome and metabolome analyses reveal differential responses in tomato-*Verticillium dahliae*-interactions. *J. Proteom.* **2019**, *207*, 103449. [CrossRef] [PubMed]
25. Subbarao, K. *Verticillium dahliae* (Verticillium wilt). In *Invasive Species Compendium*; CABI: Wallingford, UK, 2020. [CrossRef]
26. Tava, A.; Mella, M.; Avato, P.; Argentieri, M.P.; Bialy, Z.; Jurzysta, M. Triterpenoid glycosides from the leaves of *Medicago arborea* L. *J. Agric. Food Chem.* **2005**, *53*, 9954–9965. [CrossRef] [PubMed]
27. Tava, A.; Pecetti, L.; Romani, M.; Mella, M.; Avato, P. Triterpenoid glycosides from the leaves of two cultivars of *Medicago polymorpha* L. *J. Agric. Food Chem.* **2011**, *59*, 6142–6149. [CrossRef]
28. Tava, A.; Avato, P. Chemical and biological activity of triterpene saponins from *Medicago* species. *Nat. Prod. Commun.* **2006**, *1*, 1159–1180. [CrossRef]
29. Tava, A.; Biazzi, E.; Mella, M.; Quadrelli, P.; Avato, P. Artefact formation during acid hydrolysis of saponins from *Medicago* spp. *Phytochemistry* **2017**, *238*, 116–127. [CrossRef]
30. Abdul-Baki, A.A.; Anderson, J.D. Vigor Determination in Soybean Seed by Multiple Criteria. *Crop Sci.* **1973**, *13*, 630–633. [CrossRef]
31. Crombie, W.M.L.; Crombie, L. Distribution of avenacins A-1, A-2, B-1 and B-2 in oat roots: Their fungicidal activity towards "take-all" fungus. *Phytochemistry* **1984**, *25*, 2069–2073.
32. Hu, C.; Sang, S. Triterpenoid saponins in oat bran and their levels in commercial oat products. *J. Agric. Food Chem.* **2020**, *68*, 6381–6389. [CrossRef]
33. Yang, J.; Wang, P.; Wu, W.; Zhao, Y.; Idehen, E.; Sang, S. Steroidal Saponins in Oat Bran. *J. Agric. Food Chem.* **2016**, *64*, 1549–1556. [CrossRef] [PubMed]
34. Tschesche, R.; Tauscher, M.; Fehlhaber, H.W.; Wulff, G. Avenacosid A, ein bisdesmosidisches Steroidsaponin aus Avena sativa. *Chem. Ber.* **1969**, *102*, 2072–2082. [CrossRef]
35. Pecio, Ł.; Wawrzyniak-Szołkowska, A.; Oleszek, W.; Stochmal, A. Rapid analysis of avenacosides in grain and husks of oats by UPLC-TQ-MS. *Food Chem.* **2013**, *141*, 2300–2304. [CrossRef] [PubMed]
36. Osbourn, A.; Goss, R.J.M.; Field, R.A. The saponins: Polar isoprenoids with important and diverse biological activities. *Nat. Prod. Rep.* **2011**, *28*, 1261–1268. [CrossRef] [PubMed]
37. Avato, P.; Bucci, R.; Tava, A.; Vitali, C.; Rosato, A.; Bialy, Z.; Jurzysta, M. Antimicrobial activity of saponins from Medicago sp.: Structure activity relationship. *Phytother. Res.* **2006**, *20*, 454–457. [CrossRef]
38. Abbruscato, P.; Tosi, S.; Crispino, L.; Biazzi, E.; Menin, B.; Picco, A.M.; Pecetti, L.; Avato, P.; Tava, A. Triterpenoid glycosides from *Medicago sativa* as antifungal agents against *Pyricularia oryzae*. *J. Agric. Food Chem.* **2014**, *62*, 11030–11036. [CrossRef]
39. Raguindin, P.F.; Itodo, O.A.; Stoyanov, J.; Dejanovic, G.M.; Gamba, M.; Asllanaj, E.; Minder, B.; Bussler, W.; Metzger, B.; Muka, T.; et al. A systematic review of phytochemicals in oat and buckwheat. *Food Chem.* **2021**, *338*, 127982. [CrossRef]
40. Kemen, A.C.; Honkanen, S.; Melton, R.E.; Findlay, K.C.; Mugford, S.T.; Hayashi, K.; Haralampidis, K.; Rosser, S.J.; Osbourna, A. Investigation of triterpene synthesis and regulation in oats reveals a role for β-amyrin in determining root epidermal cell patterning. *Proc. Natl. Acad. Sci. USA* **2014**, *111*, 8679–8684. [CrossRef] [PubMed]

41. Leveau, A.; Reed, J.; Qiao, X.; Stephenson, M.J.; Mugford, S.T.; Melton, R.E.; Rant, J.C.; Vickerstaff, R.; Langdon, T.; Osbourn, A. Towards take-all control: A C-21β oxidase required for acylation of triterpene defence compounds in oat. *New Phytol.* **2019**, *221*, 1544–1555. [CrossRef]
42. Coleman, J.J.; Okoli, I.; Tegos, G.P.; Holson, E.B.; Wagner, F.F.; Hamblin, M.R.; Mylonakis, E. Characterization of plant-derived saponin natural products against *Candida albicans*. *ACS Chem. Biol.* **2010**, *5*, 321–332. [CrossRef]
43. Wojciechowski, K.; Jurek, I.; Góral, I.; Campana, M.; Geue, T.; Gutberlet, T. Surface-active extracts from plants rich in saponins—Effect on lipid mono-and bilayers. *Surf. Interfaces* **2021**, *27*, 101486. [CrossRef]
44. Singh, S.; Kaur, I.; Kariyat, R. The multifunctional roles of polyphenols in plant-herbivore interactions. *Int. J. Mol. Sci.* **2021**, *22*, 1442. [CrossRef]
45. Schaefer, H.M.; Rentzsch, M.; Breuer, M. Anthocyanins reduce fungal growth in fruits. *Nat. Prod. Commun.* **2008**, *3*, 1267–1272. [CrossRef]
46. Acharya, B.; Ingram, T.W.; Oh, Y.; Adhikari, T.B.; Dean, R.A.; Louws, F.J. Opportunities and challenges in studies of host-pathogen interactions and management of *Verticillium dahliae* in tomatoes. *Plants* **2020**, *9*, 1622. [CrossRef] [PubMed]
47. Mazzotta, E.; Muzzalupo, R.; Chiappetta, A.; Muzzalupo, I. Control of the Verticillium Wilt on tomato plants by means of olive leaf extracts loaded on chitosan nanoparticles. *Microorganisms* **2022**, *10*, 136. [CrossRef] [PubMed]

Article

Multi-Parameter Characterization of Disease-Suppressive Bio-composts from Aromatic Plant Residues Evaluated for Garden Cress (*Lepidium sativum* L.) Cultivation

Catello Pane [1,*], Riccardo Spaccini [2], Michele Caputo [1], Enrica De Falco [3] and Massimo Zaccardelli [1]

[1] Centro di Ricerca Orticoltura e Florovivaismo, Consiglio per la Ricerca in Agricoltura e l'Analisi dell'Economia Agraria (CREA), Via Cavalleggeri 25, 84098 Pontecagnano Faiano, Italy; caputomichele1986@libero.it (M.C.); massimo.zaccardelli@crea.gov.it (M.Z.)

[2] Dipartimento di Agraria (DIA), Università di Napoli Federico II, Via Università 100, 80055 Portici, Italy; riccardo.spaccini@unina.it

[3] Dipartimento di Farmacia, Università degli Studi di Salerno, Via Giovanni Paolo II 132, 84084 Fisciano, Italy; edefalco@unisa.it

* Correspondence: catello.pane@crea.gov.it

Citation: Pane, C.; Spaccini, R.; Caputo, M.; De Falco, E.; Zaccardelli, M. Multi-Parameter Characterization of Disease-Suppressive Bio-composts from Aromatic Plant Residues Evaluated for Garden Cress (*Lepidium sativum* L.) Cultivation. *Horticulturae* **2022**, *8*, 632. https://doi.org/10.3390/horticulturae8070632

Academic Editor: Miguel de Cara-García

Received: 22 June 2022
Accepted: 12 July 2022
Published: 13 July 2022

Publisher's Note: MDPI stays neutral with regard to jurisdictional claims in published maps and institutional affiliations.

Copyright: © 2022 by the authors. Licensee MDPI, Basel, Switzerland. This article is an open access article distributed under the terms and conditions of the Creative Commons Attribution (CC BY) license (https://creativecommons.org/licenses/by/4.0/).

Abstract: Garden cress is a vegetable crop in the Brassicaceae family that is appreciated for its nutraceutical and taste-giving components in minimally processed food chains. Due to its very short cycle, which depends on the range of production from microgreens to baby-leaf vegetables, this crop is threatened by soil-borne pathologies developing within the initial stages of germination and emergence. This study aims to evaluate the suppressive bio-compost as an innovative means to counteract the main telluric diseases of garden cress and reduce the risks of yield loss by adopting sustainable remedies and decreasing the dependence on synthetic fungicides. Therefore, eleven green composts obtained using both previously distilled and raw aromatic plant residues were analyzed for suppressive properties against *Rhizoctonia solani* and *Sclerotinia sclerotiorum* on sown garden cress. The biological active component of the composts, detected by CO_2-release, FDA-hydrolysis and microbial counts, proved to be indispensable for pathogen control in vitro and in vivo, as demonstrated by the loss of suppressiveness after sterilization. Cross-polarization magic angle spinning ^{13}C-nuclear magnetic resonance (CP-MAS-^{13}C-NMR) was used to analyze the molecular distribution of organic C in composts. The results indicated the suitability of the feedstock used to make quality compost. The suppression levels shown by composts P1 (40% wood chips, 30% escarole and 30% a mixture of sage, basil, mint and parsley) and P2 (40% wood chips, 30% escarole and 30% a mixture of essential oil-free sage, basil and rosemary) are promising for the sustainable, non-chemical production of garden cress vegetables.

Keywords: biological control; NMR; on-farm composts; organic carbon; *Rhizoctonia solani*; *Sclerotinia sclerotiorum*

1. Introduction

Garden cress (*Lepidium sativum* L.) is an annual cruciferous herb cultivated in many temperate areas of the world and prized as a leafy vegetable and a microgreen for the ready-to-eat functional food sector [1,2]. The interest in this herb is due to its distinctive taste and richness in health-beneficial phytonutrients, including antioxidants, phenolic compounds and glucosinolates [3]. Its cultivation is characterized by very short cycles that end in the order of 10–20 days after sowing with the cutting of the fresh product, which can be followed by minimal processing, bagging and distribution [4,5]. In the microgreens chain, the time is shortened even further [6]. With a view to immediate consumption, it is therefore necessary to adopt cultivation systems and management protocols, particularly for phytosanitary aspects, that minimize or even exclude the use of synthetic chemicals. On the other hand, the cultivation of garden cress at the ground level is very vulnerable to polyphagous telluric pathogens that can proliferate in sick soils and/or intensive systems [7].

The use of suppressive compost can be a very advantageous strategy in the cultivation of this vegetable, both in terms of sustainability and quality. A quality compost with its biological (microbial community and excreta) and/or physicochemical (supramolecular structures and compounds) components can hinder disease development by interfering in pathogenesis through some key mechanisms schematically referred to as biological control, the induction of resistance and direct antifungal activity [8,9]. Therefore, the use of a highly suppressive compost in the production of baby-leaf vegetables and/or microgreens can drastically reduce the dependence on synthetic fungicides and promote healthy plant development [10]. Suppressiveness adds to the advantages of using compost in cress cultivation, e.g., nutrient supply and the stimulation of the synthesis of functional metabolites that improve organoleptic quality [11].

Over the years, many studies have been conducted to investigate the main mechanisms underlying compost suppressiveness, establishing the primacy of the microbiological component that is selected during the long composting process through the continuous feedback between the transformation of organic matrices by the microorganisms and the conditioning of the microbiota structure induced by the physicochemical nature of the organic matter [12]. Many of these studies have used garden cress as a model plant to characterize compost suppressiveness [12,13] (see Table 1), and, even earlier, it proved to be particularly suitable for testing phytotoxicity [14], demonstrating the potential for use in productive systems as well.

Table 1. Percentage of control efficacy showed by many composts tested against soil-borne diseases of cress in previous studies. The asterisk indicates statistical significance ($p < 0.05$); ns indicates not significant, as compared to the infected reference.

Compost Source	Turning	Pathogen	Control (%)	Reference
Viticulture and enological factory residues		Pythium ultimum	23 *	[13]
		Rhizoctonia solani	42 *	
		Sclerotinia minor	0	
Organic fraction of differentiated municipal bio-waste		P. ultimum	53 *	[13]
		R. solani	0	
		S. minor	54 *	
Organic fraction of undifferentiated municipal bio-waste		P. ultimum	26 *	[13]
		R. solani	0	
		S. minor	38 *	
Cow manure		P. ultimum	47 *	[13]
		R. solani	60 *	
		S. minor	36 *	
Organic fraction of differentiated biowaste + peat 50% v/v		P. ultimum	53 *	[13]
		R. solani	0	
		S. minor	19 ^{ns}	
Tomato, escarole, woodchip, compost 17.5:15.5:65:2 w/w	FV	R. solani	77 *	[12]
		S. minor	61 *	
Tomato, woodchip, compost 50:48:2 w/w	FV	R. solani	55 *	[12]
		S. minor	29 ^{ns}	
Artichoke, woodchip, compost 78:20:2 w/w	FV	R. solani	46 *	[12]
		S. minor	48 *	
Artichoke, fennel, escarole, woodchip, compost 43.5:23.5:11:20:2 w/w	FV	R. solani	39 *	[12]
		S. minor	36 ^{ns}	

Table 1. Cont.

Compost Source	Turning	Pathogen	Control (%)	Reference
Urban waste		R. solani S. minor	0 0	[12]
Chestnut leaves, branches, bark, and hulls		R. solani S. minor	45 * 62 *	[15]
Solid digestate		R. solani S. minor	1 ^{ns} 53 *	[16]
Escarole, crop cardoon, compost 80:18:2 w/w	PA	R. solani S. minor	36 * 66 *	[17]
Escarole, crop cardoon, compost 80:18:2 w/w	FV	R. solani S. minor	42 * 69 *	[17]
Escarole, crop cardoon, compost 80:18:2 w/w	TH	R. solani S. minor	15 ^{ns} 32 *	[17]
Leafy vegetables, fennel, woodchip	MT→FV	R. solani S. minor	56 * 84 *	[18]
Maize, livestock waste, woodchip	MT→FV	R. solani S. minor	0 2 ^{ns}	[18]
Leafy vegetables, basil, tomato, watermelon, woodchip	MT→FV	R. solani S. minor	39 * 65 *	[18]
Leafy vegetables, basil, watermelon, woodchip	MT→FV	R. solani S. minor	36 * 95 *	[18]
Leafy vegetables, basil, pumpkin, woodchip	MT→FV	R. solani S. minor	0 11 ^{ns}	[18]
Leafy vegetables, basil, woodchip	MT→FV	R. solani S. minor	0 36 ^{ns}	[18]
Leafy vegetables, basil, watermelon, woodchip	MT→FV	R. solani S. minor	43 * 18 ^{ns}	[18]
Leafy vegetables, basil, woodchip	MT→FV	R. solani S. minor	48 * 95 *	[18]
Leafy vegetables, basil, woodchip	MT→FV	R. solani S. minor	57 * 0	[18]
Leafy vegetables, basil, pumpkin, woodchip	MT→FV	R. solani S. minor	0 72 *	[18]
Leafy vegetables, artichoke, woodchip	MT→FV	R. solani S. minor	52 * 60 *	[18]
Leafy vegetables, cabbage, walnut husk, woodchip	MT→FV	R. solani S. minor	40 * 72 *	[18]
Leafy vegetables, basil, sorghum, tomato, pumpkin, woodchip	MT→FV	R. solani S. minor	45 * 86 *	[18]
Solid digestate + chips 15:83.3 w/w		R. solani S. minor	12 ^{ns} 59 *	[19]

Composting is proposed as a circular economy process aimed at valorizing agro-industrial organic waste through recycling and reuse in new crop cycles [20]. The systematic production of multifunctional composts with suppressive and biostimulating properties is a new perspective for the development of the sector, which can be found in plant materials with a well-defined phytochemical potential, a valuable resource. This may be the case in the aromatic plant sector, which can produce a large amount of noble waste for this purpose, both from the cultivation and dressing stages. This residual aromatic biomass can continue its productive life as a co-product of the distillation of essential oils, the production of aromatic waters and, then, through just composting [21].

The present study aims to evaluate the suppression in a collection of 11 composts produced from aromatic plant waste against two telluric diseases of cress: Rhizoctonia and Sclerotinia damping-off. By characterizing the microbiological and chemical properties and the molecular distribution of organic C in the composts, inferences are made about the mechanisms underlying disease suppression.

2. Materials and Methods

2.1. Bio-composts

In this study, eleven composts (P1–11) obtained by on-farm composting (singly or in complex) different feedstocks available in the aromatic plant chain—in a static pile periodically turned by hand, as previously illustrated by [22]—were used. The starting materials are listed in Table 2; they include both raw and previously hydrodistilled feedstock [21]. These composts were available as representative of a possible combination of biomasses from the herb sector included in valuable circular economy processes.

Table 2. Main compost feedstock used to produce P1–P11 composts.

Compost	Feedstock			
	Wood Chips	Vegetable Residues	Aromatic Plant Material	
P1	40%	30% Escarole (*Cichorium endivia* L.)	30% Mixture of Sage (*Salvia officinalis* L.), Basil (*Ocimum basilicum* L.), Mint (*Mentha x piperita* L.) and Parsley (*Petroselinum crispum* (Mill.) Fuss)	
P2	40%	30% Escarole	30% Mixture of Sage, Basil and Rosemary (*Rosmarinus officinalis* L.)	Distilled
P3	20%	50.5% Parsley 6.2% Rocket (*Diplotaxis tenuifolia* L.) 3.6% Red radish (*Raphanus sativus* L.)	29.9% Basil, 6.2% Thyme (*Thymus vulgaris* L.), 2% Laurel (*Laurus nobilis* L.), 1.6% Mint	Distilled
P4	-		100% Basil	
P5	-		100% Basil	Distilled
P6	20%		50% Parsley, 28% Thyme, 8.1% Rosemary, 5.5% Mint, 3.8% Oregano (*Origanum vulgare* L.), 2.5% Sage, 0.9% Laurel, 0.7% Tarragon (*Artemisia dracunculus* L.), 0.5% Basil	
P7	20%	17.6% Parsley	39.3% Basil, 19.7% Rosemary, 16.5% Sage, 6.9% Mint	Distilled
P8	-		100% Rosemary	
P9	-		100% Rosemary	Distilled
P10	-		100% Sage	
P11	-		100% Sage	Distilled

2.2. In Planta Compost Suppressiveness Assay

The compost suppression bioassay was conducted against the Rhizoctonia and Sclerotinia diseases of *L. sativum*, as was described previously by Pane et al. [17]. In particular, the experimental setup is summarized here:

- The eleven bio-composts, both sterile (twice autoclaved) and not sterile, were supplied at a rate of 30% (vol.) to a standard peat-based growing medium. Non-amended peat was used as a control.
- Isolates of *Rhizoctonia solani* (AG-4) and *Sclerotinia sclerotiorum* from the CREA microbial collection (at the Research Center for Vegetables and Ornamental Crops in

Pontecagnano Faiano, Italy), maintained on a potato dextrose agar medium (PDA, Oxoid Ltd., Basingstoke, UK), were grown for 21 days on common millet seeds saturated with potato dextrose broth (PDB, Oxoid Ltd., Basingstoke, UK) (1/10 w/w) to prepare the pathogen inoculum to be incorporated into the substrate at a final concentration of 1% (w/w, dry weight).

- The experimental unit consisted of a mini plastic pot (7 cm diam., ~0.1 L vol.) filled with the different substrates, sown with 20 seeds of garden cress cv. Comune (Blumen, Milan, Italy) each and replicated five times for each treatment. Ultimately, there were 11 composts ×2 conditions (autoclaved and non-autoclaved) +2 controls (healthy and infected peat) for a total of 24 treatments, resulting in 120 pots and 2400 seeds. The experiment was repeated.
- The sown pots were placed in a climate chamber (25 °C) for 7 days to allow for the emergence of cress seedlings and the development of Sclerotinia and Rhizoctonia damping off (Figure 1).

Figure 1. Photograph of healthy (**A**) and Rhizoctonia- (**B**) and Sclerotinia-diseased (**C**) cress seedlings sown in pots on autoclaved peat. Details of the damping-off by *Rhizoctonia solani* (**D**) and the wilting of cress seedlings covered by *Sclerotinia sclerotiorum* mold (**E**).

The incidence of Sclerotinia and Rhizoctonia diseases on the cress was recorded as the percentage of damping-off (DO%), according to Equation (1):

$$\text{DO\%} = \frac{HSo - HSi}{HSo} \times 100 \qquad (1)$$

where HSo and HSi are the number of healthy seedlings in the non-amended control and in the ith compost mix, respectively.

2.3. *Analysis of the Main Physico-Chemical and Biological Components of Compost*

The electrical Conductivity and pH of the compost were determined according to the official methods of the Italian National Society of Soil Science [23]. The nitrate content in the compost was assessed by a colorimetric technique using Reflectoquant® strips read by a RQflex® 10 reflectometer (Merck, Darmstadt, Germany).

The phytotoxicity assay was carried out by assessing the germination rate of the cress after the exposure of 20 seeds to 4 mL of aqueous compost extracts at three different concentrations (50, 16.6 and 5 g L^{-1} of compost, dry weight). They were placed onto blotting paper in Petri plates and incubated at 25 °C for 5 days. After incubation, the number and root elongation of the seedlings were calculated in the germination index (GI%) according to Formula (1).

$$\text{GI\%} = \frac{(N° \text{ } Si)}{(N° \text{ } So)} \times \frac{(RL \text{ } Si)}{(RL \text{ } So)} \times 100 \qquad (2)$$

where the number ($N°$) and the mean root length (RL) of the germinated seedlings in both the water control (So) and the ith compost eluate (Si) are taken in account, respectively.

The population levels of the filamentous fungi and the total, spore-forming and pseudomonads-like bacteria in the composts were assessed by the plate counting of tenfold (10^{-1} to 10^{-7}) serial dilutions of water suspensions. Fungal colonies were grown on PDA pH 6 and supplemented with 150 mg L^{-1} of nalidixic acid and 150 mg L^{-1} of streptomycin. The total bacteria were counted on a selective medium (glucose 1 g L^{-1}, proteose peptone 3 g L^{-1}, yeast extract 1 g L^{-1}, K_2PO_4 1 g L^{-1}, agar 15 g L^{-1}) supplemented with actidione 100 mg L^{-1}. *Peudomonads* were counted on a selective agar medium without iron that was supplemented with actidione [24]. Finally, spore-forming bacteria were counted on Nutrient Agar [25] previously heated at 90 °C for 10 min.

Basal respiration was expressed as the CO_2 release rate of 10 g (dry weight) of compost at an 80% water holding capacity in a sealed 50 mL sterile plastic tube (Falcon, Oxnard, CA, USA), as measured with the CO_2 Analyser IRGA SBA-4 OEM (PP Systems, Haverhill, MA, USA).

The rate of hydrolysis of the Fluorescein diacetate (FDA) from the compost (2.5 g) mixed with 0.2 M potassium phosphate buffered at pH 7.6 (15 mL) and then supplemented with 0.5 mL FDA solution (2 mg mL^{-1}) was assessed after 2 h of dynamic incubation and after stopping the reaction by the addition of 15 mL $CHCl_3/CH_3OH$ (2:1 vol.) with a UV-Vis spectrophotometer (model UV-1800, Shimadzu, Canby, OR, USA) at 490 nm.

2.4. Analysis of the Molecular Carbon Components of Compost

The finely powdered compost samples were analyzed by solid-state NMR spectroscopy (^{13}C CPMAS NMR) on a Bruker AV300 Spectrometer equipped with a 4 mm wide-bore MAS probe by packing the homogenized organic substrates in 4 mm zirconium rotors with Kel-F caps. The technical parameters for the NMR acquisition were set as follows: a rotor spin rate of 13,000 Hz, 2 s of recycle time, 1 ms of contact time, 30 ms of acquisition time and 4000 scans. A conventional composite shaped "ramp" pulse on the 1H channel was used for the cross-polarization pulse sequence to account for the inhomogeneity of the Hartmann–Hann condition at a high rotor spin frequency. The Fourier transform was performed with a 4k data point and an exponential apodization of 200 Hz of line broadening.

Although the solid-state ^{13}C CPMAS NMR is a powerful technique for the direct investigation of natural organic matter, the analysis of complex solid matrices involves the *occurrence of unavoidable technical drawbacks, such as chemical shield anisotropy and* dipolar coupling effects, with a loss in signal resolution. Solid-state NMR spectra are therefore characterized by a large signal broadening and an overlapping of carbon functionalities, partly addressed by the application of magic angle spinning and high-power ^1H decoupling. Therefore, for the interpretation of solid state ^{13}C NMR spectra, the different signals are conventionally grouped into six extended chemical shift regions that are representative of the main types of carbon functional groups: Alkyl-C: 0–45 ppm; Methoxyl-C: 45–60 ppm; O-Alkyl-C: 60–110 ppm; Aryl-C: 110–145 ppm; Phenol-C: 145–160 ppm; and Carboxyl-C: 190–160 ppm.

The relative contribution of each spectral region was determined by integration (MestreNova 6.2.0 software, [26]) and expressed as a percentage of the total area. The molecular features of organic substrates can be summarized by calculating the dimensionless structural index [17,27]. The Alkyl ratio (A/OA) compares the relative intensity between Alkyl-C and O-Alkyl-C (3):

$$A/OA = \frac{(0-45 \text{ ppm})}{(60-110 \text{ ppm})} \quad (3)$$

The Hydrophobic Index (HB) corresponds to the comparison of hydrophobic apolar C functionalities and potentially more hydrophilic polar functional groups (4):

$$HB = \frac{\left[(0-45 \text{ ppm}) + \frac{(45-60 \text{ ppm})}{2} + (110-160 \text{ ppm})\right]}{\left[\frac{(45-60 \text{ ppm})}{2} + (60-110 \text{ ppm}) + (160-190 \text{ ppm})\right]} \quad (4)$$

The Lignin ratio (LigR) is based on the relation between the C distribution in Methoxyl-C and the C-N region, as related to the O-Aryl-C components (5):

$$LigR = \frac{(45-60 \text{ ppm})}{(145-160 \text{ ppm})} \quad (5)$$

The A/OA and HB indices are mainly used as references to determine the biochemical stability of organic matrices and correlate the structural composition with the stabilization processes of organic materials [28,29]. The Lig R ratio is a useful indicator to discriminate between NMR signals due to lignin and other phenolic compounds (lower LigR) versus the prevalent contribution of peptidic moieties (larger LigR) in the 45–60 ppm range of the NMR spectra [17,30].

2.5. Statistical Analysis

All of the measured parameters were subjected to descriptive statistics. A two-way ANOVA was applied to the results of the *in planta* suppression assays for each pathogen to test the effects of the compost sample and sterilization treatment on the damping-off percentage and to the phytotoxicity assay to test the effects of the compost sample and eluate concentration on the cress germination index. The percentage data were arcsine transformed to satisfy the normality assumption of distribution. A one-way ANOVA was used to test for differences in the physico-chemical, chemical and microbiological characteristics among the samples.

3. Results

3.1. Compost Suppressiveness

The composts showed varying levels of suppressiveness against the two target soil-borne diseases (Figure 2).

All of the raw composts were able to significantly ($p < 0.001$) reduce the damping-off caused by *S. sclerotiorum* on the cress in pot trials compared to the non-amended pots. Furthermore, with the exception of P4 and P5, the composts kept the disease severity below 50%; P1 was the most suppressive, sharing a statistically comparable level of suppression with a large group including P2 and P6 to P11. On the other hand, although Rhizoctonia damping-off was significantly ($p < 0.001$) reduced in 7 out of 11 cases compared to the control peat, the suppression was very low: only about 25% of the disease control capacity, except for P1 and P2, which proved to be the best performing raw composts, achieving over 60% disease control. The sterilization of the P1 and P2 composts drastically cancelled out suppression. Similarly, the sterilized composts P4, P5, P6, P8 and P9 significantly ($p < 0.01$) lost the suppressiveness that was expressed when they were raw. Indeed, the two-way ANOVA showed significant ($p < 0.01$) compost × sterilization interaction.

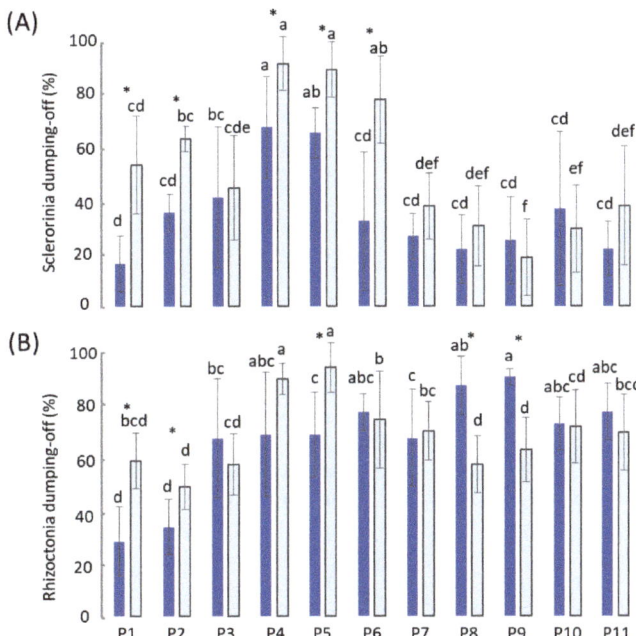

Figure 2. Percentage of Sclerotinia (**A**) and Rhizoctonia (**B**) damping-off on cress seedlings in raw (dark blue bars) and autoclaved (light blue bars) composts diluted into peat. Lowercase lettering indicates significant differences ($p \leq 0.05$), according to the Least Significant Difference post hoc test, among both the raw and autoclaved samples. Asterisk indicates significant differences ($p \leq 0.05$), according to the Least Significant Difference post hoc test, between the raw and autoclaved samples of the same compost.

3.2. Phyitotoxicity and Biological Properties of Composts

The cress germination test on the compost eluates showed the profiles of their phytotoxicity potential with, on average, a concentration-dependent behavior (Figure 3). Statistics showed that the effect of the single factor, compost sample or concentration was significant ($p < 0.001$), whereas no significant interaction was found between them (compost × concentration). Composts P4 and P5 showed the lowest toxicity, as expressed by the germination index percentage, at the highest concentration analyzed. In parallel, the eluates of P3, P6 and P7 showed intermediate values of the parameter between 50 and 100%. Passing to the remaining composts, on average, lower germination index percentages were observed.

Table 3 shows the values of the main biological parameters measured on the composts. The largest levels of FDA hydrolysis were recorded in P4 and P6, followed by P1, P2, P5 and P7, which, together, form the top cluster regarding this enzymatic activity. These behaviors were not confirmed by the assessment of basal respiration, as the rate of CO_2 release showed no significant differences among the samples. The population levels of total fungi were statistically highest in P8 and P11, followed by P9, while the remaining composts showed lower enumerations of culturable colonies. P9 showed the highest values of the total bacterial population, followed by P5, P10 and P11. Heat-resistant bacteria were significantly more numerous in composts P1, P2 and P4, while Pseudomonas-like bacteria were more numerous in P8, P9 and P10.

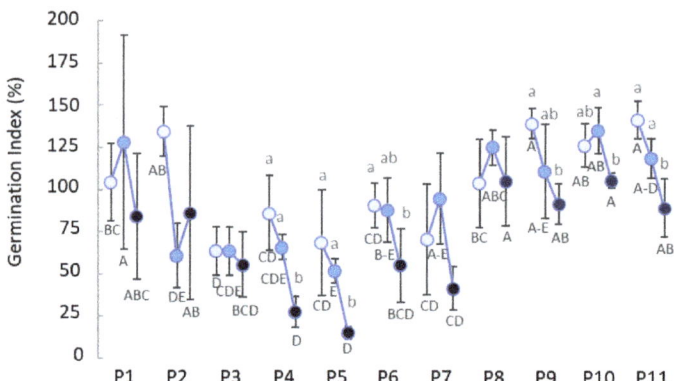

Figure 3. Germination index of cress seeds exposed to compost water extracts (P1 to P11) at 50 (dark blue), 16.6 (medium blue) and 5 g L^{-1} (light blue). Lower and uppercase lettering indicate significant differences ($p \leq 0.05$), according to the Least Significant Difference post hoc test, among the concentration within the same sample and the sample within the same concentration, respectively.

Table 3. Some biological, microbiological, physico-chemical and chemical properties of the composts (P1–P11). Asterisks indicate that the differences indicated by lowercase lettering are statistically significant ($p < 0.001$). Non-significance is reported as ns.

Compost	Fluoresceine-Diacetate Hydrolysis (μg FDA g^{-1} 12h^{-1})	Basal Respiration (μL CO$_2$ s^{-1} g^{-1})	Total Fungi (Log CFU g^{-1})	Total Bacteria (Log CFU g^{-1})	Thermal-Resistant Bacteria (Log CFU g^{-1})	*Pseudomonas*-Like Bacteria (Log CFU g^{-1})	EC	pH	NO$_3^-$ (ppm)
P1	8.22 ± 0.90 b	2.73 ± 1.85	5.27 ± 0.12 d	7.68 ± 0.05 bcd	7.51 ± 0.49 a	7.95 ± 0.00 cd	8434.00 b	8.52 d	5.87 ± 0.50 def
P2	8.59 ± 0.33 b	1.01 ± 0.18	5.61 ± 0.06 c	7.34 ± 0.12 d	7.26 ± 0.00 ab	8.30 ± 0.49 bc	4612.00 e	9.00 bc	8.60 ± 0.53 c
P3	7.03 ± 0.11 c	1.06 ± 0.00	3.56 ± 0.00 g	7.62 ± 0.31 cd	6.54 ± 0.27 cd	7.26 ± 0.30 e	5123.00 d	10.23 a	7.13 ± 1.21 cde
P4	9.40 ± 0.25 a	0.77 ± 0.00	5.31 ± 0.05 d	7.61 ± 0.16 cd	7.10 ± 0.21 ab	6.75 ± 0.18 f	11837.00 a	8.57 d	5.60 ± 1.93 ef
P5	8.73 ± 0.13 b	1.21 ± 1.24	5.25 ± 0.09 d	7.88 ± 0.24 bc	6.80 ± 0.06 bc	7.76 ± 0.13 d	7258.67 c	10.30 a	4.80 ± 0.60 f
P6	8.81 ± 0.36 ab	0.94 ± 0.00	4.86 ± 0.00 e	7.36 ± 0.17 d	6.10 ± 0.21 de	7.27 ± 0.28 e	4331.00 f	9.40 b	13.07 ± 1.10 a
P7	8.41 ± 0.19 b	2.26 ± 0.00	4.56 ± 0.00 f	5.93 ± 0.16 e	5.45 ± 0.20 g	6.75 ± 0.10 f	4630.00 e	8.48 d	7.60 ± 0.87 cd
P8	0.94 ± 0.18 de	0.65 ± 0.27	6.16 ± 0.15 ab	8.94 ± 0.32 a	4.87 ± 0.12 h	8.79 ± 0.21 a	2282.00 h	8.85 cd	11.07 ± 2.34 b
P9	0.54 ± 0.26 e	0.58 ± 0.34a	6.01 ± 0.00 b	7.57 ± 0.28 cd	5.57 ± 0.15 fg	9.01 ± 0.00 a	8.99 i	8.94 c	2.13 ± 0.61 g
P10	1.28 ± 0.29 d	0.37 ± 0.43	5.26 ± 0.21 d	7.86 ± 0.13 bc	5.42 ± 0.30 g	8.73 ± 0.00 ab	3646.00 g	9.16 bc	7.47 ± 0.42 cde
P11	1.32 ± 0.27 d	1.00 ± 0.22	6.35 ± 0.09 a	7.98 ± 0.13 b	5.88 ± 0.04 ef	7.72 ± 0.32 d	3567.67 g	9.26 b	12.13 ± 0.99 ab
Sign.	***	ns	***	***	***	***	***	***	***

3.3. Chemical and Molecular Properties of the Composts

The composts showed sub-alkaline pH values (>8.0) and variable levels of electrical conductivity. Similarly, nitrate availability showed values ranging from 2.13 ppm in P9 to 13.7 ppm in P6 (Table 3).

The ^{13}C NMR spectra of the compost samples were characterized by an overall composition dominated by the aliphatic molecules of either alkyl-C or O-alkyl-C components (Figure 4).

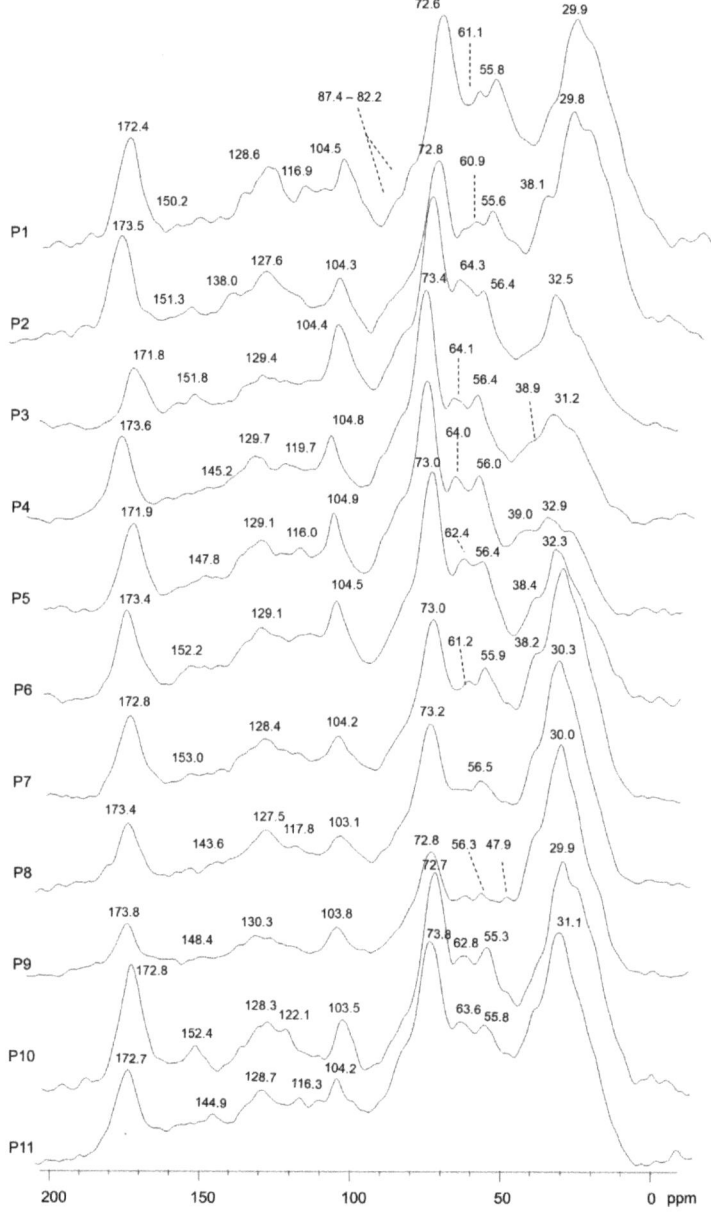

Figure 4. ^{13}C-CPMAS-NMR of compost samples (P1–P11).

The carbon distribution ranged from 21.6 to 43.6% and from 26.8 to 42.4% of the total spectral area, respectively (Table 4). The various bands grouped beneath the 110–145 ppm spectral interval are related to the unsubstituted and C-substituted nuclei of phenyl carbons pertaining to the aromatic units of terpenoid, polyphenols, flavonoids and the lignin components of plant tissues [31–33]. Moreover, the signals shown in the alkyl-C interval (0–45 ppm) also include the various lipid compounds derived from plant tissues and the microbial biomass [17]. The intense and broad bands centered around 30–33 ppm (Figure 4) are related to the overlapping of different bulk methylene (CH2) segments of predominantly straight-chain molecules of wax and cutin components [27,34], as well as those of cyclic aliphatic components such as terpene derivatives and sterols derived from the aromatic plant used in compost-starting biomasses [32,35].

Table 4. Relative C distribution (%) over the chemical shift regions (ppm) and structural index in the NMR spectra of compost samples (P1–P11).

Compost	Carboxyl-C (190–160 ppm)	O-Aryl-C (160–140 ppm)	Aryl-C (140–110 ppm)	O-Alkyl-C (110–60 ppm)	CH$_3$O/C_N (60–45 ppm)	Alkyl-C (45–0 ppm)	HB [a]	A/OA [a]	LigR [a]
P1	7.2	2.1	12.6	31.3	13.0	33.9	1.2	1.1	6.2
P2	8.2	2.9	11.5	26.8	12.3	38.2	1.4	1.4	4.3
P3	5.1	3.7	12.1	44.0	11.7	23.4	0.8	0.5	3.1
P4	8.1	3.8	14.3	41.1	11.8	21.0	0.8	0.5	3.1
P5	7.9	3.8	15.5	42.4	11.4	19.0	0.8	0.4	3.0
P6	7.4	3.3	15.1	38.7	10.9	24.6	0.9	0.6	3.3
P7	7.2	2.5	11.8	31.0	11.8	35.7	1.3	1.2	4.7
P8	6.5	2.2	11.4	30.1	12.0	37.8	1.3	1.3	5.4
P9	6.3	2.2	9.5	28.0	10.5	43.6	1.5	1.6	4.8
P10	9.5	3.1	10.4	30.7	11.5	34.7	1.2	1.1	3.8
P11	7.7	5.3	12.0	33.6	11.3	30.2	1.1	0.9	2.1

The shoulders evident at 39–40 ppm (Figure 4) can be also associated with CH_2, tertiary (CH) and quaternary (C-R) carbons in the assembled rings of terpenoid compounds [33,36].

The subsequent chemical shift region between 46 and 60 ppm, marked by the signal at 56 ppm (Figure 4), includes typical resonances found in plant-derived organic materials consisting of the contribution of various groups represented by methoxyl substituents on the aromatic rings of guaiacyl and syringyl units in lignin, or by C-N bonds in peptidic moieties [17,27].

The various peaks in the O-alkyl-C region (60–110 ppm) are currently assigned to monomeric units in the oligo and polysaccharide chains of plant tissues [28,30]. The less intense shoulders at 62/64 ppm represent the out-of-plane carbon 6 of cyclic carbohydrate conformation, followed by the sharp signal around 72 ppm due to the coalescence of the chemical shift of hydroxylated carbons in position 2, 3 and 5 in the pyranoside structure, whereas the band shift at 104 ppm is the di-O-alkyl frequency of anomeric carbons (Figure 4). In addition to the monomers of the saccharide chains, the O-alkyl regions of the NMR spectra may also include a contribution from the carbohydrate moieties of the glycosidic components in the structures of the flavonoids of aromatic plants [31,37]. The reduced evidence of carbon 4 involved in the β 1$\rightarrow$4 glycosidic bond at 82/88 ppm suggested the decomposition of the ether bond of the polysaccharides chain in composting processes, as well as the incorporation of mannan, xylan and arabinan derivatives into the hemicellulose structures of the initial fresh plant biomass [29,30].

The phenolic aromatic region (140–160 ppm) indicates the presence of O-substituted ring carbon derived from various aromatic structures [17,28]. Finally, the wide peak at 174 ppm (Figure 4) indicated the presence of carboxyl groups in aliphatic and aromatic acids, amide groups in amino acid moieties and acetyl substituents in the carbohydrate components of hemicelluloses and flavonoids.

4. Discussion

The potential application of composts and organic materials in garden cress cultivation has long been investigated with the aims to reduce peat in growing media [38], improve soil properties [39] and upgrade the nutritional quality of yield [40]. On the other hand, cress is also plentifully used in studies carried out to test the disease suppressiveness, maturity and agronomic suitability of compost. The current paper presents a comprehensive examination of bio-composts from aromatic plant residues applied to the soilless cultivations of cress as a protective means against the most common soil-borne pathogens. The study revealed that the feedstocks may affect compost features and functionality and provide benefits for crop protection.

In view of composting, aromatic plant waste differs from other crop residues, mainly because of its significant content of essential oil. This plant constituent is made up of a rich mixture of compounds that generally exhibit marked antimicrobial and phytotoxic activity, amplified by their volatility, lipophilicity and very low lethal concentration [41,42]. Essential oils contained in aromatic plants may interfere with the dynamics of microorganisms devoted to the composting, causing potential start-up difficulties, process slowdowns, negative impacts on microbial development and, secondarily, prolonged phytotoxic effects when the compost is applied to the plants [43]. However, the aware composting of these vegetable matrices well assembled with other complementary ingredients, as a sound remedy to modulate the cited microbial conditioning aspects, can lead to a high agronomic and functional value-end product. From a circular economy perspective, however, the removal/recovery of the essential oils from the raw material by means of hydrodistillation is a further way of valorizing aromatic plant leftovers, drawing attention to the concept of by-products [21]. This study, by examining the composts obtained from the raw and oil-free aromatic matrices of different botanical origins, sheds light on the consequences for the potential composts' suppressive functions.

The bioassays unequivocally demonstrated a high ability to suppress Rhizoctonia and Sclerotinia diseases on the cress of the raw composts P1 and P2, respectively obtained from the following two groups of ingredients: wood chips, escarole and a mixture of sage, basil, mint and parsley and wood chips, 30% escarole and a mixture of essential oil-free sage, basil and rosemary. Furthermore, after generating the biological vacuum in them by autoclaving, the two high-performing composts lost their ability to stop the pathogenesis dynamics, indicating the crucial role played by resident microorganisms in conferring biological control properties. This is corroborated by the high levels of the general enzymatic activity recorded by FDA hydrolysis, the respiration rate (although without statistical significance) and microbial counting profiles. Interestingly, these two composts showed the highest levels of thermal-resistant bacteria populations, including *Bacillus*-like cells, which are widely reported among the main functional microbial groups conferring suppressiveness to composts [44]. As matter of the fact, the composts assessed for suppressiveness in this study were previously used as a useful source for the stepwise selection of new strains of *Bacillus* spp. antagonistic to soil-borne pathogens [22]. In addition, the phytotoxicity of P1 and P2 was low, and the cress germination assay showed irregular behavior with the dose, suggesting an antagonistic toxic/nutritional effect of eluates. The pot assays indicated that *R. solani* is more difficult to control with suppressive composts than *S. sclerotiorum*, likely due to the insidious nature of the pathogen and the residual phytotoxicity of the non-suppressive composts that weaken the plant. This could correspond to a specific mechanism—the suppressive activity slightly related to the population levels of the *Bacillus*-like bacteria. On the contrary, against Sclerotinia damping-off, the success in biocontrol by composts has been broader, although keeping the same pattern of the Rhizoctonia bioassay regarding the most and the least suppressive ones.

There is no evidence about the modulating effects of the essential oil extraction from feedstock on the levels of disease suppressiveness; rather, the ability to counteract pathogens seems to be more influenced by the botanical diversity of the ingredients mixed into the pile. For example, both the mono-matrix composts of basil leftovers showed the lowest suppressiveness. In contrast, both the raw and essential oil-extracted rosemary-based composts showed a particularity to significantly increase suppressiveness after sterilization, possibly indicating the release or formation of highly antifungal compounds after heating. While the use of vapor heating may enhance the release of antimicrobial components in aromatic plants [27], the metabolic products of rosemary, such as flavonoids and phenolic derivates, show consistent thermal stability and bioactive performance upon thermal treatment [45,46]. It is therefore conceivable that the preliminary autoclave sterilization process of fresh biomass may have promoted the release of aromatic oligomers during the composting of rosemary-based matrices, thus enhancing the suppressive activity of the final processed composts.

Interestingly, on average, the greatest reduction in suppressiveness passing from the raw to the corresponding autoclaved compost is observed in the multi-matrix composts, suggesting a role of the carbon source complexity in the structure and functionality of the developing microbial communities. From this perspective, although the EC, pH and nitrate concentrations of the composts were in the typical range for green composts, the molecular distribution of organic carbon along the samples reflect these considerations.

Despite the broadening of the signal produced by the CPMAS technique, the NMR spectra of the compost samples allowed for an adequate evaluation of specific functional groups related to the characteristic plant components in the different chemical shift regions. The relative distribution of C in the chemical shift regions and the structural indices in the NMR spectra of the compost samples suggest a differentiation of maturity into two main groups based on the initial composition of the starting substrates. The inclusion of a higher number of poplar residues (P1) and the exclusive use of the rosemary (P8, P9) and sage (P10, P11) biomass promoted a significant incorporation of lipid compounds. On the contrary, the halving of the structural fraction and the increase in fresh biomass, consisting mainly in basil and parsley (P3, P4, P6), enabled the final prevalence of the carbohydrates and polysaccharides. This is highlighted by the distinct values found for the comparison of the aliphatic compositions carried out with the Alkyl ratio, while no significant differences were found in the overall relative distribution of the aromatic compounds. The larger residual presence of carbohydrates in the basil and/or parsley-based composts, likely due to the proliferative effects on harmful microorganisms [47], may explain the complete absence and very low suppressiveness, respectively, in the P4 and P5 composts and in the P6 and P3 composts, which also exhibited a lower germination index than the other compost samples, as an indication of their low degradation.

The bioactive properties of composts in supporting crop development through biostimulation or suppressive effects are usually associated with the combination of microbial composition and molecular characteristics [12,17,27]. Regarding chemical components, although an unambiguous relationship between structure and activity has not yet been clarified, the bioactive functionalities of compost and its derivates are mainly related to the stage of humification and the content of humified fraction. These structural characteristics are evidenced by increased stabilization properties summarized in the NMR analysis by the Hydrophobic Index and Alkyl ratio [34]. The relative preservation of apolar compounds during the dynamics of natural organic materials such as litter, soil organic matter, compost, digestates, etc. are in fact closely correlated with the advancement of carbon stabilization and the so-called humification process [29,34]. Bioactive properties are thus triggered by depolymerized organic molecules released by the intense humification process, mainly represented by aromatic and phenolic compounds and decomposed peptide moieties [48]. In contrast, the retention of more intact lignocellulose residues indicated the lower intensity of humification and the reduced availability of less decomposed active fragments [30,49]. A remarkable shared effect on the composting process was also shown by the biomasses subjected to the preliminary removal of essential oils. Almost all the extracted final composts showed a final composition characterized by an increase in the HB and A/OA indexes and/or a decrease in the Lignin ratio compared to the corresponding non-extracted samples. The trend of a decreasing Lignin ratio is related to the improved organic stabilization of the composted biomass. In fact, despite the extraction of essential oils and the consequent release of aromatic and phenolic compounds, which constitute the denominator of LigR, almost all of the de-oiled mature composts showed a steady decrease in this structural parameter. This change suggests the better fragmentation of the larger proteinaceous material underlying the peak area at 45–60 ppm combined with the constant maintenance or relative increase of the O-aryl-C signals of lignin fragments in the 145–160 ppm region, thus leading to final LigR values in the extracted biomasses comparable to those usually found in the humified fraction of stable mature composts [17,30]

Essential oils from aromatic plants are recognized as effective antioxidant and antimicrobial agents with suppressive properties against a wide range of microorganisms [21,22,27]. It is therefore conceivable that the preliminary extraction of the oil fraction may have favored the development of microbial activity in processed biomass, thus promoting an organic matter dynamic associated with compost humification.

5. Conclusions

The composting of aromatic plant residues is a suitable method to improve the circularity of this production chain and agricultural systems in general. The extraction of the essential oil does not affect the suppressive functions, so feedstock without essential oil, as well as raw materials, can potentially promote the formation of quality composts, especially in a complex combination with ingredients of different botanical origins.

Cress cultivation can benefit from the application of suppressive composts to reduce the dependence on fungicides for disease management and improve rapid soilless producing systems such as ready-to-eat salads and microgreens.

Author Contributions: Conceptualization, C.P. and M.Z.; methodology, C.P., R.S. and M.Z.; formal analysis, C.P. and R.S.; investigation, C.P., R.S. and M.C.; data curation, C.P. and R.S.; writing—original draft preparation, C.P. and R.S.; writing—review and editing, C.P., R.S., M.C., E.D.F. and M.Z.; project administration and funding acquisition, E.D.F. and M.Z. All authors have read and agreed to the published version of the manuscript.

Funding: This research was funded by the Campania Region via the EU FEASR funding program PSR 2007–2014 measure 124 through the project "Gestione innovative degli scarti di coltivazione e lavorazione nella filiera delle erbe aromatiche" (acronym: "Polieco 2").

Institutional Review Board Statement: Not applicable.

Informed Consent Statement: Not applicable.

Data Availability Statement: The data presented in this study are available on request from the corresponding author.

Conflicts of Interest: The authors declare no conflict of interest.

References

1. Nicola, S.; Hoeberechts, J.; Fontana, E. Ebb-and-flow and floating systems to grow leafy vegetables: A review for rocket, corn salad, garden cress and purslane. *Acta Hortic.* **2007**, *747*, 585–593. [CrossRef]
2. Rouphael, Y.; Colla, G.; De Pascale, S. Sprouts, microgreens and edible flowers as novel functional foods. *Agronomy* **2021**, *11*, 2568. [CrossRef]
3. Santos, J.; Oliveira, M.B.P.P.; Ibáñez, E.; Herrero, M. Phenolic profile evolution of different ready-to-eat baby-leaf vegetables during storage. *J. Chromatog. A* **2014**, *1327*, 118–131. [CrossRef] [PubMed]
4. Jain, T.; Grover, K.; Grewal, I.S. Development and sensory evaluation of ready to eat supplementary food using garden cress (*Lepidium sativum*) seeds. *J. Appl. Nat. Sci.* **2016**, *8*, 1501–1506. [CrossRef]
5. Ciesielska, K.; Ciesielski, W.; Kulawik, D.; Oszczęda, Z.; Tomasik, P. Cultivation of cress involving water treated under different atmospheres with low-temperature, low-pressure glow plasma of low frequency. *Water* **2020**, *12*, 2152. [CrossRef]
6. Keutgen, N.; Hausknecht, M.; Tomaszewska-Sowa, M.; Keutgen, A.J. Nutritional and sensory quality of two types of cress microgreens depending on the mineral nutrition. *Agronomy* **2021**, *11*, 1110. [CrossRef]
7. Ogórek, R. Enzymatic activity of potential fungal plant pathogens and the effect of their culture filtrates on seed germination and seedling growth of garden cress (*Lepidium sativum* L.). *Eur. J. Plant Pathol.* **2016**, *145*, 469–481. [CrossRef]
8. Mehta, C.M.; Palni, U.; Franke-Whittle, I.H.; Sharma, A.K. Compost: Its role, mechanism and impact on reducing soil-borne plant diseases. *Waste Manag.* **2014**, *34*, 607–622. [CrossRef]
9. Pane, C.; Zaccardelli, M. Principles of compost-based plant diseases control and innovative new developments. In *Composting for Sustainable Agriculture*; Maheshwari, D., Ed.; Sustainable Development and Biodiversity; Springer: Cham, Switzerland, 2014; Volume 3. [CrossRef]
10. Giménez, A.; Fernández, J.A.; Egea-Gilabert, C.; Santísima-Trinidad, A.B.; Ros, M.; Pascual, J.A. Agro-industry composts as growing medium for growing baby-leaf lettuces in a floating system—Added-value to suppress *Pythium irregulare*. *Acta Hortic.* **2019**, *1242*, 791–798. [CrossRef]
11. Morau, A.; Piepho, H.-P.; Fritz, J. Growth responses of garden cress (*Lepidium sativum* L.) to biodynamic cow manure preparation in a bioassay. *Biol. Agric. Hortic.* **2020**, *36*, 16–34. [CrossRef]
12. Pane, C.; Piccolo, A.; Spaccini, R.; Celano, G.; Villecco, D.; Zaccardelli, M. Agricultural waste-based composts exhibiting suppressivity to diseases caused by the phytopathogenic soil-borne fungi *Rhizoctonia solani* and *Sclerotinia minor*. *Appl. Soil Ecol.* **2013**, *65*, 43–51. [CrossRef]
13. Pane, C.; Spaccini, R.; Piccolo, A.; Scala, F.; Bonanomi, G. Compost amendments enhance peat suppressiveness to *Pythium ultimum*, *Rhizoctonia solani* and *Sclerotinia minor*. *Biol. Control* **2011**, *56*, 115–124. [CrossRef]
14. Aslam, D.N.; Horwath, W.; Vander Gheynst, J.S. Comparison of several maturity indicators for estimating phytotoxicity in compost-amended soil. *Waste Manag.* **2008**, *28*, 2070–2076. [CrossRef] [PubMed]
15. Ventorino, V.; Parillo, R.; Testa, A.; Viscardi, S.; Espresso, F.; Pepe, O. Chestnut green waste composting for sustainable forest management: Microbiota dynamics and impact on plant disease control. *J. Environ. Manag.* **2016**, *166*, 168–177. [CrossRef] [PubMed]
16. Ronga, D.; Francia, E.; Allesina, G.; Pedrazzi, S.; Zaccardelli, M.; Pane, C.; Tava, A.; Bignami, C. Valorization of vineyard by-products to obtain composted digestate and biochar suitable for nursery grapevine (*Vitis vinifera* L.) production. *Agronomy* **2019**, *9*, 420. [CrossRef]
17. Pane, C.; Spaccini, R.; Piccolo, A.; Celano, G.; Zaccardelli, M. Disease suppressiveness of agricultural greenwaste composts as related to chemical and bio-based properties shaped by different on-farm composting methods. *Biol. Control* **2019**, *137*, 104026. [CrossRef]
18. Pane, C.; Sorrentino, R.; Scotti, R.; Molisso, M.; Di Matteo, A.; Celano, G.; Zaccardelli, M. Alpha and beta-diversity of microbial communities associated to plant disease suppressive functions of on-farm green composts. *Agriculture* **2020**, *10*, 113. [CrossRef]
19. Bignami, C.; Melegari, F.; Zaccardelli, M.; Pane, C.; Ronga, D. Composted solid digestate and vineyard winter prunings partially replace peat in growing substrates for micropropagated highbush blueberry in the nursery. *Agronomy* **2022**, *12*, 337. [CrossRef]
20. De Corato, U. Agricultural waste recycling in horticultural intensive farming systems by on-farm composting and compost-based tea application improves soil quality and plant health: A review under the perspective of a circular economy. *Sci. Tot. Environ.* **2020**, *738*, 139840. [CrossRef]

21. Zaccardelli, M.; Roscigno, G.; Pane, C.; Celano, G.; Di Matteo, M.; Mainente, M.; Vuotto, A.; Mencherini, T.; Esposito, T.; Vitti, A.; et al. Essential oils and quality composts sourced by recycling vegetable residues from the aromatic plant supply chain. *Ind. Crop. Prod.* **2021**, *162*, 113255. [CrossRef]
22. Zaccardelli, M.; Pane, C.; Caputo, M.; Durazzo, A.; Lucarini, M.; Silva, A.M.; Severino, P.; Souto, E.B.; Santini, A.; De Feo, V. Sage species case study on a spontaneous Mediterranean plant to control phytopathogenic fungi and bacteria. *Forests* **2020**, *11*, 704. [CrossRef]
23. Violante, P. *Metodi di Analisi Chimica del Suolo*; Angeli, F., Ed.; Italian Ministry of Agriculture: Milan, Italy, 2000; p. 536.
24. Scher, F.M.; Baker, R. Effect of *Pseudomonas putida* and a synthetic iron chelator on induction of soil suppressiveness to Fusarium Wilt pathogens. *Phytopathology* **1982**, *72*, 1567–1573. [CrossRef]
25. Sadfi, N.; Cherif, M.; Fliss, I.; Boudabbous, A.; Antoun, H. Evaluation of bacterial isolates from salty soils and *Bacillus thuringiensis* strains for the biocontrol of Fusarium dry rot of potato tubers. *J. Plant Pathol.* **2001**, *83*, 101–118.
26. Mestrelab Research, MestReNova Manual 2010. Manual of MestreNova 6.2. Available online: https://mestrelab.com/ (accessed on 19 June 2022).
27. Verrillo, M.; Cozzolino, V.; Spaccini, R.; Piccolo, A. Humic substances from green compost increase bioactivity and antibacterial properties of essential oils in Basil leaves. *Chem. Biol. Technol. Agric.* **2021**, *8*, 28. [CrossRef]
28. Bento, L.R.; Spaccini, R.; Cangemi, S.; Mazzei, P.; de Freitas, B.B.; de Souza, A.E.O.; Moreira, A.B.; Ferreira, O.P.; Piccolo, A.; Bisinoti, M.C. Hydrochar obtained with by-products from the sugarcane industry: Molecular features and effects of extracts on maize seed germination. *J. Environ. Manag.* **2021**, *281*, 111878. [CrossRef] [PubMed]
29. Fregolente, L.G.; Dos Santos, J.V.; Vinci, G.; Piccolo, A.; Moreira, A.B.; Ferreira, O.P.; Bisinoti, M.C.; Spaccini, R. Insights on molecular characteristics of hydrochars by ^{13}C-NMR and off-line TMAH-GC/MS and assessment of their potential use as plant growth promoters. *Molecules* **2021**, *26*, 1026. [CrossRef]
30. de Aquino, A.M.; Canellas, L.P.; da Silva, A.P.S.; Canellas, N.O.; da S Lima, L.; Olivares, F.L.; Piccolo, A.; Spaccini, R. Evaluation of molecular properties of humic acids from vermicompost by ^{13}C-CPMAS-NMR spectroscopy and thermochemolysis–GC–MS. *J. Anal. Appl. Pyrol.* **2019**, *141*, 104634. [CrossRef]
31. Yang, F.; Qi, Y.; Liu, W.; Li, J.; Wang, D.; Fang, L.; Zhang, Y. Separation of five flavonoids from aerial parts of *Salvia Miltiorrhiza* bunge using HSCCC and their antioxidant activities. *Molecules* **2019**, *24*, 3448. [CrossRef]
32. Kadir, A.; Zheng, G.; Zheng, X.; Jin, P.; Maiwulanjiang, M.; Gao, B.; Aisa, H.A.; Yao, G. Structurally diverse diterpenoids from the roots of *Salvia deserta* based on nine different skeletal types. *J. Nat. Prod.* **2021**, *84*, 1442–1452. [CrossRef]
33. Aydin, S.K.; Ertaş, A.; Boğa, M.; Erol, E.; Toraman, G.O.A.; Saygı, T.K.; Halfon, B.; Topçu, G. Di-, and triterpenoids isolation and lc-ms analysis of salvia marashica extracts with bioactivity studies. *Rec. Nat. Prod.* **2021**, *15*, 463–475. [CrossRef]
34. Martinez-Balmori, D.; Spaccini, R.; Aguiar, N.O.; Novotny, E.H.; Olivares, F.L.; Canellas, L.P. Molecular characteristics of humic acids isolated from vermicomposts and their relationship to bioactivity. *J. Agr. Food Chem.* **2014**, *62*, 11412–11419. [CrossRef] [PubMed]
35. Bustos-Brito, C.; Joseph-Nathan, P.; Burgueño-Tapia, E.; Martínez-Otero, D.; Nieto-Camacho, A.; Calzada, F.; Yépez-Mulia, L.; Esquivel, B.; Quijano, L. Structure and absolute configuration of abietane diterpenoids from *Salvia clinopodioides*: Antioxidant, antiprotozoal, and antipropulsive activities. *J. Nat. Prod.* **2019**, *82*, 1207–1216. [CrossRef] [PubMed]
36. Yaris, E.; Balur Adsız, L.; Yener, I.; Tuncay, E.; Yilmaz, M.A.; Akdeniz, M.; Kaplaner, E.; First, M.; Ertas, A.; Kolak, U. Isolation of secondary metabolites of two endemic species: *Salvia rosifolia* Sm. and *Salvia cerino-pruinosa* Rech. f. var. *elazigensis* (Lamiaceae). *J. Food Meas. Charact.* **2021**, *15*, 4929–4938. [CrossRef]
37. Slimestad, R.; Fossen, T.; Brede, C. Flavonoids and other phenolics in herbs commonly used in Norwegian commercial kitchens. *Food Chem.* **2020**, *309*, 125678. [CrossRef]
38. Gajdoš, R. Effects of two composts and seven commercial cultivation media on germination and yield. *Compost Sci. Util.* **1997**, *5*, 16–37. [CrossRef]
39. Shao-qi, Z.; Wei-dong, L.; Xiao, Z. Effects of heavy metals on planting watercress in kailyard soil amended by adding compost of sewage sludge. *Process Saf. Environ.* **2010**, *88*, 263–268.
40. Tuncay, Ö.; Esiyok, D.; Yamur, B.; Okur, B. Yield and quality of garden cress affected by different nitrogen sources and growing period. *Afr. J. Agric. Res.* **2011**, *6*, 608–617.
41. Pauli, A. Antimicrobial properties of essential oil constituents. *Int. J. Arom.* **2001**, *11*, 126–133. [CrossRef]
42. De Almeida, L.F.R.; Frei, F.; Mancini, E.; De Martino, L.; De Feo, V. Phytotoxic activities of mediterranean essential oils. *Molecules* **2010**, *15*, 4309–4323. [CrossRef]
43. Greff, B.; Lakatos, E.; Szigeti, J.; Varga, L. Co-composting with herbal wastes: Potential effects of essential oil residues on microbial pathogens during composting. *Crit. Rev. Environ. Sci. Technol.* **2021**, *51*, 457–511. [CrossRef]
44. Pane, C.; Villecco, D.; Campanile, F.; Zaccardelli, M. Novel strains of *Bacillus*, isolated from compost and compost-amended soils, as biological control agents against soil-borne phytopathogenic fungi. *Biocontrol Sci. Technol.* **2012**, *22*, 1373–1388. [CrossRef]
45. Escriche, I.; Kadar, M.; Juan-Borrás, M.; Domenech, E. Suitability of antioxidant capacity, flavonoids and phenolic acids for floral authentication of honey. Impact of industrial thermal treatment. *Food Chem.* **2014**, *142*, 135–143. [CrossRef] [PubMed]
46. Doudin, K.; Al-Malaika, S.; Sheena, H.H.; Tverezovskiy, V.; Fowler, P. New genre of antioxidants from renewable natural resources: Synthesis and characterisation of rosemary plant-derived antioxidants and their performance in polyolefins. *Polym. Degrad. Stab.* **2016**, *130*, 126–134. [CrossRef]

47. Palese, A.M.; Pane, C.; Villecco, D.; Zaccardelli, M.; Altieri, G.; Celano, G. Effects of organic additives on chemical, microbiological and plant pathogen suppressive properties of aerated municipal waste compost teas. *Appl. Sci.* **2021**, *11*, 7402. [CrossRef]
48. Scaglia, B.; Nunes, R.R.; Rezende, M.O.O.; Tambone, F.; Adani, F. Investigating organic molecules responsible of auxin-like activity of humic acid fraction extracted from vermicompost. *Sci. Tot. Environ.* **2016**, *562*, 289–295. [CrossRef]
49. Castaño, R.; Borrero, C.; Avilés, M. Organic matter fractions by SP-MAS 13C NMR and microbial communities involved in the suppression of Fusarium wilt in organic growth media. *Biol. Cont.* **2011**, *58*, 286–293. [CrossRef]

Article

A Lignin-Rich Extract of Giant Reed (*Arundo donax* L.) as a Possible Tool to Manage Soilborne Pathogens in Horticulture: A Preliminary Study on a Model Pathosystem

Stefania Galletti [1], Stefano Cianchetta [1], Hillary Righini [2,*] and Roberta Roberti [2]

[1] Research Centre for Agriculture and Environment, Council for Agricultural Research and Economics, 40128 Bologna, Italy; stefania.galletti@crea.gov.it (S.G.); stefano.cianchetta@crea.gov.it (S.C.)
[2] Department of Agriculture and Food Sciences, Alma Mater Studiorum, University of Bologna, 40127 Bologna, Italy; roberta.roberti@unibo.it
* Correspondence: hillary.righini2@unibo.it; Tel.: +39-051-2096582

Abstract: Finding new sustainable tools for crop protection in horticulture has become mandatory. Giant reed (*Arundo donax* L.) is a tall, perennial, widely diffuse lignocellulosic grass, mainly proposed for bioenergy production due to the fact of its high biomass yield and low agronomic requirements. Some studies have already highlighted antimicrobial and antifungal properties of giant reed-derived compounds. This study aimed at investigating the potential of a lignin-rich giant reed extract for crop protection. The extract, obtained by dry biomass treatment with potassium hydroxide at 120 °C, followed by neutralization, was chemically characterized. A preliminary in vitro screening among several pathogenic strains of fungi and oomycetes showed a high sensitivity by most of the soilborne pathogens to the extract; thus, an experiment was performed with the model pathosystem, *Pythium ultimum*–zucchini in a growth substrate composed of peat or sand. The adsorption by peat and sand of most of the lignin-derived compounds contained in the extract was also observed. The extract proved to be effective in restoring the number of healthy zucchini plantlets in the substrate infected with *P. ultimum* compared to the untreated control. This study highlights the potential of the lignin-rich giant reed extract to sustain crop health in horticulture.

Keywords: giant reed; potassium hydroxide; lignin; *Pythium ultimum*; zucchini; *Cucurbita pepo*; extract; soilborne; polyphenol; crop protection

1. Introduction

Soilborne plant pathogens are disease-causing agents belonging to several species of fungi, oomycetes, bacteria, and nematodes that live in the soil as resting structures on plant residues or also in the absence of the host for brief or extended periods. Once established, soilborne pathogens accumulate in the soil leading to high yield losses, proving difficult to control [1]. The management options of soilborne pathogens are limited [2]. Highly effective synthetic pesticides for soil treatment, such as the fumigant methyl bromide, are no longer allowed, both in European countries and elsewhere, due to their negative impact on the environment. In Europe, there are also restrictions under Reg. (EC) No. 1107/2009 for the use of other soil fumigants: methyl isothiocyanate is not approved, while the status of chloropicrin and 1,3-dichloropropene is currently pending. Moreover, the fungicides for soil treatment, carbendazim and thiophanate methyl, are not approved because of their toxic effect on human and animal health. Currently, soilborne disease management is directed towards sustainable strategies including microbial products based on antagonistic microorganisms, crop rotation, soil solarization, or breeding for resistant plant varieties. However, none of these methods alone can effectively contain soilborne diseases. Among other control means, plant extract and their essential oils have shown some efficacy against a number of plant pathogens. Indeed, plants contain several bioactive compounds, such as phenols, flavonoid, quinones, and terpenes, have a defensive function against biotic stresses [3]. In vitro studies showed that neem, garlic, and turmeric extracts reduced the colony growth of *Fusarium oxysporum* and *Rhizoctonia solani* [4], while extracts of plants belonging to the *Asteraceae* and *Rubiaceae* families reduced *Fusarium solani* colony growth [5]. On muskmelon, the Fusarium wilt disease was suppressed by soil treatment with extracts from pepper/mustard, cassia, and clove [6], while on cucumber plants, leaf treatment with essential tea tree oil derived from *Melaleuca alternifolia* controlled *Sphaerotheca fuliginea*, the agent of powdery mildew [7]. A combination of both foliar spray and root drenching with a marine plant extract formulation from *Ascophyllum nodosum* effectively reduced the cucumber fungal pathogens *Alternaria cucumerinum*, *Didymella applanata*, *F. oxysporum*, and *Botrytis cinerea* [8]. Among sustainable alternative strategies for the control of soilborne pathogens, Armorex and Fungastop are examples of natural-based products that are commercially available in the USA [4].

Giant reed (*Arundo donax* L.) is a tall, rhizomatous, spontaneous perennial grass that is probably native to the Mediterranean environment but also widely diffused in many subtropical and temperate zones, with an invasive behavior [9]. Its use dates to 5000 BC by Egyptians who utilized the leaves as lining for grain storage and for wrapping mummies. More recently, it was utilized to produce paper/cellulose/viscose, musical instruments, stakes for plants or fishing rods, in addition to being considered an ornamental species [10,11]. For its wide pedo-climatic adaptability and high productivity, even when cultivated with low agronomic inputs [12], over the last decades this lignocellulosic species has gained interest for bioenergy production through direct combustion of the biomass, or, more recently, as feedstock for biorefinery to produce second-generation biofuel, biogas, and other biobased products [13–15]. Notably, the rhizomes, leaves, and stems were being utilized as traditional herbal remedies against several human pathologies; therefore, that giant reed can also be considered a medicinal plant [11]. Interestingly, the scientific literature reports an antimicrobial activity of giant reed extracts: an aqueous extract was able to contrast the biofilm formation of *Staphylococcus aureus* [16], while a methanolic extract showed the maximum effect among other extracts from medicinal plants against *Escherichia coli* and *Pseudomonas aeruginosa* [17].

Few studies have investigated the possible antifungal activity of giant reed lignocellulose-derived compounds: bio-oil obtained from giant reed pyrolysis showed preservative properties when used to treat Scots pine wood against *Basidiomycetes* [18]. In another study, water-soluble compounds recovered from steam-exploded giant reed during the bioethanol production process showed activity against fungal pathogens of horticultural interest [19].

The biological activity of lignin and lignin-derived compounds has been further studied: kraft lignin obtained from the delignification process of the paper industry was able to inhibit the fungal plant pathogen, *Aspergillus niger*, as well as several microbial human pathogens [20]. In addition, alkali-extracted lignin from maize residues of the bioethanol industry showed an antimicrobial effect against Gram+ bacteria and yeasts [21]. A recent review illustrated the possibility to exploit lignin with its polyhydroxyphenol network structure to produce innovative and sustainable agrochemicals, highlighting the ability to promote seedling emergence and plant growth, improve soil quality, and enhance plant resistance against phytopathogens and abiotic stresses [22].

This study aimed at investigating the potential of an innovative extract of giant reed for possible use in horticulture in the context of sustainable plant pathogen management. The extract, obtained after alkali treatment of the dry giant reed biomass, would contain lignin and lignin-derived compounds that could exert biological activity towards microbial pathogens of horticultural interest. This study moved from the chemical characterization of the extract to the evaluation of its effectiveness when applied to a model pathosystem, namely, the soilborne pathogen *Pythium ultimum* and zucchini (*Cucurbita pepo* L.), after a preliminary in vitro screening among several fungi and oomycetes.

2. Materials and Methods

2.1. Isolates of Fungi and Oomycetes

A total of 22 fungal and oomycetes isolates were preliminary tested in vitro for their sensitivity to the giant reed extract. Table 1 reports the list of the species and the origin of the different isolates used in the experiments. They were selected among soilborne and airborne pathogens, wood decay fungi, and pathogenic or nonpathogenic *Trichoderma* species. Most of them were previously isolated from infected tissues or soil samples at the laboratories of the Council for Agricultural Research and Economics (CREA), Research Centre for Agriculture and Environment of Bologna, Italy, or the Department of Agriculture and Food Sciences (DISTAL) of the University of Bologna, and identified based on morphological characteristics or molecular analysis. Wood decay isolates were purchased from public collections or private providers.

Table 1. List of the fungal and *oomycetes* isolates assayed in the preliminary in vitro experiments.

Category	Isolate	Origin
Soilborne pathogenic oomycetes and fungi	*Pythium ultimum* 16	Sugar beet
	Pythium ultimum 22	Sugar beet
	Phytophthora cactorum	Strawberry
	Ceratobasidium sp. RH3	Strawberry
	Rhizoctonia solani DAF3001	Green bean
	Sclerotium rolfsii 1	Sugar beet
	Sclerotium rolfsii 2	Potato
	Sclerotinia sclerotiorum MA3	Alfalfa
	Phoma betae	Sugar beet
	Fusarium oxysporum L1	Tomato
	Fusarium oxysporum 11.22	Tomato
Airborne pathogenic fungi	*Botrytis cinerea* 1	Strawberry
	Botrytis cinerea 2	Green bean
	Alternaria alternata 1	Potato
	Alternaria alternata 2	Tomato
Wood decay fungi	*Phanerochaete chrysosporium* D85242T	VTT, Finland
	Trametes versicolor D83211	VTT, Finland
	Pleurotus ostreatus	Funghi Mara, Italy
Pathogenic *Trichoderma* species	*Trichoderma pleuroticola* 488	*P. ostreatus* substrate
	Trichoderma pleuroti 498	*P. ostreatus* substrate
Nonpathogenic *Trichoderma* species	*Trichoderma gamsii* IMO5	Peach rhizosphere
	Trichoderma afroharzianum B75	Sugar beet rhizosphere

All isolates were maintained on potato dextrose agar (PDA) medium in tubes at 4 °C until use. The PDA medium was prepared as follows: filtered potato broth 1 L (obtained boiling 200 g/L of potatoes in distilled water), supplemented with sucrose 10 g/L, and technical agar nr 3 (Oxoid™, Thermo Fisher, Waltham, MA, USA) 12 g/L. The medium was autoclaved at 120 °C, for 20 min.

2.2. Giant Reed Meal, Seeds, and Growth Substrate

The giant reed meal was obtained starting from winter-harvested plants collected at the CREA experimental farm located at Anzola dell'Emilia (Bologna, northern Italy) (Figure 1a,b); Table 2 shows the composition of the meal, obtained after milling and sieving (<1.5 mm) the aboveground oven-dried parts, as reported in a previous paper [15].

Seeds of zucchini "Bolognese" (*Cucurbita pepo* L., Blumen Group S.p.A., Milano, Italy) were used for the experiments.

As growth substrates, peat (Terraricca, Cifo srl, S. Giorgio di Piano, Bologna, Italy, 60% humidity, 0.5 g/cm^3 density) and sterilized river sand (1.5 g/cm^3 density) were utilized.

Figure 1. (a) Giant reed plants at the growing stage; (b) plants at the end of the growth cycle. Pictures were taken at the experimental farms of the Council for Agricultural Research and Economics (CREA), Anzola dell'Emilia, Bologna, northern Italy (courtesy of Enrico Ceotto).

Table 2. Composition of the giant reed meal used in the experiments as reported in a previous paper [15] [1]. Standard deviations (sd) are reported in brackets ($n = 3$).

Composition Parameters	Mean Values (sd)
Total solids (%DW) [2]	95.51 (0.02)
Cellulose (%DW)	41.2 (1.5)
Hemicellulose (%DW)	22.5 (0.8)
Lignin (%DW)	10.9 (0.5)
Ash (%DW)	5.19 (0.01)
Total C (%DW)	44.6 (0.60)
Total N (%DW)	0.37 (0.05)
Total P (%DW)	0.05 (0.00)
C/N (mol/mol) [3]	142 (21)
pH in water	5.4 (0.5)

[1] Adapted with permission from Ref. [15]. 2022, Elsevier; [2] Percentage of dry weight at 60 °C; [3] C/N mol/mol can be converted into C/N w/w by multiplying the former by 12/14 (C/N molar mass ratio).

2.3. Experimental Design

The workflow can be summarized as follows:

(A) Preparation and chemical characterization of a lignin-rich extract from giant reed dry biomass;
(B) Preliminary tests:
- Toxicity tests of the extract towards fungi and *oomycetes* on poisoned PDA plates;
- Toxicity test of the extract towards *P. ultimum* in peat;
- Sensitivity tests of zucchini to the extract on filter paper, in peat, and in sand;
- Pathogenicity test of *P. ultimum* on zucchini in sand.
(C) Distribution analysis of the lignin-derived polyphenols in peat and sand treated with the extract;
(D) Evaluation of the extract efficacy in the *P. ultimum*–zucchini pathosystem in a growth substrate.

2.4. Giant Reed Extract Preparation and Characterization

The extract was prepared by treating a slurry of the giant reed meal at 10% weight/weight (w/w) in water with alkali (KOH 1.6% w/w in water) at 120 °C for 20 min; then, the liquid fraction was collected by filtration under vacuum and stored at 4 °C. Just before use, the extract was neutralized (pH 7) with appropriate amounts of H_2SO_4 72% (approximately 0.035 M SO_4^{2-} final) operating under sterile conditions.

To calculate the dry weights of the insoluble and soluble fractions, samples of the alkaline slurry were centrifuged (10 min, 2000 rcf, Allegra X-22 Beckman Coulter); then, the insoluble fraction (three-fold washed) and supernatants (liquid fraction) were dried at 105 °C until constant weight.

Water and total solids were determined according to ISO 12880; ashes and volatile solids according to ISO 12879; pH according to ISO 13037; electrical conductivity according to ISO 13038; total P and K according to Reg CE 2003/2003, met. 3.1.1 and met. 8.1, respectively.

Total organic C was determined according to national regulation DM 21/12/00, Suppl n. 6 [23]. C and N content were determined after elemental analysis by a LECO Truspec® CHN Analyzer (LECO Corporation; Saint Joseph, MI, USA), and the C/N ratio was calculated accordingly.

Free reducing sugars were quantified by the 3,5-dinitrosalicylic acid (DNS) method [24] adapted for 96-well microplates in duplicates [25]. A mix of glucose and xylose (1:1) was included as standards. Briefly, the assay conditions were citrate buffer 50 mM, pH 4.8, 5 min, 95 °C.

The hydrolysable holocellulose content was calculated by dividing by a correction factor of 1.12 of the amount of reducing sugars released in solution after enzymatic saccharification of the giant reed extract to account for the addition of water during hydrolysis. In detail, a sample of 10 mL of extract was saccharified in 15 mL tubes (in triplicate) using the following conditions: citrate buffer 50 mM, pH 4.8, sodium azide (0.2 mg/mL), and enzymes (SAE0020, Sigma-Aldrich, St. Louis, MO, USA) at a cellulase load of 25 FPU/g DW of the substrate. Tubes were continuously mixed in an orbital shaker (200 rpm, 2 mm radius, 50 °C, 120 h). Reducing sugars were quantified by the DNS method described above. Dilutions of the enzyme mix were included as a control.

Lignin-derived polyphenols were quantified by a spectrophotometric analysis as alkali lignin equivalents. In detail, a calibration curve was constructed using dilutions containing known amounts of alkali lignin (Lignin, alkali 370959, Sigma-Aldrich) dissolved in a strongly alkaline solution (KOH 16 g/L). Absorbance values were recorded between 280 and 420 nm, with 2 nm intervals to calculate the area under the curve (AUC). The AUC is directly proportional to the concentration (g/L); thus, the polyphenol concentration in the diluted samples (DPC) could be determined. Immediately before analysis, liquid samples were opportunely diluted with an alkaline solution (KOH 16 g/L), recording the dilution factor (DF), then briefly centrifuged, and analyzed by a spectrophotometer (Infinite 200 PRO series, Tecan) in 96-well microplates (EIA/RIA Clear Flat Bottom Polystyrene Not Treated Microplate, 3591, Corning, Corning, NY, USA) to determine the polyphenol concentration in terms of alkali lignin equivalents. The same KOH solution used to prepare the dilutions was included as a blank. The polyphenol content in undiluted samples (UPC) as g/L was calculated according to the formula:

$$UPC = DPC \times DF \qquad (1)$$

Data reported are the mean of 3 experiments.

2.5. Toxicity Tests of the Extract towards Fungi and Oomycetes

A dose-response assay was performed in Petri dishes with five doses (i.e., 0, 10, 20, 40, and 80%, volume/volume, v/v) of the giant reed extract towards 22 fungal and oomycetes strains. After neutralization (pH 7) with H_2SO_4, the extract was slowly warmed up to

40 °C and mixed, under sterile conditions, to PDA medium, previously autoclaved and then cooled to 60 °C, to obtain the extract doses of 0, 10, 20, 40% v/v. To prepare the 80% dose, a SeaKem LE agarose (Cambrex, East Rutherford, NJ, USA)-based medium, supplemented with an appropriate amount of concentrated sucrose solution, was used instead of PDA to guarantee proper gelling. The final sucrose and agarose concentrations in the 80% dose were 10 and 12 g/L, respectively, as in the PDA. The amended substrates were poured into Petri plates (90 mm in diameter) and centrally inoculated with one agar-mycelium plug taken from the edge of actively growing colonies of the isolates (3 replicates). PDA and PDA supplemented with 5 g/L K_2SO_4 were included as controls, since the extract after neutralization with sulphuric acid contained an equivalent amount of sulphate.

Plates were incubated at room temperature (22 °C). For each plate, two perpendicular colony diameters were measured and averaged after 3–7 days of incubation, depending on the isolate, just before the colony reached the plate edge in the corresponding control (PDA). The growth inhibition percentage was calculated according to the formula:

$$I\ (\%) = \frac{C - T}{C} \times 100 \qquad (2)$$

where I (%) is the inhibition percentage of the colony growth, C is the averaged diameter (mm) of the pathogen colony in the control, and T is the averaged diameter (mm) in the amended plate.

For selected isolates, the half-maximal inhibitory concentration (IC_{50}) was calculated after probit linearization of the growth inhibition values I (%). Data were obtained from 2 independent experiments, with 3 replicates for each dose.

The plugs of the isolates showing 100% growth inhibition were transferred onto PDA plates and further incubated for up to 15 days to verify fungistatic/fungicidal effects.

For *Sclerotium rolfsii* and *Sclerotinia sclerotiorum*, after 20 days of incubation, the number of sclerotia produced in the amended and control plates (3 replicates) as well as their sensitivity were also recorded. To test the sensitivity, the sclerotia were collected, dried onto sterilized filter paper, and rehydrated again on filter paper imbibed with 2 mL of extract at the doses of 0 (distilled water, control), 20, 40, and 80% for 24 h. The imbibed sclerotia were transferred into 96-well microplates on PDA supplemented with streptomycin 500 μg/mL (1 sclerotium per well, 28 sclerotia per each dose and control). After 48 h incubation at room temperature, the percentage of germinated sclerotia was counted by visual observations under a stereomicroscope. The test was repeated.

2.6. Toxicity Test of the Extract towards Pythium ultimum 22 in Peat

The experiment was conducted in a greenhouse at DISTAL, in plastic trays (16 × 14 × 10 cm) filled with 500 g of peat (60% humidity), at temperatures of 21–24 °C, and under artificial lightning (16/8 h light/dark). The pathogen inoculum was prepared by finely mincing 72 h old colonies grown on a PDA medium with sterile distilled water [26]. Peat in each tray was evenly inoculated with 100 mL of the pathogen suspension at 2% w/w (4 replicates) and covered with a plastic bag. Three days after incubation in the dark, the peat was treated with the extract by irrigation (250 mL per tray, undiluted dose), then covered with plastic bags and further incubated for 48 h in the dark. Nontreated trays (control) received the same volumes of distilled water. In order to check the fungitoxic effect of the extract against *P. ultimum*, portions of treated-inoculated and untreated-inoculated (control) peat were sampled (3 g per tray, 12 g per treatment pooled together). For each treatment, 10 g of peat were utilized to determine the dry weight at 105 °C, while the remaining 2 g were stirred in sterile distilled water. Then, 1 mL of the dilutions 1:100 and 1:1000 (w/w) was plated on a semi-selective medium (PARP) [27] in Petri dishes. The characteristic pathogen colonies were visually counted after 48 h of incubation at 19 °C to determine the number of colony-forming units per gram of peat on a dry weight basis (CFU/g DW) (3 replicates).

2.7. Sensitivity Test of Zucchini to the Extract

Considering that the effectiveness of the extract would be evaluated in the pathosystem *P. ultimum*–zucchini, it was first examined if the extract could have any effects on zucchini seed germination and seedling growth. The sensitivity of zucchini to the extract was tested on filter paper, peat, and sand.

Test on filter paper. The test was conducted on Petri plates (90 mm in diameter). Ten seeds per plate were positioned between two disks of sterile filter paper, each imbibed with 5 mL of distilled water (dose 0%, control) or the extract at the doses of 20, 40, 60, and 80% v/v, 4 replicates. Plates were incubated at 25 °C for 7 days in the dark/light (8/16 h), when the number of normally germinated seeds according to standard procedures [28] and the rootlet length were recorded. The normal seedlings were oven-dried (105 °C, 48 h) up to constant weight to determine the dry weight.

Test in peat. This test was conducted in aluminum trays (10.5 × 8 × 4.5 cm) with holes in the bottom, filled with 50 g of peat (100 mL final volume), soaked with 25 mL of undiluted extract (prepared as reported in Section 2.4) or water (control), and covered with plastic film, 3 replicates. The 25 mL volume represented the highest dose that could be absorbed by 50 g of peat. After 48 h incubation at room temperature (22 °C), the trays were watered once with 25 mL of distilled water with the aim of partially removing the extract to reduce its potential phytotoxicity. Then, ten seeds per tray were sown, the trays were covered with transparent plastic film and incubated at 25 °C for 11 days in the dark/light (8/16 h). Finally, the percentage of normal seedlings according to standard procedures [29] was recorded, and the root apparatus development was visually compared to the untreated control.

Test in sand. This assay was performed in trays as above, filled with 300 g of sterile dry river sand moistened with 50 mL of water (200 mL final volume). The trays were treated with 40 mL of undiluted extract (prepared as reported in Section 2.4) or water (control), 3 replicates, and covered with plastic film. The 40 mL volume represented the highest dose that could be absorbed by 300 g of sand. After 48 h incubation at room temperature (22 °C), the trays were watered 4 times with 25 mL of water, aimed at partially removing the extract as specified above for peat. Ten seeds were sown in each tray, incubated at 20 °C for 14 days in the dark/light (8/16 h), and covered with transparent plastic film. Finally, the percentage of normal seedlings was recorded according to standard procedures [29], and the root apparatus development was visually compared to the untreated control.

For both the tests in peat and in sand, the substrate and the liquid percolated from the trays after watering were sampled for lignin-derived polyphenol analysis as described below (Section 2.9).

2.8. Pathogenicity Test of P. ultimum 22 towards Zucchini

The isolate *P. ultimum* 22 was checked for pathogenicity towards zucchini "Bolognese". The inoculum was prepared as described above (Section 2.6). Aluminum trays (10 × 8 × 4.5 cm) filled with autoclaved river sand (300 g) were inoculated with the pathogen (0.7% w/w) suspended in 50 mL of water, 3 replicates. Non-inoculated trays were included as the control. Each tray was sown with ten seeds and incubated at 20 °C in a growth chamber (16/8 h light/dark) for 11 days. After incubation, damping-off symptoms were checked, and the percentage of non-symptomatic plantlets was recorded.

2.9. Determination of the Lignin-Derived Polyphenols in Treated Peat and Sand

To determine the fate of water-soluble polyphenols in peat and sand treated with the giant reed extract (Section 2.7) a spectrophotometric analysis was performed (Section 2.4), quantifying the input amount (i.e., endogenously present in the substrate (END) plus exogenously added with the extract (EXO)) and its distribution after watering (i.e., percolated out of the tray (PER), free in solution (FREE), and adsorbed to the substrate (ADS)). This latter was calculated by difference according to Equation (2):

$$ADS = END + EXO - PER - FREE. \qquad (3)$$

Calculations were performed on a fresh weight (FW) basis, accounting for each fraction's total weight per tray. Then, the data obtained were converted and presented per liter of substrate (accounting for density). This choice allowed for a better comparison between peat and sand substrates characterized by very different density values.

Sampling and analysis. A total of 5 g FW of each substrate (peat or sand) treated with the extract or untreated (Section 2.7) was randomly collected soon after tray watering, then diluted 1:5 in water and centrifuged. Supernatants were filtered through 1 μm syringe filter and stored at 5 °C before analysis. The extract and the liquid percolated from treated and untreated trays after tray watering were carefully collected and weighted. Then, convenient samples were centrifuged, filtered, and stored at 5 °C before analysis. Samples were opportunely diluted with an alkaline solution (KOH 16 g/L), recording the dilution factor (DF), briefly centrifuged, and analyzed for the polyphenol content (DPC) as described in Section 2.4. The lignin-derived polyphenol content (grams of alkali lignin equivalents/grams of sample) in undiluted sampled materials (UPCMs) was thus calculated according to the Formula (4):

$$UPCM = \text{Undiluted sample specific volume} \times DPC \times DF, \qquad (4)$$

where the undiluted sample specific volume corresponds to the reciprocal of its density in g/L (previously measured).

2.10. Assessment of the Extract Efficacy in the P. ultimum 22–Zucchini Pathosystem

The experiment was performed in a growth chamber at CREA using river sand as a model growth substrate, because in this substrate, the partial removal of the extract resulted in more efficiency. The sand was distributed in aluminum trays, inoculated with the pathogen (0.7% w/w), and then treated with the extract utilizing the protocols described above (Section 2.6 for inoculum preparation and Section 2.7 for the extract treatment). Four treatments were compared according to a completely randomized block design with 4 replicates (trays): (1) inoculated with the pathogen and treated with the extract; (2) non-inoculated and non-treated control; (3) non-inoculated and treated control; (4) inoculated and nontreated control. Three days after the treatment, the trays were watered with the aim of partially removing the extract and then sown with ten seeds each. All trays received the same amount of liquids.

Eleven days after sowing, samples of 40 g of sand per thesis were collected (10 g per replicate, then pooled) to determine the extract's fungitoxic effect on the pathogen CFU. For each thesis, the dry weight at 105 °C was determined on 38 g of sand, while the remaining 2 g were suspended in distilled water and utilized for serial dilutions on PARP medium in Petri dishes (3 replicates) to determine the number of CFU/g of the pathogen, as already described (Section 2.6).

Thirteen days after sowing, the number of healthy plantlets was recorded, the root apparatus was checked, and the total dry weight was determined after oven-drying at 105 °C. The experiment was repeated.

2.11. Statistical Analysis

Depending on the experimental design, data were submitted to one-way or two-way analysis of variance, after arcsine transformation of percentage data. Means were separated by Tukey's test at $p \leq 0.05$ significance level. All analyses were performed with PAST 4.09 program.

3. Results

3.1. Extract Characterization

The results of the neutralized extract of giant reed were rich in lignin-derived polyphenols and potassium but low in nitrogen. It was also characterized by a C/N ratio of 63, the presence of phosphorus, and small amounts of holocellulose and free reducing sugars. (Table 3).

Table 3. Characterization of the giant reed extract obtained by treating a slurry of dry meal (10% weight/weight in water) with alkali (KOH 1.6% weight/weight in water) at 120 °C for 20 min after filtering and neutralization with H_2SO_4. The standard deviation (sd) of the mean is reported in brackets (n = 3 experiments).

Composition Parameters [2]	Mean Values (sd)
Water (%FM [1])	94.7 (0.2)
Total solids (TS [2]) (%FM)	5.3 (0.2)
Volatile solids (%FM)	2.6 (0.1)
Ashes (%FM)	2.7 (0.1)
pH in water	6.5–7.2
Electrical conductivity (S/m)	0.30 (0.01)
Total P (%FM)	0.009 (0.002)
Total K (%FM)	0.84 (0.02)
Total organic C (%FM)	2.0 (0.1)
C (% TS)	36 (0.5)
N (% TS)	0.57 (0.1)
C:N ratio (weight/weight)	63 (10)
Hydrolysable holocellulose (%FM)	0.42 (0.02)
Free reducing sugars (%FM)	0.14 (0.04)
Lignin-derived polyphenols (%FM) [3]	1.7 (0.15)

[1] FM: fresh matter; [2] TS: total solids; [3] obtained by dividing the measured concentration (g/L) of alkali lignin equivalents by 10.

Moreover, the presence of sulphate equivalent to 5 g/L K_2SO_4 was estimated based on the amount of sulphuric acid used for the extract neutralization.

The electrical conductivity value (0.3 S/m) was largely determined by K and sulphate ions.

3.2. Toxicity of the Extract towards Fungi and Oomycetes

The fungal and oomycetes isolates tested in vitro showed a differential response towards the doses of the giant reed extract, variable from sensitivity to stimulation, depending on the species and sometimes also on the isolate (Figure 2a). No significant differences in colony growth were observed between PDA and PDA amended with K_2SO_4 (Figure 2a); thus, all the data are expressed as the percentage of the control on PDA. Figure 2b shows a test of growth-recovery for different isolates performed transferring on unamended medium mycelium plugs that appeared completely inhibited on substrate amended with the giant reed extract.

Among the soilborne pathogens (Figure 3a,c), *Phoma betae* and *Fusarium oxysporum* L1 and 11.22 grew well, even in the presence of the highest extract dose, and the former was also highly stimulated compared to the control, up to a 40% dose (Figure 3c). Apart from these three isolates, the growth of the other soilborne isolates was gradually inhibited, partially or totally, with the increasing extract doses (Figure 3a). At an 80% dose, *Rhizoctonia solani*, *Ceratobasidium* sp., and *S. sclerotiorum* still showed growth values of 29, 29, and 13% of the control, respectively (Figure 3a,c). *Phytophthora cactorum* was the unique isolate showing a complete growth reduction from a 20% dose. Both *P. ultimum* isolates were highly inhibited at a 40% dose and did not grow at all at 80%, with the isolate 22 appearing more sensitive than 16 at the intermediate doses (Figure 3a). A strong effect was observed with *S. rolfsii* since the colony growth of both isolates was highly inhibited from the 20% and stopped from the 40% dose (Figure 3a). Two weeks later, sclerotia production by these two isolates was also significantly reduced at a 40% dose; differently, *S. sclerotiorum* continued to produce sclerotia up to the maximum tested dose, even with a significant reduction with respect to the other doses (Figure 3b). However, all the sclerotia produced by *S. rolfsii* 1 and 2 and *S. sclerotiorum* grown on extract-amended PDA were able to germinate when transferred on the unamended substrate.

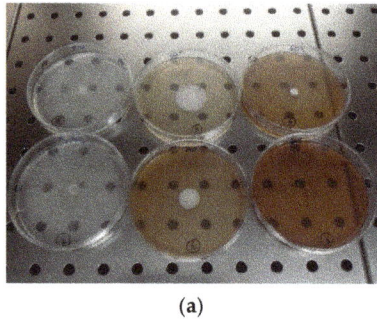 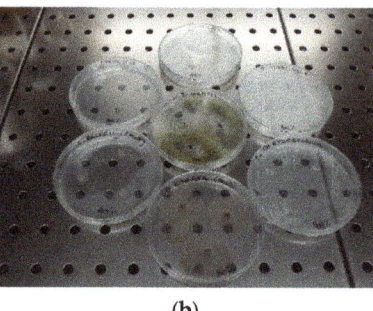

(a) (b)

Figure 2. (a) Effect of the increasing doses of the giant reed extract on the colony growth of a representative isolate (*Sclerotium rolfsii*); (b) test of the growth-recovery of different isolates on unamended medium after the occurrence of total inhibition on substrate amended with the giant reed extract.

Among the airborne pathogens, the two isolates of *Botrytis cinerea* behaved differently: the growth of the isolate 1 was gradually inhibited, while the isolate 2 was highly stimulated at the 20 and 40% doses, compared to control; nevertheless, both were almost completely inhibited at the highest dose (Figure 3d). Only one of the two isolates of *Alternaria alternata* (2) showed sensitivity to the extract, with a gradual growth inhibition up to 58% at the 80% dose. On the contrary, isolate 1 of *A. alternata* did not show any inhibition, even at the maximum dose, compared to the control, and it was also highly stimulated at the intermediate doses (Figure 3d).

The two nonpathogenic *Trichoderma* species (i.e., *T. gamsii* and *T. afroharzianum*) were gradually inhibited up to a complete stopping at the 80% dose (Figure 3e). Among the two pathogenic *Trichoderma* species, *T. pleuroti* behaved similarly to the nonpathogenic ones, while *T. pleuroticola* showed a growth reduction at the 40% dose, but it was never completely inhibited even at the maximum dose, still showing a growth of 31% compared to control (Figure 3e).

The white-rot fungi *Trametes versicolor* and *Phanerochaete chrysosporium* showed high sensitivity to the extract since their growth was dramatically reduced already at the 20% dose and stopped at 40% (Figure 3f). By contrast, the other wood decay agent, *P. ostreatus*, showed a moderate inhibition at this dose (33%), but it was completely inhibited at the dose of 80% (Figure 3f).

To further determine the extract's effect, the mycelium plugs that did not show any growth were collected and transferred on unamended PDA as shown in Figure 2b. Interestingly, almost all of the mycelium plugs were able to regrow, except those of *S. rolfsii* 2, *P. ultimum* 16, and *P. ultimum* 22, meaning that the inhibitory effect was irreversible for these isolates. Thus, *P. ultimum* 22 was selected for further experiments, also due to the higher sensitivity to the intermediate doses of the extract compared to isolate 16 (Figure 3a). This was also confirmed by the calculated IC_{50} values, which corresponded to a dose of 17.2% ± 0.1 and 23.5% ± 1.3% v/v, respectively.

The sensitivity of *P. ultimum* 22 to the extract was also confirmed in peat: 48 h after the treatment, the number of CFU/g peat DW was significantly reduced by 62% in infected + treated trays, while it did not vary significantly in the untreated trays (only infected) (Table 4). This is in accordance with the statistically significant interaction time × treatment found after two-way ANOVA.

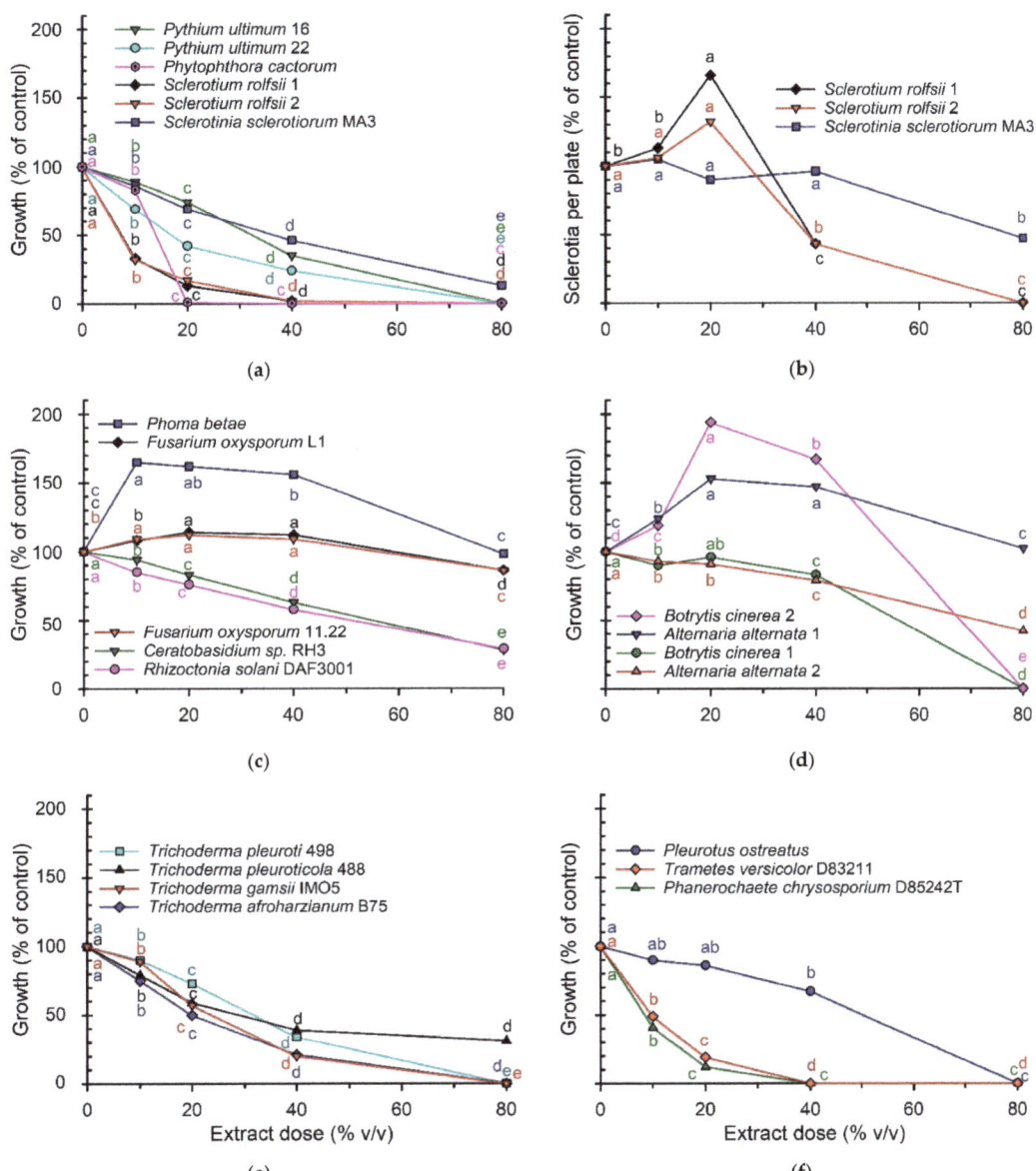

Figure 3. Screening of a collection of fungi and oomycetes for sensitivity to the giant reed extract in PDA plates amended with increasing doses of the extract (% volume/volume, v/v). Colony growth of (**a**,**c**) soilborne pathogens; (**d**) airborne fungi; (**e**) pathogenic and nonpathogenic *Trichoderma* spp.; (**f**) wood decay fungi. Panel (**b**) shows the sclerotia production by *Sclerotium rolfsii* and *Sclerotinia sclerotiorum*. Both the parameters (i.e., colony growth and number of sclerotia per plate) are expressed as a percentage of the control (0% dose). For each isolate, mean values sharing the same letter were not different at $p < 0.05$ according to Tukey's test after one-way ANOVA (n = 3). See also Tables S1 and S2 in the Supplementary Materials reporting mean and standard deviation values.

Table 4. Number of colony-forming units (CFUs) of *Pythium ultimum* 22 per gram of peat, on a dry weight basis, retrieved from the trays infected and treated with the undiluted giant reed extract, before and after the treatment (at 0 and 48 h) in comparison with the infected and nontreated trays.

Time from Treatment (h)	Treatment (CFU $\times 10^5$/g Peat Dry Weight) [1]	
	Infected	Infected + Treated
0	0.91 ab [2]	1.22 a
48	1.09 a	0.47 b

[1] Two-way ANOVA results: time ($p < 0.05$); treatment (not significant); time × treatment ($p < 0.05$); [2] Means in rows or columns followed by the same letter did not differ significantly according to Tukey's test ($p < 0.05$).

3.3. Sensitivity Test of Zucchini to the Giant Reed Extract

Test on filter paper. The extract negatively affected the germination of zucchini seeds on filter paper (Figure 4). The percentage of normal seedlings gradually decreased from 90% in the control down to 42% at the 40% dose. Complete inhibition was observed at the 60% dose. The root length and the dry weight were significantly reduced, as well, with the increasing doses (Figure 4).

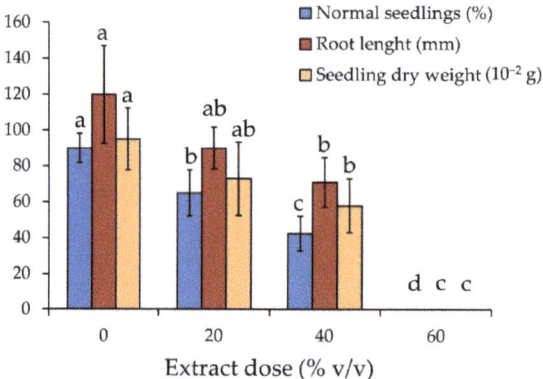

Figure 4. Effects of increasing doses (% volume/volume, v/v) of the giant reed extract on the germination of zucchini seeds, on filter paper, measured as a percentage of normal seedling, rootlet length, and total dry weight. Means labelled with the same letter within the same parameter did not differ for $p \leq 0.05$ (Tukey's test after one-way ANOVA). Bars represent standard deviation values.

Test in peat and sand. The use of the undiluted extract in peat and sand, followed by watering, did not reduce the emergence percentage of zucchini compared to untreated control, even if a slight but not statistically significant ($p \geq 0.05$) decrease was observed from 87 to 70% and from 97 to 83% in peat and in sand, respectively (Table S3). Thus, the watering step in peat and sand before sowing was effective in minimizing the inhibitory effect of the extract on the germinating seeds previously observed on filter paper.

3.4. Lignin-Derived Polyphenol Distribution in Peat and Sand

Figure 5 shows the distribution of lignin-derived polyphenols in peat compared to sand, both treated with the giant reed extract in tray experiments.

The analysis showed the presence of endogenous water-soluble polyphenols in the peat (shaded orange bar), while endogenous polyphenols were not detected in sand, as expected. The extract provided 3.3–3.6 g polyphenols per L substrate (exogenous). Watering did not remove completely the extract, but unexpectedly, both in the case of peat and sand, most of the polyphenol amount remained adsorbed to the substrate and only a relatively small amount remained free in solution. However, watering determined a relevant removal of polyphenols in sand only (percolated), corresponding to 33% of the added extract (exogenous) (Figure 5).

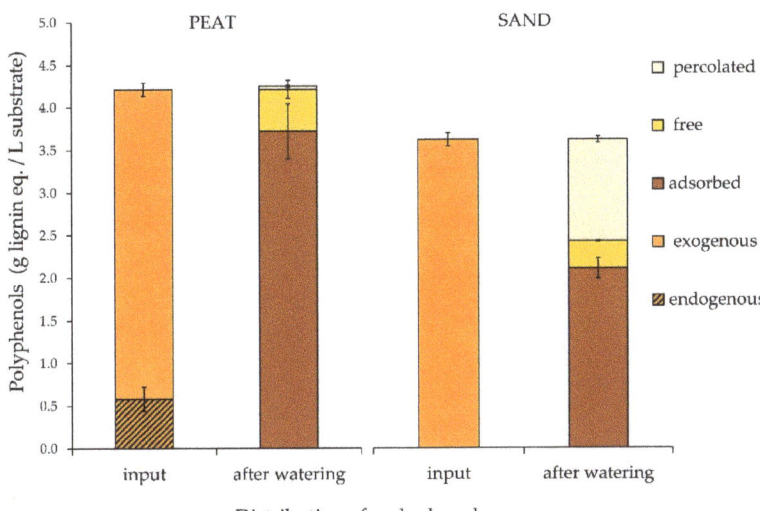

Figure 5. Distribution of lignin-derived polyphenols in peat and sand treated with undiluted giant reed extract in tray experiments. Columns represent the quantity of polyphenols in input (endogenously present in the substrate or exogenously added with the extract) and after watering (adsorbed into the substrate, free in solution, or percolated out of the tray). Bars represent standard deviation values ($n = 3$).

3.5. Extract Efficacy in the P. ultimum 22–Zucchini Patho-System

In a preliminary pathogenicity test, *P. ultimum* 22 was particularly aggressive towards zucchini, being able to attack the germinating seeds and reduce the emergence by 63.4% compared to the control. In addition, the development of the root apparatus in the inoculated trays appeared markedly reduced (Figure 6).

Figure 6. Pathogenicity test of *Pythium ultimum* 22 towards zucchini plantlets. Effect of development reduction of the root apparatus in plantlets from the inoculated trays (right) compared to the non-inoculated control (left).

In the experiment with the pathosystem, the treatment with the extract did not reduce the percentage of healthy plantlets compared to the control in the non-inoculated trays (Figure 7a). The pathogen infection instead halved the percentage of healthy plantlets (from 85 to 39%), while the treatment with the extract restored the percentage of healthy plants up to a value of 73%, which was not statistically different from the noninfected controls (Figure 7a). Moreover, the treatment with the extract significantly reduced the pathogen CFU/g by 90% compared to the control, measured 14 days after the treatment (Figure 7b).

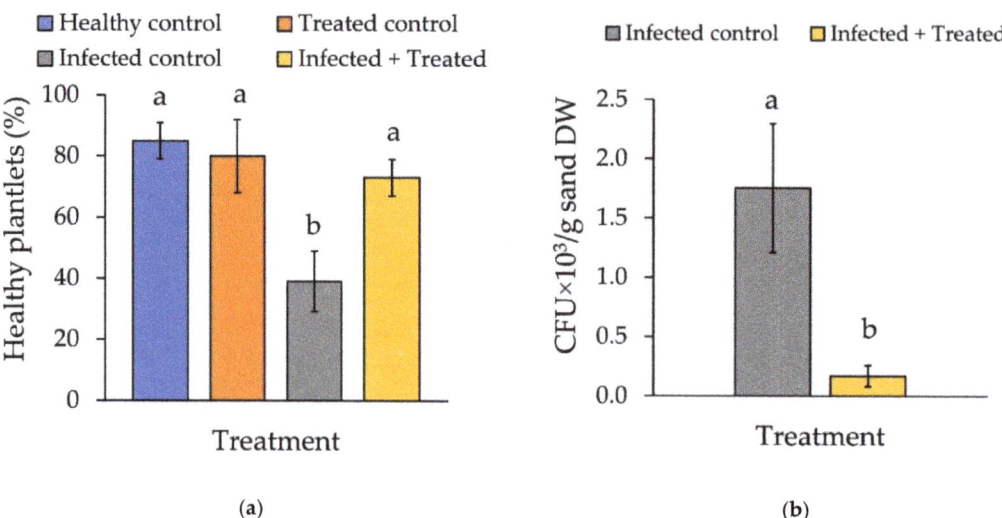

Figure 7. Effects of the treatment with the undiluted giant reed extract on the *Pythium ultimum* 22–zucchini pathosystem in sand: (**a**) percentage of healthy plantlets of zucchini in the infected and treated trays (yellow column) compared to the different controls (i.e., healthy, infected, or treated) 16 days after the treatment; (**b**) number of colony-forming units (CFUs) of *P. ultimum* 22 per g of sand dry weight (DW) in the infected and treated trays (yellow column) in comparison with the infected control, 14 days after the treatment. Bars represent the standard error of the mean. Columns labelled with the same letter did not differ for $p \leq 0.05$ (Tukey's test after one-way ANOVA).

Finally, the root apparatus of plants collected from the treated trays showed a development similar to that of the control, while it appeared reduced in the infected trays (Figure 8). Notably, the infected and treated plantlets showed a well-developed root apparatus if compared to that of the controls (Figure 8).

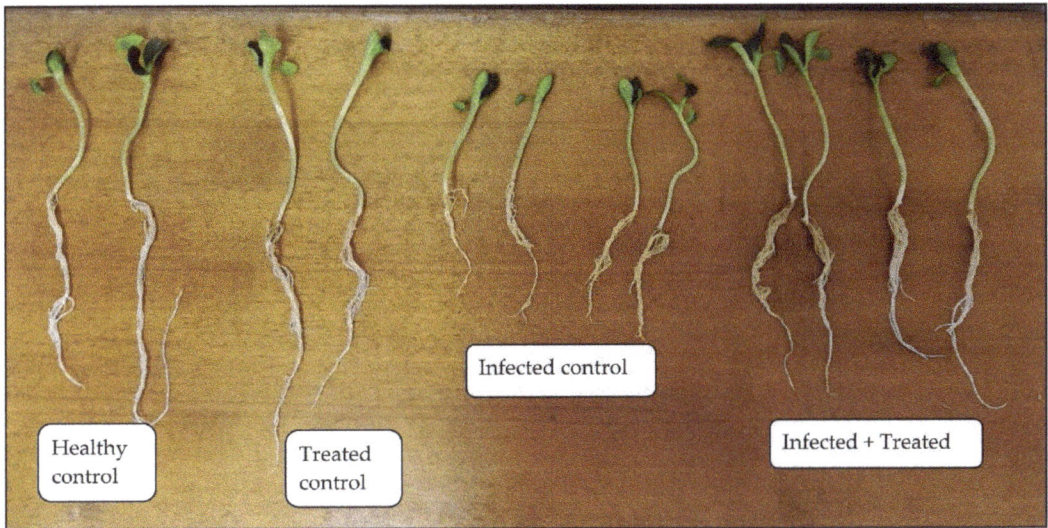

Figure 8. Effects of the treatment with the undiluted giant reed extract on the *Pythium ultimum* 22–zucchini pathosystem in sand. Plantlet development in the infected and treated trays (on the right) compared to the different controls (i.e., healthy, treated, or infected) 16 days after the treatment.

4. Discussion

4.1. Extract Characterization

The giant reed extract obtained by treating the giant reed dry biomass with potassium hydroxide can be considered a liquid organic amendment with possible ameliorative effects on the physicochemical characteristics of the soil.

Due to the high content of soluble lignin-derived compounds, the extract's application to the soil could improve soil physical characteristics while increasing C sequestration. Application of lignin provided a film-forming liquid mulch with the effect of improving soil aggregates, warming, preserving soil moisture, and protecting seedlings. [22]. In addition, soil pH, organic matter, cation exchange capacity, and available P and K can be increased by lignin amendment. Lignin application to arable soils also reduced the accumulation of heavy metals Cu, Zn, Cd, Pb, Cr, and Ni in wheat plants [30].

As a soil amendment, the use of an extract with low N content like this would not cause N leaching in areas where mandatory limits exist [31]; such a low N level was due to the plant harvesting at the end of the growth cycle when most of the nitrogen contained in the aboveground biomass had been translocated to the sink (rhizomes) for regrowth in the following season [32]. If needed, the N content of the extract can be easily increased by harvesting younger giant reed plants, which contain a high N level [12]. Thus, it is possible to modulate the fertilizing properties of the extract by utilizing plants harvested at different growth stages.

The electrical conductivity value of the extract (0.3 S/m) was far below the values of other organic fluids used as a soil amendment, for instance, the olive mill wastewaters (1 S/m) [33]. This confirms the possible use of this extract as a soil amendment.

The use of potassium hydroxide for the alkaline extraction, instead of another alkali (i.e., NaOH), enriched the extract with potassium, providing the extract with additional potential fertilizing properties. The use of KOH to pretreat the lignocellulosic biomass before anaerobic digestion for biogas production has been already proposed in the literature aiming to obtain a coproduct (digestate) with a high fertilization value [13,34].

Since nutrients can also affect disease resistance, K and P provided with the extract may also exert additional beneficial activities; K may promote the development of thicker outer walls in epidermal cells, thus preventing disease attack, while P may promote a vigorous root development permitting seedlings to escape the disease [35].

4.2. Toxicity of the Extract towards Fungi and Oomycetes

The giant reed extract contains biologically active compounds as shown by the broad fungitoxic activity observed in the screening of different species of fungi and oomycetes of agricultural interest. This biological activity could be related to the presence of lignin or soluble lignin-derived phenolic compounds, generated by the biomass treatment with potassium hydroxide. Indeed, the alkali used to obtain the extract removes the cross-links among xylan, lignin, and other hemicelluloses by solvation and saponification of the ester bonds and also further degrades lignin into different phenolic compounds [36,37].

The antimicrobial activity of lignin obtained from alkali-treated lignocellulosic feedstocks has already been reported in the literature: kraft lignin obtained from the paper industry by alkali treatment of spruce and eucalyptus showed strong activity against a range of human microbial pathogens and against the fungal plant pathogen *Aspergillus niger* [20]. Antimicrobial effects on Gram+ bacteria and yeasts have also been reported for alkali-extracted lignin from lignocellulosic maize biomass [21]. The antifungal activity of lignin-related phenolic compounds was investigated by a large in vitro screening that highlighted a higher toxicity towards true fungi (excluding *Zygomycetes*) than mitosporic ones [38].

In addition to lignin and derived compounds, the giant reed extract utilized in this work may also contain other bioactive molecules with antifungal properties. Indeed, high-temperature treatment of giant reed can generate a mixture of water-soluble compounds (such as furfural, 5-hydroxymethylfurfural, acetic, and formic acid) that have shown inhibitory activity towards pathogenic fungi of agricultural interest such as *A. alternata, B. cinerea, Colletotrichum acutatum, Cladosporium fulvum, F. oxysporum* f. sp. *lycopersici, F. oxysporum* f. sp. *melonis, F. solani* f. sp. *pisi*, and *V. dahliae* [19].

Surprisingly, the wood pathogens tested in our study (i.e., *P. chrysosporium, T. versicolor*, and *P. ostreatus*) resulted among the most sensitive isolates tested, while a good tolerance to lignin-derived compounds was expected since these strains can utilize lignin as a carbon source [39]. This finding suggests the possible presence of different kinds of inhibitory compounds in the giant reed extract.

Soilborne pathogens also appeared highly sensitive to the extract in the in vitro experiments; thus, we decided to verify, firstly, the giant reed extract's toxicity against *P. ultimum* 22 inoculated in a peat substrate and then its efficacy in the pathosystem model, *P. ultimum*–zucchini, in view of possible use in horticulture.

The two nonpathogenic *Trichoderma* isolates, representing the beneficial soil microflora, showed tolerance at low-extract doses. Notably, the ability of *Trichoderma* to tolerate and/or recover after soil treatment with fungitoxic compounds is well supported [40,41]. Therefore, it is possible that a soil treatment with the giant reed extract should not have a detrimental effect on soil *Trichoderma* species, in principle, but dedicated experiments in bulk soil should be performed to verify this effect.

4.3. Sensitivity of Zucchini to the Giant Reed Extract

The high sensitivity shown by zucchini in the filter paper test suggested the presence of anti-germinative compounds in the extract, as several potential allelochemicals have been characterized in giant reed including indoles, ketones, esters, and alcohols [42]. An anti-germinative effect of an aqueous extract of giant reed leaves reported for lentil growing on filter paper was attributed to a delay in the germination process and a reduction in the seedling vigor index [43]. The same extract applied to peat caused growth inhibition on lentils with a higher sensitivity of the rootlets compared to the shoots. These authors speculated that the phytotoxic effect could be attributed to a direct and more intensive contact between roots and allelochemicals and to a higher permeability of roots compared to shoot tissues [43]. Consequently, in our study on zucchini, the treatment of peat and sand with the extract was followed by a watering step to remove possible allelochemicals before sowing, and this actually allowed to obtain good emergence values. Lignin or lignin-derived compounds in the extract may cause anti-germinative effects or even turn to be biostimulants. Accordingly, ammonium lignosulfonate, derived from the paper industry, exerted an anti-germinative effect on beans, while a biostimulant effect was observed at low doses [44]. Moreover, in maize, very low doses (10 ppm) of water-soluble lignin isolated from miscanthus and giant reed biomass stimulated emergence and early growth [45]. A positive effect on tomato emergence and growth after field application of alkali-extracted lignin from flax was also reported [46].

4.4. Lignin-Derived Polyphenol Distribution in Peat and Sand

As already reported, among the possible allelochemicals contained in the giant reed extract, there are lignin-derived polyphenols generated from the lignocellulosic biomass treatment with alkali [37]. For this reason, we studied the distribution of polyphenols in peat and sand after treatment with the extract and watering to detect possible interactions between the substrates and the extract. The results showed that the watering step did not completely remove the polyphenols, which instead remained largely adsorbed by the substrates. This phenomenon was less marked in sand than in peat, also because leaching from sand is easier than from peat. However, the amounts of free polyphenols in the solution and available for seed imbibition and rootlet uptake resulted sufficiently diluted for good zucchini growth, both in peat and in sand, as observed in the sensitivity tests. The peat's capability to adsorb environmental pollutants, including polyphenols, is well recognized [47,48]. Recently, low-cost filters based on peat have been proposed to remove wood-derived polyphenols from water [49]. Various mechanisms are thought to be involved in the biosorption process; however, it is now recognized that ion exchange is the most prevalent mechanism [50]. Similarly, the capability to adsorb anionic, cationic, and nonionic compounds has been reported in the literature for various types of sand; thus, sand has been proposed as a low-cost adsorbing material to be used to remove coloring substances from industrial effluents [51].

4.5. Extract Efficacy in the P. ultimum–Zucchini Pathosystem

The experiment with the pathosystem *P. ultimum*–zucchini in sand confirmed the activity of the extract observed in the preliminary experiments. The giant reed extract applied at presowing in the infected substrate proved to be effective both in restoring the number of healthy plantlets, compared to nontreated control, and in reducing the pathogen CFU, as two weeks after the treatment a significant CFU reduction was still observed. Thus, the watering step after the treatment did not negatively interfere with the extract's activity. This was probably due to the adsorption phenomenon described above [51] that allowed to keep a reservoir of bioactive compounds in the system and maybe also allowed the treatment's efficacy to last over time.

While the literature reports some examples of in vitro bioactivity of lignin and lignin-related compounds towards microorganisms [20,21,38], few studies have been concerned with in vivo evaluations of these compounds for crop protection purposes. The present research reports for the first time the use of an extract of alkali-treated giant reed in a pathosystem model. De Corato et al. [19] highlighted the antifungal activity in a greenhouse of liquid wastes (SELWs) recovered during the bioethanol production process from different steam-exploded lignocellulosic biomass (*Miscanthus sinensis*, wheat straw and giant reed), even if giant reed SELW was the least effective. This harsh and highly energy-demanding technology is very effective in delignifying plant biomass but also partially degrades the holocellulose-generating toxic residues, such as furfurals and acetic and formic acids, which are responsible for the antifungal activity. Zschiegner [52] invented a plant growth regulator based on lignosulfonate that can enhance the natural resistance of plants against viruses, fungi, and bacterial pests. It cannot be excluded that our giant reed extract could also have the similar ability; thus, dedicated experiments could help to investigate possible priming or induced resistance mechanisms such as the activation of specific plant metabolic pathways and related marker genes [53].

This study reports preliminary findings on the bioactivity of the giant reed extract on a model pathosystem in tray experiments. With regard to possible application in the field, the extract dose used corresponded to 3–4.8 L m^{-2}, which is technically feasible. In addition, dose optimization or/and localized application in the field could also be considered with a reduction in the volume applied per surface unit. The technology necessary to obtain the extract is simple; moreover, the alkaline extraction could be carried out at room temperature, with obvious economic advantages [13].

Overall, the results of this study highlight an interesting potential for the use of the giant reed extract as a liquid soil improver with biological activity against soilborne pathogens such as *P. ultimum*. This pathogen affects the first crop phases and causes a pre- and post-emergence damping-off that is particularly insidious in horticulture because few chemical products are available for soil application. In addition, the repeated use of the same chemical product can increase the risk of developing resistance in the pathogen population. The phasing out of many chemicals for soil disinfestation, such as methyl bromide, has increased the demand for research regarding alternatives to controlling soilborne diseases and their recrudescence under climate change [54,55].

5. Conclusions

The awareness of the need to safeguard human and environmental health suggests the setting up of bio-based strategies in crop protection. Thus, this lignin-rich extract of alkali-treated giant reed could represent an innovative product to sustain crop health, with possible adjunctive fertilizing and amending properties. The results obtained with the model pathosystem, *P. ultimum*–zucchini, suggest both testing the extract's efficacy at a larger scale and enlarging the study to other soilborne pathogens that are particularly sensitive to the extract, such as *S. rolfsii*, or towards wood decay fungi. Further studies could regard the definition of the appropriate extract doses for practical use, the durability of the effectiveness of the extract's application in preventing disease, as well as a complete chemical characterization of the extract to highlight other possible bioactive compounds that may have a role in sustaining plant health.

Giant reed requires low pedo-climatic and agronomic inputs to produce huge biomass amounts; thus, the availability of biomass to produce the extract would not represent a limiting factor. The solvent-free process utilized for the extraction would also meet the demand for the adoption of environmentally friendly technologies. Finally, the utilization of potassium hydroxide for the alkali extraction would present an advantage to enrich the extract with K and would thus confer fertilizing properties.

Supplementary Materials: The following supporting information can be downloaded at: https://www.mdpi.com/article/10.3390/horticulturae8070589/s1, Table S1: Mean values, as a percentage of the control (0% dose), of the colony growth of different isolates of oomycetes and fungi in potato dextrose agar plates, amended with increasing doses of the giant reed extract. Standard deviations are reported in parentheses; Table S2: Mean values, as a percentage of the control (0% dose), of the number of sclerotia per plate produced by *Sclerotium rolfsii* and *Sclerotinia sclerotiorum* in potato dextrose agar plates, amended with increasing doses of the giant reed extract. Standard deviations are reported in parentheses; Table S3: Mean values of the emergence (%) of zucchini seedlings in trays filled with peat or sand treated with the undiluted giant reed extract in comparison with the untreated control. Standard deviations are reported in parentheses.

Author Contributions: Conceptualization, S.G. and R.R.; methodology, S.G., S.C., H.R. and R.R.; software, S.G. and S.C.; validation, S.G., S.C., H.R. and R.R.; formal analysis, S.G. and S.C.; investigation, S.G., S.C. and H.R.; resources, S.G.; data curation, S.G. and S.C.; writing—original draft preparation, S.G. and S.C.; writing—review and editing, S.G., S.C., H.R. and R.R.; visualization, S.C.; supervision, R.R.; project administration, S.G. and R.R.; funding acquisition, S.G. All authors have read and agreed to the published version of the manuscript.

Funding: This research was funded by the Italian Ministry of Agricultural, Food and Forestry Policies (MiPAAF) under the AGROENER project (D.D. no. 26329, 1 April 2016) http://agroener.crea.gov.it (accessed on 29 June 2022).

Institutional Review Board Statement: Not applicable.

Data Availability Statement: The data presented in this study are available upon request from the corresponding author.

Acknowledgments: The authors wish to acknowledge Enrico Ceotto for providing the giant reed biomass and the student Paolo Terzo Timoncini for their collaboration.

Conflicts of Interest: The authors declare no conflict of interest. The funders had no role in the design of the study; in the collection, analyses, or interpretation of data; in the writing of the manuscript, or in the decision to publish the results.

References

1. Haas, D.; Défago, G. Biological control of soil-borne pathogens by fluorescent pseudomonads. *Nat. Rev. Microbiol.* **2005**, *3*, 307–319. [CrossRef]
2. Chellemi, D.O.; Gamliel, A.; Katan, J.; Subbarao, K.V. Development and deployment of systems-based approaches for the management of soilborne plant pathogens. *Phytopathology* **2016**, *106*, 216–225. [CrossRef] [PubMed]
3. Gwinn, K.D. Bioactive natural products in plant disease control. In *Studies in Natural Products Chemistry*; Rahman, A., Ed.; Elsevier: Amsterdam, The Netherlands, 2018; Volume 56, pp. 229–246. [CrossRef]
4. Zaker, M. Natural plant products as eco-friendly fungicides for plant diseases control- a review. *Agriculturists* **2016**, *14*, 134–141. [CrossRef]
5. Niño, J.; Narváez, D.M.; Mosquera, O.M.; Correa, Y.M. Antibacterial, antifungal and cytotoxic activities of eight *Asteraceae* and two *Rubiaceae* plants from Colombian biodiversity. *Braz. J. Microbiol.* **2006**, *37*, 566–570. [CrossRef]
6. Bowers, J.H.; Locke, J.C. Effect of botanical extracts on the population density of *Fusarium oxysporum* in soil and control of Fusarium wilt in the greenhouse. *Plant Dis.* **2000**, *84*, 300–305. [CrossRef] [PubMed]
7. Reuveni, M.; Sanches, E.; Barbier, M. Curative and suppressive activities of essential tea tree oil against fungal plant pathogens. *Agronomy* **2020**, *10*, 609. [CrossRef]
8. Jayaraman, J.; Norrie, J.; Punja, K.Z. Commercial extract from the brown seaweed *Ascophyllum nodosum* reduces fungal diseases in greenhouse cucumber. *J. Appl. Phycol.* **2011**, *23*, 353–361. [CrossRef]
9. Lewandowski, I.; Scurlock, J.M.; Lindvall, E.; Christou, M. The development and current status of perennial rhizomatous grasses as energy crops in the US and Europe. *Biomass Bioenerg.* **2003**, *25*, 335–361. [CrossRef]
10. Antal, G. Giant reed (*Arundo donax* L.) from ornamental plant to dedicated bioenergy species: Review of economic prospects of biomass production and utilization. *Int. J. Hortic. Sci.* **2018**, *24*, 39–46. [CrossRef]
11. Al-Snafi, E.A. The constituents and biological effects of *Arundo donax*—A review. *Int. J. Phytopharm. Res.* **2015**, *6*, 34–40. Available online: https://api.semanticscholar.org/CorpusID:202637327 (accessed on 1 April 2022).
12. Ceotto, E.; Vasmara, C.; Marchetti, R.; Cianchetta, S.; Galletti, S. Biomass and methane yield of giant reed (*Arundo donax* L.) as affected by single and double annual harvest. *Glob. Change Biol. Bioenerg.* **2021**, *13*, 393–407. [CrossRef]
13. Vasmara, C.; Cianchetta, S.; Marchetti, R.; Ceotto, E.; Galletti, S. Potassium hydroxyde pre-treatment enhances methane yield from giant reed (*Arundo donax* L.). *Energies* **2021**, *14*, 630. [CrossRef]
14. Cianchetta, S.; Nota, M.; Polidori, N.; Galletti, S. Alkali pre-treatment and enzymatic hydrolysis of *Arundo donax* for single cell oil production. *Environ. Eng. Manag. J.* **2019**, *18*, 1693–1701.
15. Cianchetta, S.; Polidori, N.; Vasmara, C.; Ceotto, E.; Marchetti, R.; Galletti, S. Single cell oil production from hydrolysates of alkali pre-treated giant reed: High biomass-to-lipid yields with selected yeasts. *Ind. Crops Prod.* **2022**, *178*, 114596. [CrossRef]
16. Quave, L.C.; Plano, R.W.L.; Pantuso, T.; Bennet, C.B. Effects of extracts from Italian medicinal plants on planktonic growth, biofilm formation and adherence of methicillin-resistant *Staphylococcus aureus*. *J. Ethnopharmacol.* **2008**, *118*, 418–428. [CrossRef]
17. Shirkani, A.; Mozaffari, M.; Zarei, M. Antimicrobial effects of 14 medicinal plant species of Dashti in Bushehr province. *Iran. S. Med. J.* **2014**, *17*, 49–57. Available online: http://ismj.bpums.ac.ir/article-1-504-en.html (accessed on 1 April 2022).
18. Temiz, A.; Akbas, S.; Panov, D.; Terziev, N.; Alma, M.H.; Parlak, S.; Kose, G. Chemical composition and efficiency of bio-oil obtained from giant cane (*Arundo donax* L.) as a wood preservative. *Bioresources* **2013**, *8*, 2084–2098. [CrossRef]
19. De Corato, U.; Viola, E.; Arcieri, G.; Valerio, V.; Cancellara, A.F.; Zimbardi, F. Antifungal activity of liquid waste obtained from the detoxification of steam-exploded plant biomass against plant pathogenic fungi. *Crop Prot.* **2014**, *55*, 109–111. [CrossRef]
20. Gordobil, O.; Herrera, R.; Yahyaoui, M.; Ilk, S.; Kaya, M.; Labidi, J. Potential use of kraft and organosolv lignins as a natural additive for healthcare of products. *RSC Adv.* **2018**, *8*, 24525–24533. [CrossRef]
21. Dong, X.; Dong, M.; Lu, Y.; Turley, A.; Jin, T.; Wu, C. Antimicrobial and antioxidant activities of lignin from residue of corn stover to ethanol production. *Ind. Crops Prod.* **2011**, *34*, 1630–1631. [CrossRef]
22. Ahmad, U.M.; Ji, N.; Li, H.; Wu, Q.; Song, C.; Liu, Q.; Ma, D.; Lu, X. Can lignin be transformed into agrochemicals? Recent advances in the agricultural applications of lignin. *Ind. Crops Prod.* **2021**, *170*, 113646. [CrossRef]

23. Gazzettaufficiale.it. Available online: https://www.gazzettaufficiale.it/eli/gu/2001/01/26/21/sg/pdf (accessed on 24 November 2021).
24. Miller, G.L. Use of dinitrosalicylic acid reagent for determination of reducing sugar. *Anal. Chem.* **1959**, *31*, 426–428. [CrossRef]
25. Cianchetta, S.; Galletti, S.; Burzi, P.L.; Cerato, C. A novel microplate-based screening strategy to assess the cellulolytic potential of *Trichoderma* strains. *Biotechnol. Bioeng.* **2010**, *107*, 461–468. [CrossRef]
26. Righini, H.; Francioso, O.; Di Foggia, M.; Prodi, A.; Martel Quintana, A.; Roberti, R. Tomato seed biopriming with water extracts from *Anabaena minutissima*, *Ecklonia maxima* and *Jania adhaerens* as a new agro-ecological option against *Rhizoctonia solani*. *Sci. Hortic.* **2021**, *281*, 109921. [CrossRef]
27. Kannwischer, M.E.; Mitchell, D.J. The influence of a fungicide on the epidemiology of black shank of tobacco. *Phytopathology* **1978**, *68*, 1760–1765. [CrossRef]
28. ISTA. *International Rules for Seed Testing*; International Seed Testing Association: Zürich, Switzerland, 2010.
29. Don, R. *ISTA Handbook on Seedling Evaluation*; International Seed Testing Association: Zürich, Switzerland, 2006.
30. Zhang, S.; Wang, S.; Shan, X.-Q.; Mu, H. Influences of lignin from paper mill sludge on soil properties and metal accumulation in wheat. *Biol. Fertil. Soils* **2004**, *40*, 237–242. [CrossRef]
31. Wang, Z.H.; Li, S.X. Nitrate N loss by leaching and surface runoff in agricultural land: A global issue (a review). *Adv. Agron.* **2019**, *156*, 159–217. [CrossRef]
32. Nassi o Di Nasso, N.; Roncucci, N.; Triana, F.; Tozzini, C.; Bonari, E. Seasonal nutrient dynamics and biomass quality of giant reed (*Arundo donax* L.) and miscanthus (*Miscanthus* × *giganteus* Greef et Deuter) as energy crops. *Ital. J. Agron.* **2011**, *6*, 152–158. [CrossRef]
33. Dakhli, R.; Khatteli, H.; Ridha, L.; Taamallah, H. Agronomic application of olive mill waste water: Short-term effect on soil chemical properties and barley performance under semiarid Mediterranean conditions. *Int. J. Environ. Qual.* **2018**, *27*, 1–17. [CrossRef]
34. Jaffar, M.; Pang, Y.; Yuan, H.; Zou, D.; Liu, Y.; Zhu, B.; Korai, R.M.; Li, X. Wheat straw pretreatment with KOH for enhancing biomethane production and fertilizer value in anaerobic digestion. *Chin. J. Chem. Eng.* **2016**, *24*, 404–409. [CrossRef]
35. Dordas, C. Role of nutrients in controlling plant diseases in sustainable agriculture. A review. *Agron. Sustain. Dev.* **2008**, *28*, 33–46. [CrossRef]
36. Sun, Y.; Cheng, J.J. Hydrolysis of lignocellulosic materials for ethanol production: A review. *Bioresour. Technol.* **2002**, *83*, 1–11. [CrossRef]
37. McIntosh, S.; Vancov, T. Optimisation of dilute alkaline pretreatment for enzymatic saccharification of wheat straw. *Biomass Bioenerg.* **2011**, *35*, 3094–3103. [CrossRef]
38. Guiraud, P.; Steiman, R.; Seiglemurandi, F.; Benoitguyod, J.L. Comparison of the toxicity of various lignin-related phenolic compounds toward selected fungi perfecti and fungi imperfecti. *Ecotoxicol. Environ. Saf.* **1995**, *32*, 29–33. [CrossRef]
39. Cianchetta, S.; Di Maggio, B.; Burzi, P.L.; Galletti, S. Evaluation of selected white-rot fungal isolates for improving the sugar yield from wheat straw. *Appl. Biochem. Biotechnol.* **2014**, *173*, 609–623. [CrossRef]
40. Galletti, S.; Sala, E.; Leoni, O.; Burzi, P.L.; Cerato, C. *Trichoderma* spp. tolerance to *Brassica carinata* seed meal for a combined use in biofumigation. *Biol. Control* **2008**, *45*, 319–327. [CrossRef]
41. Madhavi, G.B.; Devi, G.U. Effect of combined application of biofumigant, *Trichoderma harzianum* and *Pseudomonas fluorescens* on *Rhizoctonia solani* f. sp. sasakii. *Indian Phytopathol.* **2018**, *71*, 257–263. [CrossRef]
42. Hong, Y.; Hu, H.Y.; Sakoda, A.; Sagehashi, M. Isolation and characterization of antialgal allelochemicals from *Arundo donax* L. *Allelopath. J.* **2010**, *25*, 357–368.
43. Abu-Romann, S. Allelopathic effect of *Arundo donax*, a mediterranean invasive grass. *Plant Omics J.* **2015**, *8*, 287–288. Available online: https://www.pomics.com/aburomman_8_4_2015_287_291.pdf (accessed on 1 April 2022).
44. Popa, V.I.; Dumitru, M.; Volf, I.; Anghel, N. Lignin and polyphenols as allelochemicals. *Ind. Crops Prod.* **2008**, *27*, 144–149. [CrossRef]
45. Savy, D.; Cozzolino, V.; Nebbioso, A.; Drosos, M.; Nuzzo, A.; Mazzei, P.; Piccolo, A. Humic-like bioactivity on emergence and early growth of maize (*Zea mays*, L.) of water-soluble lignins isolated from biomass for energy. *Plant Soil* **2016**, *402*, 221–233. [CrossRef]
46. Balas, A.; Popa, V.I. The influence of natural aromatic compounds on the development of *Lycopersicon esculentum* plantlets. *Bioresources* **2007**, *2*, 363–370. [CrossRef]
47. Mckay, G. Peat for environmental applications: A review. *Dev. Chem. Eng. Miner. Process.* **1996**, *4*, 127–155. [CrossRef]
48. Ahmaruzzaman, M. Adsorption of phenolic compounds on low-cost adsorbents: A review. *Adv. Colloid Interface Sci.* **2008**, *143*, 48–67. [CrossRef]
49. Svensson, H.; Marques, M.; Svensson, B.M.; Mårtensson, L.; Bhatnagar, A.; Hogland, W. Treatment of wood leachate with high polyphenols content by peat and carbon-containing fly ash filters. *Desalin. Water Treat.* **2015**, *53*, 2041–2048. [CrossRef]
50. Crini, G. Non-conventional low-cost adsorbents for dye removal: A review. *Bioresour. Technol.* **2006**, *97*, 1061–1085. [CrossRef]
51. Bello, O.S.; Bello, I.A.; Adegoke, K.A. Adsorption of dyes using different types of sand: A review. *S. Afr. J. Chem.* **2013**, *66*, 117–129. Available online: https://journals.co.za/doi/pdf/10.10520/EJC134491 (accessed on 1 April 2022).
52. Zschiegner, H. Plant Tonic for Induction and Improvement of Resistance. DE4404860A1, 31 August 1995.

53. Mauch-Mani, B.; Baccelli, I.; Luna, E.; Flors, V. Defense priming: An adaptive part of induced resistance. *Annu. Rev. Plant Biol.* **2017**, *68*, 485–512. [CrossRef]
54. Martin, F.N. Development of alternative strategies for management of soilborne pathogens currently controlled with methyl bromide. *Annu. Rev. Phytopathol.* **2003**, *41*, 325–350. [CrossRef]
55. Panth, M.; Hassler, S.C.; Baysal-Gurel, F. Methods for management of soilborne diseases in crop production. *Agriculture* **2020**, *10*, 16. [CrossRef]

Article

Foliar Applications of *Bacillus subtilis* HA1 Culture Filtrate Enhance Tomato Growth and Induce Systemic Resistance against *Tobacco mosaic virus* Infection

Hamada El-Gendi [1], Abdulaziz A. Al-Askar [2], Lóránt Király [3], Marwa A. Samy [4], Hassan Moawad [5] and Ahmed Abdelkhalek [4,*]

[1] Bioprocess Development Department, Genetic Engineering and Biotechnology Research Institute, City of Scientific Research and Technological Applications, Alexandria 21934, Egypt; elgendi1981@gmail.com
[2] Department of Botany and Microbiology College of Science, King Saud University, P.O. Box 2455, Riyadh 11451, Saudi Arabia; aalaskara@ksu.edu.sa
[3] Centre for Agricultural Research, Plant Protection Institute, ELKH, 15 Herman Ottó Str., H-1022 Budapest, Hungary; kiraly.lorant@atk.hu
[4] Plant Protection and Biomolecular Diagnosis Department, ALCRI, City of Scientific Research and Technological Applications, Alexandria 21934, Egypt; smarwa201291@gmail.com
[5] Agriculture Microbiology Department, National Research Centre, Cairo 12622, Egypt; hassan.moawad@gmail.com
* Correspondence: aabdelkhalek@srtacity.sci.eg; Tel.: +20-10-0755-6883

Citation: El-Gendi, H.; Al-Askar, A.A.; Király, L.; Samy, M.A.; Moawad, H.; Abdelkhalek, A. Foliar Applications of *Bacillus subtilis* HA1 Culture Filtrate Enhance Tomato Growth and Induce Systemic Resistance against *Tobacco mosaic virus* Infection. *Horticulturae* 2022, 8, 301. https://doi.org/10.3390/horticulturae8040301

Academic Editors: Hillary Righini, Roberta Roberti and Stefania Galletti

Received: 28 February 2022
Accepted: 30 March 2022
Published: 31 March 2022

Publisher's Note: MDPI stays neutral with regard to jurisdictional claims in published maps and institutional affiliations.

Copyright: © 2022 by the authors. Licensee MDPI, Basel, Switzerland. This article is an open access article distributed under the terms and conditions of the Creative Commons Attribution (CC BY) license (https://creativecommons.org/licenses/by/4.0/).

Abstract: The application of microbial products as natural biocontrol agents for inducing systemic resistance against plant viral infections represents a promising strategy for sustainable and eco-friendly agricultural applications. Under greenhouse conditions, the efficacy of the culture filtrate of *Bacillus subtilis* strain HA1 (Acc# OM286889) for protecting tomato plants from *Tobacco mosaic virus* (TMV) infection was assessed. The results showed that the dual foliar application of this culture filtrate (HA1-CF) 24 h before and 24 h after TMV inoculation was the most effective treatment for enhancing tomato plant development, with substantial improvements in shoot and root parameters. Furthermore, compared to non-treated plants, HA1-CF-treated tomato had a significant increase in total phenolic and flavonoid contents of up to 27% and 50%, respectively. In addition, a considerable increase in the activities of reactive oxygen species scavenging enzymes (PPO, SOD, and POX) and a significant decrease in non-enzymatic oxidative stress markers (H_2O_2 and MDA) were reported. In comparison to untreated control plants, all HA1-CF-treated plants showed a significant reduction in TMV accumulation in systemically infected tomato leaves, up to a 91% reduction at 15 dpi. The qRT-PCR results confirmed that HA1-CF stimulated the transcription of several defense-related tomato genes (*PR-1*, *PAL*, *CHS*, *and HQT*), pointing to their potential role in induced resistance against TMV. GC–MS analysis showed that phenol, 2,4-bis (1,1-dimethylethyl)-, Pyrrolo [1,2-a] pyrazine-1,4-dione, hexahydro-3-(2-methylpropyl)- and eicosane are the primary ingredient compounds in the HA1-CF ethyl acetate extract, suggesting that these molecules take part in stimulating induced systemic resistance in tomato plants. Our results imply that HA1-CF is a potential resistance inducer to control plant viral infections, a plant growth promoter, and a source of bioactive compounds for sustainable disease management.

Keywords: *Bacillus subtilis*; *Tobacco mosaic virus*; oxidative stress; antioxidative enzymes; gene expression; GC–MS

1. Introduction

The growing global population, as well as urbanization and global climate changes, generate an ever-increasing demand for high crop yields and improved food quality. Several strategies are being dedicated to accelerate and improve the rate of agricultural production. However, plant diseases account for significant crop losses and delay the ongoing progress of crop management [1]. Viral infections of plants represent a great threat to plant biosecurity, causing tremendous crop losses worldwide [2]. *Tobacco mosaic virus* (TMV) is one of the most infectious plant pathogens, attributed to its wide range of plant hosts (about 66 families with more than 900 plant species) and the severe consequences of its infection [3]. TMV is transmitted mechanically by rubbing with infected plants, contaminated agriculture tools, and/or contaminated seeds [4]. In addition, the extraordinary stability of viral particles ensures their persistence for years in e.g., fallen infected leaves or soil with full capacity for re-infection [5]. In tomatoes, TMV infection is associated with different morphological symptoms, including a systemic leaf mosaic and/or necrosis, in addition to leaf chlorosis [6]. Furthermore, the infection could develop into systemic changes in flowering organs that delay fruit ripening, affect crop productivity, and result in crop loss [7].

Controlling TMV infection is challenging and relies upon the use of resistant plant cultivars or prevention of vector spreading through intensive application of insecticides, which may cause adverse complications for human health and the environment [8]. Furthermore, pesticides eventually make their way into the surface–water system, complicating and exacerbating environmental and ecosystem issues [9,10]. For sustainability, in either agriculture or the environment, interest is growing in the application of biocontrol agents as eco-friendly substitutes for the hazardous chemicals currently applied in plant disease management strategies [11].

Plant growth-promoting rhizobacteria (PGPR) are natural rhizosphere microbiota that promote plant growth and resistance to various infections. Numerous studies have found that PGPR can improve plant growth and increase resistance to viral infection [11,12]. PGPRs enhance plant growth through promoting nutrient acquisition and the production of biomolecules involved in stress tolerance. These processes may either indirectly decrease the susceptibility of plants to infection and/or directly fight pathogens through antibiotic production or competition for essential nutrients [13,14]. Among others, strains of *Bacillus* sp. are widely applied as PGPRs with significant application outcomes attributed to their multiple mechanisms for infection control and plant growth stimulation mediated by numerous secondary metabolites [15,16]. However, under field conditions, the persistence and adaption of PGPR inocula to the natural microbiota are strain-dependent and heavily affected by application practices that usually lead to inconclusive outcomes [15]. Furthermore, the influence of PGPR inocula on the natural microbiota community at the site of application is unknown, with the possible potential of increasing antagonistic resistance [17–19]. Due to these limitations, the application of microbial culture filtrates has emerged as an eco-friendly, dependable alternative to PGPRs [15].

Biocontrol agents protect plants through the stimulation of induced systemic resistance (ISR) by activating a variety of cellular processes. These processes include (i) the up- or down-regulation of specific genes and the overexpression of specific transcription factors; (ii) changes in the levels of various compounds implicated in defense pathways and enhanced plant growth; (iii) activation of defense genes encoding reactive oxygen species (ROS) scavenging enzymes such as POX, SOD, CAT, and PPO, as well as the accumulation of extracellular pathogenesis-related (PR) proteins; and (iv) improved transport of macromolecules, phytohormones, and enzymes involved in defensive signaling [3,20–23].

The goals of the present study were to evaluate the ability of *Bacillus subtilis* strain HA1-culture filtrate (HA1-CF) to promote tomato growth and confer protection to TMV infection, either through direct antiviral activity or the induction of systemic resistance in tomato plants. To investigate the resistance induction mechanism, the activity of reactive oxygen species scavenging enzymes (PPO, SOD, and POX) and non-enzymatic oxidative stress markers (H_2O_2 and MDA), as well as total phenolic content and flavonoid content were assessed. Furthermore, the expression levels of several defense-related tomato genes, including *PR-1, PR-2, PAL, CHS,* and *HQT*, along with TMV accumulation inside tomato tissues, were also estimated. Finally, the gas chromatography–mass spectrometry (GC–MS) technique was used to screen and identify the bioactive constituents of HA1-CF.

2. Materials and Methods

2.1. Plant Material and Viral Source

Tomato (*Solanum lycopersicum* L.) plants of the GS 12 cultivar used during this work were purchased as virus-free seeds from the Agriculture Research Center, Egypt. The TMV strain KH1 (accession number MG264131) previously isolated from TMV-infected tomato plants, was used as a source of viral inoculum [6].

2.2. Bacterial Isolation

Bacterial isolation was conducted on nutrient agar (NA) plates with the following composition: peptone 5 g/L, yeast extract 3 g/L, NaCl 5 g/L, and agar 15 g/L [24]. Five soil samples were collected from the rhizosphere of tomato plants that appeared healthy and vigorous in different open fields in Alexandria governorate, Egypt. Soil samples were taken at a 5–15 cm root depth after removing approximately 3 cm of the soil surface. Each sample (10 g) was a mix of five samples (2 g) collected from the rhizosphere of various tomato plants at the same site. Subsequently, each sample (10 g) was suspended in 100 mL of saline solution (0.9% NaCl) for 30 min under shaking. Serial dilutions (1×10^{-4}, 1×10^{-5}, and 1×10^{-6}) were prepared, and 100 µL of each dilution were streaked aseptically on triplicate NA plates and incubated at 30 °C for 24 h. Different bacterial colonies of 10^{-6} dilution were chosen, based on colony shape, colony color, and antagonistic activity against growing fungi on the NA plates. To ensure purity, the newly developed separate colonies were streaked on the same medium. To obtain culture filtrates, all isolates were cultivated separately in nutrient broth (Agar-free nutrient medium) for 48 h under shaking at 200 rpm at 30 °C. The bacterial culture filtrate (CF) was obtained as follows: centrifugation (10 min, 10,000 rpm), collection of supernatant, and filtration with a 0.45 µm pore-size syringe filter. Using the half-leaf method [25], the antiviral potency of the purified isolates was tested in *Chenopodium amaranticolor* plants, which serve as a local lesion host for TMV. Briefly, the upper right half of the leaves were treated with 100 µL of bacterial CF, while the left half of the leaves were treated with 100 µL of sterilized nutrient broth media. After 24 h, both halves of the leaves were mechanically inoculated with TMV. The experiment was performed in three biological replicates. According to the inhibition percentage in relation to the number of local lesions, the isolate that exhibited the most potent antiviral activity was selected for further experiments.

2.3. Isolate Identification through 16S rRNA Methodology

The isolate showing the maximum antiviral potency was subjected to molecular identification through 16S rRNA sequencing. Total genomic DNA was isolated from an overnight culture, using the Wizard Genomic DNA Purification Kit (Promega, Fitchburg, WI, USA) according to the manufacturer's instructions. The PCR amplification of the 16S rRNA gene was conducted using two universal primers (Table 1). The amplified 16S gene was purified by a PCR purification kit (QIAGEN, Hilden, Germany) and sequenced through a Genetic Analyzer system (3130xl, Applied Biosystems, Bedford, MA, USA) using a BigDye Terminator v3.1 Cycle Sequencing kit. The annotated nucleotide sequence was analyzed using NCBI-BLAST and deposited in GenBank. The phylogenetic relationships of the potent isolate were elucidated through MEGA software (ver. 11) using an unweighted pair group method with arithmetic mean (UPGMA) and a bootstrap method with 2000 replications.

Table 1. List of the specific primer sequences of the different genes used in qRT PCR analysis.

Primer Name	Abbreviation	Nucleotide Sequence	References
16S ribosomal RNA	16S rRNA	Forward: AGAGTGATCCTGGCTCAG Reverse: GGTTACCTTGTTACGACTT	[26]
Tobacco mosaic virus-coat protein	TMV-CP	Forward: ACGACTGCCGAAACGTTAGA Reverse: CAAGTTGCAGGACCAGAGGT	[27]
Pathogenesis related protein-1	PR-1	Forward: GTTCCTCCTTGCCACCTTC Reverse: TATGCACCCCCAGCATAGTT	[28]
Endoglucanase	PR-2	Forward: TATAGCCGTTGGAAACGAAG Reverse: CAACTTGCCATCACATTCTG	[28]
Phenylalanine Ammonia-Lyase	PAL	Forward: ACGGGTTGCCATCTAATCTGACA Reverse: CGAGCAATAAGAAGCCATCGCAAT	[29]
Chalcone Synthase	CHS	Forward: CACCGTGGAGGAGTATCGTAAGGC Reverse: TGATCAACACAGTTGGAAGGCG	[29]
Hydroxycinnamoyl Co A: quinate (break)hydroxycinnamoyl transferase	HQT	Forward: CCCAATGGCTGGAAGATTAGCTA Reverse: CATGAATCACTTTCAGCCTCAACAA	[29]
β-actin	β-actin	Forward: TGGCATACAAAGACAGGACAGCCT Reverse: ACTCAATCCCAAGGCCAACAGAGA	[30]

2.4. Greenhouse Experimental Design and Assessment of Growth Parameters

Under greenhouse conditions, the tomato seeds were grown in plastic pots (30 cm diameter, 29.9 cm height). Each pot was provided with 4 kg of mixed sand and clay (1:1), previously sterilized through autoclaving. Tomato plants were incubated at a day/night temperature of 28 °C/16 °C with a relative humidity of 70%. The seedlings were transplanted into new pots on the 28th day after sowing, and one week later, the leaves (two upper true leaves) of each plant were mechanically inoculated with semi-purified TMV (1 mL) as described before [31]. The experiment was carried out with five different treatments, each of which included five repetitions and five tomato plants in each pot. Each treatment had five biological replicates. Each biological replicate was a pool of 15 tomato leaves collected from the five plants (3 leaves from new systemically infected leaves/plant) in each pot. Each biological replicate was run in three technical replicates for each analysis evaluation. The five treatments were as follows: the first treatment (NT) was the mock (control) group; the second group (TMV) was the viral-infected group; the third group (T1) was tomato plants sprayed with bacterial CF 24 h before viral infection; the fourth group (T2) was tomato plants sprayed with bacterial CF 24 h after viral infection; and the final group (T3) was allocated to plants sprayed with bacterial CF two times, 24 h before and 24 h after the viral inoculation. The whole plant shoots were foliar sprayed with a handheld pressure sprayer until runoff occurred and the leaves seemed to be coated with the CF. All plants were kept under insect-proof greenhouse conditions and were daily observed for the development of mosaic symptoms over the course of 3 weeks. At 22 days post-TMV inoculation (dpi), plants from each group were collected, rinsed several times with water, and evaluated for their fresh weight (g), shoot length (cm/plant), and root length (cm/plant). The plant's dry weight (g) was determined after drying at 50 °C for a constant weight.

2.5. Oxidative Stress Markers

2.5.1. Malondialdehyde (MDA) Determination

Malondialdehyde (MDA) levels were assessed in all treatments by using thiobarbituric acid (TBA), as in Heath and Packer [32]. Briefly, 100 mg of tomato leaf samples was ground in 1 mL of 0.1% trichloroacetic acid (TCA) and centrifuged at 10,000 rpm for 30 min. Sample supernatants (1 mL) were mixed separately with 4 mL of TBA solution (0.5% TBA: 20% TCA) and incubated at 95 °C for 30 min. The reaction was terminated through immediate immersion in ice, where the developed color was measured at 600 nm, indicating the malondialdehyde concentration (μM/g of fresh weight).

2.5.2. Hydrogen Peroxide Determination

Hydrogen peroxide (H_2O_2) was determined in the fresh plant samples using KI as described by Junglee et al. [33], with a few modifications. The fresh plant samples (100 mg) were separately homogenized in 0.1% TCA and centrifuged to obtain a clear homogenate. The H_2O_2 reaction was conducted by adding 1 mL of plant homogenate to 2 mL of KI solution (1 M KI in 10 mM phosphate buffer, pH 7.0). After 20 min, the reaction absorbance was measured at 390 nm, with results deducted using the H_2O_2 extinction coefficient (0.28 M^{-1} cm^{-1}) and expressed as μM/g fresh weight.

2.6. Determination of Antioxidant Enzymatic Activities

2.6.1. Polyphenol Oxidase (PPO)

The PPO activity was determined by quinone methods [34]. In brief, 500 μL of crude plant extract was added to 1 mL of 50 mM quinone (in 100 mM Tris-HCl buffer pH 6.0) and incubated for 10 min at 25 °C. The reaction absorbance was measured at 420 nm where a 0.001 increase in the absorbance was equivalent to one unit of enzyme activity/min and expressed as μM/g fresh weight.

2.6.2. Superoxide Dismutase (SOD)

The SOD activity was determined through the nitroblue tetrazolium (NBT) photoreduction inhibition method with minor modifications [35]. Crude plant extract (100 μL in phosphate buffer pH 7.0) was added to 50 μM NBT, 10 μM riboflavin, 0.1 mM EDTA, 50 mM sodium carbonate, and 12 mM L-methionine. A 50 mM phosphate buffer, pH 7.6, was added to adjust the final reaction volume to 3 mL. The reaction mixtures without plant extract were considered as controls. The mixtures were exposed to fluorescent lamps for 15 min to initiate the photochemical reaction, before being placed in the dark and then measured at 560 nm. The inhibition of photochemical reduction (50%) was defined as one unit of enzyme activity [36]. The activity of SOD was expressed as μmol/g fresh weight.

2.6.3. Peroxidase (POX)

The POX activity was evaluated according to the method of Angelini et al. [37]. In brief, crude plant extract (80 μL in phosphate buffer pH 7.0) was added to 500 μL of 5 mM guaiacol and 120 μL of 1 mM hydrogen peroxide to a final volume of 1200 μL adjusted by 100 mM phosphate buffer pH 7.0. After that, the reaction was incubated for 10 min at 30 °C and the absorbance was measured at 480 nm. The extinction coefficient of $\varepsilon = 26{,}600$ M^{-1} cm^{-1} was used to calculate the results.

2.7. Determination of Total Phenolic Contents

The total phenolic content in the dried plant samples was assayed according to the Folin–Ciocalteau method [38]. Dried plant samples (0.5 g) were extracted with 80% methanol (25 mL) after shaking for 24 h. After extraction, 400 μL of the clear plant extract was added to 2 mL of Folin–Ciocalteau reagent for 5 min, and then 1.6 mL of Na_2CO_3 (7.5%) was added at room temperature with a vortex. The reaction mixture was incubated in the dark for 1 h and the developed color was measured at 760 nm, with the results expressed using a gallic acid standard curve.

2.8. Determination of Total Flavonoid Contents

The total flavonoid content in the plant samples was evaluated through an aluminium chloride colorimetric approach adapted from Ghosh et al. [39] as follows: 500 μL of plant extract (in phosphate buffer pH 7.0) was added to 100 μL of aluminium chloride (10%), 100 μL of potassium acetate (1 M), and 1500 μL of methanol. The final reaction volume was adjusted to 5 mL with distilled water, and incubated for 30 min at room temperature. After incubation, the absorbance was measured at 415 nm, where results were expressed using a standard curve for quercetin.

2.9. Quantitative RT-PCR (qRT-PCR) Assay and Data Analysis

2.9.1. Total RNA Extraction and cDNA Synthesis

Total RNA was extracted from around 100 mg of tomato leaves using the RNeasy plant mini kit (Qiagen, Hilden, Germany). The RNA integrity was examined by visualizing the quality of 28S and 18S rRNA bands separated by 1.2% agarose gels by electrophoresis, whereas purity (A_{260}/A_{280}) and concentration were determined by the SPECTRO Star Nano instrument (BMG Labtech, Ortenberg, Germany). For each sample, 1 µg of DNase I-treated RNA was used as a template to synthesize cDNA in a reverse transcription reaction using oligo (dT) and random hexamer primers, as described in previous studies [40,41]. The final cDNA product was stored at −20 °C until employment as a qRT-PCR template.

2.9.2. TMV-CP Accumulation and Defense Genes Expression Levels

The transcriptional levels of two tomato genes encoding for pathogenesis-related proteins (*PR-1* and *PR-2*) and three genes involved in polyphenol metabolism (*PAL*, *CHS*, and *HQT*), as well as transcript accumulation of the TMV-coat protein gene (*TMV-CP*), were evaluated in all treatments and compared to controls using the qRT-PCR technique. Based on the expression levels of the *TMV-CP* gene and the housekeeping gene (*β-actin*) in the control treatment, viral accumulation levels were determined [42]. The gene expression levels were also normalized by using *β-actin* gene expression as a housekeeping gene. The nucleotide sequences of the primers are listed in Table 1. PCR reactions for each biological treatment were separately performed in a real-time thermocycler (Rotor-Gene 6000, QIAGEN, Germantown, MD, USA) using a SYBR Green Mix (Thermo, Foster, CA, USA) as previously described [43,44]. For each tested gene, relative expression levels were accurately calculated according to the $2^{-\Delta\Delta CT}$ method [45].

2.10. Assessment of Active Biomolecules in the Bacterial Culture Filtrate through Gas Chromatography–Mass Spectrometry

The active biomolecules in the bacterial CF were identified by gas chromatography-mass spectrometry (GC–MS) upon extraction with ethyl acetate. Ethyl acetate was added to the CF in a ratio of 1:1 and vigorously shaken for 15 min. The ethyl acetate phase was separated and concentrated in a rotatory evaporator at 50 °C. The concentrated extract was analyzed through GC–MS (TRACE 1300 Series, Thermo, Waltham, MA, USA), equipped with a split mode mass detector, with helium gas as a carrier at a flow rate of 1 mL/min. The injector was set to 250 °C and the oven was set to 60 °C for 2 min, with a scan time of 0.2 s, a mass range of 50–650 amu, and a 20-min ramp to 250 °C. During the running time of 53 min, mass spectra were obtained at 70 eV. The CF components were identified by comparing them to data in the literature and the GC–MS library.

2.11. Statistical Analysis

Using the GraphPad Prism software (version 6.01, San Diego, CA, USA), all of the obtained data was statistically evaluated using a one-way ANOVA. Significant differences were calculated according to the least significant difference (LSD) method at $p \leq 0.05$ level of probability. Standard deviation (±SD) is represented numerically in tables and as a column bar in histograms. When compared to mock-inoculated tomato tissues (NT treatment), the relative expression values of more than one indicated an increase in gene accumulation, whereas values of less than one indicated a drop in expression levels.

3. Results and Discussion

3.1. Bacterial Isolation and Molecular Characterization

Among the 25 selected purified bacterial isolates, the CF of the isolate coded HA1 displayed maximum antiviral activity and hence was selected for further application in the following experiments. An NCBI-BLAST analysis revealed that the nucleotide sequence of the full length 16S RNA of HA1 has a similarity of 100% to isolates of *Bacillus subtilis*. *B. subtilis* is a naturally occurring soil microorganism that is commonly isolated from the rhizosphere of various plants [46,47]. Based on the homology results of the HA1 isolate, the isolate was putatively defined as *B. subtilis* and the annotated sequence was deposited in the GenBank under the accession number OM286889. In addition, a phylogenetic tree analysis, constructed through MEGA 11, revealed that HA1 is closely related to other *B. subtilis* strains, especially the Egyptian isolate (Acc# MT222787, strain SE05); therefore, HA1 belongs to the evolutionary lineage of *B. subtilis* (Figure 1).

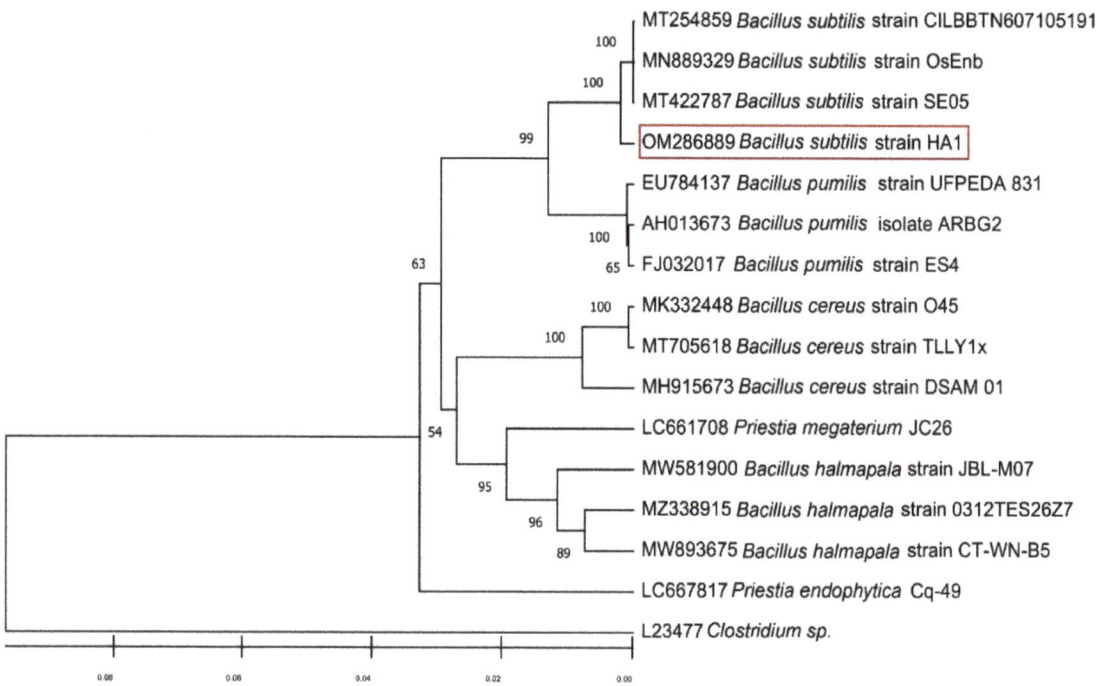

Figure 1. Phylogenetic tree showing the relationship of the locally isolated *B. subtilis* strain HA1 (shows in red rectangle) and other closely related isolates based on 16s rRNA (GenBank) nucleotide sequences. The tree was developed by the MEGA 11 program, based on the UPGMA method, with a bootstrap method of 2000 replications.

3.2. Effect of HA1-CF on Development of TMV Symptoms and Growth Parameters

In a greenhouse experiment, the foliar application of *B. subtilis* strain HA1-CF significantly reduced disease severity following TMV infection, enhanced tomato plant growth, and decreased viral accumulation in all treated tomato plants (T1, T2, and T3) compared to non-treated plants (TMV treatment). The obtained results showed that the TMV treatment (i.e., tomato plants inoculated with TMV only) developed chlorosis patterns and severe mosaic symptoms at 14 dpi (Figure 2), similar to those previously described [3,8]. On the other hand, and compared to TMV plants, the symptom development of T1 (tomato plants treated with HA1-CF 24 h before TMV inoculation) and T2 (tomato plants treated with HA1-CF 24 h after TMV inoculation) plants was delayed by five and three days, respectively. Moreover, dual foliar application of HA1-CF (T3), 24 h before and after TMV inoculation, showed a 7-day delay in symptoms' appearance. The slow increase in symptomatic plant numbers or delayed symptom development in HA1-CF-treated plants may be attributed to an obstruction of virus movement or replication. Many authors report that the foliar application of bacterial CF is associated with delays in the development of plant virus-elicited symptoms [3,11,48,49]. The NT (tomato plants treated with virus-free inoculation buffer and foliar sprayed with bacterium-free broth medium) plants showed no symptoms (Figure 2).

Figure 2. Effect of foliar application of *B. subtilis* strain HA1-CF on the development of disease symptoms in tomato leaves at 14 days post-TMV inoculation. NT: mock plants; TMV: virus-infected plants; T1: tomato plants sprayed with bacterial culture filtrate 24 h before virus infection; T2: tomato plants sprayed with bacterial culture filtrate 24 h after virus infection; T3: tomato plants sprayed with bacterial culture filtrate two times, 24 h before and 24 h after virus inoculation.

The growth parameter assays revealed that the T3 tomato plants gave the best results, exhibiting a significant increase in the plants' fresh weight, up to 44% and 64% compared to NT and TMV plants, respectively (Table 2). Moreover, the T1 treatment came second, in increasing the plants' fresh weight by 30.1% and 49% compared to NT and TMV, respectively. No significant differences were reported regarding plant dry weight between HA1-CF treatments (T1, T2, and T3) and the control treatment. However, compared to TMV plants, the three treatments enhanced dry weight by up to 28%, as reported in T3 treatment plants (Table 2). The shoot and root lengths were also improved in the three treatments (T1, T2, and T3) compared to the non-treated TMV-infected plants, with a notable enhancement in shoot and root lengths in T3, compared to control plants. Together, the results indicated stimulation and enhancement in all measured growth parameters in all treatments involving CF (T1, T2, and T3) and the superiority of the T3 treatment strategy for growth enhancement, even over non-infected plants (NT). The results are in accordance with several studies that have reported the efficacy of *B. subtilis* in enhancing plant growth of various plant species [50,51], mediated by a vast number of growth-promoting secondary metabolites and phytohormones [52]. Posada et al. [53] found that CFs of *B. subtilis* EA-CB0575, derived from vegetative cells or spores, significantly enhanced the shoot length as well as the dry weight of Musa plants, as compared to controls.

3.3. Evaluation of Oxidative Stress Markers

Elevated levels of reactive oxygen species (ROS) are a defining feature of plant virus infections [54–56], so the concentration of these species may be directly proportional to infection severity. In this regard, two oxidative stress markers (MDA and H_2O_2) were evaluated in all treatments, including those treated with HA1-CF or those not treated (Figure 3). Compared to NT plants (112 ± 5.4 and 6.8 ± 0.4 μM/g f.wt. for MDA and H_2O_2, respectively), the elevation of MDA (151 ± 3.7) and H_2O_2 (8.1 ± 0.3) levels by 35% and 19%, respectively, was obtained in TMV treatment plants (Figure 3). The findings are in accordance with results reported from many viral infections in different plants [8,57–59].

The elevation of H_2O_2 levels at the early stages of pathogenesis is a process contributing to plant resistance against viral infection [60,61]. However, unbalanced levels of ROS during e.g., the late stages of plant virus infections, lead to oxidation of vital plant cell components, such as proteins, DNA, and unsaturated fatty acids, adversely affecting the whole plant [62,63]. It was reported that the reduction of MDA maintained cell membrane integrity and stability [64]. Duan et al. [65] reported that the cell-free culture filtrate of *B. amyloliquefaciens* QSB-6 enhanced plant tolerance by reducing MDA accumulation inside plant tissues. The treatment with HA1-CF revealed a varied reduction in the two stress markers, related mainly to the application time. The T3 treatment exhibited the maximum reduction in MDA (119 ± 4.1 µM/g f.wt.) and H_2O_2 (6.4 ± 0.4 µM/g f.wt.) levels compared to that of the non-infected group levels (NT), indicating the potency of the T3 strategy in alleviating oxidative stress and lipid peroxidation in virus-infected plants.

Table 2. Effect of foliar application of *B. subtilis* strain HA1-CF on growth parameters of tomato plants at 22 days post-TMV inoculation.

Treatment *	Fresh Weight g/Plant	Dry Weight g/Plant	Shoot Length cm/Plant	Root Length cm/Plant
NT	7.15 ± 0.37 bc	2.24 ± 0.03 ab	27.67 ± 3.30	14.17 ± 1.55 ab
TMV	6.28 ± 0.61 c	1.89 ± 0.07 c	26.17 ± 0.24	9.33 ± 2.63 b
T1	9.36 ± 2.20 ab	2.27 ± 0.11 ab	28.33 ± 4.50	14 ± 4.55 ab
T2	6.47 ± 0.87 bc	2.14 ± 0.15 bc	27.33 ± 2.49	11.33 ± 1.25 ab
T3	10.35 ± 1.19 a	2.43 ± 0.16 a	29.67 ± 2.06	16.33 ± 2.87 a

* NT: mock plants; TMV: virus-infected plants; T1: tomato plants sprayed with bacterial culture filtrate 24 h before viral infection; T2: tomato plants sprayed with bacterial culture filtrate 24 h after viral infection; T3: tomato plants sprayed with bacterial culture filtrate two times, 24 h before and 24 h after viral inoculation. Each column value represents the mean result obtained from five biological replicates. Significant differences were calculated using a one-way ANOVA according to the least significant difference (LSD) method at a $p \leq 0.05$ level of probability with the GraphPad Prism software package. The mean values of each column with the same letter do not differ significantly.

3.4. Antioxidant Enzymatic Activities

Due to the crucial role of antioxidant enzymes in plant defense mechanisms against several plant pathogens, the current work aimed to evaluate the activities of the antioxidant enzymes PPO, SOD, and POX in HA1-CF-treated and -untreated tomato plants upon TMV infection (Figure 4). The enzyme activity assay results indicated a remarkable reduction in PPO activities of more than 50% in the TMV treatment plants (0.13 ± 0.01 µM/g f.wt. min^{-1}) compared to the NT plants (0.25 ± 0.01 µM/g f.wt. min^{-1}). Upon treatment of tomato plants with HA1-CF, PPO activities increased in treatments T1 (0.19 ± 0.01 µM/g f.wt. min^{-1}) as compared to the TMV treatment group. The maximum PPO activity was detected in the T3 treatment (0.32 ± 0.02 µM/g f.wt. min^{-1}) reporting 28% increases in PPO activities compared to the NT control group (Figure 4). The measured PPO activities in the T3 plants were 2.5-fold higher than those assayed in the TMV plants, demonstrating the role of HA1-CF in up-regulation of PPO genes and/or enzymatic activities as a part of the defense mechanism against TMV infection in tomatoes. As a result, it has been reported that over-expression of PPO in various plants has defense properties against bacterial infection in tomato plants [66] and fungal infection in strawberry fruits [67]. Regarding SOD, the results indicated a slight enhancement of about 27% (0.14 ± 0.02 µM/g f.wt. min^{-1}) in SOD activity in the TMV treatment when compared to the NT treatment (0.11 ± 0.01 µM/g f.wt. min^{-1}), which could be attributed to an initial response of the plant defense system to oxidative stress. SOD enzymes have a major role in detoxifying reactive superoxide ($O_2^{\cdot-}$) species into H_2O_2, which is subsequently degraded through catalases. The SOD activity was also enhanced in all HA-CF treated plants, where the maximum level was shown in T3 (0.22 ± 0.02 µM/g f.wt. min^{-1}), followed by T1 (0.17 ± 0.07 µM/g f.wt. min^{-1}), displaying 2- and 1.5-fold increases, respectively, compared to non-treated plants. Concerning

POX activities, the results showed a significant reduction (55%) in POX activity upon TMV infection in TMV treatment (0.15 ± 0.01 μM/g f.wt. min^{-1}), when compared to the NT plants (0.27 ± 0.02 μM/g f.wt. min^{-1}). Compared to TMV treatment plants, the treatment of tomato plants with HA1-CF elevated the POX activities in all treatments at varying concentrations depending on the treatment strategy. The maximum POX activity (0.24 ± 0.02 μM/g f.wt. min^{-1}) was detected in T1 plants, followed by T3, with a level of 0.19 ± 0.01 μM/g f.wt. min^{-1} (Figure 4). Although the highest activity of POX observed in T1 treatment was less than that reported in the control treatment, no significant changes were reported. Moreover, the increasing POX activity, which is 60% higher than that reported in the TMV treatment plants, demonstrates the efficacy of HA1-CF in improving the POX activities, resulting in boosting tomato plant tolerance under the TMV challenge. POX remarkably enhances the plant's defense against infection by lignin synthesis, using ROS as a substrate. Lignin deposition increases the physical barrier against viral infection [68]. The obtained results are consistent with other studies that have reported the ability of *Bacillus* sp. to increase the activity of plant enzymes against ROS [69–71]. Moreover, the application of the culture filtrate of *B. amyloliquefaciens* considerably increased the enzyme activities of SOD, POX, and CAT of *Malus hupehensis* plants [65].

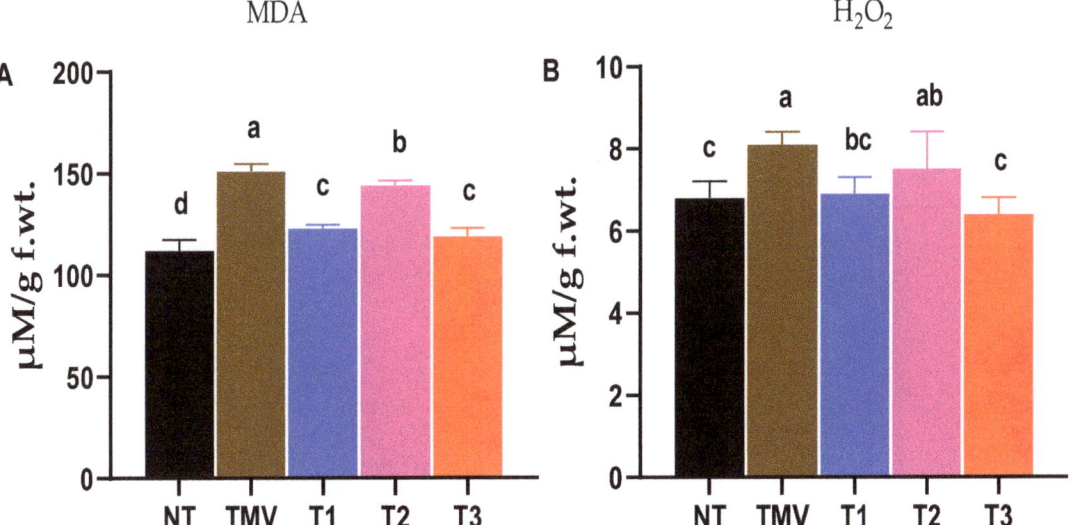

Figure 3. Effect of foliar application of *B. subtilis* strain HA1-culture filtrate on the two oxidative stress markers: MDA (**A**) and H_2O_2 (**B**) of tomato plants at 22 days post-TMV inoculation. NT: mock plants; TMV: virus-infected plants; T1: tomato plants sprayed with bacterial culture filtrate 24 h before viral infection; T2: tomato plants sprayed with bacterial culture filtrate 24 h after viral infection; T3: tomato plants sprayed with bacterial culture filtrate two times, 24 h before and 24 h after viral inoculation. Each column value represents the mean result obtained from five biological replicates. Significant differences were calculated using a one-way ANOVA according to the least significant difference (LSD) method at a $p \leq 0.05$ level of probability with the GraphPad Prism software package. The mean values of each column with the same letter do not differ significantly.

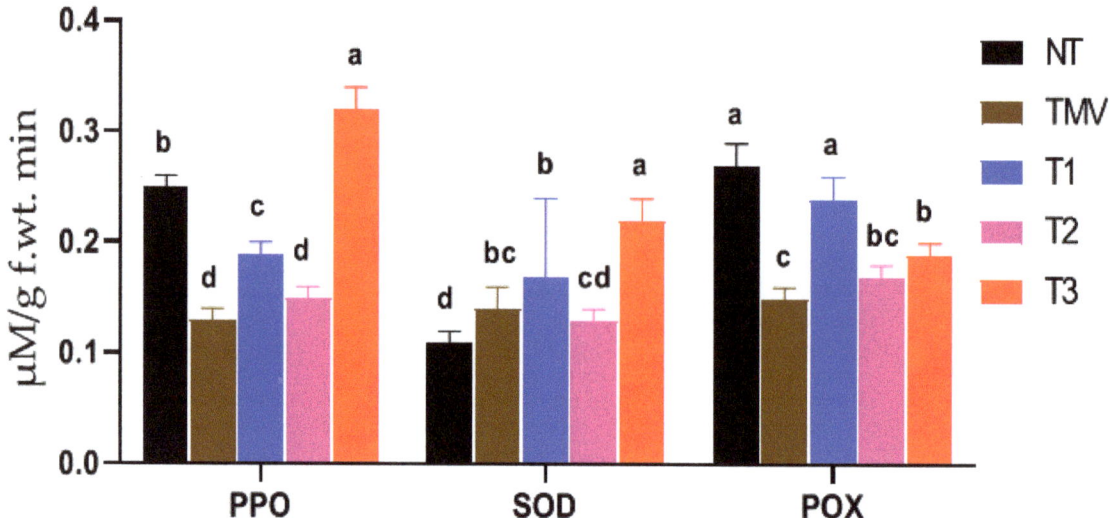

Figure 4. Effect of foliar application of *B. subtilis* strain HA1-culture filtrate on activities of antioxidant enzymes PPO, SOD, and POX of tomato plants at 22 days post-TMV inoculation. NT: mock plants; TMV: virus-infected plants; T1: tomato plants sprayed with bacterial culture filtrate 24 h before viral infection; T2: tomato plants sprayed with bacterial culture filtrate after 24 h of viral infection; T3: tomato plants sprayed with bacterial culture filtrate two times, 24 h before and 24 h after the viral inoculation. Each column value represents the mean result obtained from five biological replicates. Significant differences were calculated using a one-way ANOVA according to the least significant difference (LSD) method at a $p \leq 0.05$ level of probability with the GraphPad Prism software package. The mean values of each column with the same letter do not differ significantly.

3.5. Total Phenolic and Total Flavonoid Contents

It is well known that the accumulation of polyphenolic phytochemicals, including phenolic and flavonoid compounds, is a main defense mechanism for plants to tolerate biotic and abiotic stresses, including virus infections [8,44,72]. The obtained results in the present study indicated a noticeable reduction (about 35%) in total phenolic contents in the TMV treatment plants (74 ± 14.3 mg/g d.wt.) when compared to control plants (114 ± 2.0 mg/g d.wt.). Interestingly, the T3 treatment exhibited a significant increase in tomato total phenolic contents of 27% (94 ± 6.1 mg/g d.wt.) when compared to the TMV plants (Figure 5A). No significant difference was found between TMV, T1, and T2 treatments (Figure 5A). Consequently, the dual treatment of tomato plants with HA1-CF enhanced the accumulation levels of total phenolic contents inside TMV-infected tissues. On the other hand, the total flavonoid content estimation revealed a significant reduction (18%) in flavonoid contents of the TMV plants (10.1 ± 0.53 mg/g d.wt.), compared to the NT plants (11.4 ± 2.0 mg/g d.wt.), as indicated in Figure 5B. The three HA1-CF-treated groups revealed higher flavonoid contents as compared to the TMV treatment plants. Among them, the T3 plants displayed flavonoid contents (15.0 ± 0.92 mg/g d.wt.) which were 50% and 20% higher than that of the TMV and NT groups, respectively. Notably, no significant differences were detected in total flavonoid contents in the T1 and T2 treatments when compared to the control (NT) treatment. The significant flavonoid accumulation in the T3 treatment plants indicates the importance of treatment doses (two doses in the T3 group) with respect to the treatment time (before or after infection) in the accumulation of such defense compounds.

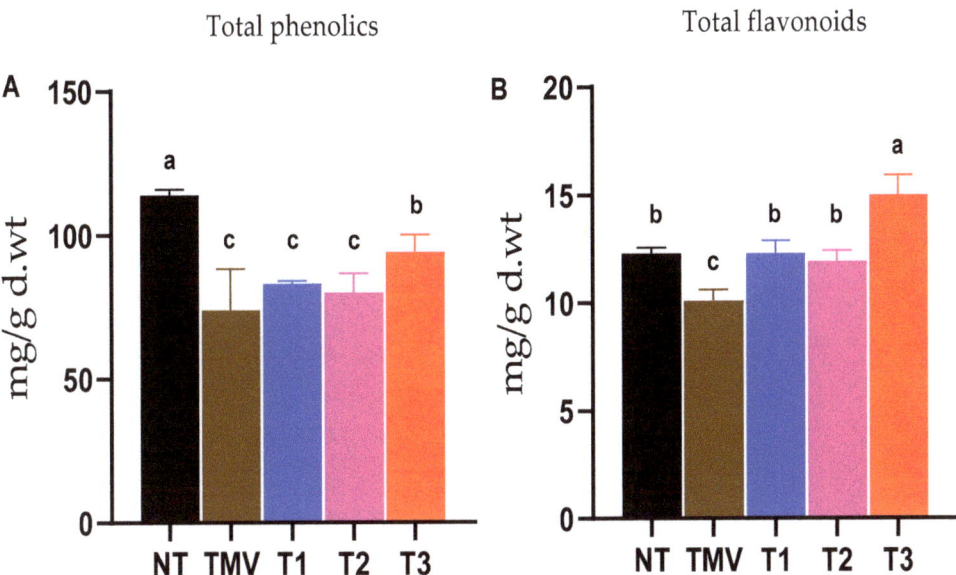

Figure 5. Effect of foliar application of *B. subtilis* strain HA1-culture filtrate on the total phenolic (**A**) and total flavonoid (**B**) contents of tomato plants at 22 days post-TMV inoculation. NT: mock plants; TMV: virus-infected plants; T1: tomato plants sprayed with bacterial culture filtrate 24 h before virus infection; T2: tomato plants sprayed with bacterial culture filtrate 24 h after virus infection; T3: tomato plants sprayed with bacterial culture filtrate two times, 24 h before and 24 h after virus inoculation. Each column value represents the mean result obtained from five biological replicates. Significant differences were calculated using a one-way ANOVA according to the least significant difference (LSD) method at a $p \leq 0.05$ level of probability with the GraphPad Prism software package. The mean values of each column with the same letter do not differ significantly.

3.6. Effect of HA1-CF on Systemic Accumulation of TMV

In line with the previously obtained results in this study, the application of HA1-CF, either before or after TMV inoculation, resulted in a considerable reduction in viral accumulation inside tomato tissues. The TMV content was calculated using the *TMV-CP* gene cycle threshold (Ct) value and the Ct value of the internal control *β-actin* gene of the tomato control treatment. The qRT-PCR results revealed that the TMV treatment exhibited the highest levels of *TMV-CP* transcripts, representing a 28.2-fold change, indicating the plant's viral infection. On the other hand, the HA1-CF-treated plants exhibited relative expression levels of 3.23-, 4.35-, and 2.48-fold changes for T1, T2, and T3 treatments, respectively. These results corresponded with the appearance of the symptoms in terms of the highest and lowest concentrations of the virus inside tomato tissues. Consequently, the significant decline in TMV accumulation levels in T1, T2, and T3 tomato leaves by 88.55, 84.57, and 91.20% compared to TMV treatment plants, confirmed the biocontrol activity of the HA1-CF against TMV infection. In line with our results, the foliar application of the culture filtrate of *B. licheniformis* and *Streptomyces* sp. caused a significant reduction in the accumulation of AMV and PVY in potato plants [11,73]. Moreover, the application of *Streptomyces cellulosae* and *B. amyloliquefaciens* reduced TMV and CMV severity, and decreased viral accumulation levels in the treated leaves [3,74]. Thus, the application of HA1-CF could protect the tomato plants from TMV infection by preventing the accumulation of viral particles as well as activating the plant defense responses [3,11,48].

3.7. Effect of HA1-CF on Transcriptional Levels of Pathogenesis-Related Protein Genes

According to several research reports, the induction of different groups of PR proteins plays a vital role in SAR activation and is also efficient at preventing pathogen formation, multiplication, and/or spread. Among them, the genes encoding for PR-1 and PR-2 (evaluated in this study) may be the most significant markers for plant viral infection [3,8,75]. For *PR-1*, it was shown that the TMV treatment plants showed a significant down-regulation, with a relative expression level of 0.56-fold change lower than control treatment plants (Figure 6). Interestingly, all HA1-CF-treated plants exhibited a considerable increase in the relative expression level of *PR-1*, with relative expression levels of 2.43-, 3.30-, and 5.14-fold change in T2, T1, and T3, respectively, greater than control (Figure 6). It was reported that the application of the cell-free filtrate of *B. velezensis* increased the resistance of *Datura stramonium* and tomato plants against *Cucumber mosaic virus* and *Tomato yellow leaf curl virus*, respectively, by significantly elevating the expression of the *PR-1* gene [48,76]. It is well known that salicylic acid (SA) plays a crucial role in stimulating plant systemic resistance through activating the plant defense system [77]. Upon viral infection, the accumulation of SA is frequently accompanied by the induction of *PR-1*, an SA marker gene. Meanwhile, the induction of *PR-1* is associated with plant immunity activation and increasing plant resistance against pathogens [21,78]. Consequently, we propose that HA1-CF contains some elicitor secondary metabolite compounds that play a vital role in the induction of *PR-1*, resulting in activating systemic acquired resistance (SAR) and boosting plant resistance to viral infection.

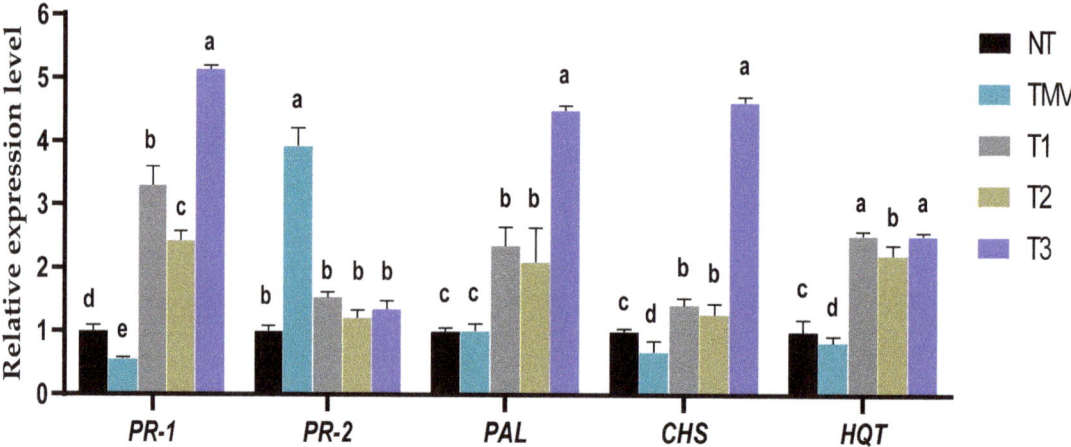

Figure 6. The relative expression levels of genes encoding pathogenesis-related proteins (*PR-1*, and *PR-2*) and polyphenol biosynthetic enzymes (*PAL*, *CHS*, and *HQT*) in tomato plants at 22 days post-TMV inoculation. NT: mock plants; TMV: virus-infected plants; T1: tomato plants sprayed with bacterial culture filtrate 24 h before virus infection; T2: tomato plants sprayed with bacterial culture filtrate after 24 h of virus infection; T3: tomato plants sprayed with bacterial culture filtrate two times, 24 h before and 24 h after virus inoculation. Each column value represents the mean result obtained from five biological replicates. Significant differences were calculated using a one-way ANOVA according to the least significant difference (LSD) method at a $p \leq 0.05$ level of probability with the GraphPad Prism software package. The mean values of each column with the same letter do not differ significantly.

Regarding the *PR-2* transcript, the results indicated a significant induction of *PR-2*, observed only in TMV plants, with a relative expression level of 3.93-fold change higher than control (Figure 6). Despite HA1-CF-treated plants showing a minor up-regulation of *PR-2*, with relative expression levels of 1.54-, 1.21-, and 1.35-fold change in T1, T2, and T3, respectively, no significant differences were detected when compared to the control (Figure 6).

The *PR-2* gene encodes a protein conferring β-1,3-glucanase activity that facilitates viral translocation between plant cells across plasmodesmata, and hence TMV may induce this gene to facilitate its movement and spread through plant cells [6,79]. Previous research has shown that *PR-2* is clearly induced during viral infections in potato, Arabidopsis, onion, tobacco, and tomato plants [3]. Furthermore, a lack of tobacco *PR-2* expression reduced viral infection susceptibility, whereas overexpression accelerated the transmission of PVY across cells [80–82]. Consequently, the foliar application of HA1-CF may reduce TMV infection via lowering *PR-2* expression and inhibiting long-distance movement between cells.

3.8. Effect of HA1-CF on the Transcript Levels of Polyphenolic Biosynthesis Genes

It is well known that the accumulation of polyphenolic compounds is a crucial plant defense mechanism against several biotic and abiotic stressors. They are brought to infection sites, elicit hypersensitive responses, and promote programmed cell death when plants are infected [72,83]. In higher plants, biosynthesis of polyphenolic compounds occurs mainly via the shikimate pathway, with three major routes, including the phenylpropanoid, chlorogenic, and flavonoid pathways [29,84]. Besides being the key enzyme in the first step of the phenylpropanoid pathway that is responsible for the conversion of phenylalanine to cinnamic acid, PAL plays a vital role in SA biosynthesis regulation [85,86]. Compared to the mock-inoculated (NT) tomato plants in the present study, the expression levels of *PAL* were significantly induced in HA1-CF-treated plants, with relative transcriptional levels of 2.36-, 2.10-, and 4.50-fold change in T1, T2, and T3, respectively, higher than the control (Figure 6). On the other hand, no significant difference was reported between TMV treatment and NT treatment plants. The obtained data revealed that the treatment of tomato plants with HA1-CF triggered the expression of *PAL*, which may potentially increase SA accumulation. The obtained results agree with previous reports showing that the application of the culture filtrate of *B. licheniformis* and *B. velezensis* significantly increased the expression levels of *PAL* and resulted in increased plant resistance against viral infections [11,48]. Consequently, we suggest that HA1-CF could be applied as an efficient ISR elicitor, boosting secondary metabolite biosynthesis, polyphenolic compounds, and SA accumulation in treated plants.

CHS is the initial enzyme in the flavonoid pathway, catalyzing the conversion of p-coumaroyl CoA to naringenin chalcones, which are the major precursor and key intermediates for the synthesis of a wide variety of flavonoids in different plant tissues [84,87]. Likewise, HQT is a strategic enzyme in the biosynthesis of chlorogenic acid. It is one of the most significant polyphenolic molecules that directly enhances plant defense, being involved in various pathogen resistances as well as displaying antioxidant properties [88–90]. Compared to the NT treatment plants, the untreated leaves challenged with TMV only showed a significant down-regulation of both CHS and HQT, with relative expression levels of 0.68- and 0.83-fold change, respectively (Figure 6). Thus, the down-regulation of transcriptional levels of CHS and HQT in TMV treatment plants reflected the decrease in the total flavonoid contents detected in this treatment. In this context, many previous studies have shown that the biosynthesis of flavonoid compounds, including chlorogenic acid, is suppressed inside virally infected plant tissues [8,11,75,91,92]. Notably, HA1-CF-treated leaves of T1, T2, and T3 treatments exhibited significant elevations of both *CHS* and *HQT* expression levels when compared to the control. (Figure 6). The highest expression level of *CHS* (4.63-fold) was reported in T3 plants, while the maximum transcriptional level of *HQT* was observed in both T1 and T3, with the same expression level of 2.52-fold change higher than control (Figure 6). In line with the obtained results in this study, the induction of *CHS* and *HQT* genes indicated the accumulation of the flavonoid compounds that play a role in plant resistance against pathogen infection [8,11,22]. Overall, results obtained in the present study show that the foliar application of HA1-CF activates ISR and up-regulates *PR-1*, *PAL*, *CHS*, and *HQT*. In addition, the accumulation of polyphenolic compounds in response to HA1-CF may contribute to the development of induced resistance and the suppression of TMV infection.

3.9. Identification of Bioactive Metabolites of HA1-CF by GC–MS

Because microbial secondary metabolites are the primary precursors for many biological activities, investigating such metabolites using various analytical approaches is a prerequisite for gaining a deep and comprehensive understanding of biological control toward novel applications [93]. GC–MS analysis is currently a reliable and robust analytical technique, widely applied in bioactive compound analysis and identification [94]. In the present study, the identification of bioactive compounds in the ethyl acetate extract of HA1-CF was conducted by using the mass spectrum of a GC–MS instrument. Results of our GC–MS analyses indicated the presence of 15 different components in HA1-CF; the most abundant compounds are presented in Table 3. The dominant compound detected was phenol, 2,4-bis(1,1-dimethylethyl)- at a retention time of 12.19. Interestingly, phenol, 2,4-bis (1,1-dimethylethyl)- has been reported to accumulate in plant cells under fungal and bacterial attack, and represents a major compound contributing to disease resistance in avocado and Malaysian mango kernel through inhibition of reactive oxygen species (ROS) produced by the pathogen [95,96]. TMV infection is usually associated with elevated levels of ROS [55,56], therefore, inhibition of ROS could alleviate the symptoms of viral infection. Furthermore, several amino acid residues, such as L-proline, N-valeryl, and heptadecyl ester, were detected in our analyses at a retention time of 15.54. Some additional compounds were also detected in the HA1-CF, as indicated in Table 3. The GC–MS results indicated the presence of two fatty acids, eicosane (a long-chain fatty acid) and pentadecanoic acid (asaturated fatty acid) in the HA1-CF at a retention time of 13.7 and 15.48 min, respectively. Eicosane is a biologically active compound derived from *Streptomyces* sp., shown to be a potent antifungal in treating *Rhizoctonia solani* spot disease in tobacco leaves [97]. The presence of eicosane has also been reported in other microbes [98], as well as plant extracts that exhibit major antimicrobial activities [99]. Another compound, Pyrrolo [1,2-a] pyrazine-1,4-dione, hexahydro-3-(2-methylpropyl)- was also detected at a retention time of 15.435 min (Table 3). Pyrrole and pyrrolizidines are well-known heterocyclic compounds with diverse biological activities such as antimicrobial, anticancer, antiviral, and anti-inflammatory effects [100]. The unique structures of pyrrole and pyrrolizidines with at least two different elements broaden their biological activities and applications, accounting for more than 75% of the currently applied drugs in clinical use [101]. The findings are consistent with the previously reported efficacy of *B. velezensis* PEA1 as a *Fusarium oxysporum* inhibitor, as well as an enhancer of systemic resistance to *Cucumber mosaic virus* in *Datura stramonium*, which was attributed to the presence of several different pyrrolo[1,2-a]pyrazine-1,4-dione compounds in the culture filtrate extract [48]. In addition, pyrrole and pyrrolizidines have been detected in various microbial extracts displaying potent biological activities, including antimicrobial [102], antioxidant [103], and antiviral effects [104]. Accordingly, our GC–MS analysis results revealed the presence of several biologically active compounds with reported antifungal and antibacterial activities in the HA1-CF supernatant. Importantly, the current study demonstrated a significant antiviral activity against TMV, plus the growth stimulation properties of HA1-CF in tomatoes. However, further investigations are required to directly associate the above-mentioned biological activities to the compounds we have detected in HA1-CF.

Table 3. The highest six compounds detected in HA1-CF as revealed with GC–MS analysis.

Peak	R. Time (min.)	Area	Name	Chemical Formula	Molecular Structure
2	11.960	560.11	Nonane, 5-(2-methylpropyl)-	$C_{13}H_{28}$	
3	12.187	1.592.35	Phenol, 2,4-bis(1,1-dimethylethyl)-	$C_{14}H_{22}O$	
6	13.689	475.26	Eicosane	$C_{20}H_{42}$	
12	15.435	535.02	Pyrrolo[1,2-a]pyrazine-1,4-dione, hexahydro-3-(2-methylpropyl)-	$C_{10}H_{16}N_2O_2$	
13	15.481	549.58	Pentadecanoic acid	$C_{15}H_{30}O_2$	
14	15.542	835.29	L-Proline, N-valeryl-, heptadecyl ester	$C_{26}H_{49}NO_3$	

4. Conclusions

We have demonstrated that, under greenhouse conditions, the foliar application of the culture filtrate of *B. subtilis* strain HA1 (HA1-CF) significantly enhances tomato growth parameters and antioxidant enzyme activities (PPO, SOD, and POX) that alleviate the oxidative stress resulting from TMV infection in all treated plants compared to non-treated plants. Furthermore, HA1-CF treatments resulted in the induction of different defense-related genes (*PR-1*, *PAL*, *CHS*, and *HQT*) and a significant decline in TMV accumulation, indicating the activation of induced resistance, effective against TMV. Certain compounds detected in the ethyl acetate extract of HA1-CF (e.g., phenol, 2,4-bis (1,1-dimethylethyl)-, Pyrrolo[1,2-a]pyrazine-1,4-dione, hexahydro-3-(2-methylpropyl)-, and eicosane) could potentially be used as plant growth promoters and defense modulatory agents to protect tomato plants against TMV infection. However, further investigations are required to elucidate the different antiviral properties of HA1-CF before it can be used in open-field applications or for commercial purposes. Future research should primarily focus on optimizing and testing the culture filtrate in different plant-virus systems.

Author Contributions: Conceptualization, A.A.; methodology, H.E.-G., M.A.S. and A.A.; software, H.E.-G. and A.A.; validation, A.A.A.-A., A.A. and H.M.; formal analysis, A.A.; investigation, H.E.-G.; resources, A.A.A.-A.; data curation, H.E.-G.; writing—original draft preparation, H.E.-G. and A.A.; writing—review and editing, H.E.-G., A.A.A.-A., A.A., L.K. and H.M.; visualization, A.A.;

supervision, H.M.; project administration A.A.A.-A.; funding acquisition, A.A.A.-A. and A.A. All authors have read and agreed to the published version of the manuscript.

Funding: This work was partially funded by the Science and Technology Development Fund (STDF), Egypt, Grant No 30102. This research was financially supported by the Researchers Supporting Project number (RSP2022R505), King Saud University, Riyadh, Saudi Arabia.

Institutional Review Board Statement: Not applicable.

Informed Consent Statement: Not applicable.

Data Availability Statement: Not applicable.

Acknowledgments: This paper is based upon work supported by Science, Technology & Innovation Funding Authority (STDF) under grant (30102). The authors would like to extend their appreciation to the Researchers Supporting Project number (RSP2022R505), King Saud University, Riyadh, Saudi Arabia.

Conflicts of Interest: The authors declare no conflict of interest.

References

1. Abdelkhalek, A.; Hafez, E. Plant Viral Diseases in Egypt and Their Control. In *Cottage Industry of Biocontrol Agents and Their Applications*; Springer: Berlin/Heidelberg, Germany, 2020; pp. 403–421.
2. Mumford, R.A.; Macarthur, R.; Boonham, N. The role and challenges of new diagnostic technology in plant biosecurity. *Food Secur.* **2016**, *8*, 103–109. [CrossRef]
3. Abo-Zaid, G.A.; Matar, S.M.; Abdelkhalek, A. Induction of Plant Resistance against Tobacco Mosaic Virus Using the Biocontrol Agent Streptomyces cellulosae Isolate Actino 48. *Agronomy* **2020**, *10*, 1620. [CrossRef]
4. McDaniel, L.; Maratos, M.; Farabaugh, J. Infection of plants by tobacco mosaic virus. *Am. Biol. Teach.* **1998**, *60*, 434–439. [CrossRef]
5. Peng, J.; Song, K.; Zhu, H.; Kong, W.; Liu, F.; Shen, T.; He, Y. Fast detection of Tobacco Mosaic Virus infected tobacco using laser-induced breakdown spectroscopy. *Sci. Rep.* **2017**, *7*, 44551. [CrossRef]
6. Abdelkhalek, A. Expression of tomato pathogenesis related genes in response to Tobacco mosaic virus. *J. Anim. Plant Sci.* **2019**, *29*, 1596–1602.
7. Bazzini, A.A.; Hopp, H.E.; Beachy, R.N.; Asurmendi, S. Infection and coaccumulation of Tobacco Mosaic Virus proteins alter microRNA levels, correlating with symptom and plant development. *Proc. Natl. Acad. Sci. USA* **2007**, *104*, 12157–12162. [CrossRef]
8. Abdelkhalek, A.; Al-Askar, A.A.; Alsubaie, M.M.; Behiry, S.I. First Report of Protective Activity of Paronychia argentea Extract against Tobacco Mosaic Virus Infection. *Plants* **2021**, *10*, 2435. [CrossRef] [PubMed]
9. Sharma, A.; Kumar, V.; Shahzad, B.; Tanveer, M.; Sidhu, G.P.S.; Handa, N.; Kohli, S.K.; Yadav, P.; Bali, A.S.; Parihar, R.D.; et al. Worldwide pesticide usage and its impacts on ecosystem. *SN Appl. Sci.* **2019**, *1*, 1446. [CrossRef]
10. Alengebawy, A.; Abdelkhalek, S.T.; Qureshi, S.R.; Wang, M.-Q. Heavy Metals and Pesticides Toxicity in Agricultural Soil and Plants: Ecological Risks and Human Health Implications. *Toxics* **2021**, *9*, 42. [CrossRef] [PubMed]
11. Abdelkhalek, A.; Al-Askar, A.A.; Behiry, S.I. Bacillus licheniformis strain POT1 mediated polyphenol biosynthetic pathways genes activation and systemic resistance in potato plants against Alfalfa mosaic virus. *Sci. Rep.* **2020**, *10*, 16120. [CrossRef] [PubMed]
12. Sorokan, A.; Cherepanova, E.; Burkhanova, G.; Veselova, S.; Rumyantsev, S.; Alekseev, V.; Mardanshin, I.; Sarvarova, E.; Khairullin, R.; Benkovskaya, G.; et al. Endophytic *Bacillus* spp. as a Prospective Biological Tool for Control of Viral Diseases and Non-vector Leptinotarsa decemlineata Say. in *Solanum tuberosum* L. *Front. Microbiol.* **2020**, *11*, 1–13. [CrossRef] [PubMed]
13. Jiao, X.; Takishita, Y.; Zhou, G.; Smith, D.L. Plant Associated Rhizobacteria for Biocontrol and Plant Growth Enhancement. *Front. Plant Sci.* **2021**, *12*, 420. [CrossRef] [PubMed]
14. Wang, H.; Liu, R.; You, M.P.; Barbetti, M.J.; Chen, Y. Pathogen biocontrol using plant growth-promoting bacteria (PGPR): Role of bacterial diversity. *Microorganisms* **2021**, *9*, 1988. [CrossRef] [PubMed]
15. Pellegrini, M.; Pagnani, G.; Bernardi, M.; Mattedi, A.; Spera, D.M.; Del Gallo, M. Cell-free supernatants of plant growth-promoting bacteria: A review of their use as biostimulant and microbial biocontrol agents in sustainable agriculture. *Sustainability* **2020**, *12*, 9917. [CrossRef]
16. Radhakrishnan, R.; Hashem, A.; Abd Allah, E.F. Bacillus: A biological tool for crop improvement through bio-molecular changes in adverse environments. *Front. Physiol.* **2017**, *8*, 1–14. [CrossRef] [PubMed]
17. Vejan, P.; Abdullah, R.; Khadiran, T.; Ismail, S.; Nasrulhaq Boyce, A. Role of plant growth promoting rhizobacteria in agricultural sustainability—A review. *Molecules* **2016**, *21*, 573. [CrossRef] [PubMed]
18. Saeed, Q.; Xiukang, W.; Haider, F.U.; Kučerik, J.; Mumtaz, M.Z.; Holatko, J.; Naseem, M.; Kintl, A.; Ejaz, M.; Naveed, M.; et al. Rhizosphere bacteria in plant growth promotion, biocontrol, and bioremediation of contaminated sites: A comprehensive review of effects and mechanisms. *Int. J. Mol. Sci.* **2021**, *22*, 10529. [CrossRef] [PubMed]
19. Trabelsi, D.; Mhamdi, R. Microbial inoculants and their impact on soil microbial communities: A review. *BioMed Res. Int.* **2013**, *2013*, 863240. [CrossRef]

20. Vitti, A.; Pellegrini, E.; Nali, C.; Lovelli, S.; Sofo, A.; Valerio, M.; Scopa, A.; Nuzzaci, M. Trichoderma harzianum T-22 induces systemic resistance in tomato infected by Cucumber mosaic virus. *Front. Plant Sci.* **2016**, *7*, 1520. [CrossRef] [PubMed]
21. Abo-Zaid, G.; Abdelkhalek, A.; Matar, S.; Darwish, M.; Abdel-Gayed, M. Application of Bio-Friendly Formulations of Chitinase-Producing Streptomyces cellulosae Actino 48 for Controlling Peanut Soil-Borne Diseases Caused by Sclerotium rolfsii. *J. Fungi* **2021**, *7*, 167. [CrossRef]
22. Heflish, A.A.; Abdelkhalek, A.; Al-Askar, A.A.; Behiry, S.I. Protective and Curative Effects of Trichoderma asperelloides Ta41 on Tomato Root Rot Caused by Rhizoctonia solani Rs33. *Agronomy* **2021**, *11*, 1162. [CrossRef]
23. Kandan, A.; Ramiah, M.; Vasanthi, V.J.; Radjacommare, R.; Nandakumar, R.; Ramanathan, A.; Samiyappan, R. Use of Pseudomonas fluorescens-based formulations for management of tomato spotted wilt virus (TSWV) and enhanced yield in tomato. *Biocontrol Sci. Technol.* **2005**, *15*, 553–569. [CrossRef]
24. Deng, Z.S.; Zhao, L.F.; Kong, Z.Y.; Yang, W.Q.; Lindström, K.; Wang, E.T.; Wei, G.H. Diversity of endophytic bacteria within nodules of the *Sphaerophysa salsula* in different regions of Loess Plateau in China. *FEMS Microbiol. Ecol.* **2011**, *76*, 463–475. [CrossRef] [PubMed]
25. El-Dougdoug, K.A.; Ghaly, M.F.; Taha, M.A. Biological control of Cucumber mosaic virus by certain local Streptomyces isolates: Inhibitory effects of selected five Egyptian isolates. *Intl. J. Virol.* **2012**, *8*, 151–164. [CrossRef]
26. Weisburg, W.G.; Barns, S.M.; Pelletier, D.A.; Lane, D.J. 16S ribosomal DNA amplification for phylogenetic study. *J. Bacteriol.* **1991**, *173*, 697–703. [CrossRef] [PubMed]
27. Zhao, L.; Dong, J.; Hu, Z.; Li, S.; Su, X.; Zhang, J.; Yin, Y.; Xu, T.; Zhang, Z.; Chen, H. Anti-TMV activity and functional mechanisms of two sesquiterpenoids isolated from Tithonia diversifolia. *Pestic. Biochem. Physiol.* **2017**, *140*, 24–29. [CrossRef]
28. Kavroulakis, N.; Ehaliotis, C.; Ntougias, S.; Zervakis, G.I.; Papadopoulou, K.K. Local and systemic resistance against fungal pathogens of tomato plants elicited by a compost derived from agricultural residues. *Physiol. Mol. Plant Pathol.* **2005**, *66*, 163–174. [CrossRef]
29. André, C.M.; Schafleitner, R.; Legay, S.; Lefèvre, I.; Aliaga, C.A.A.; Nomberto, G.; Hoffmann, L.; Hausman, J.-F.; Larondelle, Y.; Evers, D. Gene expression changes related to the production of phenolic compounds in potato tubers grown under drought stress. *Phytochemistry* **2009**, *70*, 1107–1116. [CrossRef] [PubMed]
30. Sagi, M.; Davydov, O.; Orazova, S.; Yesbergenova, Z.; Ophir, R.; Stratmann, J.W.; Fluhr, R. Plant respiratory burst oxidase homologs impinge on wound responsiveness and development in Lycopersicon esculentum. *Plant Cell* **2004**, *16*, 616–628. [CrossRef]
31. Hafez, E.E.; El-Morsi, A.A.; El-Shahaby, O.A.; Abdelkhalek, A.A. Occurrence of iris yellow spot virus from onion crops in Egypt. *VirusDisease* **2014**, *25*, 455–459. [CrossRef] [PubMed]
32. Heath, R.L.; Packer, L. Photoperoxidation in isolated chloroplasts: I. Kinetics and stoichiometry of fatty acid peroxidation. *Arch. Biochem. Biophys.* **1968**, *125*, 189–198. [CrossRef]
33. Junglee, S.; Urban, L.; Sallanon, H.; Lopez-Lauri, F. Optimized Assay for Hydrogen Peroxide Determination in Plant Tissue Using Potassium Iodide. *Am. J. Anal. Chem.* **2014**, *05*, 730–736. [CrossRef]
34. Cho, Y.K.; Ahn, H.K. Purification and characterization of polyphenol oxidase from potato: II. Inhibition and catalytic mechanism. *J. Food Biochem.* **1999**, *23*, 593–605. [CrossRef]
35. Beauchamp, C.; Fridovich, I. Superoxide dismutase: Improved assays and an assay applicable to acrylamide gels. *Anal. Biochem.* **1971**, *44*, 276–287. [CrossRef]
36. Kumar, A.; Dutt, S.; Bagler, G.; Ahuja, P.S.; Kumar, S. Engineering a thermo-stable superoxide dismutase functional at sub-zero to >50 °C, which also tolerates autoclaving. *Sci. Rep.* **2012**, *2*, srep00387. [CrossRef]
37. Angelini, R.; Manes, F.; Federico, R. Spatial and functional correlation between diamine-oxidase and peroxidase activities and their dependence upon de-etiolation and wounding in chick-pea stems. *Planta* **1990**, *182*, 89–96. [CrossRef] [PubMed]
38. Singleton, V.L.; Orthofer, R.; Lamuela-Raventós, R.M. [14]Analysis of total phenols and other oxidation substrates and antioxidants by means of folin-ciocalteu reagent. In *Oxidants and Antioxidants Part A*; Academic Press: San Diego, CA, USA, 1999; Volume 299, pp. 152–178, ISBN 0076-6879.
39. Ghosh, S.; Derle, A.; Ahire, M.; More, P.; Jagtap, S.; Phadatare, S.D.; Patil, A.B.; Jabgunde, A.M.; Sharma, G.K.; Shinde, V.S.; et al. Phytochemical analysis and free radical scavenging activity of medicinal plants Gnidia glauca and Dioscorea bulbifera. *PLoS ONE* **2013**, *8*, e82529. [CrossRef] [PubMed]
40. Abdelkhalek, A.; Ismail, I.A.I.A.; Dessoky, E.S.E.S.; El-Hallous, E.I.E.I.; Hafez, E. A tomato kinesin-like protein is associated with Tobacco Mosaic Virus infection. *Biotechnol. Biotechnol. Equip.* **2019**, *33*, 1424–1433. [CrossRef]
41. Abdelkhalek, A.; Qari, S.H.S.H.; Hafez, E. Iris yellow spot virus–induced chloroplast malformation results in male sterility. *J. Biosci.* **2019**, *44*, 142. [CrossRef] [PubMed]
42. Abdelkhalek, A.; Al-Askar, A.A.; Arishi, A.A.; Behiry, S.I. Trichoderma hamatum Strain Th23 Promotes Tomato Growth and Induces Systemic Resistance against Tobacco Mosaic Virus. *J. Fungi* **2022**, *8*, 228. [CrossRef]
43. Hafez, E.E.; Abdelkhalek, A.A.; Abd El-Wahab, A.S.E.-D.; Galal, F.H. Altered gene expression: Induction/suppression in leek elicited by Iris Yellow Spot Virus infection (IYSV) Egyptian isolate. *Biotechnol. Biotechnol. Equip.* **2013**, *27*, 4061–4068. [CrossRef]
44. Abdelkhalek, A.; Qari, S.H.; Abu-Saied, M.A.A.-R.; Khalil, A.M.; Younes, H.A.; Nehela, Y.; Behiry, S.I. Chitosan Nanoparticles Inactivate Alfalfa Mosaic Virus Replication and Boost Innate Immunity in Nicotiana glutinosa Plants. *Plants* **2021**, *10*, 2701. [CrossRef] [PubMed]

45. Livak, K.J.; Schmittgen, T.D. Analysis of relative gene expression data using real-time quantitative PCR and the 2− ∆∆CT method. *Methods* **2001**, *25*, 402–408. [CrossRef] [PubMed]
46. Hashem, A.; Tabassum, B.; Abd_Allah, E.F. Bacillus subtilis: A plant-growth promoting rhizobacterium that also impacts biotic stress. *Saudi J. Biol. Sci.* **2019**, *26*, 1291–1297. [CrossRef] [PubMed]
47. Blake, C.; Christensen, M.N.; Kovacs, A.T. Molecular aspects of plant growth promotion and protection by bacillus subtilis. *Mol. Plant-Microbe Interact.* **2021**, *34*, 15–25. [CrossRef] [PubMed]
48. Abdelkhalek, A.; Behiry, S.I.; Al-Askar, A.A. Bacillus velezensis PEA1 Inhibits Fusarium oxysporum Growth and Induces Systemic Resistance to Cucumber Mosaic Virus. *Agronomy* **2020**, *10*, 1312. [CrossRef]
49. Kloepper, J.W.; Ryu, C.-M.; Zhang, S. Induced systemic resistance and promotion of plant growth by *Bacillus* spp. *Phytopathology* **2004**, *94*, 1259–1266. [CrossRef] [PubMed]
50. Qiao, J.; Yu, X.; Liang, X.; Liu, Y.; Borriss, R.; Liu, Y. Addition of plant-growth-promoting Bacillus subtilis PTS-394 on tomato rhizosphere has no durable impact on composition of root microbiome. *BMC Microbiol.* **2017**, *17*, 131. [CrossRef] [PubMed]
51. Bokhari, A.; Essack, M.; Lafi, F.F.; Andres-Barrao, C.; Jalal, R.; Alamoudi, S.; Razali, R.; Alzubaidy, H.; Shah, K.H.; Siddique, S.; et al. Bioprospecting desert plant Bacillus endophytic strains for their potential to enhance plant stress tolerance. *Sci. Rep.* **2019**, *9*, 18154. [CrossRef]
52. Saxena, A.K.; Kumar, M.; Chakdar, H.; Anuroopa, N.; Bagyaraj, D.J. *Bacillus* species in soil as a natural resource for plant health and nutrition. *J. Appl. Microbiol.* **2020**, *128*, 1583–1594. [CrossRef] [PubMed]
53. Posada, L.F.; Ramírez, M.; Ochoa-Gómez, N.; Cuellar-Gaviria, T.Z.; Argel-Roldan, L.E.; Ramírez, C.A.; Villegas-Escobar, V. Bioprospecting of aerobic endospore-forming bacteria with biotechnological potential for growth promotion of banana plants. *Sci. Hortic.* **2016**, *212*, 81–90. [CrossRef]
54. Arena, G.D.; Ramos-González, P.L.; Nunes, M.A.; Ribeiro-Alves, M.; Camargo, L.E.A.; Kitajima, E.W.; Machado, M.A.; Freitas-Astúa, J. Citrus leprosis virus C infection results in hypersensitive-like response, suppression of the JA/ET plant defense pathway and promotion of the colonization of its mite vector. *Front. Plant Sci.* **2016**, *7*, 1757. [CrossRef]
55. Zhu, F.; Deng, X.-G.; Xu, F.; Jian, W.; Peng, X.-J.; Zhu, T.; Xi, D.-H.; Lin, H.-H. Mitochondrial alternative oxidase is involved in both compatible and incompatible host-virus combinations in Nicotiana benthamiana. *Plant Sci.* **2015**, *239*, 26–35. [CrossRef] [PubMed]
56. Zhu, F.; Zhang, Q.; Che, Y.; Zhu, P.; Zhang, Q.; Ji, Z. Glutathione contributes to resistance responses to TMV through a differential modulation of salicylic acid and reactive oxygen species. *Mol. Plant Pathol.* **2021**, *22*, 1668–1687. [CrossRef] [PubMed]
57. Jaiswal, N.; Singh, M.; Dubey, R.S.; Venkataramanappa, V.; Datta, D. Phytochemicals and antioxidative enzymes defence mechanism on occurrence of yellow vein mosaic disease of pumpkin (Cucurbita moschata). *Biotech* **2013**, *3*, 287–295. [CrossRef] [PubMed]
58. Madhusudhan, K.N.; Srikanta, B.M.; Shylaja, M.D.; Prakash, H.S.; Shetty, H.S. Changes in antioxidant enzymes, hydrogen peroxide, salicylic acid and oxidative stress in compatible and incompatible host-tobamovirus interaction. *J. Plant Interact.* **2009**, *4*, 157–166. [CrossRef]
59. Radwan, D.E.M.; Ismail, K.S. The impact of hydrogen peroxide against cucumber green mottle mosaic virus infection in watermelon plants. *Polish J. Environ. Stud.* **2020**, *29*, 3771–3782. [CrossRef]
60. Rui, R.; Liu, S.; Karthikeyan, A.; Wang, T.; Niu, H.; Yin, J.; Yang, Y.; Wang, L.; Yang, Q.; Zhi, H. Fine-mapping and identification of a novel locus Rsc15 underlying soybean resistance to Soybean mosaic virus. *Theor. Appl. Genet.* **2017**, *130*, 2395–2410. [CrossRef] [PubMed]
61. Yoda, H.; Yamaguchi, Y.; Sano, H. Induction of hypersensitive cell death by hydrogen peroxide produced through polyamine degradation in tobacco plants. *Plant Physiol.* **2003**, *132*, 1973–1981. [CrossRef]
62. de Dios Alché, J. A concise appraisal of lipid oxidation and lipoxidation in higher plants. *Redox Biol.* **2019**, *23*, 101136. [CrossRef] [PubMed]
63. Hernández, J.A.; Gullner, G.; Clemente-Moreno, M.J.; Künstler, A.; Juhász, C.; Díaz-Vivancos, P.; Király, L. Oxidative stress and antioxidative responses in plant–virus interactions. *Physiol. Mol. Plant Pathol.* **2016**, *94*, 134–148. [CrossRef]
64. Balal, R.M.; Khan, M.M.; Shahid, M.A.; Mattson, N.S.; Abbas, T.; Ashfaq, M.; Garcia-Sanchez, F.; Ghazanfer, U.; Gimeno, V.; Iqbal, Z. Comparative studies on the physiobiochemical, enzymatic, and ionic modifications in salt-tolerant and salt-sensitive citrus rootstocks under NaCl stress. *J. Am. Soc. Hortic. Sci.* **2012**, *137*, 86–95. [CrossRef]
65. Yanan, D.; Ran, C.; Rong, Z.; Jiang, W.; Xuesen, C.; Chengmiao, Y.; Zhiquan, M. Isolation, identification, and antibacterial mechanisms of Bacillus amyloliquefaciens QSB-6 and its effect on plant roots. *Front. Microbiol.* **2021**, *12*, 746799.
66. Li, L.; Steffens, J.C. Overexpression of polyphenol oxidase in transgenic tomato plants results in enhanced bacterial disease resistance. *Planta* **2002**, *215*, 239–247. [CrossRef] [PubMed]
67. Jia, H.; Zhao, P.; Wang, B.; Tariq, P.; Zhao, F.; Zhao, M.; Wang, Q.; Yang, T.; Fang, J. Overexpression of Polyphenol Oxidase Gene in Strawberry Fruit Delays the Fungus Infection Process. *Plant Mol. Biol. Rep.* **2016**, *34*, 592–606. [CrossRef]
68. Mohammadi, M.; Kazemi, H. Changes in peroxidase and polyphenol oxidase activities in susceptible and resistant wheat heads inoculated with Fusarium graminearum and induced resistance. *Plant Sci.* **2002**, *162*, 491–498. [CrossRef]
69. Khan, M.S.; Gao, J.; Chen, X.; Zhang, M.; Yang, F.; Du, Y.; Moe, T.S.; Munir, I.; Xue, J.; Zhang, X. Isolation and Characterization of Plant Growth-Promoting Endophytic Bacteria *Paenibacillus polymyxa* SK1 from *Lilium lancifolium*. *Biomed Res. Int.* **2020**, *2020*, 8650957. [CrossRef]

70. Miljaković, D.; Marinković, J.; Balešević-Tubić, S. The Significance of *Bacillus* spp. in Disease Suppression and Growth Promotion of Field and Vegetable Crops. *Microorganisms* **2020**, *8*, 1037. [CrossRef]
71. Rais, A.; Jabeen, Z.; Shair, F.; Hafeez, F.Y.; Hassan, M.N. *Bacillus* spp., a bio-control agent enhances the activity of antioxidant defense enzymes in rice against Pyricularia oryzae. *PLoS ONE* **2017**, *12*, e0187412. [CrossRef] [PubMed]
72. Akyol, H.; Riciputi, Y.; Capanoglu, E.; Caboni, M.; Verardo, V. Phenolic compounds in the potato and its byproducts: An overview. *Int. J. Mol. Sci.* **2016**, *17*, 835. [CrossRef] [PubMed]
73. Nasr-Eldin, M.; Messiha, N.; Othman, B.; Megahed, A.; Elhalag, K. Induction of potato systemic resistance against the potato virus Y (PVY NTN), using crude filtrates of *Streptomyces* spp. under greenhouse conditions. *Egypt. J. Biol. Pest Control* **2019**, *29*, 62. [CrossRef]
74. Lee, G.H.; Ryu, C.-M. Spraying of leaf-colonizing Bacillus amyloliquefaciens protects pepper from *Cucumber mosaic virus*. *Plant Dis.* **2016**, *100*, 2099–2105. [CrossRef] [PubMed]
75. Abdelkhalek, A.; Al-Askar, A.A. Green Synthesized ZnO Nanoparticles Mediated by Mentha Spicata Extract Induce Plant Systemic Resistance against Tobacco Mosaic Virus. *Appl. Sci.* **2020**, *10*, 5054. [CrossRef]
76. Guo, Q.; Li, Y.; Lou, Y.; Shi, M.; Jiang, Y.; Zhou, J.; Sun, Y.; Xue, Q.; Lai, H. Bacillus amyloliquefaciens Ba13 induces plant systemic resistance and improves rhizosphere microecology against tomato yellow leaf curl virus disease. *Appl. Soil Ecol.* **2019**, *137*, 154–166. [CrossRef]
77. Dempsey, D.M.A.; Vlot, A.C.; Wildermuth, M.C.; Klessig, D.F. Salicylic acid biosynthesis and metabolism. *Arab. Book/Am. Soc. Plant Biol.* **2011**, *9*, e0156. [CrossRef]
78. Breen, S.; Williams, S.J.; Outram, M.; Kobe, B.; Solomon, P.S. Emerging Insights into the Functions of Pathogenesis-Related Protein 1. *Trends Plant Sci.* **2017**, *22*, 871–879. [CrossRef]
79. Iglesias, V.A.; Meins, F.; Meins, F., Jr. Movement of plant viruses is delayed in a β-1, 3-glucanase-deficient mutant showing a reduced plasmodesmatal size exclusion limit and enhanced callose deposition. *Plant J.* **2000**, *21*, 157–166. [CrossRef]
80. Otulak-Kozieł, K.; Kozieł, E.; Lockhart, B. Plant cell wall dynamics in compatible and incompatible potato response to infection caused by Potato virus Y (PVYNTN). *Int. J. Mol. Sci.* **2018**, *19*, 862. [CrossRef]
81. Bucher, G.L.; Tarina, C.; Heinlein, M.; Di Serio, F.; Meins, F., Jr.; Iglesias, V.A. Local expression of enzymatically active class I β-1, 3-glucanase enhances symptoms of TMV infection in tobacco. *Plant J.* **2001**, *28*, 361–369. [CrossRef]
82. Dobnik, D.; Baebler, Š.; Kogovšek, P.; Pompe-Novak, M.; Štebih, D.; Panter, G.; Janež, N.; Morisset, D.; Žel, J.; Gruden, K. β-1, 3-glucanase class III promotes spread of PVY NTN and improves in planta protein production. *Plant Biotechnol. Rep.* **2013**, *7*, 547–555. [CrossRef]
83. Beckman, C.H. Phenolic-storing cells: Keys to programmed cell death and periderm formation in wilt disease resistance and in general defence responses in plants? *Physiol. Mol. Plant Pathol.* **2000**, *57*, 101–110. [CrossRef]
84. Abdelkhalek, A.; Dessoky, E.S.; Hafez, E. Polyphenolic genes expression pattern and their role in viral resistance in tomato plant infected with Tobacco mosaic virus. *Biosci. Res.* **2018**, *15*, 3349–3356.
85. Huang, J.; Gu, M.; Lai, Z.; Fan, B.; Shi, K.; Zhou, Y.-H.; Yu, J.-Q.; Chen, Z. Functional analysis of the Arabidopsis PAL gene family in plant growth, development, and response to environmental stress. *Plant Physiol.* **2010**, *153*, 1526–1538. [CrossRef] [PubMed]
86. Su, H.; Song, S.; Yan, X.; Fang, L.; Zeng, B.; Zhu, Y. Endogenous salicylic acid shows different correlation with baicalin and baicalein in the medicinal plant Scutellaria baicalensis Georgi subjected to stress and exogenous salicylic acid. *PLoS ONE* **2018**, *13*, e0192114. [CrossRef]
87. Kang, J.-H.; McRoberts, J.; Shi, F.; Moreno, J.E.; Jones, A.D.; Howe, G.A. The flavonoid biosynthetic enzyme chalcone isomerase modulates terpenoid production in glandular trichomes of tomato. *Plant Physiol.* **2014**, *164*, 1161–1174. [CrossRef]
88. Sonnante, G.; D'Amore, R.; Blanco, E.; Pierri, C.L.; De Palma, M.; Luo, J.; Tucci, M.; Martin, C. Novel hydroxycinnamoyl-coenzyme A quinate transferase genes from artichoke are involved in the synthesis of chlorogenic acid. *Plant Physiol.* **2010**, *153*, 1224–1238. [CrossRef]
89. Moglia, A.; Lanteri, S.; Comino, C.; Hill, L.; Knevitt, D.; Cagliero, C.; Rubiolo, P.; Bornemann, S.; Martin, C. Dual catalytic activity of hydroxycinnamoyl-coenzyme A quinate transferase from tomato allows it to moonlight in the synthesis of both mono- and dicaffeoylquinic acids. *Plant Physiol.* **2014**, *166*, 1777–1787. [CrossRef]
90. Niggeweg, R.; Michael, A.J.; Martin, C. Engineering plants with increased levels of the antioxidant chlorogenic acid. *Nat. Biotechnol.* **2004**, *22*, 746. [CrossRef]
91. Abdelkhalek, A.; Al-Askar, A.A.; Hafez, E. Differential induction and suppression of the potato innate immune system in response to Alfalfa mosaic virus infection. *Physiol. Mol. Plant Pathol.* **2020**, *110*, 101485. [CrossRef]
92. Bazzini, A.A.; Manacorda, C.A.; Tohge, T.; Conti, G.; Rodriguez, M.C.; Nunes-Nesi, A.; Villanueva, S.; Fernie, A.R.; Carrari, F.; Asurmendi, S. Metabolic and miRNA profiling of TMV infected plants reveals biphasic temporal changes. *PLoS ONE* **2011**, *6*, e28466. [CrossRef]
93. Ullah, A.; Bano, A.; Janjua, H.T. Microbial Secondary Metabolites and Defense of Plant Stress. *Microb. Serv. Restor. Ecol.* **2020**, *11*, 37–46. [CrossRef]
94. Nas, F.; Aissaoui, N.; Mahjoubi, M.; Mosbah, A.; Arab, M.; Abdelwahed, S.; Khrouf, R.; Masmoudi, A.-S.; Cherif, A.; Klouche-Khelil, N. A comparative GC-MS analysis of bioactive secondary metabolites produced by halotolerant *Bacillus* spp. isolated from the Great Sebkha of Oran. *Int. Microbiol. Off. J. Spanish Soc. Microbiol.* **2021**, *24*, 455–470. [CrossRef] [PubMed]
95. María Teresa, R.-C.; Rosaura, V.-G.; Elda, C.-M.; Ernesto, G.-P. The avocado defense compound phenol-2,4-bis (1,1-dimethylethyl) is induced by arachidonic acid and acts via the inhibition of hydrogen peroxide production by pathogens. *Physiol. Mol. Plant Pathol.* **2014**, *87*, 32–41. [CrossRef]

96. Abdullah, A.S.H.; Mirghani, M.E.S.; Jamal, P. Antibacterial activity of Malaysian mango kernel. *African J. Biotechnol.* **2011**, *10*, 18739–18748. [CrossRef]
97. Ahsan, T.; Chen, J.; Zhao, X.; Irfan, M.; Wu, Y. Extraction and identification of bioactive compounds (eicosane and dibutyl phthalate) produced by Streptomyces strain KX852460 for the biological control of Rhizoctonia solani AG-3 strain KX852461 to control target spot disease in tobacco leaf. *AMB Express* **2017**, *7*, 54. [CrossRef]
98. Octarya, Z.; Novianty, R.; Suraya, N. Saryono Antimicrobial activity and GC-MS analysis of bioactive constituents of Aspergillus fumigatus 269 isolated from Sungai Pinang hot spring, Riau, Indonesia. *Biodiversitas* **2021**, *22*, 1839–1845. [CrossRef]
99. Naeim, H.; El-Hawiet, A.; Abdel Rahman, R.A.; Hussein, A.; El Demellawy, M.A.; Embaby, A.M. Antibacterial activity of *Centaurea pumilio* L. Root and aerial part extracts against some multidrug resistant bacteria. *BMC Complement. Med. Ther.* **2020**, *20*, 79. [CrossRef]
100. Bhardwaj, V.; Gumber, D.; Abbot, V.; Dhiman, S.; Sharma, P. Pyrrole: A resourceful small molecule in key medicinal heteroaromatics. *RSC Adv.* **2015**, *5*, 15233–15266. [CrossRef]
101. Sharma, S.; Kumar, D.; Singh, G.; Monga, V.; Kumar, B. Recent advancements in the development of heterocyclic anti-inflammatory agents. *Eur. J. Med. Chem.* **2020**, *200*, 112438. [CrossRef]
102. Kumari, N.; Menghani, E.; Mithal, R. GCMS analysis of compounds extracted from actinomycetes AIA6 isolates and study of its antimicrobial efficacy. *Indian J. Chem. Technol.* **2019**, *26*, 362–370.
103. Ser, H.-L.; Palanisamy, U.D.; Yin, W.-F.; Abd Malek, S.N.; Chan, K.-G.; Goh, B.-H.; Lee, L.-H. Presence of antioxidative agent, Pyrrolo[1,2-a]pyrazine-1,4-dione, hexahydro- in newly isolated *Streptomyces mangrovisoli* sp. nov. *Front. Microbiol.* **2015**, *6*, 854. [CrossRef] [PubMed]
104. Pooja, S.; Aditi, T.; Naine, S.J.; Devi, C.S. Bioactive compounds from marine *Streptomyces* sp. VITPSA as therapeutics. *Front. Biol.* **2017**, *12*, 280–289. [CrossRef]

Communication

Control of Seed-Borne Fungi by Selected Essential Oils

Simona Chrapačienė *, Neringa Rasiukevičiūtė * and Alma Valiuškaitė

Laboratory of Plant Protection, Institute of Horticulture, Lithuanian Research Centre for Agriculture and Forestry, Kaunas District, LT-54333 Babtai, Lithuania; alma.valiuskaite@lammc.lt
* Correspondence: simona.chrapaciene@lammc.lt (S.C.); neringa.rasiukeviciute@lammc.lt (N.R.)

Abstract: Seed-borne pathogens reduce the quality and cause infections at various growth stages of horticultural crops. Some of the best-known are fungi of genus *Alternaria*, that cause destructive vegetable and other crop diseases, resulting in significant yield losses. Over several years, much attention has been paid to environmentally-friendly solutions for horticultural disease management regarding the environmental damage caused by chemicals. For example, plant extracts and essential oils could be alternative sources for biopesticides and help to control vegetable seed-borne pathogens. This study aimed to evaluate essential oils' influence on the growth of seed-borne fungi *Alternaria* spp. The microbiological contamination of vegetable seeds (carrot, tomato, onion) was determined by the agar-plate method. The essential oils' impact on the growth of fungi was evaluated by mixing them with PDA medium at different amounts. The hydrodistillation was used for extraction of thyme and hyssop essential oils, and common juniper essential oil was purchased. The investigation revealed that the highest contamination of carrot and tomato seeds was by *Alternaria* spp. fungi. Furthermore, the highest antifungal effect on *Alternaria* spp. growth was achieved using 200–1000 µL L^{-1} of thyme essential oil. Meanwhile, the antifungal effect of other investigated essential oils differed from low to moderate. Overall, essential oils expressed a high potential for fungal pathogens biocontrol and application in biopesticides formulations.

Keywords: *Thymus vulgaris*; *Juniperus communis*; *Hyssopus officinalis*; *Alternaria* spp.; biocontrol

Citation: Chrapačienė, S.; Rasiukevičiūtė, N.; Valiuškaitė, A. Control of Seed-Borne Fungi by Selected Essential Oils. *Horticulturae* 2022, *8*, 220. https://doi.org/10.3390/horticulturae8030220

Academic Editors: Hillary Righini, Roberta Roberti and Stefania Galletti

Received: 31 December 2021
Accepted: 28 February 2022
Published: 2 March 2022

Publisher's Note: MDPI stays neutral with regard to jurisdictional claims in published maps and institutional affiliations.

Copyright: © 2022 by the authors. Licensee MDPI, Basel, Switzerland. This article is an open access article distributed under the terms and conditions of the Creative Commons Attribution (CC BY) license (https://creativecommons.org/licenses/by/4.0/).

1. Introduction

Vegetables are a crucial part of food production and are consumed worldwide. However, fungal diseases often lead to significant economic yield losses [1]. For example, horticultural production yield spoilage caused by fungal *Alternaria* species ranges from 20% to 80% [1,2]. This fungus can induce seedling death, petiole base blackening, leaf death or blight, leaf lesions, stem canker, black rot, and other symptoms depending on the host plant [3,4]. *Alternaria* spp. can also be considered a seed-borne pathogen, responsible for destructive diseases of various vegetables such as carrot, tomato, onion, etc. [1]. For example, *Alternaria radicina* Meier, Drechsler, and Eddy is known primarily as a carrot pathogen, responsible for root and crown disease and causing foliar blight under certain conditions. *Alternaria dauci* (Kühn) Groves and Skolko mainly cause carrot Alternaria leaf blight. However, *A. dauci* has also been documented to cause disease on parsnip, spinach, celery, and parsley [5]. *Alternaria solani* (Ellis and Martin) Sorauer causes early blight on foliage, collar rot on basal stems of seedlings, stem lesions on adult plants, and fruit rot of tomatoes [6]. Sources indicate that *Alternaria arborescens* Keissler also causes stem canker of tomato [7]. The purple blotch of onion is a disease caused by *Alternaria porri* (Ellis) Cif. [8]. *Alternaria alternata* (Fr.) Keissl. causes a black spot in many fruits and vegetables around the world. Some studies reported seed contamination with various *Alternaria* species, including saprotrophic *A. alternata*, *Alternaria tenuissima* Samuel Paul Wiltshire, *Alternaria longipes* (Ellis and Everh.) E.W. Mason [9–11]. In addition, due to its presence on the seeds' surface, *Alternaria* spp. can adversely affect seed germination [1,3,12]. Therefore, high-quality, fungi-free seeds are prioritised because vegetable consumption increases yearly [3]. Seed-borne diseases can be controlled by selecting resistant varieties, production technology, seed treatments and dressings, and soil disinfection [13]. Over the last several decades, seed and soil or foliar treatments with synthetic chemicals have been shown to prevent plant disease epidemics caused by seed-borne fungi [14–19].

Nevertheless, following regulation No. 1452/2003 developed by the European Commission, organic horticulture is limited to using only organic seeds [20]. Therefore, fungicides in organic production are not used for seeds to prevent the influence of micromycetes. Additionally, their non-target impact on pathogen resistance gain risks to human health and other organisms. Chemical nature and horticultural products pollution by pesticide residues has encouraged the investigation for alternative solutions to control and make horticulture more sustainable [21,22].

Essential oils, due to their broad applicability in various industries, like pharmacy or food industries, have received much attention [23,24]. Furthermore, more comprehensive studies of essential oils revealed their potential in environmental-friendly horticultural disease management as they have antiseptic, antiviral, antibacterial, and antifungal properties [25,26]. Additionally, essential oils, as secondary metabolites, exhibit high plant defence effects, are non-toxic, biodegradable, and limit pathogenic organisms [24,27]. Due to these features, they can be applied as biopesticides for alternative plant protection. For example, Karaca et al. [28] reported good inhibition of investigated fungal species growth under oregano, mint, and clove essential oils application. Muthukumar et al. [29] also stated significant results of geranium and palmarosa essential oils efficacy against rice micromycetes of genera *Cochliobolus* and *Fusarium*. According to other studies, thyme essential oil has a potent antifungal effect on the development of fungal plant pathogens [19,24,25,28,30–33].

The literature review showed that there is a lack of studies regarding environmentally friendly ways to prevent fungal infections of vegetable seeds. Hence, the aim of this study was to determine the predominant seed-born fungi in carrot, tomato, and onion seeds, then to evaluate the antifungal activity of essential oils of thyme, hyssop, and common juniper on the growth of *Alternaria* spp.

2. Materials and Methods

2.1. Seed Samples

For the research, three seed samples of carrot, onion, and tomato were obtained from the Department of Vegetable Breeding and Technology, Institute of Horticulture (IH), Lithuanian Research Centre for Agriculture and Forestry (LAMMC) (Table 1).

Table 1. Vegetable seeds used in the experiments.

Common Name	Botanical Name	Cultivar
Carrot	*Daucus carota sativus* L.	Svalia
Tomato	*Solanum lycopersicum* L.	Rutuliai
Onion	*Allium cepa* L.	Babtų didieji

Vegetable seeds were surface-sterilised by rinsing them in 70% ethanol for 3 min and then washing them three times with sterile distilled water for 5 min in total [34]. After this, seeds were left to dry for 5–10 min in laminar flow. The internal seeds infestation with fungi was determined when external microorganisms were removed during surface sterilisation.

2.2. Determination of Predominant Fungi

The microbiological contamination of seeds samples was evaluated using the agar-plate method [35]. The potato dextrose agar (PDA) medium (Sigma-Aldrich, St. Louis, MO, USA) composed of 15 g L^{-1} agar, 20 g L^{-1} dextrose, and 4 g L^{-1} potato extract was autoclaved and distributed to the Petri dishes [36]. Prepared surface-sterilised samples were arranged in a square shape (five rows and five columns) on each Petri dish (Figure 1) and kept at 22 ± 2 °C temperature in the dark [37,38].

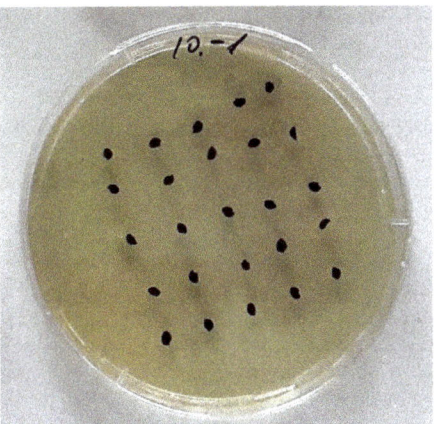

Figure 1. Arrangement of 25 seeds.

The experiment was repeated twice (four replications of each treatment). While inspecting the internal infestation of seeds, the settlements of fungi were counted to get the percentage of dominating fungi in the treatment after 2, 5, and 7 days of incubation (DOI). Visual and microscopical fungi identification was made based on morphological and cultural characteristics typical to the colonies [17,39]. Their detection frequency was determined using the detection rate of micromycetes: less than 30%—random species, more than 30%—typical species, more than 50%—dominant species [40].

2.3. Essential Oils Efficacy Assay

Three essential oils of thyme (*Thymus vulgaris* L.), hyssop *(Hyssopus officinalis* L.), and common juniper *(Juniperus communis* L.) were used in the experiment. The essential oils of thyme and hyssop were separately hydro-distilled for 2 h using Clevenger type of apparatus [41] from naturally dried herb material, harvested from the experimental fields of IH, LAMMC. The essential oil of the common juniper was purchased (Naujoji Barmune, Vilnius, Lithuania). The major compounds of thyme essential oil: thymol (41.35%), *p*-cymene (16.95%), and γ-terpinene (10.81%), were identified earlier by Morkeliūnė et al. [32]. The hyssop essential oil was characterised by cis-pinocamphone (40.16%), β-phellandrene (12.51%), and β-pinene (8.07%) and the process of chemical analysis was described previously by Šernaitė et al. [42]. The essential oil of the common juniper was mainly characterised by α-pinene (21.0–67.4%) and myrcene (7.8–18.7%) chemotypes [43].

Then, different amounts of each essential oil were added to one litre PDA medium after cooling to 45 °C, to get 200, 400, 600, 800, and 1000 μL L^{-1} concentrations, then mixed and poured into new Petri plates [44]. A control treatment was without essential oil in PDA, prepared as previously described. Treatments with thyme essential oil were coded T1, hyssop—T2, and common juniper—T3. Surface-sterilised samples of each seeds cultivars were placed in the same order as before (Figure 1) on PDA with different essential oil concentrations and incubated at 22 ± 2 °C in the dark for 7 days. There were four replicates for each vegetable seed cultivar, and the experiment was repeated twice.

The percentage of *Alternaria* spp. was calculated based on the number of grown fungal colonies in each plate after 2, 5, and 7 DOI. Fungi were identified according to cultural and morphological characteristics typical to the colonies [17,39]. Essential oils effect on *Alternaria* species was evaluated according to the disease incidence using the formula below (1) [12]:

Alternaria spp. incidence (%) = Number of seeds infected by *Alternaria* spp. × 100/Total number of infected seeds   (1)

Lower disease incidence showed effective essential oil mean activity for seed-borne fungi control.

2.4. Statistics

The experimental data were analysed using the analysis of variance (ANOVA) from the software SAS Enterprise Guide 7.1 (SAS Institute Inc., Cary, NC, USA). Duncan's multiple range test ($p < 0.05$) was used to determine differences among the treatments.

3. Results

The fungal contamination of vegetable seeds at the 7 DOI is summarised in Table 2. Carrot seeds were infected by 100%, and the predominant fungi were *Alternaria* spp. Fungi of genera *Penicillium* and *Fusarium* occurrence reached up to 4% and were considered random. However, the internal infection of tomato and onion seeds did not exceed 20%. The *Alternaria* spp. also dominated on tomato seeds. Fungi of the genera *Mucor* and *Penicillium* were typical for onion seeds and *Aspergillus* and *Mucor* for tomato seeds.

Table 2. Seeds contamination with fungi after seven days of incubation.

Seeds	Total Seeds Infected, %	Fungal Contamination, %					
		Alternaria spp.	*Fusarium* spp.	*Aspergillus* spp.	*Mucor* spp.	*Penicillium* spp.	Mycelia sterilia
Carrot	100	93.4	3.77	0	0	2.83	0
Tomato	20	50	0	15	25	0	10
Onion	15	20	0	0	40	40	0

As *Alternaria* species prevailed as the dominant fungi in vegetable seeds, it was decided to test the influence of three essential oils (T1, T2, and T3) on the growth of seed-borne fungi *Alternaria* spp. in vitro.

The incidence of *Alternaria* spp. on carrot seeds under the influence of T1, T2, and T3 treatments is presented in Figure 2.

The treatments applied to carrot seeds showed an intermittent effect. In the case of treatments with T1, all concentrations significantly suppressed ($p < 0.05$) the growth of seed-borne fungi. Furthermore, no colonies were detected at 2 DOI regardless of the amount of essential oil. Meanwhile, the emergence of *Alternaria* species reached 33% at 2 DOI, 36% at 5 DOI, and 64% at 7 DOI in the control treatment.

During the first assessment, seeds infection with *Alternaria* fungi was 1% at 400 and 1000 $\mu L\ L^{-1}$ of T2 treatment. Later, the abundance of these fungi was higher: 63% (200 $\mu L\ L^{-1}$), 68% (400 $\mu L\ L^{-1}$), 74% (600 $\mu L\ L^{-1}$), 48% (800 $\mu L\ L^{-1}$), and 64% (1000 $\mu L\ L^{-1}$) at 5 DOI. Likewise, 400 $\mu L\ L^{-1}$ of T2 cause significant decreation of *Alternaria* incidence at 7 DOI. However, the remaining T2 concentrations of 200, 600, 800, and 1000 $\mu L\ L^{-1}$ did not affect fungal growth— *Alternaria* spp. incidence increased compared with the control at 5 and 7 DOI. Thus, the opposite effect of T2 was observed than expected.

The T3 treatment performed weaker on *Alternaria* spp. on the second incubation day. Nevertheless, the 600 $\mu L\ L^{-1}$ had the best antifungal activity. The 200 $\mu L\ L^{-1}$ and 600–1000 $\mu L\ L^{-1}$ of T3 slightly controlled the prevalence of the fungi compared to controls at 5 DOI and did not differ significantly at 7 DOI. Still, the best fungal incidence suppression was exhibited by 400 $\mu L\ L^{-1}$ of this treatment at the fifth and seventh DOI.

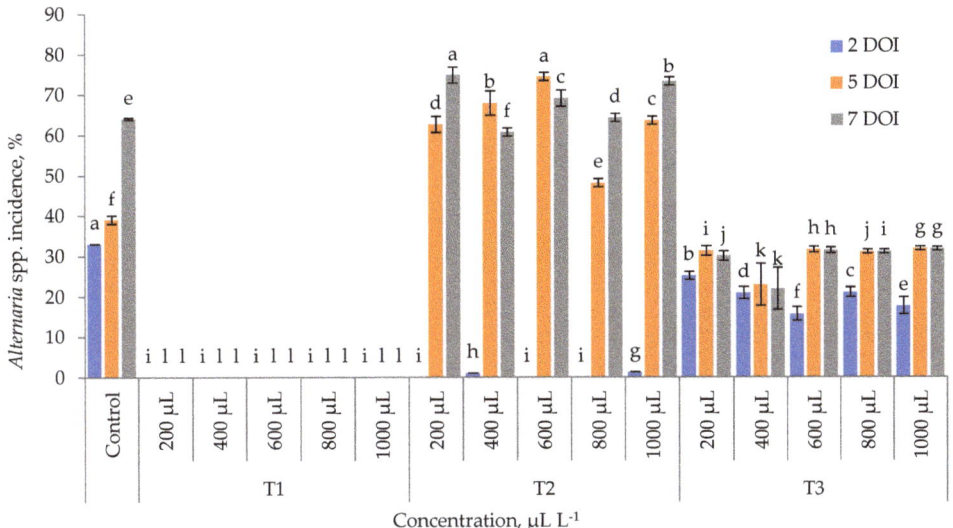

Figure 2. The incidence of *Alternaria* spp. on carrot seeds under the influence of thyme (T1), hyssop (T2), and common juniper (T3) essential oils treatments after 2, 5, and 7 days of incubation (DOI); according to Duncan's multiple range test ($p < 0.05$), the same letters demonstrate no significant differences between treatments at 2, 5, and 7 DOI.

The incidence of *Alternaria* spp. on tomato seeds under T1, T2, and T3 treatments is presented in Figure 3. Evaluating the effect of T1 concentrations from 200 to 1000 µL L^{-1}, no colonies of *Alternaria* spp. were observed at 2 and 5 DOI. However, 6% incidence was reached at 7 DOI under 200 µL L^{-1}.

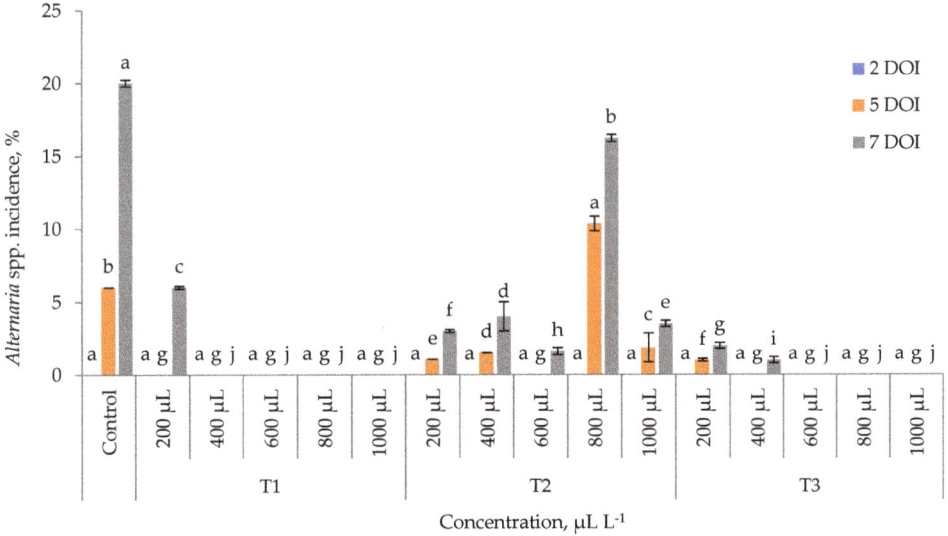

Figure 3. The incidence of *Alternaria* spp. on tomato seeds under the influence of thyme (T1), hyssop (T2), and common juniper (T3) essential oils treatments after 2, 5, and 7 days of incubation (DOI); according to Duncan's multiple range test ($p < 0.05$), the same letters demonstrate no significant differences between treatments at 2, 5, and 7 DOI.

Meanwhile, T2 performed a weaker impact on fungi occurrence, although no fungal colonies were noticed at 2 DOI. The development of micromycetes was observed: 1% under 200 µL L^{-1}, 2%—400 µL L^{-1}, 10%—800 µL L^{-1}, and 2%—1000 µL L^{-1} at 5 DOI. During the third estimation, the number of *Alternaria* spp. increased 2–6% from the previous evaluation. However, none of the T2 values exceeded those of the control incidence; the prevalence value of *Alternaria* spp. came up with 6% at 5 DOI and 20% at 7 DOI.

Estimation of tomato seeds incidence with *Alternaria* species at different concentrations of T3 revealed that fungi infected 1% of seeds under the lowest concentration used at 5 DOI. Later, the incidence increased to 2% under 200 µL L^{-1} and 1% under 400 µL L^{-1} of T3 at 7 DOI. The T3 treatment concentrations from 600 µL L^{-1} had significant antifungal activity at 2, 5, and 7 DOI.

The influence of T1, T2, and T3 treatments on the onion seeds infestation with *Alternaria* spp. is presented in Figure 4. At 5 DOI, the frequency in control was 18%, and at 7 DOI, 20%. The assay with T1 (200–1000 µL L^{-1}) revealed total development inhibition of the genus *Alternaria* fungi at all assessment days.

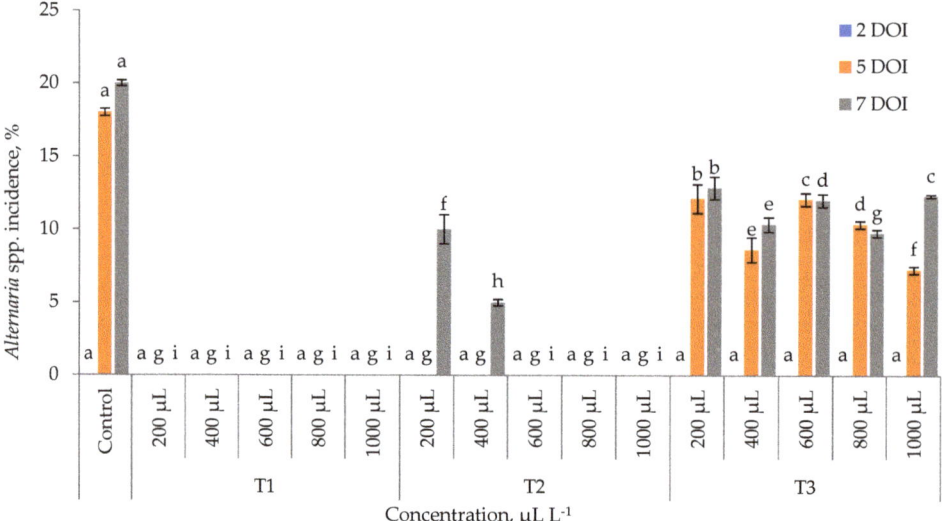

Figure 4. The incidence of *Alternaria* spp. on onion seeds under the influence of thyme (T1), hyssop (T2), and common juniper (T3) essential oils treatments after 2, 5, and 7 days of incubation (DOI); according to Duncan's multiple range test ($p < 0.05$), the same letters demonstrate no significant differences between treatments at 2, 5, and 7 DOI.

T2 treatment also gave excellent inhibition of *Alternaria* fungi growth recorded only with 200 and 400 µL L^{-1} at 7 DOI.

However, the overall antifungal effect of T3 treatment at concentrations from 200 to 1000 µL L^{-1} was average compared to the control evaluating onion seeds at 2, 5, and 7 DOI. In addition, the augmentation of *Alternaria* spp. was observed in all treatments (control, T1, T2, and T3) at 2 DOI. Therefore, the 400 and 1000 µL L^{-1} of all T3 treatment concentrations had the best antifungal activity on *Alternaria* spp. incidence.

From all T1, T2, and T3 treatments applied on the carrot, tomato, and onion seeds, the T1 at all rates most significantly inhibited the growth of *Alternaria* spp., and concentrations from 400 µL L^{-1} of T3 also showed modest antimycotic results. It is important to remember that tomato and onion seeds had less internal contamination with pathogenic fungi than carrot seeds.

4. Discussion

Innovative plant protection solutions are necessary due to the chemical fungicides' negative impact on ecology and seeds germination issues caused by infestation with *Alternaria* spp. [12,21,22]. Therefore, appropriate antifungal measures and their application strategy are crucial. The current study provides the latest findings regarding the antifungal effects of plant-based substances on the seed-borne fungi of the genus *Alternaria* isolated on tomato, carrot, and onion seeds.

Fungi incidence on vegetable seeds was significantly reduced by thyme (T1) treatment at all rates (200–1000 µL L^{-1}). There are numerous investigations about thyme essential oil antimicrobial impacts determined by the main chemotypes of thymol, γ-terpinene, *p*-cymene, β-caryophyllene, and carvacrol [19,24,25,27,28,30–33]. Some other experiments reported an effective control mechanism for *Alternaria* and other genera fungi on carrot and tomato seeds by this oil [30,31,33,45,46]. For example, Dorna and Szopińska [30] noticed that different applications of commercial oils involving thyme, possibly with 22–38% of thymol and 1–2% carvacrol, reduced the fungal contamination of the carrot seeds (cultivar 'Flakkese 2'), including *Alternaria alternata* (Fr.) Keissl. micromycetes. Contrary, the incidence of *Alternaria dauci* (J.G. Kühn), J.W. Groves and Skolko, and *Alternaria radicina* Meier, Drechsler, and E.D. Eddy increased under this oil influence in the case of seed cultivar 'Amsterdam 3' [30]. The same good effects of the treatment were seen against *Alternaria* spp. when a 'Laguna' cultivar seed sample was stirred in 1% oil emulsion for four hours. Nevertheless, the authors emphasised that choosing the optimal concentration is critical due to inherent oil phytotoxicity, and they recommended pre-testing [31]. Our results with T1 (41.35% of thymol) treatment support Riccioni and Orzali research [46], where a thyme essential oil (41% of thymol) concentration range of 0.05–1% considerably reduced the development of *A. dauci* in vitro.

Our results revealed that the effect on vegetable seeds was unequal when applying hyssop (T2) essential oil. It had the most negligible impact on carrot seeds, then better than average on tomato seeds, and the best fungi growth inhibition on onion seeds. It did not inhibit fungal development on carrot seeds and even promoted it compared to the control. Nonetheless, Fraternale et al. [47] have described various phytopathogenic fungi, like *Rhizoctonia solani* Ell. et Mart, *Botrytis cinerea* Pers., *Fusarium graminearum* Schwabe (ATCC 15624) as significantly sensitive to two hyssop essential oils. These oils were mainly characterised by pinocamphone (34% and 18.5%), β-pinene (10.5% and 10.8%), and isopinocamphone (3.2% and 29%). Their experiment also revealed that concentrations of 1400 and 1600 µL mL^{-1} of both oils inhibited 13 different fungal plant pathogens by 100%, *Alternaria solani* Ell. et Mart either. The seeds utilised in our experiments were characterised by higher contamination of *Alternaria* species. Thus, according to previously discussed results, the tested concentrations of T2 treatment were not high enough to achieve expected efficacy in our experiment. Still, the similarity between chemical compositions of T2 (cis-pinocamphone, 40.16%; β-phellandrene, 12.51%; β-pinene, 8.07%) and earlier described hyssop oil [40] prompts a potentially optimistic effect of higher T2 concentrations. Moreover, many studies investigated hyssop antimicrobial properties on other microorganisms and substances of this herb exhibited undeniable prospects [48].

Our study found that common juniper T3 inhibited *Alternaria* spp. depending on seeds, and the concentration of 400 µL L^{-1} was most effective in all seed experiments. T3 active compounds were possibly α-pinene (21.0–67.4%) and myrcene (7.8–18.7%). Other authors found that common juniper essential oil was effective against some soil- and seed-borne pathogens. For example, in Zabka et al. [49] research, 1 µL mL^{-1} concentration of this oil moderately influenced pathogens, such as *Fusarium verticillioides* (Sacc.) Nirenberg, *Fusarium oxysporum* Schlechtendahl, *Aspergillus fumigates* Fresenius, *Aspergillus flavus* Link, *Penicillium expansum* Link, and *Penicillium brevicompactum* Dierckx; the effect on *Alternaria* species was not studied. Indeed, thyme oil was described as the most robust to reduce target fungi growth [49]. Additionally, good antimicrobial activity (*A. flavus*, *A. niger*, and

Candida albicans (C.P. Robin) Berkhout) of methanolic extract of *Juniperus communis* L. was highlighted, emphasising the leading activity against *A. niger* and *A. flavus* [50].

To conclude, essential oils of hyssop and common juniper resulted in a moderate capability to control the seed-borne fungi on tested samples of vegetable seeds. Besides, results demonstrated that thyme essential oil had a significant reducing impact on the carrot, tomato, and onion fungi, affirming what is already published in the literature. Despite this, as in vitro effects do not always positively affect in vivo performances, further studies are required to prove the effectiveness in field conditions as seeds treatments and their possible phytotoxicity on the plant or seed material. Furthermore, thyme essential oil is a promising agent for vegetable seed-borne fungi *Alternaria* spp. management.

Author Contributions: Conceptualisation, S.C., N.R. and A.V.; methodology, S.C., N.R. and A.V.; software, S.C.; validation, N.R.; formal analysis, S.C.; investigation, S.C.; resources, N.R.; data curation, S.C.; writing—original draft preparation, S.C.; writing—review and editing, S.C., N.R. and A.V.; visualisation, S.C.; supervision, N.R. and A.V. All authors have read and agreed to the published version of the manuscript.

Funding: This research received no external funding.

Institutional Review Board Statement: Not applicable.

Conflicts of Interest: The authors declare no conflict of interest. The funders had no role in the design of the study; in the collection, analyses, or interpretation of data; in the writing of the manuscript, or in the decision to publish the results.

References

1. Tournas, V.H. Spoilage of vegetable crops by bacteria and fungi and related health hazards. *Crit. Rev. Microbiol.* **2005**, *31*, 33–44. [CrossRef] [PubMed]
2. Nowicki, M.; Nowakowska, M.; Niezgoda, A.; Kozik, E. Alternaria Black Spot of Crucifers: Symptoms, Importance of Disease, and Perspectives of Resistance Breeding. *J. Fruit Ornam. Plant Res.* **2012**, *76*, 5–19. [CrossRef]
3. Gaur, A.; Kumar, A.; Kiran, R.; Kumari, P. Importance of seed-borne diseases of agricultural crops: Economic losses and impact on society. In *Seed-Borne Diseases of Agricultural Crops: Detection, Diagnosis & Management*; Springer: Singapore, 2020; pp. 3–23. [CrossRef]
4. EPPO Standard. PP 2/1 (1) Guideline on good plant protection practice: Umbelliferous crops. *Bull. OEPP/EPPO Bull.* **1994**, *24*, 233–240.
5. Farrar, J.J.; Pryor, B.M.; Davis, R.M. Alternaria diseases of carrot. *Plant Dis.* **2004**, *88*, 776–784. [CrossRef] [PubMed]
6. Chaerani, R.; Voorrips, R.E. Tomato early blight (*Alternaria solani*): The pathogen, genetics, and breeding for resistance. *J. Gen. Plant Pathol.* **2006**, *72*, 335–347. [CrossRef]
7. Singh, V. Alternaria diseases of vegetable crops and its management control to reduce the low production. *Int. J. Agric. Sci.* **2015**, *7*, 834–840.
8. Mohsin, S.M.; Islam, M.R.; Ahmmed, A.N.F.; Nisha, H.A.C.; Hasanuzzaman, M. Cultural, morphological and pathogenic characterization of Alternaria porri causing purple blotch of onion. *Not. Bot. Horti Agrobot. Cluj-Napoca* **2016**, *44*, 222–227. [CrossRef]
9. Solfrizzo, M.; Girolamo, A.D.; Vitti, C.; Tylkowska, K.; Grabarkiewicz-Szczęsna, J.; Szopińska, D.; Dorna, H. Toxigenic profile of Alternaria alternata and Alternaria radicina occurring on umbelliferous plants. *Food Addit. Contam.* **2005**, *22*, 302–308. [CrossRef]
10. Addrah, M.E.; Zhang, Y.; Zhang, J.; Liu, L.; Zhou, H.; Chen, W.; Zhao, J. Fungicide treatments to control seed-borne fungi of sunflower seeds. *Pathogens* **2020**, *9*, 29. [CrossRef]
11. Siciliano, I.; Gilardi, G.; Ortu, G.; Gisi, U.; Gullino, M.L.; Garibaldi, A. Identification and characterization of Alternaria species causing leaf spot on cabbage, cauliflower, wild and cultivated rocket by using molecular and morphological features and mycotoxin production. *Eur. J. Plant Pathol.* **2017**, *149*, 401–413. [CrossRef]
12. Zhang, X.; Wang, R.; Ning, H.; Li, W.; Bai, Y.; Li, Y. Evaluation and management of fungal-infected carrot seeds. *Sci. Rep.* **2020**, *10*, 10808. [CrossRef] [PubMed]
13. Davis, R.M. Carrot diseases and their management. In *Diseases of Fruits and Vegetables*; Naqvi, S.A.M.H., Ed.; Springer: Dordrecht, The Netherlands, 2004; Volume I, pp. 397–439.
14. González, M.; Caetano, P.; Sánchez, M.E. Testing systemic fungicides for control of Phytophthora oak root disease. *For. Pathol.* **2017**, *47*, e12343. [CrossRef]
15. Lamichhane, J.R.; You, M.P.; Laudinot, V.; Barbetti, M.J.; Aubertot, J.N. Revisiting sustainability of fungicide seed treatments for field crops. *Plant Dis.* **2020**, *104*, 610–623. [CrossRef] [PubMed]

16. Singh, U.B.; Chaurasia, R.; Manzar, N.; Kashyap, A.S.; Malviya, D.; Singh, S.; Kannojia, P.; Sharma, P.K.; Imran, M.; Sharma, A.K. Chemical Management of Seed-Borne Diseases: Achievements and Future Challenges. In *Seed-Borne Diseases of Agricultural Crops: Detection, Diagnosis & Management*; Kumar, R., Gupta, A., Eds.; Springer: Singapore, 2020; pp. 665–682. [CrossRef]
17. Töfoli, J.G.; Domingues, R.J.; Tortolo, M.P.L. Effect of various fungicides in the control of Alternaria Leaf Blight in carrot crops. *Biol. São Paulo* **2019**, *81*, 1–30. [CrossRef]
18. Sharma, K.K.; Singh, U.S.; Sharma, P.; Kumar, A.; Sharma, L. Seed treatments for sustainable agriculture-A review. *J. Appl. Nat. Sci.* **2015**, *7*, 521–539. [CrossRef]
19. Mancini, V.; Romanazzi, G. Seed treatments to control seedborne fungal pathogens of vegetable crops. *Pest Manag. Sci.* **2014**, *70*, 860–868. [CrossRef]
20. The European Commission Regulation (EC) No. 1452/2003. Official Journal of the European Community. 2003. Available online: https://eur-lex.europa.eu/LexUriServ/LexUriServ.do?uri=OJ:L:2003:206:0017:0021:EN:PDF (accessed on 9 March 2021).
21. Silva, V.; Mol, H.G.; Zomer, P.; Tienstra, M.; Ritsema, C.J.; Geissen, V. Pesticide residues in European agricultural soils–A hidden reality unfolded. *Sci. Total Env.* **2019**, *653*, 1532–1545. [CrossRef]
22. Cozma, P.; Apostol, L.C.; Hlihor, R.M.; Simion, I.M.; Gavrilescu, M. Overview of human health hazards posed by pesticides in plant products. In Proceedings of the 2017 E-Health and Bioengineering Conference (EHB), Sinaia, Romania, 22–24 June 2017; IEEE: Piscataway, NJ, USA, 2017; Volume 17066084, pp. 293–296. [CrossRef]
23. Ložienė, K.; Venskutonis, P.R. Juniper (*Juniperus communis* L.) oils. In *Essential Oils in Food Preservation, Flavor and Safety*; Academic Press: Cambridge, MA, USA, 2016; pp. 495–500. [CrossRef]
24. Hanif, M.A.; Nisar, S.; Khan, G.S.; Mushtaq, Z.; Zubair, M. Essential oils. In *Essential Oil Research*; Springer: Cham, Switzerland, 2019; pp. 3–17. [CrossRef]
25. Valdivieso-Ugarte, M.; Gomez-Llorente, C.; Plaza-Díaz, J.; Gil, Á. Antimicrobial, antioxidant, and immunomodulatory properties of essential oils: A systematic review. *Nutrients* **2019**, *11*, 2786. [CrossRef]
26. Moulodi, F.; Khezerlou, A.; Zolfaghari, H.; Mohamadzadeh, A.; Alimoradi, F. Chemical Composition and Antioxidant and Antimicrobial Properties of the Essential Oil of *Hyssopus officinalis* L. *J. Kermanshah Univ. Med. Sci.* **2018**, *22*, e85256. [CrossRef]
27. Dauqan, E.M.; Abdullah, A. Medicinal and functional values of thyme (*Thymus vulgaris* L.) herb. *J. Appl. Biol. Biotechnol.* **2017**, *5*, 17–22. [CrossRef]
28. Karaca, G.; Bilginturan, M.; Olgunsoy, P. Effects of some plant essential oils against fungi on wheat seeds. *Indian J. Pharm. Educ. Res* **2017**, *51*, S385–S388. [CrossRef]
29. Muthukumar, A.; Sangeetha, G.; Naveenkumar, R. Antimicrobial activity of essential oils against seed borne fungi of rice (*Oryza sativa* L.). *J. Environ. Biol.* **2016**, *37*, 1429–1436.
30. Dorna, H.; Qi, Y.; Szopińska, D. The effect of acetic acid, grapefruit extract and selected essential oils on germination, vigour and health of carrot (*Daucus carota* L.) seeds. *Acta Sci. Pol. Hortorum Cultus* **2018**, *17*, 27–38. [CrossRef]
31. Koch, E.; Schmitt, A.; Stephan, D.; Kromphardt, C.; Jahn, M.; Krauthausen, H.J.; Forsberg, G.; Werner, S.; Amein, T.; Wright, S.A.I.; et al. Evaluation of non-chemical seed treatment methods for the control of *Alternaria dauci* and *A. radicina* on carrot seeds. *Eur. J. Plant Pathol.* **2010**, *127*, 99–112. [CrossRef]
32. Morkeliūnė, A.; Rasiukevičiūtė, N.; Šernaitė, L.; Valiuškaitė, A. The Use of Essential Oils from Thyme, Sage and Peppermint against Colletotrichum acutatum. *Plants* **2021**, *10*, 114. [CrossRef]
33. Soković, M.D.; Vukojević, J.; Marin, P.D.; Brkić, D.D.; Vajs, V.; van Griensven, L.J.L.D. Chemical composition of essential oils of Thymus and Mentha species and their antifungal activities. *Molecules* **2009**, *14*, 238–249. [CrossRef]
34. Graj, W.; Lisiecki, P.; Szulc, A.; Chrzanowski, Ł.; Wojtera-Kwiczor, J. Bioaugmentation with petroleum-degrading consortia has a selective growth-promoting impact on crop plants germinated in diesel oil-contaminated soil. *Water Air Soil Pollut.* **2013**, *224*, 1676. [CrossRef]
35. Mathur, S.B.; Kongsdal, O. *Common Laboratory Seed Health Testing Methods for Detecting Fungi*, 1st ed.; International Seed Testing Association: Bassersdorf, Switzerland, 2003; 425p, ISBN 3906549356.
36. Acumedia. A Subsidiary of NEOGEN Corporation. Neogen Food Safety. Potato Dextrose Agar (7149)-Protocol. PI7149 Rev 4. 2011. Available online: http://biotrading.com/assets/productinformatie/acumedia/tds/7149.pdf (accessed on 10 March 2021).
37. Thobunluepop, P. The inhibitory effect of the various seed coating substances against rice seed borne fungi and their shelf-life during storage. *Pak. J. Biol. Sci.* **2009**, *12*, 1102. [CrossRef]
38. Mancini, V.; Murolo, S.; Romanazzi, G. Diagnostic methods for detecting fungal pathogens on vegetable seeds. *Plant Pathol.* **2016**, *65*, 691–703. [CrossRef]
39. Mamgain, A.; Roychowdhury, R.; Tah, J. Alternaria pathogenicity and its strategic controls. *Res. J. Biol.* **2013**, *1*, 1–9.
40. Stankevičienė, A.; Lugauskas, A. Mikromicetų įvairovė augalų rizosferoje skirtingose oranžerijos sekcijose. *Vytauto Didžiojo Univ. Bot. Sodo Raštai = Scr. Horti Bot. Univ.* **2008**, *12*, 84–93.
41. AOAC. Volatile oil in spices. In *Official Methods of Analysis*, 15th ed.; Helrich, K., Ed.; Association of Official Analytical Chemists: Washington, DC, USA, 1990; Volume 1.
42. Šernaitė, L.; Rasiukevičiūtė, N.; Valiuškaitė, A. The extracts of cinnamon and clove as potential biofungicides against strawberry grey mould. *Plants* **2020**, *9*, 613. [CrossRef] [PubMed]
43. Judžentienė, A. *Juniperus communis* L.: A review of volatile organic compounds of wild and cultivated common juniper in Lithuania. *Chemija* **2019**, *30*, 184–193. [CrossRef]

44. Morkeliūnė, A.; Rasiukevičiūtė, N.; Valiuškaitė, A. Pathogenicity of *Colletotrichum acutatum* to different strawberry cultivars and anthracnose control with essential oils. *Zemdirb. Agric.* **2021**, *10*, 173–180. [CrossRef]
45. Alexa, E.; Sumalan, R.M.; Danciu, C.; Obistioiu, D.; Negrea, M.; Poiana, M.A.; Dehelean, C. Synergistic antifungal, allelopatic and anti-proliferative potential of *Salvia officinalis* L., and *Thymus vulgaris* L. essential oils. *Molecules* **2018**, *23*, 185. [CrossRef]
46. Riccioni, L.; Orzali, L. Activity of tea tree (*Melaleuca alternifolia*, Cheel) and thyme (*Thymus vulgaris*, Linnaeus.) essential oils against some pathogenic seed borne fungi. *J. Essent. Oil Res.* **2011**, *23*, 43–47. [CrossRef]
47. Fraternale, D.; Ricci, D.; Epifano, F.; Curini, M. Composition and antifungal activity of two essential oils of hyssop (*Hyssopus officinalis* L.). *J. Essent. Oil Res.* **2004**, *16*, 617–622. [CrossRef]
48. Judžentienė, A. Hyssop (*Hyssopus officinalis* L.) Oils. In *Essential Oils in Food Preservation, Flavor and Safety*; Academic Press: Cambridge, MA, USA, 2016; pp. 471–479. [CrossRef]
49. Zabka, M.; Pavela, R.; Slezakova, L. Antifungal effect of *Pimenta dioica* essential oil against dangerous pathogenic and toxinogenic fungi. *Ind. Crops Prod.* **2009**, *30*, 250–253. [CrossRef]
50. Menghani, E.; Sharma, S.K. Antimicrobial activity of *Juniperus communis* and *Solanum xanthocarpum*. *Int. J. Pharm. Sci. Res.* **2012**, *3*, 2815. [CrossRef]

Article

Effects of *Bacillus amyloliquefaciens* QSB-6 on the Growth of Replanted Apple Trees and the Soil Microbial Environment

Yanan Duan, Yifan Zhou, Zhao Li, Xuesen Chen, Chengmiao Yin * and Zhiquan Mao *

State Key Laboratory of Crop Biology, College of Horticultural Science and Engineering, Shandong Agricultural University, Taian 271018, China; 2019010063@sdau.edu.cn (Y.D.); 2020120319@sdau.edu.cn (Y.Z.); 2020120361@sdau.edu.cn (Z.L.); chenxs@sdau.edu.cn (X.C.)
* Correspondence: cmyin@sdau.edu.cn (C.Y.); mzhiquan@sdau.edu.cn (Z.M.); Tel.: +86-18653880060 (C.Y.); +86-538-8241984 (Z.M.)

Abstract: Apple replant disease (ARD), caused largely by soil-borne fungal pathogens, has seriously hindered the development of the apple industry. The use of antagonistic microorganisms has been confirmed as a low-cost and environmentally friendly means of controlling ARD. In the present study, we assessed the effects of *Bacillus amyloliquefaciens* QSB-6 on the growth of replanted apple saplings and the soil microbial environment under field conditions, thus providing a theoretical basis for the successful use of microbial biocontrol agents. Four treatments were implemented in three apple orchards: untreated replant soil (CK1), methyl bromide fumigation (CK2), blank carrier treatment (T1), and QSB-6 bacterial fertilizer treatment (T2). The plant height, ground diameter, and branch length of apple saplings treated with T2 in three replanted apple orchards were significantly higher than that of the CK1 treatment. Compared with the other treatments, T2 significantly increased the number of soil bacteria, the proportion of actinomycetes, and the activities of soil enzymes. By contrast, compared with the CK1 treatments, the phenolic acid content, the number of fungi, and the abundance of *Fusarium oxysporum*, *Fusarium moniliforme*, *Fusarium proliferatum*, and *Fusarium solani* in the soil were significantly reduced. PCoA and cluster analysis showed that soil inoculation with strain QSB-6 significantly decreased the Mcintosh and Brillouin index of soil fungi and increased the diversity of soil bacteria in T2 relative to CK1. The soil bacterial community structure in T2 was different from the other treatments, and the soil fungal communities of T2 and CK2 were similar. In summary, QSB-6 bacterial fertilizer shows promise as a potential bio-inoculum for the control of ARD.

Keywords: apple replant disease; *Bacillus amyloliquefaciens*; phenolic acid; soil microorganisms

Citation: Duan, Y.; Zhou, Y.; Li, Z.; Chen, X.; Yin, C.; Mao, Z. Effects of *Bacillus amyloliquefaciens* QSB-6 on the Growth of Replanted Apple Trees and the Soil Microbial Environment. *Horticulturae* 2022, 8, 83. https://doi.org/10.3390/horticulturae8010083

Academic Editors: Hillary Righini, Roberta Roberti and Stefania Galletti

Received: 23 November 2021
Accepted: 12 January 2022
Published: 17 January 2022

Publisher's Note: MDPI stays neutral with regard to jurisdictional claims in published maps and institutional affiliations.

Copyright: © 2022 by the authors. Licensee MDPI, Basel, Switzerland. This article is an open access article distributed under the terms and conditions of the Creative Commons Attribution (CC BY) license (https://creativecommons.org/licenses/by/4.0/).

1. Introduction

In the main apple planting regions of China, many old apple orchards are facing renewal but are limited by land resources, leading to the development of apple replant disease (ARD), which seriously affects the yield and quality of apples [1,2]. The disease is particularly harmful to young, replanted apple trees, causing slow growth, disease susceptibility, stopping root growth, below-ground necrosis, and even death in severe cases [3,4]. Previous studies have reported that replant disease does not arise from a single cause, but instead emerges from a combination of biotic and abiotic factors. The abiotic factors include allelopathy, autotoxicity, and soil physicochemical imbalances [5–8]. The biotic factors include the accumulation of soil-borne pathogens such as nematodes (*Pratylenchus* spp.), fungi (*Rhizoctonia solani*, *Fusarium* spp., and *Cylindrocarpon* spp.) and oomycetes (species of *Pythium* and *Phytophthora*). Although it is generally believed that replant disease results from a combination of biotic and abiotic factors, biotic factors play a leading role in disease development [2,3,8–10]. This has been widely demonstrated in other studies through soil pasteurization and biocide application [11–13].

Sheng et al. [14] demonstrated that the relative abundances of *F. oxysporum* and *F. solani* were significantly reduced after the replant soil was fumigated with methyl bromide. Quantitative analysis based on qPCR confirmed the wide distribution of *F. oxysporum* in replanted soil around the Bohai Gulf region, and *F. verticillioides* (formerly *F. moniliforme*), *F. solani*, *F. oxysporum*, and *F. proliferatum* isolated from apple replant soil around the Bohai Gulf region were highly pathogenic to *Malus hupehensis* seedlings. These findings demonstrate that *Fusarium* spp. are among the main soil-borne pathogenic fungi that cause ARD in China [15–18], and similar results have been reported from orchards in South Africa [10]. *Fusarium* spp. can survive saprophytically in soil or crop debris for more than six years. It infects the roots of susceptible hosts, colonizes the vascular system of plants, and produces toxins that kill plants by blocking xylem vessels and restricting water transport [19–21]. Traditional control methods, such as crop rotation and organic material application, are usually ineffective against this pathogen, and the cultivation of disease-resistant rootstocks and the application of chemical fumigants (e.g., methyl bromide) are the main measures used to control ARD. However, methyl bromide has been gradually eliminated because of its high environmental toxicity and time-consuming application [12,22–24]. The use of biological control agents (BCAs) is considered to be a safe, environmentally friendly, and sustainable means of protecting plants from soil-borne pathogenic fungi, and it has been widely used to control various fungal diseases of greenhouse and field crops. Therefore, the development of BCAs is one of the main directions for disease control research in the future [25–27].

Among the previously developed BCAs, *Pseudomonas* spp., arbuscular mycorrhizal fungi, *Bacillus* spp., and *Trichoderma* spp. have been reported to be effective antagonists against wilt caused by *Fusarium* spp. under field conditions [28–33]. *Bacillus* spp. offer several advantages over other BCAs for protection against root pathogens, including broad-spectrum activity of their antifungal secondary metabolites, plant growth promoting (PGP) properties, unique sporulation capacity, and ability to resist various stresses, which facilitates their long-term storage and commercialization [34,35]. Haddoudi et al. [36] revealed that the application of *Bacillus amyloliquefaciens*, *Bacillus velezensis*, *Bacillus subtilis*, and *Bacillus mojavensis* significantly reduced broad bean Fusarium wilt caused by *Bacillus equisetum* under greenhouse conditions and promoted plant growth. Zheng et al. [37] reported that *B. velezensis* D61-A had strong inhibitory activity against *R. solani*; its control efficacy reached 61.5% and 74.6% in the greenhouse and the field, respectively. Anusha et al. [38] obtained five strains of *Bacillus* spp. with strong antagonistic effects on *F. oxysporum* f. sp. *ciceri* from rhizosphere soils. Plants treated with these strains showed reduced disease incidence and delayed symptom development relative to non-inoculated controls under greenhouse and wilt-infested field conditions. Therefore, the introduction of disease-suppressing microorganisms is of great significance to the control of ARD [39,40].

In earlier work, we isolated a strain of *B. amyloliquefaciens* QSB-6 with broad-spectrum antifungal properties. Its fermentation broth was found to contain antibacterial substances (1,2-benzenedicarboxylic acid and benzeneacetic acid, 3-hydroxy-, methyl ester) with growth-promoting properties that significantly inhibited the mycelial growth and spore germination of *F. oxysporum*, *F. moniliforme*, *F. proliferatum*, and *F. solani*. These substances also protected the roots of *M. hupehensis* seedlings from *F. oxysporum*, *F. moniliforme*, *F. proliferatum*, and *F. solani* damage and promoted plant growth in a pot experiment [18]. Here, on the basis of these initial results, we evaluated the effects of *B. amyloliquefaciens* QSB-6 as a biological control agent of ARD under field conditions. Our specific aims were to evaluate the effects of QSB-6 on: (i) the growth of replanted apple saplings; (ii) the microbial diversity and community structure of the rhizosphere soil; and (iii) soil enzyme activities and the phenolic acid content. Lastly, we evaluated this new method in the renewal of aging apple orchards.

2. Materials and Methods

2.1. Bacterial Fertilizer Production

A strain of *B. amyloliquefaciens* QSB-6 was previously isolated from the rhizosphere soil of healthy apple trees in a replanted orchard and demonstrated to have good inhibitory effects on *F. oxysporum*, *F. moniliforme*, *F. proliferatum*, and *F. solani* [18].

QSB-6 bacterial fertilizer was produced by Chuangdi Microbial Resources Co., Ltd. (Dezhou, China). The production process was as follows: strain QSB-6 was first subjected to liquid fermentation for 12 h in an optimal flask fermentation medium that contained 20.0 g sucrose, 15.0 g yeast extract, 1.0 g $MnSO_4$, 2.0 g $NaH_2PO_4 \cdot 2H_2O$, 4.0 g $Na_2HPO_4 \cdot 2H_2O$, and 1 L distilled water. The liquid inoculum was then thoroughly and evenly mixed with sterilized, decomposed carrier (3:1 cow dung:wheat straw by weight). The mixture was stored in a cool place, covered with a plastic sheet, and maintained at a temperature of 35–38 °C and a humidity of 45%. After 12–24 h, it was placed in a sealed container and used after 15 days of fermentation when the bacterial density was 5.0×10^9 $CFU \cdot g^{-1}$. The content of organic matter was 35.57%, available nitrogen was 0.36 $mg \cdot g^{-1}$, available phosphorus was 1.49 $mg \cdot g^{-1}$, and available potassium was 1.03 $mg \cdot g^{-1}$.

2.2. Test Materials

The apple seedlings used in the experiment were two-year-old grafted seedlings. The rootstock was T337, and the scion was Yanfu 3. The grafted seedlings had a stem thickness of approximately 10 mm and a stem height of approximately 1.4 m. They were purchased from Laizhou Nature Horticultural Technology Co., Ltd. (Laizhou, China). The row spacing of the plants was 1.5 m × 4 m, and the trees were pruned to a spindle shape.

2.3. Field Experiment

The field test was carried out in Wangtou Village, Laizhou City (119.81 longitude, 37.10 latitude), Sujiadian Town, Qixia City (120.83 longitude, 37.28 latitude), and Yiyuan, Zibo City (118.43 longitude, 36.09 latitude) in Shandong Province, China. The soil textures at the Laizhou (LZ), Qixia (QX), and Yiyuan (TY) sites were clay loam, sandy loam, and loam, respectively. Physicochemical properties of the soils are presented in Table 1. In March 2021, 30-year-old trees were removed from the orchards, and replanted orchards were established simultaneously.

Table 1. Basic physical and chemical properties in the rhizosphere soil in three orchards.

Place	Ammonium-Nitrogen (mg kg^{-1})	Nitrate Nitrogen (mg kg^{-1})	Organic Matter (%)	Available Phosphorus (mg kg^{-1})	Soil Bulk Density (g cm^{-3})	Available Potassium (mg kg^{-1})	Soil pH	Soil Moisture Content (%)
LZ	1.5 ± 0.0 b	39.2 ± 1.1 b	2.2 ± 0.1 a	17.5 ± 0.0 c	1.0 ± 0.1 a	63.2 ± 8.8 c	7.1 ± 0.0 a	8.9 ± 0.2 a
QX	1.6 ± 0.0 b	42.8 ± 0.2 a	2.2 ± 0.0 a	19.2 ± 0.3 b	1.2 ± 0.1 a	108.4 ± 5.0 b	6.7 ± 0.3 a	10.3 ± 1.7 a
TY	2.3 ± 0.2 a	43.7 ± 0.2 a	2.3 ± 0.0 a	21.7 ± 0.5 a	1.1 ± 0.1 a	223.3 ± 11.7 a	7.1 ± 0.0 a	11.5 ± 1.1 a

LZ: Laizhou, QX: Qixia, TY: Yiyuan. Numbers followed by the same letter in each column are not significantly different based on the Duncan multiple range test at $p < 5\%$.

The experiment consisted of four treatments: 30-year-old orchard soil (CK1), 30-year-old orchard soil fumigated with methyl bromide (CK2), bacterial fertilizer carrier treatment (T1), and QSB-6 bacterial fertilizer treatment (T2). The planting holes (80 cm^3) were dug according to the row spacing, and the bacterial manure carrier or QSB-6 bacterial manure were mixed with the soil and backfilled. Each soil amendment was applied at a rate of 1 kg per tree, and there were 20 trees per treatment. All measurements (plant height, ground diameter, number of branches, and length of new branches) and soil sampling were performed on 20 October 2021. A Canon PowerShot G16 camera was used to photograph the saplings. The surface soil was removed, and multiple soil samples were collected within a 0.5-m radius around each sapling. Three replicate samples from each treatment were used for each measurement. First, impurities, such as roots, weeds, soil organisms, and stones, were removed from each soil sample. Next, each sample was divided into three portions: one portion was stored in a refrigerator at 4 °C for the determination of culturable microbes; one portion was air-dried for the measurement of soil enzyme activities, nutrient content, and phenolic acid content; and one portion was stored in a freezer at −80 °C for DNA extraction, real-time fluorescence quantitative PCR (qPCR), and terminal-restriction fragment length polymorphism (T-RFLP) analysis [14].

2.3.1. Soil Physical and Chemical Properties

The soil organic matter content was determined by potassium dichromate capacity-spectrophotometry as described in the *Soil Physical and Chemical Analysis Experiment Guide* [41]. Soil ammonium-nitroge and nitrate nitrogen contents were determined by colorimetric methods, soil available phosphorus content was determined by the 0.05 M $NaHCO_3$ method, and soil available potassium content was determined by flame photometry. The soil bulk density was determined by the ring knife method, and the moisture content was determined by drying at 105 °C. Soil pH was measured in a water extract (1:2.5, w/v) using a Shanghai Lei PHS-3EJ Magnetic Benchtop pH Meter (Shanghai, China). The particle size distribution (the percentages of clay, silt, and sand) was determined by hydrometry [42].

2.3.2. Microbial Culture Methods

Soil microbial populations (bacteria, fungi, and actinomycetes) were assessed by the dilution method of plate counting described by Zhang et al. [43]. Bacteria, fungi, and actinomycetes were incubated in beef broth peptone substrate, potato dextrose agar (PDA; Difco, Detroit, MI, USA), and Gause No. 1 substrate, respectively. Five plates per dilution were used for each measurement of each soil sample, and the populations of bacteria, fungi, and actinomycetes were quantified as CFU per gram of dry soil.

2.3.3. Determination of Soil Enzyme Activities

Soil urease activity was measured by the indophenol blue colorimetric method as described in the instructions of the solid-urease (S-UE) activity kit. Soil neutral phosphatase activity was measured by the disodium phenyl phosphate method as described in the instructions of the soil neutral phosphatase (S-NP) activity kit. Soil sucrase activity was measured by the 3,5-dinitrosalicylic acid (DNS) method as described in the instructions of the soil sucrase (S-SC) activity kit. The measurement of soil catalase activity used H_2O_2, which has a characteristic absorption peak at 240 nm. The specific method is described in the instructions of the solid-catalase (S-CAT) activity kit. All kits were purchased from Suzhou Keming Biotechnology Co., Ltd. (Suzhou, China).

2.3.4. Quantitative Determination of Soil Phenolic Acids by HPLC

The soil phenolic acid content was measured using the method of Yin et al. [44]. A sample of dry soil (100 g) was passed through a 12-mesh size sieve, mixed with diatomaceous earth, and placed into a 100-mL extraction tank. The ASE 350 Fast Solvent Extractor (Sunnyvale, CA, USA) was used to perform the extraction. First, absolute ethanol was used as the extraction solvent, and static extraction was performed for 5 min at 120 °C and 10.3 MPa two times, followed by a purge volume of 60% and a purge time of 90 s. Next, the same sample was extracted again under the same conditions using methanol as the extraction solvent. After the extraction was completed, the two solvents were mixed and concentrated under reduced pressure at 34 °C to near dryness, and the sample was then reconstituted with 1 mL methanol and passed through a 0.22-μm organic phase filter membrane for HPLC analysis.

The HPLC procedure followed that described by Xiang et al. [45], with some modifications. An UltiMate 3000 HPLC system (Dionex) with an Acclaim 120 C18 column (3 μm, 150 mm × 3 mm) and a column temperature of 30 °C were used for quantification. The mobile phase A was acetonitrile, and the mobile phase B was water (adjusted to pH 2.6

with acetic acid). The flow rate was 0.5 mL·min^{-1}, the automatic injection volume was 5 µL, and the detection wavelength was 280 nm. All reagents were chromatographic grade.

2.3.5. DNA Extraction from Four Species of Fusarium and Quantitative PCR

Sieved fresh soil (5.0 g) was used for DNA extraction with the DNeasy PowerMax Soil Kit (Qiagen, Hilden, Germany). Quantitative PCR amplifications for standard and environmental DNA samples were performed with a volume of 20 µL in each reaction using SYBR Premix Ex Taq (Takara, Dalian, China) and a CFX96 Touch Real-Time PCR Detection System (Bio-Rad, Hercules, CA, USA) following the method of Duan et al. [18]. Each 20 µL PCR reaction contained 2 µL of genomic DNA, 10 µL of SYBR green PCR Master Mix (TaKaRa Biotech, Dalian, China), 0.4 µL of each primer pair, and 7.2 µL of ddH$_2$O. Reactions were carried out using the following PCR cycling conditions: JR/JF and CHR/CHF primer pairs, 95 °C 30 s, 94 °C 5 s (40 cycles), 60 °C 30 s, 72 °C 1 min, followed by a 4 °C soak; CR/CF and FR/FF primer pairs, 95 °C 30 s, 94 °C 5 s (40 cycles), 65 °C 30 s, 72 °C 1 min, followed by a 4 °C soak. The primer pairs used in this experiment are as follows: JR (5′GGCCTGAGGGTTGTAATG-3′) × JF (5′CATACCACTTGTTGTCTCGGC-3′) for *F. oxysporum*; CHR (5′GACTCGCGAGTCAAATCGCGT-3′) × CHF (5′GGGGTTTAAC GGCGTGGCC-3′) for *F. moniliforme*; CR (5′GATCGGCGAGCCCTTGCGGCAAG-3′) × CF (5′CGCCGCGTACCAGTTGCGAGGGT-3′) for *F. proliferatum*; FR (5′CGAGTTATACAACT CATCAACC-3′) × FF (5′GGCCTGAGGGTTGTAATG-3′) for *F. solani*. Four species of *Fusarium* DNAs were used as templates, and specific primers were used for PCR amplification. The purified PCR products were connected to *pMD*18-T vector (TaKaRa Biotech, China), and transformed into *E. coli* competent cells DH5α (TaKaRa Biotech, Dalian, China). After sequencing verification, the plasmid was extracted according to the method of the plasmid extraction kit (TaKaRa Biotech, Dalian, China), its concentration was measured and a gradient dilution with sterile water was carried out. The standard curves were generated by plotting the cycle threshold (Cq) values obtained for each specific DNA concentration versus the log of the initial concentration of species DNA. The standard curve is as follows: *F. oxysporum*: y = −2.291x + 36.396, R^2 = 0.993; *F. moniliforme*: y = −3.495x + 12.421, R^2 = 0.998; *F. proliferatum*: y = −3.675x + 9.128, R^2 = 0.999; *F. solani*: y = −2.352x + 26.941, R^2 = 0.994. The concentration of plasmid DNA was measured and converted to copy concentration using the following equation from Whelan et al. [46]: DNA (copy) = [6.02 × 10^{23} (copies mol^{-1}) × DNA amount (g)]/[DNA length (bp) × 660 (g mol^{-1} bp^{-1})]. Sterile water was used as the negative control instead of the template. All real-time PCR reactions were performed in triplicate with three biological replicates, so that each treatment was analyzed nine times.

2.3.6. Terminal-Restriction Fragment Length Polymorphism (T-RFLP) Analysis

The DNA was amplified using the universal primers 27F-FAM/1492R and ITS1F-FAM/ITS4R that target the bacterial 16S rRNA gene and the fungal ITS region between 18S and 28S rRNA regions, respectively. The forward primers were labeled at the 5′ end with 6-carboxyfluorescein (FAM), which was synthesized by Sangon Biotech (Shanghai, China). The specific steps are described in Xu et al. [47]. The 50-µL PCR mixture contained 0.6 µL of 5 U/µL Ex Taq (TaKaRa), 5 µL of 10 × Ex Taq Buffer, 1 µL of 2.5 mM dNTP mixture, 2 µL of 0.5 mM forward and reverse primers, 12.6 µL of ddH$_2$O, and 2.0 µL containing 100 ng of the extracted DNA template. The primer pairs used in this experiment are as follows: 27F-FAM (5′AGAGTTTGATCCTGGCTCAG-3′) × 1492R (5′GTTACCTTGTTACGACTT-3′) for bacteria; ITS1F-FAM (5′CTTGGTCATTTAGAGGAAGTAA-3′) × ITS4R (5′CAGGAGACTTGTAC ACGGTCCAG-3′) for fungi [48,49]. All PCR amplifications were performed on an Applied Biosystems 2720 Thermal Cycler (Applied Biosystems, Foster City, CA, USA). For bacteria, PCR conditions consisted of 94 °C for 3 min, followed by 30 cycles at 94 °C for 45 s, 52 °C for 45 s, and 72 °C for 1 min. A final extension was performed at 72 °C for 10 min. For fungi, the PCR conditions consisted of 95 °C for 5 min, followed by 30 cycles at 94 °C for 30 s, 50 °C for 30 s, and 72 °C for 1 min. A final extension was performed at 72 °C for

10 min. Prior to digestion, PCR products were cleaned with the EZNA PCR Purification Kit (OMEGA Bio-tek Inc., Doraville, GA, USA) following the manufacturer's instructions and quantified using a DNAmaster Nucleic Acid and Protein Analyzer (Dynamica, Scientific Ltd., Newport Pagnell, UK). The purified PCR product (500 ng) was digested with *MspI* (TaKaRa, Tokyo, Japan) for 16S rRNA gene amplicons and *HinfI* (TaKaRa, Japan) for ITS amplicons in two separate reactions according to their protocols [50,51]. T-RFLP analysis was run on an ABI 3730 DNA Analyzer (Applied Biosystems, Melbourne, Australia) by Sangon Biotech Co., Ltd. (Shanghai, China) using LIZ-labeled GS500 (−250) as the internal size standard.

2.4. Statistical Analysis

All statistical analyses were performed with IBM SPSS 20.0 (IBM SPSS Statistics, IBM Corporation, Armonk, NY, USA). Different lowercase letters represent significant differences between treatments (one-way ANOVA, $p < 0.05$) according to Duncan's multiple range test. The figures were plotted with Microsoft Excel 2013 (Microsoft Corporation, Redmond, WA, USA) and GraphPad Prism 7.0 (GraphPad software, Inc., San Diego, CA, USA). TBtools software was used for cluster analysis; similarities and differences among treatments are indicated by the color gradient, and the color intensity is directly proportional to substance content.

T-RFLP profiles were analyzed using Peak Scanner Software v1.0 (Thermo Fisher Scientific, Wilmington, NC, USA); T-RFs < 50 bp in length and T-RFs that contributed to <0.5% of total peak area in each sample were excluded from subsequent analyses. The apparent T-RF sizes in capillary electrophoresis were compared against the MiCA database to determine the phylotype [52]. The R statistical platform (v.4.1.1) was used for principal component analysis (PCoA) and cluster analysis to study differences in community composition among samples. Differences among samples were calculated by Bray–Curtis dissimilarity, and analysis of similarity (ANOSIM) was performed to identify significant differences among the fungal communities [53,54]. The richness index (SR) and evenness index (E) were calculated using Bio-Dap software [18].

3. Results

3.1. Effect of Strain QSB-6 on the Biomass of Replanted Apple Saplings

The application of QSB-6 bacterial fertilizer (T2) increased the plant height, ground diameter, and branch length of young apple trees relative to CK1 (Table 2). In Yiyuan, Qixia, and Laizhou, T2 treatment increased plant height by 31.30%, 29.98%, and 24.78%, the ground diameter by 37.84%, 42.28%, and 46.94%, and the average branch length by 34.62%, 43.46%, and 30.34% relative to the replant soil control (CK1). In Yiyuan, T2 treatment increased the plant height and the ground diameter by 9.74% and 23.19% relative to the blank carrier (T1), and their growth was close to that of seedlings planted in methyl bromide-fumigated soil. In Qixia, T2 treatment increased the plant height, the ground diameter, and the average branch length by 9.05%, 10.85%, and 27.99% relative to T1.

3.2. Effect of Strain QSB-6 on Soil Microorganisms

In Laizhou, Qixia, and Yiyuan, the number of soil bacteria was significantly higher in the QSB-6 bacterial fertilizer treatment (T2) than in the other treatments. The number of soil bacteria was 2.21-, 2.82-, and 2.0-fold higher in T2 than in the replant soil control (CK1), and 1.91-, 1.69-, and 1.74-fold higher in T2 than in the blank carrier treatment (T1) (Table 3). The number of soil fungi was significantly higher in T1 than in T2. The T1 treatment also significantly increased the soil fungus in Laizhou (1.31-fold), Qixia (1.81-fold), and Yiyuan (1.53-fold) relative to the T2 treatment. In Laizhou, Qixia, and Yiyuan, the number of soil fungi was reduced by 40.38%, 68.45%, and 49.07% in CK2 relative to CK1 and by 34.62%, 53.40%, and 41.61% in T2 relative to CK1. The effects of T2 were similar to those of methyl bromide fumigation treatment in Laizhou and Yiyuan. The number of soil actinomycetes at all three sites could be ranked as: T2 > T1 > CK2. The T2 treatment also significantly increased the soil bacteria/fungi and actinomycetes/fungi ratios in Laizhou (2.55- and 2.05-fold), Qixia (3.06- and 2.73-fold), and Yiyuan (2.66- and 2.26-fold) relative to the T1 treatment.

Table 2. Effects of different treatments on the biomass of apple replanting saplings in three areas.

Place	Soil Treatments	Plant Height (cm)	Ground Diameter (mm)	Numbers of Branches	Branch Length (cm)
LZ	CK1	150.7 ± 3.0 c	13.8±0.1 c	7.7 ± 0.9 b	44.5 ± 6.6 c
	CK2	206.0 ± 7.6 a	24.3±1.1 a	11.0 ± 0.6 a	73.8 ± 2.4 a
	T1	179.3 ± 0.9 b	18.4±0.9 b	7.7 ± 0.7 b	54.3 ± 17 bc
	T2	188.0 ± 0.6 b	20.2±0.9 b	10.0 ± 0.6 ab	58.0 ± 3.1 b
QX	CK1	139.0 ± 5.9 c	18.0 ± 0.1 d	8.7 ± 1.5 b	34.0 ± 1.7 b
	CK2	191.3 ± 6.3 a	28.4 ± 0.6 a	15.0 ± 1.0 a	53.6 ± 2.7 a
	T1	165.7 ± 2.3 b	23.1 ± 0.0 c	11.7 ± 1.2 ab	38.1 ± 1.6 b
	T2	180.7 ± 1.7 a	25.6 ± 1.0 b	12.0 ± 1.5 ab	48.8 ± 3.0 a
TY	CK1	191.7 ± 12.1 c	17.0 ± 1.5 b	10.0 ± 1.5 b	49.1 ± 4.7 b
	CK2	254.3 ± 4.3 a	25.8 ± 0.1 a	16.3 ± 0.7 a	68.0 ± 5.6 a
	T1	229.3 ± 0.7 b	19.1 ± 0.2 b	12.0 ± 1.0 ab	55.7 ± 3.8 ab
	T2	251.7 ± 4.9 a	23.5 ± 0.6 a	15.0 ± 2.1 a	66.1 ± 1.1 a

LZ: Laizhou, QX: Qixia, TY: Yiyuan. CK1: replant control, CK2: methyl bromide fumigation, T1: blank carrier treatment, T2: QSB-6 bacterial fertilizer treatment. Numbers followed by the same letter in the columns for each place are not significantly different based on the Duncan multiple range test at $p < 5\%$.

Table 3. Effects of different treatments on the density of microorganisms in the rhizosphere of young apple trees in three places.

Place	Soil Treatments	The Number of Soil Bacteria (×10⁵ CFU/g Soil)	The Number of Soil Fungi (×10³ CFU/g Soil)	The Number of Soil Actinomycete (×10⁴ CFU/g Soil)	The Ratio of Bacteria and Fungi	The Ratio of Actinomycete and Fungi
LZ	CK1	40.3 ± 1.5 b	52.0 ± 2.1 a	60.7 ± 3.5 c	78.0 ± 5.8 b	11.7 ± 1.0 c
	CK2	25.7 ± 1.2 c	31.0 ± 0.6 c	52.0 ± 1.0 c	82.7 ± 2.4 b	16.8 ± 0.4b c
	T1	46.7 ± 1.5 b	44.7 ± 1.8 b	91.0 ± 1.2 b	104.7 ± 3.7 b	20.4 ± 0.9 b
	T2	89.0 ± 3.8 a	34.0 ± 2.5 c	140.3 ± 4.2 a	266.4 ± 31.7 a	41.9 ± 4.2 a
QX	CK1	38.0 ± 1.5 c	68.7 ± 0.9 a	92.3 ± 1.2 b	55.4 ± 3.0 c	13.5 ± 0.1 c
	CK2	27.0 ± 3.0 d	21.7 ± 0.9 d	75.0 ± 2.3 c	123.9 ± 9.6 b	34.6 ± 0.4 b
	T1	63.6 ± 1.8 b	58.0 ± 2.1 b	102.7 ± 4.1 b	110.2 ± 6.0 b	17.78 ± 1.2 c
	T2	107.3 ± 1.9 a	32.0 ± 1.2 c	154.7 ± 6.4 a	336.7 ± 17.9 a	48.54 ± 3.2 a
TY	CK1	56.0 ± 0.6 c	53.7 ± 2.0 a	85.0 ± 2.5 c	104.6 ± 3.7 c	15.9 ± 0.3 b
	CK2	24.3 ± 0.9 d	27.3 ± 0.9 c	65.3 ± 2.7 c	89.0 ± 0.4 d	24.0 ± 1.7 b
	T1	66.0 ± 2.1 b	48.0 ± 1.2 b	114.3 ± 7.7 b	137.5 ± 3.1 b	23.9 ± 2.2 b
	T2	114.7 ± 3.7 a	31.3 ± 0.9 c	169.0 ± 9.3 a	365.9 ± 2.5 a	54.1 ± 4.0 a

LZ: Laizhou, QX: Qixia, TY: Yiyuan. CK1: replant control, CK2: methyl bromide fumigation, T1: blank carrier treatment, T2: QSB-6 bacterial fertilizer treatment. Numbers followed by the same letter in the columns for each place are not significantly different based on the Duncan multiple range test at $p < 5\%$.

3.3. Effect of Strain QSB-6 on Soil Enzyme Activities

The T2 treatment significantly increased the soil activities of urease, phosphatase, invertase, and catalase in Qixia, Yiyuan, and Laizhou (Figure 1). Urease activity was 1.86-, 1.74-, and 1.55-fold higher in T2 than in CK1. Phosphatase activity was 2.17-, 2.16-, and 2.16-fold higher in T2 than in CK1. Sucrase activity was 2.94-, 3.84-, and 3.79-fold higher in T2 than in CK1. Catalase activity was 1.90-, 1.85-, and 1.73-fold higher in T2 than in CK1. Compared with the T1 treatment, urease activity was 22.24%, 21.20%, and 20.38% higher in T2; phosphatase activity was 40.47%, 27.42%, and 23.02% higher in T2; sucrase activity was 49.21%, 35.86%, and 73.91% higher in T2; and catalase activity was 40.80%, 27.39%, and 26.97% higher in T2.

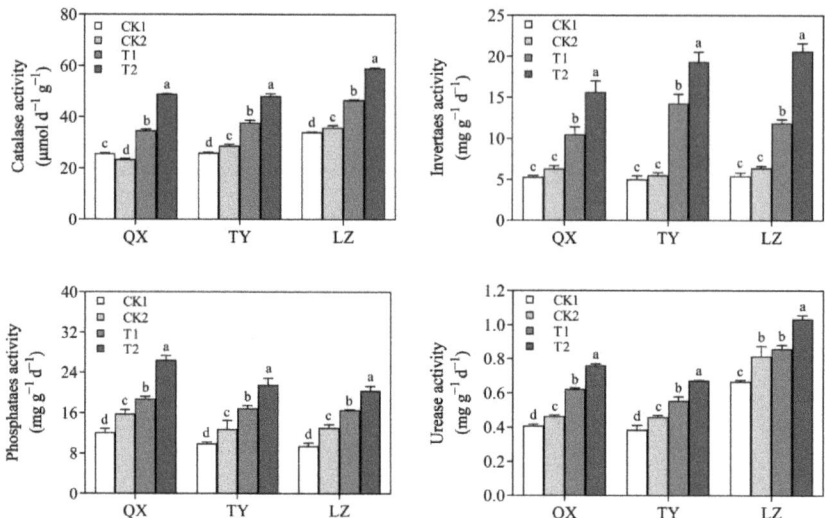

Figure 1. Effect of different treatments on soil enzyme activity in three places. LZ: Laizhou, QX: Qixia, TY: Yiyuan. CK1: replant control, CK2: methyl bromide fumigation, T1: blank carrier treatment, T2: QSB-6 bacterial fertilizer treatment. Columns with the same letter for each place are not significantly different based on the Duncan multiple range test at $p < 5\%$.

3.4. Effect of Strain QSB-6 on Soil Phenolic Acids

The contents of several phenolic acids were highest in the replant soil control (CK1) at all three field sites, whereas the phenolic acid contents were significantly reduced in soil treated with QSB-6 bacterial fertilizer (T2) (Figure 2). In Laizhou, the soil contents of cinnamic acid, phlorizin, benzoic acid, ferulic acid, vanillin, p-hydroxybenzoic acid, and caffeic acid were reduced by 53.58%, 67.64%, 61.45%, 62.62%, 71.31%, 60.50%, and 71.31% in T2 relative to CK1 (Table S1). In Qixia, the contents of these phenolic acids were reduced by 53.15–83.32% in T2 relative to CK1, and in Yiyuan, they were reduced by 52.65–72.73%. In Laizhou, Qixia, and Yiyuan, the soil total phenolic acid contents were higher in the T1 treatment than in the T2 treatment: cinnamic acid (1.59-, 1.98-, and 1.72-fold higher in T1), phenophyllin (1.66-, 1.11-, and 1.20-fold), benzoic acid (2.11-, 3.20-, and 2.03-fold), ferulic acid (2.21-, 1.99-, and 1.80-fold), vanillin (2.59-, 2.86-, and 1.54-fold), p-hydroxybenzoic acid (2.15-, 2.42-, and 2.62-fold), and caffeic acid (2.76-, 3.29-, and 1.50-fold).

3.5. Inhibitory Effect of Strain QSB-6 on Four Species of Fusarium in Rhizosphere Soil

The qPCR results showed that *F. proliferatum*, *F. solani*, *F. verticillioides*, and *F. oxysporum* abundance was significantly reduced in the CK2 and T2 treatments compared with the CK1 treatment at all three field sites (Figure 3). In Qixia, the abundances of *F. proliferatum*, *F. solani*, *F. verticillioides*, and *F. oxysporum* were 35.23%, 32.79%, 57.48%, and 36.60% lower in T2 than in CK1. In Yiyuan, the abundances of *F. proliferatum*, *F. solani*, *F. verticillioides*, and *F. oxysporum* were 32.05%, 33.16%, 32.62%, and 49.63% lower in T2 than in CK1. In Laizhou, the abundances of *F. proliferatum*, *F. solani*, *F. verticillioides*, and *F. oxysporum* were 28.26%, 43.56%, 55.86%, and 38.04% lower in T2 than in CK1. In Qixia, Yiyuan, and Laizhou, the abundances of *F. proliferatum* were 47.22%, 43.11%, and 31.33% lower in CK2 than in CK1. The abundances of *F. solani* were 41.94%, 50.27%, and 50.60% lower in CK2 than in CK1. The abundances of *F. verticillioides* were 64.94%, 37.88%, and 61.05% lower in CK2 than in CK1. The abundances of *F. oxysporum* were 45.49%, 59.51%, and 49.74% lower in CK2 than in CK1. The gene copy number of four species of *Fusarium* was significantly higher in T1 soil than T2 soil. In Qixia, Yiyuan, and Laizhou, the relative abundance of *F. proliferatum* was 1.33-, 1.30-, and 1.26-fold higher in T1 soil than in T2 soil. The relative abundance of

F. solani was 1.26-, 1.25-, and 1.41-fold higher; that of *F. verticillioides* was 1.48-, 1.22-, and 1.45-fold higher; and that of *F. oxysporum* was 1.25-, 1.41-, and 1.40-fold higher.

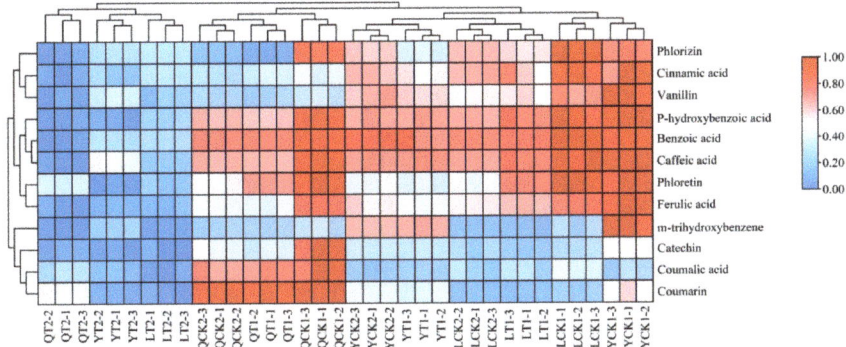

Figure 2. Effects of different treatments on soil phenolic acids in three places. LCK1: replant control in Laizhou, LCK2: methyl bromide fumigation in Laizhou, LT1: blank carrier treatment in Laizhou, LT2: QSB-6 bacterial fertilizer treatment in Laizhou, QCK1: replant control in Qixia, QCK2: methyl bromide fumigation in Qixia, QT1: blank carrier treatment in Qixia, QT2: QSB-6 bacterial fertilizer treatment in Qixia, YCK1: replant control in Yiyuan, YCK2: methyl bromide fumigation in Yiyuan, YT1: blank carrier treatment in Yiyuan, YT2: QSB-6 bacterial fertilizer treatment in Yiyuan. The color intensity was proportional to the total phenolic acid content. Taxa relative abundances were log10-transformed, and the scale method (from zero to one) was used for the heatmap representation. Each treatment included three repetitions.

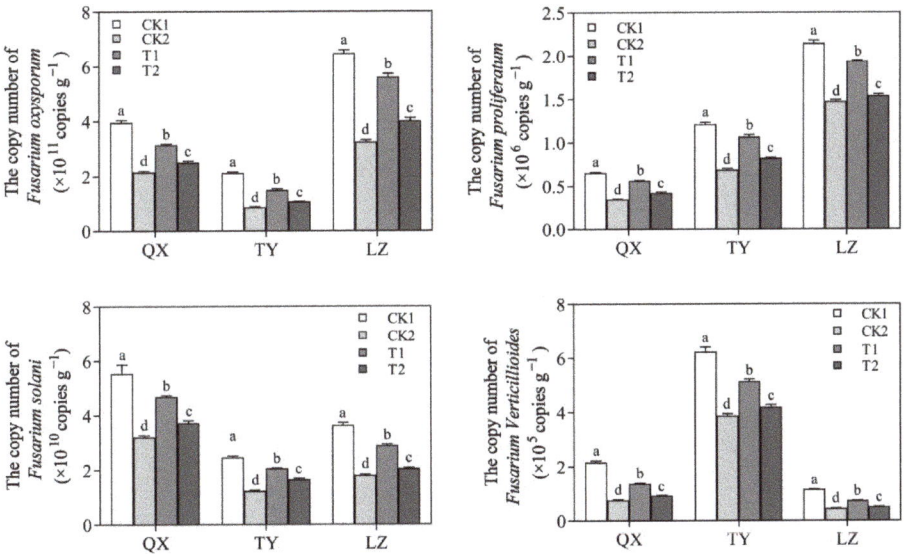

Figure 3. Effect of different treatments on the copy number of four species of *Fusarium* in three places. LZ: Laizhou, QX: Qixia, TY: Yiyuan. CK1: replant control, CK2: methyl bromide fumigation, T1: blank carrier treatment, T2: QSB-6 bacterial fertilizer treatment. Columns with the same letter for each place are not significantly different based on the Duncan multiple range test at $p < 5\%$.

3.6. Effect of Strain QSB-6 on the Soil Microbial Community

Principal component analysis and cluster analysis showed that the soil microbial community structure in T2 and CK2 was significantly different from that in CK1 (Figure 4).

The soil bacterial community structure in T2 was different from the other treatments, and the soil fungal and bacterial communities of T1 and CK1 were similar. Margalef, Mcintosh, Brillouin, and Shannon indices reflect the richness and diversity of soil microbial communities (Table 4). The Margalef index reflects the abundance of soil microbial communities; the Mcintosh index reflects the number of different types of carbon sources utilized; and the Brillouin and Shannon indices reflect the diversity of the soil microbial community [55]. Compared with CK1 treatment, the abundance of the soil fungal community was significantly increased after treatment with strain QSB-6. The Brillouin index of the soil fungal community and the utilized carbon sources were significantly reduced, whereas the bacterial community showed the opposite pattern.

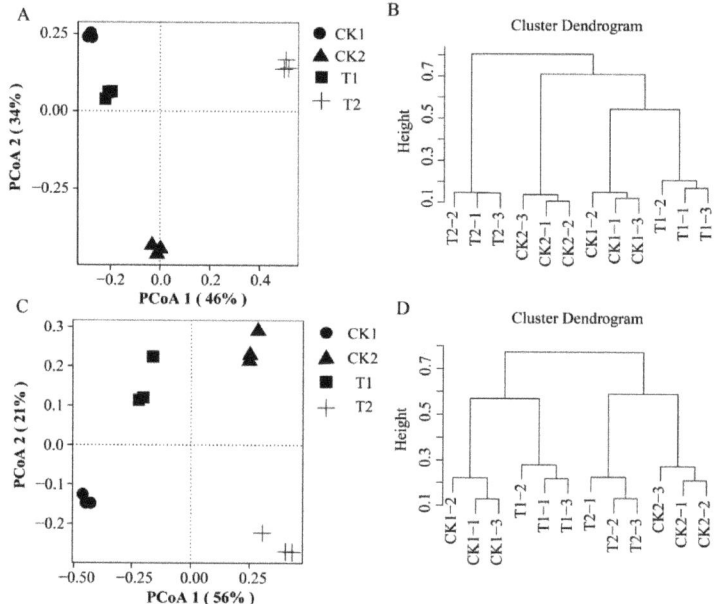

Figure 4. Principal component analysis (**A**,**C**) and cluster analysis (**B**,**D**) of T-RFLP patterns of bacteria (**A**,**B**) and fungi (**C**,**D**) in the soil relative to treatment based on the Bray–Curtis method. CK1: replant control, CK2: methyl bromide fumigation, T1: blank carrier treatment, T2: QSB-6 bacterial fertilizer treatment. Each treatment includes three repetitions.

Table 4. Effects of different treatments on microbial diversity in the rhizosphere of young apple trees in three places.

Microorganism	Treatment	Margalef's Index	Shannon's Index	Mcintosh's Index	Brillouin's Index
Fungi	CK1	10.3 ± 0.1 b	3.6 ± 0.0 ab	16.6 ± 0.8 a	3.0 ± 0.0 a
	CK2	9.9 ± 0.7 b	3.5 ± 0.1 ab	8.7 ± 0.9 c	2.7 ± 0.1 bc
	T1	12.2 ± 0.2 a	3.7 ± 0.0 a	11.9 ± 0.4 b	2.9 ± 0.1 ab
	T2	11.7 ± 0.3 a	3.4 ± 0.1 b	8.5 ± 0.6 c	2.5 ± 0.1 c
Bacterial	CK1	11.0 ± 0.2 b	3.5 ± 0.0 b	12.420.6 b	2.8 ± 0.0 c
	CK2	9.6 ± 0.1 d	3.3 ± 0.0 c	12.6 ± 0.5 b	2.6 ± 0.0 d
	T1	12.1 ± 0.1 a	3.7 ± 0.0 a	12.6 ± 0.4 b	3.0 ± 0.0 b
	T2	10.5 ± 0.0 c	3.7 ± 0.0 a	25.5 ± 1.2 a	3.3 ± 0.0 a

CK1: replant control, CK2: methyl bromide fumigation, T1: blank carrier treatment, T2: QSB-6 bacterial fertilizer treatment. Numbers followed by the same letter in the columns for each microorganism are not significantly different based on the Duncan multiple range test at $p < 5\%$.

4. Discussion

Replant disease is a complex, multifaceted disease that often occurs in crops and orchards and strongly constrains the sustainable development of global agriculture and fruit production [56,57]. In this study, strain QSB-6 acted as a biocontrol agent that significantly improved the growth and health of replanted young apple trees under field conditions. Multiple aspects of plant vegetative growth (height, ground diameter, and length of branches) were measured, and all were markedly increased by QSB-6 fertilizer application relative to the replanted soil control. The biomass of young apple trees was higher in Yiyuan than in Laizhou and Qixia, perhaps because of the better soil physicochemical properties and management practices [58,59]. These plant growth results indicate that treatment with strain QSB-6 was effective for the control of ARD under field conditions.

4.1. Effect of QSB-6 Fertilizer on the Soil Microbial Community

Bacteria, actinomyces, and fungi are three categories of the soil microbial community that constitute the majority of the soil microbial biomass; their community structure and abundances play important roles in plant growth and disease control [60,61]. Previous studies have found that long-term continuous cropping leads to a transformation of the soil microbial community structure from a "bacterial type" to a "fungal type". The soil microbial community structure becomes unbalanced, leading to an increase in harmful pathogens and a decrease in beneficial microbiota, ultimately leading to poor soil health [2,62–65]. Soil microbial diversity is considered to be a key factor in disease prevention [66]. High microbial diversity and appropriate microbial composition play key roles in preventing pathogen invasion, maintaining soil health, and promoting plant growth [67,68]. Gadhave et al. [69] reported that the soil bacterial diversity and richness index increased significantly after application of *Bacillus* spp., and we reached a similar conclusion in this study. The application of QSB-6 bacterial fertilizer to three apple orchards significantly increased the diversity and number of bacteria. The results of the plate count test revealed that most of the cultured bacterial colonies showed morphologies consistent with that of strain QSB-6. A single colony with the same morphology was randomly selected and sequenced, and its similarity to strain QSB-6 was 100%. These results indicate that strain QSB-6 was able to reproduce normally after being inoculated into the soil. At the same time, the addition of vectors can also support the survival of the strain QSB-6 and improve its performance in preventing and controlling plant diseases [70,71]. The raw materials (cow dung and wheat straw) in the formula are cheap and easy to obtain, and the fermentation level is high, which can provide a good foundation for large-scale industrial production [72,73].

A decrease in the quantity of soil fungi may be related to fungistatic compounds (1,2-benzenedicarboxylic acid and benzeneacetic acid, 3-hydroxy-, methyl ester) produced by strain QSB-6. These compounds can significantly inhibit the reproduction of harmful *F. oxysporum*, *F. moniliforme*, *F. proliferatum*, and *F. solani* in rhizosphere soil, and they can also inhibit other harmful fungi to some extent [18]. Our results were consistent with those of Cao et al. [74]. *Bacillus subtilis* SQR 9 was able to survive well in the rhizosphere of cucumber, where it suppressed the growth of *F. oxysporum* and protected the host from the pathogen. Tao et al. [40] reported that the population densities of *Bacillus* spp. and *Pseudomonas* spp. were correlated with one another and negatively correlated with *F. oxysporum* density and wilt disease. Several well-known plant growth-promoting bacteria belong to the genus *Pseudomonas* spp. [75–77]. The increase in soil bacterial diversity reported here after QSB-6 inoculation may be due to the enrichment of some specific beneficial bacteria such as *Pseudomonas* spp., *Gemmatimonas* spp., and *Sphingomonas* spp., resulting in a significant decrease in bacterial richness [78], but a significant increase in the bacteria/fungi ratio, thus transforming the soil into a "bacterial type" and improving the soil microbial community structure, which may promote the healthy growth of plants [79,80]. Sheng et al. [14] demonstrated that soil fumigation treatment can effectively kill most fungi, especially *F. oxysporum* and *F. solani*. The relative abundances of *F. solani* and *F. oxysporum* are significantly reduced, but with prolonged time, some fungi are recruited to form new microbial communities, such as *Ohtaekwangia* spp., *Opitutus* spp., *Mortierella* spp., and *Synchytrium* spp., which improve the soil fungal community, thus effectively controlling plant diseases [81]. These results were consistent with the findings of this study. Methyl bromide fumigation treatment significantly reduced the relative abundance of *F. oxysporum*, *F. moniliforme*, *F. proliferatum*, and *F. solani* and promoted the growth of replanted apple trees. Cluster analysis showed that the soil fungal communities of T2 and CK2 were similar, indicating that the addition of strain QSB-6 may also promote the growth of replanted apple trees by improving the soil fungal community structure.

4.2. Effect of QSB-6 Fertilizer on Soil Phenolic Acids

There is evidence that allelochemicals, such as phenolic acids, in root exudates or decomposing residues contribute to apple replanting obstacles [23]. Yin et al. [17] showed that phlorizin can promote a rapid increase in the number of *F. moniliforme* and accelerate the speed of mycelial division, thus increasing apple replant challenges. The replanted soil contains a variety of phenolic acids such as p-hydroxybenzoic acid, phloroglucinol, syringic acid, benzoic acid, caffeic acid, and ferulic acid. Above a certain concentration, these compounds can impair the root antioxidant system of apple seedlings and inhibit their growth [44,66,82,83]. Here, we found that strain QSB-6 may have the ability to degrade phenolic acids. The addition of strain QSB-6 significantly reduced the contents of major phenolic acids in the soil, including phlorizin, cinnamic acid, benzoic acid, and p-hydroxybenzoic acid. At the same time, the abundance of *F. oxysporum*, *F. moniliforme*, *F. proliferatum*, and *F. solani* also declined significantly in the rhizosphere soil, consistent with the results of Yin et al. [17]. Bai et al. [65] demonstrated that phenolic acids significantly affected the biomass, diversity, and community structure of soil microbes, selectively increasing specific microbial species, such as soil-borne pathogenic microorganisms, and thus, increasing morbidity [84]. It is possible that the addition of strain QSB-6 increases the populations of some beneficial bacteria that can degrade phenolic acid, increase bacterial diversity, and improve the bacterial community structure of the rhizosphere soil, thus accelerating the decomposition and transformation of phenolic acids and promoting plant growth [85,86].

4.3. Effect of QSB-6 Fertilizer Treatment on Soil Enzyme Activity

Soil enzymes are an important component of the soil, catalyzing various reactions and processes of organic matter metabolism (e.g., soil organic matter formation and degradation; C, N, and P cycling; and plant nutrient transformation) and generating energy for microorganisms and plants. Therefore, the extent of soil enzyme activity can be used to characterize the extent of soil maturation and fertility [87,88]. Soil urease is involved in the soil N cycle. Phosphatases hydrolyze P-containing organic compounds into inorganic P that is required by plants. Invertase hydrolyzes sucrose to glucose and fructose for use as plant and microbial energy sources, whereas catalase is an oxidoreductase that protects organisms from H_2O_2 toxicity [89–92]. Sabaté et al. [93] showed that inoculation with *Bacillus* sp. P12 significantly increased soil enzyme activities and the number of beneficial microorganisms, improved soil quality, and reduced the incidence of *Macrophomina phaseolina*. Our results were similar, that is, inoculation with strain QSB-6 at three field sites significantly increased the rhizosphere activities of urease, phosphatase, invertase, and catalase, thus presumably promoting nitrogen, phosphorus, and potassium transformation in the soil, improving soil fertility, and strengthening plant resistance to abiotic and biotic stresses [94,95]. Yadav et al. [96] reported that the micro-organisms associated with rhizosphere soil also have profound impacts on plant health and soil fertility, as they strongly influence nutrient mineralization and soil organic matter decomposition. Thus, increases in soil enzyme activity may be related to increases in soil microbial richness and diversity, as well as changes in microbial community composition [97].

5. Conclusions

The results of our field experiments were similar to those of our previous pot study [18]. The addition of strain QSB-6 can significantly promote the height, ground diameter, and length of branches of apple plants under field conditions, reduce the abundance of *F. oxysporum*, *F. moniliforme*, *F. proliferatum*, and *F. solani* and the content of phenolic acids in rhizosphere soil, increase soil bacterial diversity and activity, improve the structure of the soil microbial community, and increase soil enzyme activities. Together, these effects help to mitigate ARD. In summary, QSB-6 bacterial fertilizer appears to offer a more sustainable approach to enhancing apple growth and soil health under replant conditions, thereby advancing the development of the apple industry.

Supplementary Materials: The following supporting information can be downloaded at: https://www.mdpi.com/article/10.3390/horticulturae8010083/s1, Table S1: Effects of different treatments on soil phenolic acids in three places.

Author Contributions: Conceptualization, Z.M. and C.Y.; methodology, Y.D.; software, Z.L.; validation, Y.D., Y.Z. and Z.L.; investigation, Y.D.; data curation, Y.D. and Y.Z.; writing—original draft preparation, Y.D.; writing—review and editing, C.Y. and X.C. All authors have read and agreed to the published version of the manuscript.

Funding: This research was funded by China Agriculture Research System of MOF and MARA (CARS-27), the National Natural Science Foundation of China (32072510), Shandong Agricultural Major Applied Technology Innovation Project (SD2019ZZ008); Taishan Scholar Funded Project (NO.ts20190923); Qingchuang Science and Technology Support Project of Shandong Colleges and Universities (2019KJF020); and the Natural Science Foundation of Shandong Province (ZR2020MC131).

Institutional Review Board Statement: Not applicable.

Informed Consent Statement: Not applicable.

Data Availability Statement: The study did not report any data.

Acknowledgments: We thank all the colleagues that helped with the development of different parts of this manuscript.

Conflicts of Interest: The authors declare no conflict of interest.

References

1. Chen, X.S.; Han, M.Y.; Su, G.L.; Liu, F.Z.; Guo, G.N.; Jiang, Y.M.; Mao, Z.Q.; Peng, F.T.; Shu, H.R. Discussion on today's world apple industry trends and the suggestions on sustainable and efficient development of apple industry in China. *J. Fruit Sci.* **2021**, *27*, 598–604. (In Chinese)
2. Tewoldemedhin, Y.T.; Mazzola, M.; Labuschagne, I.; McLeod, A. A multi-phasic approach reveals that apple replant disease is caused by multiple biological agents, with some agents acting synergistically. *Soil Biol. Biochem.* **2011**, *43*, 1917–1927. [CrossRef]
3. Tilston, E.L.; Deakin, G.; Bennett, J.; Passey, T.; Harrison, N.; Fernández, F.; Xu, X. Effect of fungal, oomycete and nematode interactions on apple root development in replant soil. *CABI Agric. Biosci.* **2020**, *1*, 14. [CrossRef]
4. Balbín-Suárez, A.; Jacquiod, S.; Rohr, A.D.; Liu, B.; Flachowsky, H.; Winkelmann, T.; Smalla, K. Root exposure to apple replant disease soil triggers local defense response and rhizoplane microbiome dysbiosis. *FEMS Microbiol. Ecol.* **2021**, *97*, fiab031. [CrossRef] [PubMed]
5. Bezemer, T.M.; Lawson, C.S.; Hedlund, K.; Edwards, A.R.; Brook, A.J.; Igual, J.M.; Van der Putten, W.H. Plant species and functional group effects on abiotic and microbial soil properties and plant-soil feedback responses in two grasslands. *J. Ecol.* **2006**, *94*, 893–904. [CrossRef]
6. Zhang, B.; Li, X.; Wang, F.; Li, M.; Zhang, J.; Gu, L.; Zhang, Z. Assaying the potential autotoxins and microbial community associated with *Rehmannia glutinosa* replant problems based on its 'autotoxic circle'. *Plant Soil* **2016**, *407*, 307–322. [CrossRef]
7. Cavael, U.; Lentzsch, P.; Schwärzel, H.; Eulenstein, F.; Tauschke, M.; Diehl, K. Assessment of Agro-Ecological Apple Replant Disease (ARD) Management Strategies: Organic Fertilisation and Inoculation with Mycorrhizal Fungi and Bacteria. *Agronomy* **2021**, *11*, 272. [CrossRef]
8. Hofmann, A.; Wittenmayer, L.; Arnold, G.; Schieber, A.; Merbach, W. Root exudation of phloridzin by apple seedlings (*Malus* × *domestica* Borkh.) with symptoms of apple replant disease. *J. Appl. Bot. Food Qual.* **2009**, *82*, 193–198.
9. Kelderer, M.; Manici, L.M.; Caputo, F.; Thalheimer, M. Planting in the 'inter-row' to overcome replant disease in apple orchards: A study on the effectiveness of the practice based on microbial indicators. *Plant Soil* **2012**, *357*, 381–393. [CrossRef]

10. Van Schoor, L.; Denman, S.; Cook, N.C. Characterisation of apple replant disease under South African conditions and potential biological management strategies. *Sci. Hortic.* **2009**, *119*, 153–162. [CrossRef]
11. Singh, N.; Sharma, D.P.; Kaushal, R. Controlling replant disease of apple in Himachal Pradesh, India by rootstocks and soil agro-techniques. *Pharma Innov. J.* **2017**, *6*, 288–293.
12. Leinfelder, M.M.; Merwin, I.A.; Fazio, G.; Robinson, T. Resistant rootstocks, preplant compost amendments, soil fumigation, and row repositioning for managing apple replant disease. *HortScience* **2004**, *39*, 841. [CrossRef]
13. Garbeva, P.V.; Van Veen, J.A.; Van Elsas, J.D. Microbial diversity in soil: Selection of microbial populations by plant and soil type and implications for disease suppressiveness. *Annu. Rev. Phytopathol.* **2004**, *42*, 243–270. [CrossRef] [PubMed]
14. Sheng, Y.F.; Wang, H.Y.; Wang, M.; Li, H.; Xiang, L.; Pan, F.B.; Mao, Z.Q. Effects of soil texture on the growth of young apple trees and soil microbial community structure under replanted conditions. *Hortic. Plant J.* **2020**, *6*, 123–131. [CrossRef]
15. Wang, G.S.; Yin, C.M.; Pan, F.B.; Wang, X.B.; Xiang, L.; Wang, Y.F.; Mao, Z.Q. Analysis of the fungal community in apple replanted soil around Bohai Gulf. *Hortic. Plant J.* **2018**, *4*, 175–181. [CrossRef]
16. Liu, Z. The Isolation and Identification of Fungi Pathogen from Apple Replant Disease in Bohai Bay Region and the Screening of the Antagonistic *Trichoderma* Strains. Master's Thesis, Shandong Agricultural University, Tai'an, China, 2013.
17. Yin, C.M.; Xiang, L.; Wang, G.S.; Wang, Y.F.; Shen, X.; Chen, X.S.; Mao, Z.Q. Phloridzin promotes the growth of *Fusarium moniliforme* (*Fusarium verticillioides*). *Sci. Hortic.* **2017**, *214*, 187–194. [CrossRef]
18. Duan, Y.N.; Chen, R.; Zhang, R.; Jiang, W.T.; Chen, X.S.; Yin, C.M.; Mao, Z.Q. Isolation, identification, and antibacterial mechanisms of *Bacillus amyloliquefaciens* QSB-6 and its effect on plant roots. *Front. Microbiol.* **2021**, *12*, 2727. [CrossRef]
19. Trapero Casas, A.; Jimenez-Diaz, R.M. Fungal wilt and root rot disease of chickpea in Southern Spain. *Phytopathology* **1985**, *75*, 1146–1151. [CrossRef]
20. Anjaiah, V.; Cornelis, P.; Koedam, N. Effect of genotype and root colonization in biological control of Fusarium wilts in pigeonpea and chickpea by *Pseudomonas aeruginosa* PNA1. *Can. J. Microbiol.* **2003**, *49*, 85–91. [CrossRef]
21. Gopalakrishnan, S.; Beale, M.H.; Ward, J.L.; Strange, R.N. Chickpea wilt: Identification and toxicity of 8-O-methyl-fusarubin from *Fusarium acutatum*. *Photochemistry* **2005**, *66*, 1536–1539. [CrossRef] [PubMed]
22. Clements, J.; Cowgill, W.; Embree, C.; Gonzalez, V.; Hoying, S.; Kushad, M.; Autio, W. Rootstock Tolerance to Apple Replant Disease for Improved Sustainability of Apple Production. In Proceedings of the XXVIII International Horticultural Congress on Science and Horticulture for People (IHC2010): International Symposium on the Challenge for a Sustainable Production, Protection and Consumption of Medit, Lisbon, Portugal, 22–27 August 2010; pp. 521–528.
23. Wang, Y.F.; Pan, F.B.; Wang, G.S.; Zhang, G.; Wang, Y.; Chen, X.S.; Mao, Z.Q. Effects of biochar on photosynthesis and antioxidative system of *Malus hupehensis* Rehd. seedlings under replant conditions. *Sci. Hortic.* **2014**, *175*, 9–15. [CrossRef]
24. Winkelmann, T.; Smalla, K.; Amelung, W.; Baab, G.; Grunewaldt-Stöcker, G.; Kanfra, X.; Schloter. M. Apple replant disease: Causes and mitigation strategies. *Curr. Issues Mol. Biol.* **2018**, *30*, 89–106.
25. Raymaekers, K.; Ponet, L.; Holtappels, D.; Berckmans, B.; Cammue, B.P. Screening for novel biocontrol agents applicable in plant disease management—A review. *Biol. Control* **2020**, *144*, 104240. [CrossRef]
26. Chowdhury, S.K.; Majumdar, S.; Mandal, V. Application of *Bacillus* sp. LBF-01 in Capsicum annuum plant reduces the fungicide use against *Fusarium oxysporum*. *Biocatal. Agric. Biotechnol.* **2020**, *27*, 101714. [CrossRef]
27. Azabou, M.C.; Gharbi, Y.; Medhioub, I.; Ennouri, K.; Barham, H.; Tounsi, S.; Triki, M.A. The endophytic strain *Bacillus velezensis* OEE1: An efficient biocontrol agent against Verticillium wilt of olive and a potential plant growth promoting bacteria. *Biol. Control* **2020**, *142*, 104168. [CrossRef]
28. Bubici, G.; Kaushal, M.; Prigigallo, M.I.; Gómez-Lama Cabanás, C.; Mercado-Blanco, J. Biological control agents against Fusarium wilt of banana. *Front. Microbiol.* **2019**, *10*, 616. [CrossRef] [PubMed]
29. Hua, G.K.H.; Wang, L.; Chen, J.; Ji, P. Biological control of Fusarium wilt on watermelon by *fluorescent pseudomonads*. *Biocontrol Sci. Technol.* **2020**, *30*, 212–227. [CrossRef]
30. Patil, S.; Sriram, S. Biological control of Fusarium wilt in crop plants using non-pathogenic isolates of *Fusarium* species. *Indian Phytopathol.* **2020**, *73*, 11–19. [CrossRef]
31. Moutassem, D.; Belabid, L.; Bellik, Y. Efficiency of secondary metabolites produced by *Trichoderma* spp. in the biological control of Fusarium wilt in chickpea. *J. Crop Prot.* **2020**, *9*, 217–231.
32. Khedher, S.B.; Mejdoub-Trabelsi, B.; Tounsi, S. Biological potential of *Bacillus subtilis* V26 for the control of Fusarium wilt and tuber dry rot on potato caused by *Fusarium* species and the promotion of plant growth. *Biol. Control* **2021**, *152*, 104444. [CrossRef]
33. Wang, M.; Zhang, R.; Zhao, L.; Wang, H.; Chen, X.; Mao, Z.; Yin, C. Indigenous arbuscular mycorrhizal fungi enhance resistance of apple rootstock 'M9T337' to apple replant disease. *Physiol. Mol. Plant Pathol.* **2021**, *116*, 101717. [CrossRef]
34. Ab Rahman, S.F.S.; Singh, E.; Pieterse, C.M.; Schenk, P.M. Emerging microbial biocontrol strategies for plant pathogens. *Plant Sci.* **2018**, *267*, 102–111. [CrossRef] [PubMed]
35. Kumbar, B.; Mahmood, R.; Nagesha, S.N.; Nagaraja, M.S.; Prashant, D.G.; Kerima, O.Z.; Chavan, M. Field application of *Bacillus subtilis* isolates for controlling late blight disease of potato caused by *Phytophthora infestans*. *Biocatal. Agric. Biotechnol.* **2019**, *22*, 101366. [CrossRef]
36. Haddoudi, I.; Cabrefiga, J.; Mora, I.; Mhadhbi, H.; Montesinos, E.; Mrabet, M. Biological control of Fusarium wilt caused by *Fusarium equiseti* in Vicia faba with broad spectrum antifungal plant-associated *Bacillus* spp. *Biol. Control* **2021**, *160*, 104671. [CrossRef]

37. Zheng, T.W.; Liu, L.; Nie, Q.W.; Hsiang, T.; Sun, Z.X.; Zhou, Y. Isolation, identification and biocontrol mechanisms of endophytic bacterium D61-A from *Fraxinus hupehensis* against *Rhizoctonia solani*. *Biol. Control* **2021**, *158*, 104621. [CrossRef]
38. Anusha, B.G.; Gopalakrishnan, S.; Naik, M.K.; Sharma, M. Evaluation of *Streptomyces* spp. and *Bacillus* spp. for biocontrol of Fusarium wilt in chickpea (*Cicer arietinum* L.). *Arch. Phytopathol. Plant Prot.* **2019**, *52*, 417–442. [CrossRef]
39. Toju, H.; Peay, K.G.; Yamamichi, M.; Narisawa, K.; Hiruma, K.; Naito, K.; Kiers, E.T. Core microbiomes for sustainable agroecosystems. *Nat. Plants* **2018**, *4*, 247–257. [CrossRef] [PubMed]
40. Tao, C.; Li, R.; Xiong, W.; Shen, Z.; Liu, S.; Wang, B.; Kowalchuk, G.A. Bio-organic fertilizers stimulate indigenous soil Pseudomonas populations to enhance plant disease suppression. *Microbiome* **2020**, *8*, 137. [CrossRef]
41. MAFF; ADAS. The analysis of agricultural materials. In *Reference Book 427: Methods 32 and 56*; Ministry of Agriculture, Fisheries and Food: London, UK; Her Majesty's Stationery Office: London, UK, 1986.
42. Avery, B.W. Soil classification in the Soil Survey of England and Wales. *J. Soil Sci.* **1973**, *243*, 324–338. [CrossRef]
43. Zhang, Q.; Zhu, L.; Wang, J.; Xie, H.; Wang, J.; Wang, F.; Sun, F. Effects of fomesafen on soil enzyme activity, microbial population, and bacterial community composition. *Environ. Monit. Assess.* **2014**, *186*, 2801–2812. [CrossRef]
44. Yin, C.M.; Wang, G.S.; Li, Y.Y.; Che, J.S.; Shen, X.; Chen, X.S.; Wu, S.J. A new method for analysis of phenolic acids in the soil-soil from replanted apple orchards was investigated. *China Agric. Sci.* **2013**, *46*, 4612–4619. (In Chinese)
45. Xiang, L.; Wang, M.; Jiang, W.T.; Wang, Y.F.; Chen, X.S.; Yin, C.M.; Mao, Z.Q. Key indicators for renewal and reconstruction of perennial trees soil: Microorganisms and phloridzin. *Ecotoxicol. Environ. Saf.* **2021**, *225*, 112723. [CrossRef]
46. Whelan, J.A.; Russell, N.B.; Whelan, M.A. A method for the absolute quantification of cDNA using real-time PCR. *J. Immunol. Methods* **2003**, *278*, 261–269. [CrossRef]
47. Xu, H.J.; Bai, J.; Li, W.Y.; Zhao, L.X.; Li, Y.T. Removal of persistent DDT residues from soils by earthworms: A mechanistic study. *J. Hazard. Mater.* **2019**, *365*, 622–631. [CrossRef]
48. Gardes, M.; Bruns, T.D. ITS primers with enhanced specificity for basidiomycetes-application to the identification of mycorrhizae and rusts. *Mol. Ecol.* **1993**, *2*, 113–118. [CrossRef] [PubMed]
49. Weisburg, W.G.; Barns, S.M.; Pelletier, D.A.; Lane, D.J. 16S ribosomal DNA amplification for phylogenetic study. *J. Bacteriol.* **1991**, *173*, 697–703. [CrossRef]
50. Yu, H.Y.; Wang, Y.K.; Chen, P.C.; Li, F.B.; Chen, M.J.; Hu, M. The effect of ammonium chloride and urea application on soil bacterial communities closely related to the reductive transformation of pentachlorophenol. *J. Hazard. Mater.* **2014**, *272*, 10–19. [CrossRef]
51. Zhang, C.; Wang, J.; Zhu, L.; Du, Z.; Wang, J.; Sun, X.; Zhou, T. Effects of 1-octyl-3-methylimidazolium nitrate on the microbes in brown soil. *J. Environ. Sci.* **2018**, *67*, 249–259. [CrossRef]
52. Shyu, C.; Soule, T.; Bent, S.J.; Foster, J.A.; Forney, L.J. MiCA: A web-based tool for the analysis of microbial communities based on terminal-restriction fragment length polymorphisms of 16S and 18S rRNA genes. *Microb. Ecol.* **2007**, *53*, 562–570. [CrossRef] [PubMed]
53. Liu, H.; Pan, F.; Han, X.; Song, F.; Zhang, Z.; Yan, J.; Xu, Y. Response of soil fungal community structure to long-term continuous soybean cropping. *Front. Microbiol.* **2019**, *9*, 3316. [CrossRef]
54. Xu, F.; Chen, Y.; Cai, Y.; Gu, F.; An, K. Distinct roles for bacterial and fungal communities during the curing of vanilla. *Front. Microbiol.* **2020**, *11*, 2342. [CrossRef] [PubMed]
55. Perkins, J.L. *A Study of Diversity and Community Comparison Indices by Bioassay of Copper Using Colonized Benthic Macroinvertebrates*; The University of Texas School of Public Health: Houston, TX, USA, 1981.
56. Li, H.L.; Qiang, S.W. Mitigation of replant disease by mycorrhization in horticultural plants: A review. *Folia Hortic.* **2018**, *30*, 269–282. [CrossRef]
57. Ma, Z.; Wang, Q.; Wang, X.; Chen, X.; Wang, Y.; Mao, Z. Effects of Biochar on Replant Disease by Amendment Soil Environment. *Commun. Soil Sci. Plant Anal.* **2021**, *52*, 673–685. [CrossRef]
58. Di Prima, S.; Rodrigo-Comino, J.; Novara, A.; Iovino, M.; Pirastru, M.; Keesstra, S.; Cerdà, A. Soil physical quality of citrus orchards under tillage, herbicide, and organic managements. *Pedosphere* **2018**, *28*, 463–477. [CrossRef]
59. Adekiya, A.O.; Ejue, W.S.; Olayanju, A.; Dunsin, O.; Aboyeji, C.M.; Aremu, C.; Akinpelu, O. Different organic manure sources and NPK fertilizer on soil chemical properties, growth, yield and quality of okra. *Sci. Rep.* **2020**, *10*, 16083. [CrossRef]
60. Hill, P.; Krištůfek, V.; Dijkhuizen, L.; Boddy, C.; Kroetsch, D.; van Elsas, J.D. Land use intensity controls actinobacterial community structure. *Microb. Ecol.* **2011**, *61*, 286–302. [CrossRef] [PubMed]
61. Bhatti, A.A.; Haq, S.; Bhat, R.A. Actinomycetes benefaction role in soil and plant health. *Microb. Pathog.* **2017**, *111*, 458–467. [CrossRef]
62. Caputo, F.; Nicoletti, F.; Picione, F.D.L.; Manici, L.M. Rhizospheric changes of fungal and bacterial communities in relation to soil health of multi-generation apple orchards. *Biol. Control* **2015**, *88*, 8–17. [CrossRef]
63. Franke-Whittle, I.H.; Manici, L.M.; Insam, H.; Stres, B. Rhizosphere bacteria and fungi associated with plant growth in soils of three replanted apple orchards. *Plant Soil* **2015**, *395*, 317–333. [CrossRef]
64. Spath, M.; Insam, H.; Peintner, U.; Kelderer, M.; Kuhnert, R.; Franke-Whittle, I.H. Linking soil biotic and abiotic factors to apple replant disease: A greenhouse approach. *J. Phytopathol.* **2015**, *163*, 287–299. [CrossRef]
65. Bai, Y.; Wang, G.; Cheng, Y.; Shi, P.; Yang, C.; Yang, H.; Xu, Z. Soil acidification in continuously cropped tobacco alters bacterial community structure and diversity via the accumulation of phenolic acids. *Sci. Rep.* **2019**, *9*, 12499. [CrossRef]

66. Liang, B.W.; Ma, C.Q.; Fan, L.M.; Wang, Y.Z.; Yuan, Y.B. Soil amendment alters soil physicochemical properties and bacterial community structure of a replanted apple orchard. *Microbiol. Res.* **2018**, *216*, 1–11. [CrossRef] [PubMed]
67. Liang, B.; Ma, C.; Fan, L.; Wang, Y.; Yuan, Y. Compost amendment alters soil fungal community structure of a replanted apple orchard. *Arch. Agron. Soil Sci.* **2021**, *67*, 739–752. [CrossRef]
68. Zhao, W.S.; Guo, Q.G.; Su, Z.H.; Wang, P.P.; Dong, L.H.; Hu, Q.; Lu, X.Y.; Zhang, X.Y.; Li, S.Z.; Ma, P. Rhizosphere soil fungal community structure of healthy potato plants and *Verticillium dahliae* and their utilization characteristics of carbon sources. *Sci. Agric. Sin.* **2021**, *54*, 296–309.
69. Gadhave, K.R.; Devlin, P.F.; Ebertz, A.; Ross, A.; Gange, A.C. Soil inoculation with *Bacillus* spp. modifies root endophytic bacterial diversity, evenness, and community composition in a context-specific manner. *Microb. Ecol.* **2018**, *76*, 741–750. [CrossRef] [PubMed]
70. Wei, Z.; Huang, J.; Yang, C.; Xu, Y.; Shen, Q.; Chen, W. Screening of suitable carriers for *Bacillus amyloliquefaciens* strain QL-18 to enhance the biocontrol of tomato bacterial wilt. *Crop Prot.* **2015**, *75*, 96–103. [CrossRef]
71. Malusá, E.; Sas-Paszt, L.; Ciesielska, J. Technologies for beneficial microorganisms inocula used as biofertilizers. *Sci. World J.* **2012**, *2012*, 491206. [CrossRef]
72. Smith, R.S. Legume inoculant formulation and application. *Can. J. Microbiol.* **1992**, *38*, 485–492. [CrossRef]
73. Vijayaraghavan, P.; Arun, A.; Al-Dhabi, N.A.; Vincent, S.G.P.; Arasu, M.V.; Choi, K.C. Novel *Bacillus subtilis* IND19 cell factory for the simultaneous production of carboxy methyl cellulase and protease using cow dung substrate in solid-substrate fermentation. *Biotechnol. Biofuels* **2016**, *9*, 1–13. [CrossRef]
74. Cao, Y.; Zhang, Z.; Ling, N.; Yuan, Y.; Zheng, X.; Shen, B.; Shen, Q. Bacillus subtilis SQR 9 can control Fusarium wilt in cucumber by colonizing plant roots. *Biol. Fertil. Soils* **2011**, *47*, 495–506. [CrossRef]
75. Weller, D.M. *Pseudomonas* biocontrol agents of soilborne pathogens looking back over 30 years. *Phytopathology* **2007**, *97*, 250–256. [CrossRef] [PubMed]
76. Mazurier, S.; Corberand, T.; Lemanceau, P.; Raaijmakers, J.M. Phenazine antibiotics produced by *fluorescent pseudomonads* contribute to natural soil suppressiveness to fusarium wilt. *ISME J* **2009**, *3*, 977–991. [CrossRef]
77. Mendes, R.; Kruijt, M.; De Bruijn, I.; Dekkers, E.; Van der Voort, M.; Schneider, J.H.; Raaijmakers, J.M. Deciphering the rhizosphere microbiome for disease-suppressive bacteria. *Science* **2011**, *332*, 1097–1100. [CrossRef]
78. Shen, Z.; Wang, D.; Ruan, Y.; Xue, C.; Zhang, J.; Li, R.; Shen, Q. Deep 16S rRNA pyrosequencing reveals a bacterial community associated with banana Fusarium wilt disease suppression induced by bio-organic fertilizer application. *PLoS ONE* **2014**, *9*, e98420.
79. Qin, S.; Yeboah, S.; Cao, L.; Zhang, J.; Shi, S.; Liu, Y. Breaking continuous potato cropping with legumes improves soil microbial communities, enzyme activities and tuber yield. *PLoS ONE* **2017**, *12*, e0175934. [CrossRef]
80. Liu, H.; Li, J.; Carvalhais, L.C.; Percy, C.D.; Prakash Verma, J.; Schenk, P.M.; Singh, B.K. Evidence for the plant recruitment of beneficial microbes to suppress soil-borne pathogens. *New Phytol.* **2021**, *229*, 2873–2885. [CrossRef]
81. Li, R.; Shen, Z.; Sun, L.; Zhang, R.; Fu, L.; Deng, X.; Shen, Q. Novel soil fumigation method for suppressing cucumber Fusarium wilt disease associated with soil microflora alterations. *Appl. Soil Ecol.* **2016**, *101*, 28–36. [CrossRef]
82. Gao, X.B.; Zhao, F.X.; Xiang, S.H.E.N.; Hu, Y.L.; Hao, Y.H.; Yang, S.Q.; Mao, Z.Q. Effects of cinnamon acid on respiratory rate and its related enzymes activity in roots of seedlings of *Malus hupehensis* Rehd. *Agric. Sci. China* **2020**, *9*, 833–839. [CrossRef]
83. Yin, C.; Xiang, L.; Wang, G.; Wang, Y.; Shen, X.; Chen, X.; Mao, Z. How to plant apple trees to reduce replant disease in apple orchard: A study on the phenolic acid of the replanted apple orchard. *PLoS ONE* **2016**, *11*, e0167347. [CrossRef] [PubMed]
84. Zhang, S.; Jin, Y.; Zhu, W.; Tang, J.; Hu, S.; Zhou, T.; Chen, X. Baicalin released from Scutellaria baicalensis induces autotoxicity and promotes soilborn pathogens. *J. Chem. Ecol.* **2011**, *36*, 329–338. [CrossRef]
85. Zhou, X.; Yu, G.; Wu, F. Responses of soil microbial communities in the rhizosphere of cucumber (*Cucumis sativus* L.) to exogenously applied p-hydroxybenzoic acid. *J. Chem. Ecol.* **2012**, *38*, 975–983. [CrossRef] [PubMed]
86. Hannula, S.E.; Morriën, E.; De Hollander, M.; Van Der Putten, W.H.; Van Veen, J.A.; De Boer, W. Shifts in rhizosphere fungal community during secondary succession following abandonment from agriculture. *ISME J.* **2017**, *11*, 2294–2304. [CrossRef]
87. Poonguzhali, S.; Madhaiyan, M.; Sa, T. Cultivation-dependent characterization of rhizobacterial communities from field grown Chinese cabbage Brassica campestris ssp pekinensis and screening of traits for potential plant growth promotion. *Plant Soil* **2006**, *286*, 167–180. [CrossRef]
88. Song, Y.; Song, C.; Yang, G.; Miao, Y.; Wang, J.; Guo, Y. Changes in labile organic carbon fractions and soil enzyme activities after marshland reclamation and restoration in the Sanjiang Plain in Northeast China. *Environ. Manag.* **2012**, *50*, 418–426. [CrossRef]
89. Stpniewska, Z.; Wolińska, A.; Ziomek, J. Response of soil catalase activity to chromium contamination. *J. Environ. Sci.* **2009**, *21*, 1142–1147. [CrossRef]
90. Hu, B.; Liang, D.; Liu, J.; Lei, L.; Yu, D. Transformation of heavy metal fractions on soil urease and nitrate reductase activities in copper and selenium co-contaminated soil. *Ecotoxicol. Environ. Saf.* **2014**, *110*, 41–48. [CrossRef]
91. Wei, K.; Chen, Z.; Zhu, A.; Zhang, J.; Chen, L. Application of ^{31}P NMR spectroscopy in determining phosphatase activities and p composition in soil aggregates influenced by tillage and residue management practices. *Soil Tillage Res.* **2014**, *138*, 35–43. [CrossRef]
92. Yu, P.; Liu, S.; Han, K.; Guan, S.; Zhou, D. Conversion of cropland to forage land and grassland increases soil labile carbon and enzyme activities in northeastern China. *Agric. Ecosyst. Environ.* **2017**, *245*, 83–91. [CrossRef]

93. Sabaté, D.C.; Petroselli, G.; Erra-Balsells, R.; Audisio, M.C.; Brandan, C.P. Beneficial effect of *Bacillus* sp. P12 on soil biological activities and pathogen control in common bean. *Biol. Control* **2020**, *141*, 104131. [CrossRef]
94. Zhang, X.; Gao, J.; Zhao, F.; Zhao, Y.; Li, Z. Characterization of a salt-tolerant bacterium *Bacillus* sp. from a membrane bioreactor for saline wastewater treatment. *J. Environ. Sci.* **2014**, *26*, 1369–1374. [CrossRef]
95. Rawat, J.; Sanwal, P.; Saxena, J. Potassium and its role in sustainable agriculture. In *Potassium Solubilizing Microorganisms for Sustainable Agriculture*; Springer: New Delhi, India, 2016; pp. 235–253.
96. Yadav, R.; Ror, P.; Rathore, P.; Kumar, S.; Ramakrishna, W. Bacillus subtilis CP4, isolated from native soil in combination with arbuscular mycorrhizal fungi promotes biofortification, yield and metabolite production in wheat under field conditions. *J. Appl. Microbiol.* **2021**, *131*, 339–359. [CrossRef] [PubMed]
97. Iovieno, P.; Morra, L.; Leone, A.; Pagano, L.; Alfani, A. Effect of organic and mineral fertilizers on soil respiration and enzyme activities of two Mediterranean horticultural soils. *Biol. Fertil. Soils* **2009**, *45*, 555–561. [CrossRef]

Review

Basic Substances and Potential Basic Substances: Key Compounds for a Sustainable Management of Seedborne Pathogens

Laura Orzali [1,*], Mohamed Bechir Allagui [2], Clemencia Chaves-Lopez [3], Junior Bernardo Molina-Hernandez [3], Marwa Moumni [4], Monica Mezzalama [5] and Gianfranco Romanazzi [4,*]

[1] Council for Agricultural Research and Economics (CREA), Research Center for Plant Protection and Certification (CREA-DC), Via C.G. Bertero 22, 00156 Rome, Italy
[2] Plant Protection Laboratory, National Institute of Agricultural Research of Tunisia (INRAT), Carthage University, Rue Hedi Karray, Ariana 2080, Tunisia; allagui.bechir@gmail.com
[3] Bioscience and Agro-Food and Environmental Technology Department, University of Teramo, Campus "Coste Sant'Agostino", Via R. Balzarini 1, 64100 Teramo, Italy; cchaveslopez@unite.it (C.C.-L.); jbmolinahernandez@unite.it (J.B.M.-H.)
[4] Department of Agricultural, Food and Environmental Sciences, Marche Polytechnic University, Via Brecce Bianche, 60131 Ancona, Italy; m.moumni@staff.univpm.it
[5] AGROINNOVA—Interdepartmental Centre for Innovation in the Agricultural and Food Sector, University of Torino, Largo Paolo Braccini 2, 10095 Grugliasco, TO, Italy; monica.mezzalama@unito.it
* Correspondence: laura.orzali@crea.gov.it (L.O.); g.romanazzi@univpm.it (G.R.)

Citation: Orzali, L.; Allagui, M.B.; Chaves-Lopez, C.; Molina-Hernandez, J.B.; Moumni, M.; Mezzalama, M.; Romanazzi, G. Basic Substances and Potential Basic Substances: Key Compounds for a Sustainable Management of Seedborne Pathogens. *Horticulturae* **2023**, *9*, 1220. https://doi.org/10.3390/horticulturae9111220

Academic Editors: Hillary Righini, Roberta Roberti and Stefania Galletti

Received: 21 September 2023
Revised: 1 November 2023
Accepted: 7 November 2023
Published: 11 November 2023

Copyright: © 2023 by the authors. Licensee MDPI, Basel, Switzerland. This article is an open access article distributed under the terms and conditions of the Creative Commons Attribution (CC BY) license (https://creativecommons.org/licenses/by/4.0/).

Abstract: Seedborne pathogens represent a critical issue for successful agricultural production worldwide. Seed treatment with plant protection products constitutes one of the first options useful for reducing seed infection or contamination and preventing disease spread. Basic substances are active, non-toxic substances already approved and sold in the EU for other purposes, e.g., as foodstuff or cosmetics, but they can also have a significant role in plant protection as ecofriendly, safe, and ecological alternatives to synthetic pesticides. Basic substances are regulated in the EU according to criteria presented in Article 23 of Regulation (EC) No 1107/2009. Twenty-four basic substances are currently approved in the EU and some of them such as chitosan, chitosan hydrochloride, vinegar, mustard seed powder, and hydrogen peroxide have been investigated as seed treatment products due to their proven activity against fungal, bacterial, and viral seedborne pathogens. Another basic substance, sodium hypochlorite, is under evaluation and may be approved soon for seed decontamination. Potential basic substances such as essential oils, plant extracts, and ozone were currently found effective as a seed treatment for disease management, although they are not yet approved as basic substances. The aim of this review, run within the Euphresco BasicS project, is to collect the recent information on the applications of basic substances and potential basic substances for seed treatment and describe the latest advanced research to find the best application methods for seed coating and make this large amount of published research results more manageable for consultation and use.

Keywords: chitosan; essential oils; phytotoxicity; seed coating; seed quality; seed treatment; sustainability

1. Introduction

The seed is an essential input for crop production, since 90% of food crops are grown from seeds. For this reason, the use of healthy seeds is an essential key to successful agricultural production and serves as the backbone for good economic harvest. Seeds can carry a heavy load of microorganisms, which can cause severe diseases and be responsible for various negative effects on yield and the spread of pathogen inoculum in the soil. Seed movement is also the main cause of pathogenic spread across international borders and the introduction of diseases into previously unaffected areas or their re-emergence [1]. There are many examples of seedborne pathogens that have spread globally, some of which can cause devastating diseases in some of the most important staple crops. Just as examples, Karnal bunt of wheat, caused by *Tilletia indica*, was introduced from India to Mexico in 1972, from Mexico to the USA in 1996 [2], and from an unknown source to South Africa in 2002 [3]; wheat blast caused by the *Magnaporthe oryzae Triticum* pathotype was spread from South America to Bangladesh [4] and to Zambia [5]; wheat streak mosaic virus was spread from Mexico to Australia [6]; and maize lethal necrosis caused by maize mottle chlorotic virus was spread from Asia to Kenya [7].

Seed treatments represent the first line of defense against seedborne (surface-borne or internally seedborne) and soilborne pests. Seed treatments are defined as "the biological, physical, and chemical agents and techniques applied to seed to provide protection and improve the establishment of healthy crops" [8], and in the last 200 years from the discovery of the Bordeaux mixture, several active ingredients have been developed to be used as coating to protect seeds and seedlings in the early stages of their growth. Munkvold [9] exhaustively reviewed the history and development of chemical control of seedborne pathogens. Over the past decade, the number of studies on seed treatment has increased significantly, reflecting the growing interest of the scientific community [10]. Lamichhane et al. [11] summarized the potential negative effects of synthetic fungicides used for seed treatments on nontarget organisms. These effects could consist of a reduction in biocontrol agents and earthworms' activities, alteration of litter decomposition rate, decline in the number of rhizobia on seeds and in the arbuscular mycorrhiza colonization, as well as a reduction in fungal endophytes of seedlings. These fungicide-induced disturbances also had negative consequences on root and shoot biomass and grain yield. Following European Community initiatives, many lines of research and scientific efforts have focused on the development of environmentally friendly alternatives to the use of pesticides for managing crop diseases, in particular seedborne diseases [12–14].

Reducing the use of synthetic pesticides is a major challenge in many countries, and the search for alternative crop protection products is a strategy for promoting more sustainable agricultural systems. Nowadays, the use of traditional environmentally friendly practices (e.g., sanitation, crop rotation, adjusting the age of planting) to control diseases is integrated with new advanced techniques or tools to avoid or at least limit the use of synthetic pesticides. Several sustainable seed treatments can be used including physical treatments such as heat treatments, with the most common being hot water, hot air, and electron treatments, biocontrol agents with species belonging to the genus *Trichoderma*, or plant growth-promoting rhizobacteria (PGPR) and the use of natural substances with antimicrobial activity and/or priming effects [10]. Alternative methods such as seed treatment using basic substances or potential basic substances to manage seedborne pathogens can be a solution to ensuring safe agricultural production, but these substances are still poorly known by researchers and growers and have not been placed on the market as plant protection products [15]. Basic substances are relatively novel compounds already approved and sold in the EU for other purposes, e.g., as foodstuff or cosmetics, which can be used in plant protection without neurotoxic or immune-toxic effects as ecofriendly, safe, and ecological alternatives to synthetic pesticides [16,17]. Among the 24 basic substances approved in the EU, five of them were approved as a seed treatment: chitosan hydrochloride, chitosan, vinegar, mustard seed powder, and hydrogen peroxide. Moreover, potential basic substances such as ozone, essential oils, and plant extracts have been used as seed treatment.

The number of studies on seed treatment with such natural/ecofriendly substances has increased over the last decade resulting in a large amount of published investigations. The aim of this review, carried out in the framework of the Euphresco BasicS project, is to provide an overview of the use of already approved basic substances and of potential basic substances as seed treatments for the control of seedborne pathogens, in order to make this large amount of published results more manageable for consultation and use. Since there are many different techniques that can be used for this purpose, the latest advanced research in finding the best application method as seed coating, dressing, or spraying is also described.

2. Methods for Seed Treatment

2.1. Seed Immersion

Seed immersion methods are those in which seeds are soaked in aqueous or solvent-based liquid for a certain length of time, depending on the nature of the seed coat and the substance used. The soaking results in partial or full hydration of both the host and pathogen and produces microscopic ruptures, making them more susceptible to the penetration of active substances compared to the dry state [12]. Not all the substances are soluble in water, so in some cases (e.g., essential oil, chitosan), it is necessary to use an emulsifier to allow for mixing and emulsion homogeneity [18]. Besides the direct antimicrobial effects that depend on the type of substance used, immersion treatment can have the following priming effects: increased germination rate and seedling vigor; induced diverse range of morphophysiological, biochemical, and molecular responses in plants; and thus improved abiotic and biotic stress tolerance and increased crop yields [19]. Immersion represents the most widely used method for treatment with elicitors for resistance induction, such as chitosan and methyl salicylate [20,21]. Timing of the treatment plays a key role in phytotoxicity, negatively influencing seed vitality [20]. Moreover, excessive imbibition during seed submersion can damage the outer seed coats, especially in the case of seeds with softer teguments such as legume seeds [22]. The challenge is to find the right combination of treatment durations for different seed types to ensure efficacy without causing phytotoxicity. Primed seeds are known to have low storage longevity, which can be partially remedied via post-storage treatments such as dehydration, heat shock, or post-storage humidification [23]. The soaking process is considered cumbersome and time-consuming when treating large quantities of seeds at a large scale, because it requires a large volume of liquid and needs subsequent drying [18].

2.2. Seed Dressing and Coating

Innovative seed coating and dressing technologies are useful as delivery systems for the application of active ingredients on the seed surface. The technique of seed dressing involves the application on the seed surface of a thin layer of the active product, such as pesticides, fertilizers, or growth promoters which can be applied both as dry or liquid formulations [12]. Seed dressing is the most widely used method for low dosages of active components onto seeds [24] and although there are many types of equipment used for coating, the most commonly used device is performed with a rotary coater [18]. Seed coating is a technique in which an external material is applied to the surface of the seed using a binder which acts as an adhesive to improve the adhesion of the active ingredients to the seed. The role of the binder is also to ensure coating integrity during and after drying and to prevent cracking and dusting off during handling and sowing [18]. The layer is applied to the seed typically from 2 to 5% of the seed weight [25]. In this context, nanotechnology could represent an innovative tool exploitable in agriculture, since nanoparticles (materials with a size ranging from 1 to 100 nm) [26,27] can be effective carriers of seed health-promoting compounds when applied as seed coatings or seed dressing material [27]. Nanoagroproducts are an upcoming technology that might be beneficial for the development of future generations of formulations for seed treatment to enhance the sustainability of agricultural systems. Among them, a wide selection of organic and natural compounds can be loaded into these nanoparticles, including essential oils, cellulose, and chitosan, making this technology suitable for sustainable and ecofriendly farming. Basic substances can take advantage of this technology to take place in adapted formulations of seed coating products. Seed coating allows for a controlled release of the substance reducing the active ingredient dosage needed, thus reducing their release into the ecosystem and soils, the possible toxicity for plants and the environment, and the treatment cost. Nanoscale materials used in seed coating technologies such as nanocapsules, nanogels, nanofibers, nanoclays, and nanosuspensions are supposed to increase the accuracy and efficiency of seed protection products, allowing for a reduction in pesticides in the field [27]. On the other hand, specific machines and equipment are required for seed dressing and coating techniques which are performed with a dry power applicator, rotary or drum machine, motor, or hand driving [18].

3. Seed Treatment with Approved Basic Substances

3.1. Activity of Approved Basic Substances against Fungi and Oomycetes

Chitosan is a naturally occurring biopolymer with antimicrobial properties explored in agriculture for many uses as a plant defense inducer, growth promoter, and carrier for delivery systems of biocontrol agents [28]. In 2014, chitosan hydrochloride was approved by the EU as one of the first basic substances for plant protection [29], and a second chitosan formulation was approved in 2022 [30]. Chitosan has shown activity against several species of seedborne pathogens (Tables 1 and 2). El-Mohamedy et al. [31] reported that soaking seeds of green bean (*Phaseolus vulgaris*) in chitosan (1 g L^{-1}) reduced the pre-emergence incidence of *Rhizoctonia solani* and *Fusarium solani* by 54.4% and 52.6%, respectively, after 40 days of plant growth in a greenhouse in soils naturally infested with either of these fungi. No sign of phytotoxicity was reported on the plants obtained by germinated seeds. Fenugreek (*Trigonella foenum-graecum*) seeds were treated with different concentrations of chitosan and then inoculated 24 h later with *F. solani* conidia. Results showed that six days post-inoculation, root rot disease incidence was reduced by 87.5% and 90.1%, with no significant difference for the seeds treated with chitosan (2 g L^{-1}) or carbendazim (0.5 g L^{-1}), respectively [32]. In this experiment, the radicle length of fenugreek seedlings due to chitosan (0.5 g L^{-1}) was significantly higher (3.76 cm) over the control (2.26 cm) and carbendazim (3.34 cm). Bhardwaj et al. [33] evaluated different pearl millet (*Pennisteum glaucum*) seed treatments including chitosan that were sown in several experiment fields in India regarding blast disease caused by *Pyricularia grisea*. The application rate of chitosan seed immersion was 0.5 g kg^{-1} per liter of water and resulted in a blast severity reduction ranging from 4.7% to 26.9%, depending on the field location and growing season. Spelt (*Triticum spelta*) seeds immersed in a conjugate complex solution of chitosan (1.5 g L^{-1}) and tyrosine (15 g L^{-1}), then inoculated with a conidia suspension of *Fusarium culmorum*, showed a 50% reduction in the incidence of root rot in the seedlings [34]. No phytotoxicity was observed. Seeds of groundnut (*Arachis hypogaea*) were coated with chitosan polymer (1 g L^{-1}), sowed in potted soil infested with *Aspergillus niger* and grown for 50 days under greenhouse conditions. Incidence of *Aspergillus* collar rot on seedlings from coated seeds was reduced by 51.8% compared to inoculated untreated seeds [28]. Similar studies were carried out by the same authors on safflower (*Carthamus tinctorius*) seeds coated and sown in infested soil with *Macrophomina phaseolina*. Chitosan coating did not affect the germination rate of either the groundnut or safflower seeds. Reduction of the pathogen was 15.7% on seedlings from seeds treated with chitosan. Chitosan was also tested on cucumber (*Cucumis sativus*) seeds against the oomycete *Phytophthora capsici*. Cucumber seedlings coming from the seeds immersed in chitosan at 500 ppm (0.05%) were grown in plastic pots in a screenhouse; chitosan treatment provided 85% disease suppression of damping off caused by seedling inoculation with zoospores of *P. capsici* injected into the rhizosphere [35]. Moreover, seed germination and root and shoot growth of cucumber were enhanced by chitosan seed treatment in a dose-dependent way up to 500 ppm. Chitosan (0.5%, w/v) seed immersion treatment was also used to reduce foot and root rot caused by *Fusarium graminearum* in durum wheat (*Triticum durum*) plants from both naturally and artificially infected seeds. This treatment caused the stimulation of a defense system as phenolic content increasing and defense-related enzyme activation in seedlings. In the field, seedlings from natural and artificial seed infection showed a reduction of foot and root rot disease by 36% and 56%, respectively. In the greenhouse, the disease reduction was 38% for seedlings from seeds that were artificially infected [20].

Table 1. In vivo and in-field activities of basic substances applied as seed treatments to control seedborne fungi and oomycetes in different crops. Effectiveness is reported as the disease incidence or symptom percentage reduction compared to the untreated control. Phytotoxicity was evaluated through germination testing and the results are compared to the untreated control.

Crop	Disease/Pathogen	Substance (Concentration)	Application	Effectiveness	Possible Phytotoxicity	Activity/Defense Response	Reference
Green bean (*Phaseolus vulgaris*)	*Rhizoctonia solani* [1,*]	Chitosan (1 g L^{-1})	Immersion	54.4%	Data not available		[31]
	Fusarium solani [1,*]	Chitosan (1 g L^{-1})	Immersion	52.6%	Rate of seed germination equal to the control		[31]
Fenugreek (*Trigonella foenum-graecum*)	*Fusarium solani* [1,*]	Chitosan (2 g L^{-1})	Immersion	87.5%	Rate of seed germination equal to the control	Radicle length improvement	[32]
Pearl millet (*Pennisetum glaucum*)	*Magnaporthe grisea* [2]	Chitosan (0.5 g L^{-1})	Immersion	4.7%–26.9%	Data not available		[33]
Spelt (*Triticum spelta*)	*F. culmorum* [2]	Chitosan (1.5 g L^{-1})	Immersion	50.0%	Rate of seed germination equal to the control	Seed germination increasing	[34]
Groundnut (*Arachis hypogaea*)	*Aspergillus niger* [2,*]	Chitosan (1 g L^{-1}) + *Trichoderma* spores	Immersion	51.8%	Rate of seed germination equal to the control		[28]
Safflower (*Carthamus tinctorius*)	*Macrophomina phaseolina* [2,*]		Immersion	15.7%			[28]
Cucumber (*Cucumis sativus*)	*Phytophthora capsici* [1,*]	Chitosan (500 ppm)	Immersion	85.0%	Increased seed germination	Seedling shoot and root growth increasing	[35]
Durum wheat (*Triticum durum*)	Fusarium foot rot *F. graminearum* [1,2]	Chitosan (0.5% v/v)	Immersion	In field [1]: 36% In field [2]: 56% In greenhouse [2]: 38%	Rate of seed germination equal to the control	Phenolic content increasing and defense-related enzyme activation	[20]
Common wheat (*Triticum aestivum*)	*F. culmorum* [2]	White mustard meal (15 g mustard + 45 mL H$_2$O per kg)	Wet and dry seed dressing	In vitro: 67% In field: 43%–78%	Rate of seed germination equal to the control	Plant development stimulation: improving grain quality and wheat plant growth	[36]
Pine (*Pinus radiata*)	*F. circinatum* [2]	Hydrogen peroxide (33% w/v)	Immersion	98.2%	Seedling emergence reduction		[37]
Carrot (*Daucus carota*)	*Alternaria radicina* [1]	Hydrogen peroxide stabilized with silver ions (0.025%)	Immersion	43.2%	Rate of seed germination equal to the control		[38]
White lupin (*Lupinus albus*)	*Colletotrichum lupini* [1]	Vinegar (5% acetic acid)	Immersion for 30 min	16.9%	Rate of seed germination equal to the control		[39]

[1] Natural contamination; [2] artificial inoculation; * soil contamination.

Besides chitosan, other compounds like mustard seed power, vinegar, and hydrogen peroxide were approved as basic substances by the European Union between 2015 and 2017 [15] and allowed for agricultural uses (Table 1). Kowalska et al. [36] recommended the dose of 15 g mustard meal per 1 kg common wheat grain (*Triticum aestivum* ssp. *vulgare*) as a seed dressing applied with 45 mL of water, to significantly reduce disease caused by *F. culmorum* on wheat during the early stage of growth. The authors reported a stimulating effect of mustard meal seed dressing on seedling development without perceiving any negative influence on the germination and development of seedlings, accompanied by a reduction in the number of infected seeds and by a 43–78% disease incidence reduction in the field, according to the type of seed dressing applied, respectively, wet or dry. Berbegal et al. [37] evaluated *Pinus radiata* seed treatments using hydrogen peroxide (33% w/v, disinfectant conc. 30%) to control *Fusarium circinatum*. Seeds artificially inoculated and treated by soaking in hydrogen peroxide were sown in peat moss and then maintained in a forest nursery. The reduction of disease incidence in seedlings from seeds treated with hydrogen peroxide ranged from 98.2% to 100% but the germination rate was also reduced compared to inoculated untreated seeds. Differently, hydrogen peroxide stabilized with silver ions applied to *Daucus carota* seeds had no phytotoxic effects, and it caused a significant decrease in the percentage of seeds infested with *Alternaria radicina* [38].

Table 2. In vivo and in-field activities of basic substances applied as seed treatments to control seedborne bacteria in different crops. Effectiveness is reported as the disease incidence or symptom percentage reduction compared to the untreated control. Phytotoxicity was evaluated through germination testing and the results are compared to the untreated control.

Basic Substances—Bacteria							
Crop	Disease/Pathogen	Substance (Concentration)	Application	Effectiveness (Disease/Symptoms Reduction)	Possible Phytotoxicity	Activity/Defense Response	Reference
Lettuce (*Lactuva sativa*)	*Xanthomonas campestris* pv. *vitians* [2]	Hydrogen peroxide (3% w/v)	Immersion	100%	Rate of seed germination equal to the control	Direct antibacterial activity	[40]
		Hydrogen peroxide (5% w/v)			Significant reductions in germination		
Cabbage (*Brassica oleracea*)	*Xanthomonas campestris* pv. *campestris* [1]	Hydrogen peroxide (10%; 20% w/v)	Immersion	Depending on the concentration up to 100%	Rate of seed germination equal to the control	Direct antibacterial activity	[41]

[1] Natural contamination; [2] artificial inoculation.

Table vinegar (pH = 3, acetic acid 5%) was also tested in order to reduce *Colletotrichum lupini* seed infection on lupin (*Lupinus albus*) [39]. Anthracnose-infected seeds from highly infected plots were soaked in vinegar and grown under field conditions. The authors reported that vinegar treatment successfully reduced disease severity (16.9%) and increased yield to levels similar to those observed for certified seeds, without significantly affecting germination rate [39].

3.2. Activity of Approved Basic Substances against Bacteria

The bactericidal action of oxygen released from peroxides is well known, and the possibility of direct horticultural benefits plus bactericidal activity make hydrogen peroxide attractive in agriculture for seed disinfection. However, there are only a few recent reports on in vivo or field applications (Table 2). Since 2002, hydrogen peroxide was investigated as a seed treatment for the control of bacterial leaf spot of lettuce (*Lactuca sativa*) caused by the seedborne bacterium *Xanthomonas campestris* pv. *vitians*. Bacteria were not detected when seeds were treated with 3 or 5% hydrogen peroxide, even if the treatments at 5% concentration reduced seed germination up to 28% compared with controls [40]. More recent works about seed treatment with hydrogen peroxide against bacterial diseases have only come after years of research: hydrogen peroxide at 3% was investigated as a seed treatment against *Xanthomonas campestris* pv. *campestris* in cabbage (*Brassica oleracea*) seeds [41]. The treatment for 30 min was the most effective, both in terms of disinfection rate and of seed viability, but the side effects on the seed coat observed when the procedure was carried out at the company facilities suggested 15 min as the maximum time of immersion without losing effectiveness.

4. Seed Treatment with Potential Basic Substances against Pathogens

4.1. Activity of Potential Basic Substances against Fungi and Oomycetes

Essential oils (EOs) are secondary metabolites accumulated by aromatics or medical plants and extracted from leaves, flowers, roots, and barks. They exhibit antifungal activity due to the presence of different bioactive ingredients (alkaloids, phenols, monoterpenes and sesquiterpenes, isoprenoids) in different concentrations, their composition may vary even within the same species, affecting antimicrobial activity [42,43]. EOs have widely demonstrated over the years their efficacy against various fungal pathogens in vitro [44,45] and in recent years, the scientific research in this field has focused primarily on in vivo and field applications (Tables 3 and 4). Immersion seed treatment with clove (*Syzygium aromaticum*) EO was able to reduce *Fusarium* spp. infection on maize and wheat seeds at different doses, but the effective rates (5×10^3 and 5×10^4 ppm, respectively, for maize and wheat) had a high phytotoxicity effect [46]. Clove oil has also been tested in field trials, both as a seed soak and as coating (spray) on wheat and field peas against, respectively, *Tilletia laevis* [47] and *Ascochyta blight* complex [22], artificially inoculated on seeds, with good effectiveness, which varied from year to year. Submersion application has demonstrated a more reliable effectiveness over the years, compared to coating application. In the tomato, eucalyptus (*Eucalyptus grandis*), caraway (*Cuminum cyminum*), and citrus (*Citrus sinensis*), EOs have been tested as seed treatments against *Fusarium oxysporum* [48], and oregano EO (*Origanum vulgare*), against *F. oxysporum* f.sp. *lycopersici* [49] artificially inoculated in soil, with a reduction in disease incidence and severity. Tomato seedlings showed no phytotoxic effects after soaking treatment at the applied rates (Table 3).

Table 3. In vivo and in- field activities of potential basic substances applied as seed treatments to control seedborne fungi and oomycetes in different crops. Effectiveness is reported as the disease incidence or symptoms percentage reduction compared to the untreated control. Phytotoxicity was evaluated through germination testing and the results are compared to the untreated control.

		Potential Basic Substances—Fungi and Oomycetes					
Crop	Target Disease/Pathogen	Substance (Concentration)	Application	Effectiveness (Disease/Symptoms Reduction)	Possible Phytotoxicity	Activity/Defense Response	Reference
Durum wheat (*Triticum durum*)	Common bunt/*Tilletia laevis* *	*Syzygium aromaticum* EO (0.3% v/v)	Immersion for 10 min	From 30% to 90%	Seed germination reduction	Reduction in pathogen incidence	[47]
		S. aromaticum formulation (2.5% v/v)		From 40% to 100%	Rate of seed germination equal to the control		
		S. aromaticum EO (1% v/v)	Coating	From 30% to 82%	Rate of seed germination equal to the control		
		S. aromaticum formulation (5% v/v)	Coating	From 30% to 85%			
Wheat (*Triticum aestivum*)	*Fusarium equiseti* [2]; *F. culmorum* [2]; *F. poae* [2]; *F. avenaceum* [2]	*S. aromaticum* EO 5×10^3 ppm	Immersion for 8 min	100%	Total inhibition of seed germination	Inhibition of pathogen development	[46]
	Alternaria spp. *Fusarium* spp. *Drechslera* spp.	*Origanum vulgare*, *Thymus vulgaris* and *Coriandrum sativum* Eos	Vapour	50%	Inhibition of seed germination at 0.4% (thyme and oregano EO)	Inhibition of deoxynivalenol (DON) occurrence	[50]
	Aspergillus spp. *Fusarium* spp. [1,2]	Ozone (60 mg L^{-1})	Ozonation for 300 min	54.3%	–		[51]

Table 3. Cont.

		Potential Basic Substances—Fungi and Oomycetes					
Crop	Target Disease/Pathogen	Substance (Concentration)	Application	Effectiveness (Disease/Symptoms Reduction)	Possible Phytotoxicity	Activity/Defense Response	Reference
Pea *Pisum sativum*	Ascochyta blight fungal complex (*Dydimella pinodes, D. pinodella, D. pisi*) [2]	*S. aromaticum*-based formulation (0.2% v/v)	Immersion for 10 and 20 min	From 68% to 71%	Rate of seed germination equal to the control but in field an excessive handling after imbibition could damage seeds	In vivo: reduction in seed infection percentage In field: seedling protection and established plants enhancement	[22]
		Thymus vulgaris EO (0.2% v/v)		86%			
		Melaleuca alternifolia EO (2% v/v)		71.5%			
		S. aromaticum-based formulation (0.4% v/v) + pinolene	Seed coating	From 6% to 80%	Rate of seed germination equal to the control		
		T. vulgaris EO (0.3% v/v) + pinolene		53%			
		M. alternifolia EO (2% v/v) + pinolene		5%			
Maize (*Zea mays*)	*F. verticillioides* [2]	*Jacaranda mimosifolia* WE (0.6% v/v)	Immersion for 1 h	Pot experiment: 75% Field experiment: 64%	–	Induction of defense-related enzymes	[52]
	F. equiseti [2]; *F. culmorum* [2]; *F. poae* [2]; *F. avenaceum* [2]	*S. aromaticum* EO (5×10^4 ppm)	Immersion for 8 min		Total inhibition of seed germination	Inhibition of pathogen development	[46]
	Aspergillus spp. [2]	Ozone (60 mg L^{-1})	Ozonation for 480 min	99.7%	–	Aflatoxins and microbial contamination reduction	[53]
	Fusarium spp. [2]			99.9%			
	Aspergillus spp. [1]	Ozone (2.14 mg L^{-1})	Ozonation for 50 h	78.5%	–	Pathogen incidence reduction	[54]
	Penicillium spp. [1]			98.0%			
Tomato (*Solanum lycopersicum*)	Fusarium wilt *F. oxysporum* *	*Artemisia absinthium* EO (0.5 mg mL^{-1})	Seed coating	Reduction in disease symptoms.	Rate of seed germination equal to the control	Induction of a long-term response (ROS production and callose deposition)	[55]
		Eucalyptus grandis EO (6% v/v)	Immersion	73.0%	Rate of seed germination equal to the control		[48]
		Cuminum cyminum EO (6% v/v)		53.1%			
		Citrus sinensis EO (6% v/v)		84.3%			
	F. oxysporum f. sp. *lycopersici* *	*Origanum vulgare* EO 1200 μg mL^{-1}	Immersion	52.0%	No phytotoxicity	Reduction in percentage disease severity and incidence	[49]
Squash (*Cucurbita maxima*)	*Stagonosporopsis cucurbitacearum* [1] and seven other fungal species	*Cymbopogon citratus* EO and six other essential oils. (0.5 mg mL^{-1})	Immersion for 6 h	From 67% to 84.4%		Seedling emergence increasing	[56]
Bean (*Phaseolus vulgaris*)	Anthracnose/ *Colletotrichum lindemuthianum* [2]	*Ocimum gratissimum* EO (80 mg kg^{-1})	Immersion	Anthracnose symptoms reduction of 73.9%	Rate of seed germination equal to the control		[57]
		S. aromaticum EO (80 mg kg^{-1})		Anthracnose symptoms reduction of 65.5%			

Table 3. Cont.

		Potential Basic Substances—Fungi and Oomycetes					
Crop	Target Disease/Pathogen	Substance (Concentration)	Application	Effectiveness (Disease/Symptoms Reduction)	Possible Phytotoxicity	Activity/Defense Response	Reference
Lettuce (*Lactuca sativa*)	*Cladosporium* sp. [1]	*Eugenia caryophyllus* EO (500 µL L^{-1})		86.0%	Seed germination reduction		[58]
	Alternaria sp. [1]			70.0%			
	Cladosporium sp. [1]	*Cymbopogon citratus* EO (500 µL L^{-1})		98.0%			
	Alternaria sp. [1]			85.0%			
	Cladosporium sp. [1]	*Rosmarinus officinalis* EO (500 µL L^{-1})		33.0%			
	Alternaria sp. [1]			7.5%			
Onion (*Allium cepa*)	*A. alternata* [1]	*Abies alba* EO (0.2 µL cm^{-3})	Immersion for 6 h	10.4%	Rate of seed germination equal to the control		[59]
	Botrytis allii [1]			80.5%			
	B. cinerea [1]			76.9%			
	Cladosporium spp. [1]			28.5%			
	Fusarium spp. [1]			84.2%			
	A. alternata [1]	*Pinus sylvestris* EO (0.2 µL cm^{-3})	Immersion for 6 h	16.3%			
	Botrytis allii [1]			55.5%			
	B. cinerea [1]			88.4%			
	Cladosporium spp. [1]			7.1%			
	Fusarium spp. [1]			84.2%			
	A. alternata [1]	*T. vulgaris* EO (0.2 µL cm^{-3})	Immersion for 6 h	10.4%			
	Botrytis allii [1]			80.5%			
	B. cinerea [1]			100%			
	Cladosporium spp. [1]			35.7%			
	Fusarium spp. [1]			94.7%			
Sunflower (*Helianthus annuus*)	*Plasmopara halstedii* [1]	*Nigella sativa* EO (0.6%)	Spray	Decrease in sporangium quantity 70.1%	–		[60]
		Sambucus nigra EO (0.6%)		87.3%			
		Hypericum perforatum EO (0.6%)		90.5%			
		Allium sativum EO (0.6%)		90.0%			
		Vitis vinifera EO (0.6%)		91.2%			
		Zingiber officinale EO (0.6%)		90.2%			

[1] Natural contamination; [2] artificial inoculation; * soil contamination; EO = essential oil; WE = water extract.

Naturally contaminated *Colletotrichum lindemuthianum* beans were treated with basil (*Ocimum gratissimum*) and clove EOs, and the treatment caused a significant reduction in anthracnose incidence without affecting the germination and the emergence speed index [57]. Lemongrass (*Cymbopogon citratus*), lavender (*Lavandula dentata*), lavandin (*Lavandula hybrida*), tea tree (*Melaleuca alternifolia*), bay laurel (*Laurus nobilis*), and two different marjoram (*Origanum majorana*) EOs were tested as seed treatments against the main *Cucurbita maxima* seedborne fungal pathogens: *Stagonosporiopsis cucurbitacearum*, *Alternaria alternata*, and *F. solani* [56]. The seed immersion treatments were carried out at a concentration of 0.5 mg mL^{-1} for 6 h, with mixing every 30 min, and the results showed that the incidence of multiple seedborne fungal pathogens was significantly reduced on squash seeds, with no negative effect on germination. In addition, the *C. citratus* EO increased seedling emergence and reduced the incidence of *S. cucurbitacearum* in plantlets.

Waureck et al. [58] found that the main fungi observed in organic and untreated lettuce seeds were *Cladosporium* sp. and *Alternaria* sp. seed, and treatments with clove, lemongrass, and rosemary EOs at a dose of 0.5% (v/v) significantly reduced their presence on seeds, but with negative effects on germination, suggesting that the application dose of these essential oils should be modulated for lettuce seeds [58].

Exogenous application of specific plant extracts can induce resistance in the host plant via higher levels of host defense enzymes and PR protein stimulation. An absinthium (*Artemisia absinthium*) EO seed coating was tested on tomato seeds and was able to protect seed germination and seedling growth, priming tolerance in tomato seedlings previously infected with *F. oxysporum* f.sp. *lycopersici* by the induction of metabolic changes responsible for the long-term tolerance of the tomato [55]. An extract of *Jacaranda mimosifolia* (1.2%) applied to maize seeds provided significant protective effects on plants compared to the inoculated control, by also inducing a systemic resistance in the host plant [52].

Silver fir (*Abies alba*), pine (*Pinus sylvestris*), and thyme EOs were tested as seed treatments on onion by immersion for 6 h, and seed health test on potato dextrose agar showed that all the oil treatments effectively controlled *Fusarium* spp. on the onion seeds and frequently reduced their infestation with *Botrytis* spp. The lowest dose tested with antifungal activity and without phytotoxic effects was 0.2 $\mu L\ cm^{-3}$, while increasing the dose led to increased phytotoxicity [59].

Commercial EOs obtained from different parts of black cumin (*Nigella sativa*), mustard (*Sambucus nigra*), St. John's wort (*Hypericum perforatum*), garlic (*Allium sativum*), grape (*Vitis vinifera*), and ginger (*Zingiber officinale*) plants were evaluated in vivo against the oomycete *Plasmopara halstedii*. The application of the above oils as a spray seed treatment was shown to provide protection against mildew in sunflower plants under in vivo conditions, assessed as a percentage reduction in the sporangium count ranging from 70.1% to 90.5% [60].

In order to obtain the best advantages from the volatile nature of active compounds, oregano, thyme (*Thymus vulgaris*), and coriander (*Coriandrum sativum*) EOs were tested in vapor form for their antifungal potential against *Alternaria* spp., *Fusarium* spp. and *Drechslera* spp. infection on wheat seeds [50]. Wheat seeds were stored in an atmosphere enriched with essential oil vapors and a selective antifungal effect was highlighted as the following: oregano EO and thyme EO significantly inhibited *Alternaria*, *Fusarium*, and *Drechslera* (that was the most sensitive). Regarding the phytotoxic effects of EO vapors on the germination of the seeds, thyme EO and oregano EO had an inhibitory effect, especially at 0.4%. This effect was cumulative over time. The EOs inhibited deoxynivalenol (DON) occurrence, and the maximum percentage of inhibition was obtained after 21 days of vapor exposure, with the most effective timing being when applied at 0.2%.

Ozone has been declared as a generally recognized as safe (GRAS) substance and its application in agriculture has increased in recent years (Table 2) [61]. Ozone gas was applied on maize and wheat seeds for fungal decontamination: ozone gas application for 300 min at a rate of 60 mg L^{-1} was able to reduce the incidence of *Aspergillus* spp. and *Penicillium* spp. (both ~ 54%) on artificially infected wheat seeds [51], while 50 h application at a rate of 2.14 mg L^{-1} reduced *Aspergillus* spp. (78.5%) and *Penicillium* spp. (98.0%) incidence on naturally infected maize seeds [54]. Thanks to its oxidizing properties, ozonation can also represent an effective method for the remediation of cereals contaminated by mycotoxins, where gaseous ozone application for 480 min at the rate of 60 mg L^{-1} reduced aflatoxins and microbial contamination in corn artificially infected with *Aspergillus* spp. and *Penicillium* spp. [53].

4.2. Activity of Potential Basic Substances against Bacteria

Several studies have investigated the effects of potential basic substances to control bacterial seedborne pathogens (Table 4). Kotan et al. [62] revealed the antibacterial effects of different extracts of *Origanum onites* (hexane, acetone, and chloroform) on tomato and lettuce seeds inoculated with *Clavibacter michiganensis* ssp. *michiganensis*, *Xanthomonas axonopodies* pv. *vesicatoria*, and *X. campestris* pv. *zinniae*. Extracts were applied by seed soaking after inoculation. The hexane extract was the most effective against *C. michiganensis* ssp. *michiganensis*, with a 75% disease severity reduction at 15 mg mL^{-1}, whereas the chloroform extract was more effective against *X. axonopodies* pv. *vesicatoria* and *X. campestris* pv. *vitians*, with a 77% reduction at 20 mg ml^{-1} and a 74% reduction at 15 mg mL^{-1}, respectively. The authors attributed this strong antibacterial activity to the presence of carvacrol and thymol, two of EO's major constituents. No phytotoxicity was found on seeds treated with all the extracts tested; indeed, different extracts even increased seed germination and plant height in tomato seedlings at concentrations of 5 and 10 mg mL^{-1}. A study by Karabüyük and Aysan [63] on the reduction in bacterial speck disease caused by *Pseudomonas syringae* pv. *tomato* demonstrated that immersion treatments of tomato seeds with aqueous extracts of *Zingiber officinale* and *Origanum vulgare* (Istanbul thyme) reduced 100% of bacterial speck disease incidence and severity on tomato seedlings. In addition, aqueous extracts of *Eucalyptus camaldulensis* and *Allium sativum* reduced disease incidence and severity by 98%–97% and 99.3%–56.8%, respectively, whereas coriander extracts only reduced disease incidence by up to 63%. All the tested extracts did not affect seed germination. The antimicrobial activity of thyme EO on soybean seeds infected with *P. savastanoi* pv. *glycinea* B076 and *P. syringae* M7-C1, causal agents of bacterial blight in soybean, was investigated at a greenhouse scale by Sotelo et al. [64]. The results obtained demonstrated that 1.76 mg mL^{-1} of the essential oil previously diluted in skim milk powder reduced the number of phytopathogenic bacteria inoculated on the seeds by about 6 logs. In addition, the germination of the treated seeds was 73%, whereas for the infected seeds it was near 50%. Similarly, the disease incidence of soybean plants from infected seeds and treated with thyme EO was reduced by 24.05% for *P. syringae* M7-C1 and by 29.76% for *P. savastanoi* pv. *glycinea* B076. Another study [65] focused on the plant pathogenic bacteria *Burkholderia glumae*, a rice seedborne pathogen that causes grain rot in rice plants, and showed that immersion treatment of rice seeds for 10 min with clove EO at 2% and 5% *v/v* and citronella (*Cymbopogon nardus*) EO at 1% and 3% *v/v* reduced by 50% the disease incidence in plants, with the 5% clove oil treatment giving the highest rice grain production. However, no phytotoxicity data were provided. *Cistus ladaniferus* subsp. *ladanifer* EO, together with its methanolic and ethanolic extracts, and *Mentha suaveolens* EO, were used for the treatment of tomato seeds infected with the phytopathogenic bacterium *C. michiganensis* subsp. *michiganensis* [66]. The results evidenced that *C. ladaniferus* subsp. *ladanifer* oil and extracts and *Mentha suaveolens* EO inhibited in vitro the growth of *C. michiganensis* with a minimal inhibitory concentration (MIC) of 0.78 mg mL^{-1}, but the in vivo treatment with such EOs at MIC and 4 × MIC showed a negative effect on tomato seed germination. On the contrary, treatment with ethanolic and methanolic extracts of *C. ladaniferus* showed no phytotoxicity, with the methanolic extract revealing the highest percentages of germination. Treatments were performed by soaking the seeds for 1 h. In another study on the tomato, two other EOs (cinnamon and oregano) were tested in vivo for their antibacterial activity against *C. michiganensis* subsp. *michiganensis* [67]. Artificially infected tomato seeds were treated by immersion with these two oils at a concentration of 0.4% and their efficacy in controlling the pathogen was evaluated using a real-time PCR molecular assay for in planta bacterial quantification at the very first stage of development: both oils significantly reduced the bacterial presence in seedlings compared to controls (untreated and water-treated), with oregano being the most effective. Oregano EO showed no phytotoxicity at the concentrations tested up to 0.4%, while cinnamon EO had little effect on germination, reducing it by one or two percentage points.

Table 4. In vivo and in-field activities of potential basic substances applied as seed treatments to control seedborne bacteria in different crops. Effectiveness is reported as the disease incidence or symptom percentage reduction compared to the untreated control. Phytotoxicity was evaluated through germination testing and the results are compared to the untreated control.

		Potential Basic Substances—Bacteria					
Crop	Target Disease/Pathogen	Substance (Concentration)	Application	Effectiveness (Disease/Symptoms Reduction)	Possible Phytotoxicity	Activity/Defense Response	Reference
Tomato (Solanum lycopersicum)	Clavibacter michiganensis subsp. michiganensis [1]	Cinnamomum zeylanicum EO (0.4% v/v)	Immersion	25%	Germination reduced by 1–2%	Bactericidal activity	[67]
		Origanum vulgare EO 0.4% (v/v)	Immersion	100%	Rate of seed germination equal to the control		
	C. michiganensis subsp. michiganensis [1]	O. onites HE (15 mg mL^{-1})	Immersion	75%	Rate of seed germination equal to the control	Different extracts increased seed germination and plant height	[62]
	Xanthomonas axonopodies pv. vesicatoria [1]	O. onites CE (20 mg mL^{-1})		76.91%			
	X. campestris pv. zinniae [1]	O. onites chloroform extract (15 mg mL^{-1})		74.22%			
	Pseudomonas syringae pv. tomato [1] (Pst)	Zingiber officinale AE	Immersion	100%	Rate of seed germination equal to the control		[63]
		O. vulgare L. AE (Istanbul thyme and Izmir thyme)		100%			
		Eucalyptus camaldulensis AE		98% (incidence) 97% (severity)			
		Allium sativum AE		99% (incidence) 57% (severity)			
		Coriandrum sativum extracts		Up to 63% (incidence)			
Soybean (Glycine max)	P. savastanoi pv. glycinea B076 [1]	Thymus vulgaris EO (1.76 mg mL^{-1})		24.05%	Seed germination increasing	Increasing seed germination	[64]
	P. syringae M7-C1 [1]			29.76%			
Rice (Oryza sativa)	Burkholderia glumae [1]	S. aromaticum EO Cymbopogon nardus		50%	Rate of seed germination equal to the control		[65]
Tomato (Solanum lycopersicum)	C. michiganensis subsp. michiganensis [1]	Cistus ladaniferus subsp. ladanifer EO	Immersion for 1 h	Minimal inhibitory concentration (MIC): 0.78 mg mL^{-1}	Rate of seed germination equal to the control	Bacterial growth inhibition	[66]
		Cistus ladaniferus subsp. ladanifer ME			Seed germination increasing		
		Mentha suaveolens EO			Rate of seed germination equal to the control		

[1] Natural contamination; EO = essential oil; AE = aqueous extract; ME = methanolic extract; CE = chloroform extract; HE = hexane extract.

4.3. Activity of Potential Basic Substances against Viruses and Phytoplasma

Basic substances or potential basic substances having a direct action on viruses or phytoplasma inside plant cells are nowadays quite unknown. Research directly targeting these pathogens inside the plant host cells by applying sustainable means of control is useful and highly recommended. Stommel and colleagues demonstrated [68] that exposure of pepper mild mottle virus to ozone resulted in viral inactivation, but at insufficient levels to prevent viral transmission from highly contaminated pepper (Capsicum annuum) seeds (Table 5). Viruses and phytoplasma are non-culturable organisms; therefore, it is not easy to verify their direct effects on pathogens and just in vivo trials can be used. However, in vivo trials are much more complex and require infected materials with a high load of the pathogen to gain significant results. Virus and phytoplasma can be controlled by physical treatments such as thermotherapy or by controlling their insect vectors. Basic substances or potentially basic substances can also effectively be used against the vectors to reduce the spread of viruses and phytoplasma.

Table 5. In vivo and in-field activities of potential basic substances applied as seed treatments to control seedborne viruses in different crops. Effectiveness is reported as the disease incidence or symptom percentage reduction compared to the untreated control. Phytotoxicity was evaluated through germination testing and results are compared to the untreated control.

			Potential Basic Substances—Viruses				
Crop	Target Disease/ Pathogen	Substance (Concentration)	Application	Effectiveness (Disease/Symptoms Reduction)	Possible Phytotoxicity	Activity/Defense Response	Reference
Pepper (*Capsicum annum*)	Pepper mild mottle virus (PMoV)[1]	Ozone (20 ppm)	Ozonation for 14 h	Inactivation of the seedborne virus; however, at high seed contamination levels, this treatment was insufficient to prevent infection	Rate of seed germination equal to the control		[68]

[1] Natural contamination.

5. Conclusions

Sowing high-quality seeds is important to reduce yield losses. Seed treatment is an essential step in the management of crops diseases. This step can play economic and environmental roles in reducing the cost and quantity of pesticides in the field. Sustainable seed treatments using basic substances and potential basic substances can be good alternatives to controlling the main seedborne pathogens and for promoting more sustainable crop systems. Basic substances have a registration cost that is much lower than the one of synthetic pesticides (EUR 50,000 versus EUR 300 million) [15], and small companies can also promote the application of a basic substance. There is relatively poor information about the effectiveness of basic substances and potential basic substances as seed treatments compared to synthetic pesticides, but in some research, their effectiveness can be considered comparable or slightly lower than the one of synthetic fungicide. The cost of the product is comparable to or slightly higher than synthetic pesticides. Conversely, by applying basic substances, there are no issues with the safety of the treated commodities and a there is a lower impact on the environment. These alternatives now need to be further developed as appropriate seed treatments for ensuring global food security in a green way.

Author Contributions: Conceptualization, L.O. and G.R.; investigation, L.O., M.B.A., C.C.-L., M.M. (Marwa Moumni), M.M. (Monica Mezzalama) and J.B.M.-H.; writing—original draft preparation, L.O. and M.M. (Marwa Moumni); writing—review and editing, L.O., M.B.A., C.C.-L., M.M. (Marwa Moumni), M.M. (Monica Mezzalama) and J.B.M.-H.; supervision, G.R. and L.O.; funding acquisition, G.R. and L.O. All authors have read and agreed to the published version of the manuscript.

Funding: This work was carried out within the "Euphresco Basic substances as an environmentally friendly alternative to synthetic pesticides for plant protection (BasicS)" project (Objective 2020-C-353). G.R. acknowledges the support of the PSR Marche Project "CleanSeed".

Data Availability Statement: All data are contained in the article.

Acknowledgments: This review is dedicated to Luca Riccioni, who greatly supported the researchers and the project activities and recently suddenly passed away prematurely. Many thanks are also expressed to Baldissera Giovani, the Euphresco Coordinator, who promoted and supported the project Euphresco BasicS along with its development.

Conflicts of Interest: The authors declare no conflict of interest.

References

1. Umesha, S. *Diversity of Seed-Borne Bacterial Phytopathogens*; Springer: Singapore, 2020. [CrossRef]
2. Bonde, M.R.; Peterson, G.L.; Schaad, N.W.; Smilanick, J.L. Karnal Bunt of Wheat. *Plant Dis.* **1997**, *81*, 1370–1377. [CrossRef]
3. Rong, I. Karnal bunt of wheat detected in South Africa. *Plant Prot. News* **2000**, *58*, 15–17.
4. Malaker, P.K.; Barma, N.C.D.; Tiwari, T.P.; Collis, W.J.; Duveiller, E.; Singh, P.K.; Joshi, A.K.; Singh, R.P.; Braun, H.-J.; Peterson, G.L.; et al. First report of wheat blast caused by *Magnaporthe oryzae* pathotype *triticum* in Bangladesh. *Plant Dis.* **2016**, *100*, 2330. [CrossRef]
5. Tembo, B.; Mulenga, R.M.; Sichilima, S.; M'siska, K.K.; Mwale, M.; Chikoti, P.C.; Singh, P.K.; He, X.; Pedley, K.F.; Peterson, G.L.; et al. Detection and characterization of fungus (*Magnaporthe oryzae* pathotype *triticum*) causing wheat blast disease on rain-fed grown wheat (*Triticum aestivum* L.) in Zambia. *PLoS ONE* **2020**, *15*, e0238724. [CrossRef] [PubMed]
6. Dwyer, G.I.; Gibbs, M.J.; Gibbs, A.J.; Jones, R.A.C. Wheat streak mosaic virus in Australia: Relationship to isolates from the Pacific Northwest of the USA and its dispersion via seed transmission. *Plant Dis.* **2007**, *91*, 164–170. [CrossRef] [PubMed]
7. Wangai, A.W.; Redinbaugh, M.G.; Kinyua, Z.M.; Miano, D.W.; Leley, P.K.; Kasina, M.; Mahuku, G.; Scheets, K.; Jeffers, D. First Report of Maize chlorotic mottle virus and Maize Lethal Necrosis in Kenya. *Plant Dis.* **2012**, *96*, 1582. [CrossRef]
8. International Seed Federation. Seed Treatment. Available online: https://worldseed.org/our-work/seed-treatment (accessed on 15 September 2023).
9. Munkvold, G.P. Seed pathology progress in Academia and industry. *Annu. Rev. Phytopathol.* **2009**, *47*, 285–311. [CrossRef]
10. Moumni, M.; Brodal, G.; Romanazzi, G. Recent innovative seed treatment methods in the management of seedborne pathogens. *Food Secur.* **2023**, *15*, 1365–1382. [CrossRef]
11. Lamichhane, J.R.; You, M.P.; Laudinot, V.; Barbetti, M.J.; Aubertot, J.N. Revisiting sustainability of fungiside seed treatments for field crops. *Plant Dis.* **2020**, *104*, 610–623. [CrossRef]
12. Sharma, K.K.; Singh, U.S.; Sharma, P.; Kumar, A.; Sharma, L. Seed treatments for sustainable agriculture—A review. *J. Appl. Nat. Sci.* **2015**, *7*, 521–539. [CrossRef]
13. Kesraoui, S.; Andrès, M.F.; Berrocal-Lobo, M.; Soudani, S.; Gonzales-Coloma, A. Direct and indirect effects of essential oils for sustainable crop protection. *Plants* **2022**, *11*, 2144. [CrossRef]
14. Chrapačienė, S.; Rasiukevičiūtė, N.; Valiuškaitė, A. Control of seed-borne fungi by selected essential oils. *Horticulturae* **2022**, *8*, 220. [CrossRef]
15. Romanazzi, G.; Orçonneau, Y.; Moumni, M.; Davillerd, Y.; Marchand, P.A. Basic substances, a sustainable tool to complement and eventually replace synthetic pesticides in the management of pre and postharvest diseases: Reviewed instructions for users. *Molecules* **2022**, *27*, 3484. [CrossRef] [PubMed]
16. Marchand, P.A.; Davillerd, Y.; Riccioni, L.; Sanzani, S.M.; Horn, N.; Matyjaszczyk, E.; Golding, J.; Roberto, S.R.; Mattiuz, B.-H.; Xu, D.; et al. BasicS, an Euphresco International Network on Renewable Natural Substances for Durable Crop Protection Products Chronicle of Bioresource Management An International E-magazine BasicS, an Euphresco International Network on Renewable Natural Substances for. *Chron. Bioresour. Manag.* **2021**, *2021*, 77–080.
17. EC. Commission Regulation No 1107/2009 of the European parliament and of the council of 21 October 2009 concerning the placing of plant protection products on the market and repealing council directives 79/117/EEC and 91/414/EEC. *OJ* **2009**, *L309*, 1–50.
18. Afzal, I.; Javed, T.; Amirkhani, M.; Taylor, A.G. Modern seed technology: Seed coating delivery systems for enhancing seed and crop performance. *Agriculture* **2020**, *10*, 526. [CrossRef]
19. Zulfiqar, Z.F. Effect of seed priming on horticultural crops. *Sci. Hortic.* **2021**, *286*, 110197. [CrossRef]
20. Orzali, L.; Forni, C.; Riccioni, L. Effect of chitosan seed treatment as elicitor of resistance to *Fusarium graminearum* in wheat. *Seed Sci. Technol.* **2014**, *42*, 132–149. [CrossRef]
21. Kalaivani, K.; Kalaiselvi, M.M.; Senthil-Nathan, S. Effect of methyl salicylate (MeSA), an elicitor on growth, physiology and pathology of resistant and susceptible rice varieties. *Sci. Rep.* **2016**, *6*, 34498. [CrossRef] [PubMed]
22. Riccioni, L.; Orzali, L.; Romani, M.; Annicchiarico, P.; Pecetti, L. Organic seed treatments with essential oils to control ascochyta blight in pea. *Eur. J. Plant Pathol.* **2019**, *155*, 831–840. [CrossRef]
23. Hussain, S.; Zheng, M.; Khan, F.; Khaliq, A.; Fahad, S.; Peng, S.; Huang, J.; Cui, K.; Nie, L. Benefits of rice seed priming are offset permanently by prolonged storage and the storage conditions. *Sci. Rep.* **2015**, *5*, 8101. [CrossRef]
24. Kimmelshue, C.; Goggi, A.S.; Cademartiri, R. The use of biological seed coatings based on bacteriophages and polymers against *Clavibacter michiganensis* subsp. *nebraskensis* in maize seeds. *Sci. Rep.* **2019**, *9*, 17950. [CrossRef] [PubMed]
25. Taylor, J.B.; Cass, K.L.; Armond, D.N.; Madsen, M.D.; Pearson, D.E.; St. Clair, S.B. Deterring rodent seed-predation using seed-coating technologies. *Restor. Ecol.* **2020**, *28*, 927–936. [CrossRef]
26. Shang, Y.; Hasan, K.; Ahammed, G.J.; Li, M.; Yin, H. Applications of nanotechnology in plant growth and crop protection: A review. *Molecules* **2019**, *24*, 2558. [CrossRef] [PubMed]
27. Shelar, A.; Nile, S.H.; Singh, A.V.; Rothenstein, D.; Bill, J.; Xiao, J.; Chaskar, M.; Kai, G.; Patil, R. Recent advances in nano-enabled seed treatment strategies for sustainable agriculture: Challenges, risk assessment, and future perspectives. *Nanomicro Lett.* **2023**, *15*, 54. [CrossRef] [PubMed]
28. Prasad, R.D.; Chandrika, K.S.V.P.; Godbole, V. A novel chitosan biopolymer based *Trichoderma* delivery system: Storage stability, persistence and bio efficacy against seed and soil borne diseases of oilseed crops. *Microbiol. Res.* **2020**, *237*, 126487. [CrossRef]

29. EU. Regulation (EU) 563/2014 of 23.5.2014 approving the basic substance chitosan hydrochloride. *Off. J. Eur. Union* **2014**, *L156*, 5–7.
30. EU. Commission Implementing Regulation (EU) 2022/456 of 21 March 2022 approving the basic substance chitosan in accordance with Regulation (EC) No 1107/2009 of the European Parliament and of the Council concerning the placing of plant protection products on the market, and amending the Annex to Implementing Regulation (EU) No 540/2011. *OJ* **2022**, *L93*, 138–141.
31. El-Mohamedy, R.S.R.; Shafeek, M.R.; El-Samad, E.E.D.H.A.; Salama, D.M.; Rizk, F.A. Field application of plant resistance inducers (PRIs) to control important root rot diseases and improvement growth and yield of green bean (*Phaseolus vulgaris* L.). *Aust. J. Crop Sci.* **2017**, *11*, 496–505. [CrossRef]
32. Ghule, M.R.; Ramteke, P.K.; Ramteke, S.D.; Kodre, P.S.; Langote, A.; Gaikwad, A.V.; Holkar, S.K.; Jambhekar, H. Impact of chitosan seed treatment of fenugreek for management of root rot disease caused by *Fusarium solani* under in vitro and in vivo conditions. *3 Biotech* **2021**, *11*, 290. [CrossRef] [PubMed]
33. Bhardwaj, N.R.; Atri, A.; Banyal, D.K.; Dhal, A.; Roy, A.K. Multi-location evaluation of fungicides for managing blast (*Magnaporthe grisea*) disease of forage pearl millet in India. *J. Crop Prot.* **2022**, *195*, 106019. [CrossRef]
34. Buzón-Durán, L.; Martín-Gil, J.; Marcos-Robles, J.L.; Fombellida-Villafruela, Á.; Pérez-Lebeña, E.; Martín-Ramos, P. Antifungal activity of chitosan oligomers–amino acid conjugate complexes against *Fusarium culmorum* in spelt (*Triticum spelta* L.). *Agronomy* **2020**, *10*, 1427. [CrossRef]
35. Zohara, F.; Surovy, M.Z.; Khatun, A.; Prince, M.F.R.K.; Akanda, M.A.M.; Rahman, M.; Islam, M.T. Chitosan biostimulant controls infection of cucumber by *Phytophthora capsici* through suppression of asexual reproduction of the pathogen. *Acta Agrobot.* **2019**, *72*, 1763. [CrossRef]
36. Kowalska, J.; Tyburski, J.; Krzymińska, J.; Jakubowska, M. Effects of seed treatment with mustard meal in control of *Fusarium culmorum* Sacc. and the growth of common wheat (*Triticum aestivum* ssp. *vulgare*). *Eur. J. Plant Pathol.* **2021**, *159*, 327–338. [CrossRef]
37. Berbegal, M.; Landeras, E.; Sánchez, D.; Abad-Campos, P.; Pérez-Sierra, A.; Armengol, J. Evaluation of *Pinus radiata* seed treatments to control *Fusarium circinatum*: Effects on seed emergence and disease incidence. *For. Pathol.* **2015**, *45*, 525–533. [CrossRef]
38. Górski, R.; Szopińska, D.; Dorna, H.; Rosińska, A.; Stefańska, Z.; Lisiecka, J. Effects of plant extracts and disinfectant huva-san tr 50 on the quality of carrot (*Daucus carota* L.). *Seeds. Ecol. Chem. Eng.* **2020**, *27*, 617–628. [CrossRef]
39. Alkemade, J.A.; Arncken, C.; Hirschvogel, C.; Messmer, M.M.; Leska, A.; Voegele, R.T.; Finckh, M.R.; Kölliker, R.; Groot, S.P.C.; Hohmann, P. The potential of alternative seed treatments to control anthracnose disease in white lupin. *Crop Prot.* **2022**, *158*, 106009. [CrossRef]
40. Pernezny, K.; Nagata, R.; Raid, R.N.; Collins, J.; Carroll, A. Investigation of seed treatments for management of bacterial leaf spot of lettuce. *Plant Dis.* **2002**, *86*, 151–155. [CrossRef] [PubMed]
41. Sanna, M.; Gilardi, G.; Gullino, M.L.; Mezzalama, M. Evaluation of physical and chemical disinfection methods of Brassica oleracea seeds naturally contaminated with *Xanthomonas campestris* pv. *campestris*. *J. Plant Dis. Prot.* **2022**, *129*, 1145–1152. [CrossRef]
42. Chouhan, S.; Sharma, K.; Guleria, S. Antimicrobial activity of some essential oils—Present status and future perspectives. *Medicines* **2017**, *4*, 58. [CrossRef]
43. Gonçalves, L.A.; Lorenzo, J.M.; Trindade, M.A. Fruit and agro-industrial waste extracts as potential antimicrobials in meat products: A brief review. *Foods* **2021**, *10*, 1469. [CrossRef] [PubMed]
44. M Maurya, A.; Prasad, J.; Das, S.; Dwivedy, A.K. Essential Oils and Their Application in Food Safety. *Front. Sustain. Food Syst.* **2021**, *5*, 653420. [CrossRef]
45. Moumni, M.; Romanazzi, G.; Najar, B.; Pistelli, L.; Amara, H.B.; Mezrioui, K.; Karous, O.; Chaieb, I.; Allagui, M.B. Antifungal activity and chemical composition of seven essential oils to control the main seedborne fungi of cucurbits. *Antibiotics* **2021**, *10*, 104. [CrossRef]
46. Grzanka, M.; Sobiech, Ł.; Danielewicz, J.; Horoszkiewicz-Janka, J.; Skrzypczak, G.; Sawinska, Z.; Radzikowska, D.; Świtek, S. Impact of essential oils on the development of pathogens of the *Fusarium* genus and germination parameters of selected crops. *Open Chem.* **2021**, *19*, 884–893. [CrossRef]
47. Valente, M.T.; Orzali, L.; Manetti, G.; Magnanimi, F.; Matere, A.; Bergamaschi, V.; Grottoli, A.; Bechini, S.; Riccioni, L.; Aragona, M. Rapid molecular assay for the evaluation of clove essential oil antifungal activity against wheat common bunt. *Front. Plant Sci.* **2023**, *14*, 1130793. [CrossRef] [PubMed]
48. Yousafi, Q.; Bibi, S.; Saleem, S.; Hussain, A.; Hasan, M.M.; Tufail, M.; Qandeel, A.; Khan, M.S.; Mazhar, S.; Yousaf, M.; et al. Identification of novel and safe fungicidal molecules against *Fusarium oxysporum* from plant essential oils: In vitro and computational approaches. *Biomed Res. Int.* **2022**, *2022*, 5347224. [CrossRef]
49. Gonçalves, D.C.; Queiroz, V.T.; Costa, A.V.; Lima, W.P.; Belan, L.L.; Moraes, W.B.; Iorio, N.L.P.P.; Póvoa, H.C.C. Reduction of Fusarium wilt symptoms in tomato seedlings following seed treatment with *Origanum vulgare* L. essential oil and carvacrol. *Crop Prot.* **2021**, *141*, 105487. [CrossRef]
50. Bota, V.; Sumalan, R.M.; Obistioiu, D.; Negrea, M.; Cocan, I.; Popescu, I.; Alexa, E. Study on the Sustainability Potential of Thyme, Oregano, and Coriander Essential Oils Used as Vapours for Antifungal Protection of Wheat and Wheat Products. *Sustainability* **2022**, *14*, 4298. [CrossRef]

51. Trombete, F.M.; Porto, Y.D.; Freitas-Silva, O.; Pereira, R.V.; Direito, G.M.; Saldanha, T.; Fraga, M.E. Efficacy of ozone treatment on mycotoxins and fungal reduction in artificially contaminated soft wheat grains. *J. Food Process. Preserv.* **2017**, *41*, e12927. [CrossRef]
52. Naz, R.; Bano, A.; Nosheen, A.; Yasmin, H.; Keyani, R.; Shah, S.T.A.; Anwar, Z.; Roberts, T.H. Induction of defense-related enzymes and enhanced disease resistance in maize against *Fusarium verticillioides* by seed treatment with *Jacaranda mimosifolia* formulations. *Sci. Rep.* **2021**, *11*, 59. [CrossRef]
53. Porto, Y.D.; Trombete, F.M.; Freitas-Silva, O.; de Castro, I.M.; Direito, G.M.; Ascheri, J.L.R. Gaseous ozonation to reduce aflatoxins levels and microbial contamination in corn grits. *Microorganisms* **2019**, *7*, 220. [CrossRef]
54. Brito, J.G.D.; Faroni, L.R.D.A.; Cecon, P.R.; Benevenuto, W.C.A.D.N.; Benevenuto, A.A.; Heleno, F.F. Efficacy of ozone in the microbiological disinfection of maize grains. *Brazilian J. Food Technol.* **2018**, *21*, e2017022. [CrossRef]
55. Soudani, S.; Poza-Carrión, C.; De la Cruz Gómez, N.; González-Coloma, A.; Andrés, M.F.; Berrocal-Lobo, M. Essential oils prime epigenetic and metabolomic changes in tomato defense against *Fusarium oxysporum*. *Front. Plant Sci.* **2022**, *13*, 804104. [CrossRef]
56. Moumni, M.; Allagui, M.B.; Mezrioui, K.; Ben Amara, H.; Romanazzi, G. Evaluation of seven essential oils as seed treatments against seedborne fungal pathogens of *Cucurbita maxima*. *Molecules* **2021**, *26*, 2354. [CrossRef]
57. Silva, A.A.; Pereira, F.A.C.; de Souza, E.A.; de Oliveira, D.F.; Nobre, D.A.C.; Macedo, W.R.; Silva, G.H. Inhibition of anthracnose symptoms in common bean by treatment of seeds with essential oils of *Ocimum gratissimum* and *Syzygium aromaticum* and eugenol. *Eur. J. Plant Pathol.* **2022**, *163*, 865–874. [CrossRef]
58. Waureck, A.; da Luz Coelho Novembre, A.D. Physiological and sanitary attributes of organic lettuce seeds treated with essential oils during storage. *Comun. Sci.* **2022**, *13*, e3394. [CrossRef]
59. Dorna, H.; Szopińska, D.; Rosińska, A.; Górski, R. Chemical composition of fir, pine and thyme essential oils and their effect on onion (*Allium cepa* L.) seed quality. *Agronomy* **2021**, *11*, 2445. [CrossRef]
60. Er, Y.; Özer, N.; Katırcıoğlu, Y.Z. In vivo anti-mildew activity of essential oils against downy mildew of sunflower caused by *Plasmopara halstedii*. *Eur. J. Plant Pathol.* **2021**, *161*, 619–627. [CrossRef]
61. McHugh, T. Ozone processing of foods and beverages. *Food Technol.* **2015**, *69*, 72–74.
62. Kotan, R.; Cakir, A.; Ozer, H.; Kordali, S.; Cakmakci, R.; Dadasoglu, F.; Dikbas, N.; Aydin, T.; Kazaz, C. Antibacterial effects of *Origanum onites* against phytopathogenic bacteria: Possible use of the extracts from protection of disease caused by some phytopathogenic bacteria. *Sci. Hortic.* **2014**, *172*, 210–220. [CrossRef]
63. Karabüyük, F.; Aysan, Y. Aqueous plant extracts as seed treatments on tomato bacterial speck disease. *Acta Hortic.* **2018**, *1207*, 193–196. [CrossRef]
64. Sotelo, J.P.; Oddino, C.; Giordano, D.F.; Carezzano, M.E.; Oliva, M.d.l.M. Effect of *Thymus vulgaris* essential oil on soybeans seeds infected with *Pseudomonas syringae*. *Physiol. Mol. Plant Pathol.* **2021**, *116*, 101735. [CrossRef]
65. Sari, S.P.; Safni, I.; Lubis, L. Seed treatments to control *Burkholderia glumae* on rice seeds in the screenhouse. *IOP Conf. Ser. Earth Environ. Sci.* **2022**, *977*, 012021. [CrossRef]
66. Benali, T.; Bouyahya, A.; Habbadi, K.; Zengin, G.; Khabbach, A.; Achbani, E.H.; Hammani, K. Chemical composition and antibacterial activity of the essential oil and extracts of *Cistus ladaniferus* subsp. *Ladanifer* and *Mentha suaveolens* against phytopathogenic bacteria and their ecofriendly management of phytopathogenic bacteria. *Biocatal. Agric. Biotechnol.* **2020**, *28*, 101696. [CrossRef]
67. Orzali, L.; Valente, M.T.; Scala, V.; Loreti, S.; Pucci, N. Antibacterial activity of essential oils and trametes versicolor extract against *Clavibacter michiganensis* subsp. *michiganensis* and *Ralstonia solanacearum* for seed treatment and development of a rapid in vivo assay. *Antibiotics* **2020**, *9*, 628. [CrossRef]
68. Stommel, J.R.; Dumm, J.M.; Hammond, J. Effect of ozone on inactivation of purified pepper mild mottle virus and contaminated pepper seed. *PhytoFrontiers* **2021**, *1*, 85–93. [CrossRef]

Disclaimer/Publisher's Note: The statements, opinions and data contained in all publications are solely those of the individual author(s) and contributor(s) and not of MDPI and/or the editor(s). MDPI and/or the editor(s) disclaim responsibility for any injury to people or property resulting from any ideas, methods, instructions or products referred to in the content.

 *horticulturae*

Review

Sustainable Management of Diseases in Horticulture: Conventional and New Options

Marco Scortichini

Research Centre for Olive, Fruit and Citrus Crops, Council for Agricultural Research and Economics (CREA), Via di Fioranello, 52, 00134 Roma, Italy; marco.scortichini@crea.gov.it

Abstract: To reduce the impact of chemical pesticides on the environment, there are relevant efforts to enhance the possibility of controlling plant diseases using environmentally friendly biocontrol agents or natural products that show pathogen control capacity. The European Union, FAO, and the United Nations largely promote and finance projects and programs in order to introduce crop protection principles that can attain sustainable agriculture. Preventive measures related to the choice of cultivars, soil fertility, integrated pest management (IPM), and organic farming strategies are still the basis for obtaining satisfactory crop yields and reducing classical pesticide utilisation through the application of commercially available and ecofriendly control agents. Effective pathogen detection at borders to avoid quarantine pathogens is mandatory to reduce the risk of future epidemics. New technical support for the development of sustainable pathogen control is currently being provided by forecasting models, precision farming, nanotechnology, and endotherapy. New biocontrol agents and natural products, disease management through plant nutrition, systemic resistance inducers, and gene-silencing technology will provide solutions for obtaining satisfactory disease control in horticulture. The "multi-stakeholder partnership" strategy can promote the implementation of sustainable crop protection.

Keywords: Green Deal; integrated pest management; biocontrol agents; natural products; models; precision agriculture; nanotechnology; endotherapy; systemic resistance inducers; gene silencing

Citation: Scortichini, M. Sustainable Management of Diseases in Horticulture: Conventional and New Options. *Horticulturae* **2022**, *8*, 517. https://doi.org/10.3390/horticulturae8060517

Academic Editors: Hillary Righini, Roberta Roberti, Stefania Galletti and Harald Scherm

Received: 11 April 2022
Accepted: 8 June 2022
Published: 13 June 2022

Publisher's Note: MDPI stays neutral with regard to jurisdictional claims in published maps and institutional affiliations.

Copyright: © 2022 by the author. Licensee MDPI, Basel, Switzerland. This article is an open access article distributed under the terms and conditions of the Creative Commons Attribution (CC BY) license (https://creativecommons.org/licenses/by/4.0/).

1. Introduction

The concepts that illustrate sustainable agriculture have been posed and defined decades ago and can be summarised by the principles and approaches described by F.A.O. "Building a common vision for sustainable food and agriculture" (https://www.fao.org/3/i3940e/i3940e.pdf, accessed on 22 May 2022) as "an integrated system of plant and animal production practices having a site-specific application that over the long-term will: (a) satisfy human food and fiber needs; (b) enhance environmental quality and the natural resources; (c) make the most efficient use of nonrenewable resources and on-farm resources and integrate natural biological cycles and control; (d) sustain the economic viability of farm operations; (e) enhance the quality of life for farmers and society as a whole".

Considering that the complete achievement of all such goals still requires a relevant effort [1], the success of sustainable agriculture mainly depends on the acceptance of these principles by the farmers, which should actively identify strategies for maintaining, enhancing, and developing their on-site resources (i.e., soil, water, air, biodiversity, and landscape) for future generations [2]. The successful application of such goals can be assessed by indicators that measure the percentage of the agricultural area which satisfies the specified criteria of sustainability regarding water, soil, and biodiversity, and achieving a specific level of productivity [3]. However, the need for a continuously widespread application of sustainability criteria in agriculture with less impact on the environment is also necessary in a world where food demand is increasing.

For the European Union, sustainable agriculture, through the "Farm to Fork" strategy, is one of the main objectives of the European Green Deal (Annex to the European Green

Deal, 2019). The aim of agriculture before the year 2030 should be: (a) at least 25% of agriculture in Europe being organic; (b) reduction by 50% of chemical pesticide utilisation; (c) reduction by 50% of more hazardous pesticide utilisation; (d) reduction by 20% of fertiliser utilisation; (e) reduction of soil nutrient losses by at least 50%. In addition, to diminish the utilisation of copper in agriculture, the executive regulation 2018/1981 of the European Commission reduced the maximum limit of usable copper to 4 kg per hectare, and a maximum of 28 kg per hectare in seven years to minimise the potential accumulation of copper in soil and the exposure of non-target organisms. Similarly, the United Nations 2030 agenda promotes sustainability in agriculture through Sustainable Development Goals (SDG) (https://sdgs.un.org/goals, accessed on 22 May 2022). Within this scenario, there are already examples of communities that, upon a referendum, decided to ban the use of chemical pesticides to protect the local environment and obtain pesticide-free food [4].

It should be noted that the use of traditional pesticides showed a 2% decrease per year owing to the application of regulatory restriction laws, compared to the 15% increase per year in favour of biopesticide utilisation [5]. Within this context, the future of agriculture will be based on environmentally friendly agronomical techniques that, at the same time, can assure a profit to the farmer and the sustainability of the farm itself [6].

The success of obtaining satisfactory pathogen management according to sustainable agriculture principles requires parallel actions to prevent the spread of phytopathogens. From this perspective, effective quarantine measures are necessary to avoid the introduction of destructive plant pathogens into new areas of cultivation. Currently, this aspect is particularly relevant because of the extensive global circulation of plant materials and climate change [7]. Modern diagnostic tools should be implemented at the points of plant material circulation (i.e., airports and ports) and the local entry points (i.e., regional phytosanitary services) [8]. Local quarantine agencies can be assisted by climate-matching tools and geographical information systems that can predict the possibility of pathogen spread in a new area [9].

Many reviews have been published on the different aspects of sustainable agriculture, including basic knowledge on the control of phytopathogens [10–14]. The principles that rule out agro-ecology and organic farming are not discussed. This review attempts to provide a broad overview of sustainable agriculture and integrated pest management (IPM) principles applied to achieve pesticide reduction, with a focus on disease management under the regulatory framework of the European Union. There is a focus on the main strategies based on the utilisation of well-known and new biocontrol agents and products or compounds with a low impact on the environment that are already developed or undergoing achievements regarding the control of some diseases of woody and herbaceous crops. New technologies to augment the efficacy of disease control in sustainable agriculture are also presented and discussed.

A synoptic panel of current control strategies in relation to sustainability principles and policies is shown in Figure 1.

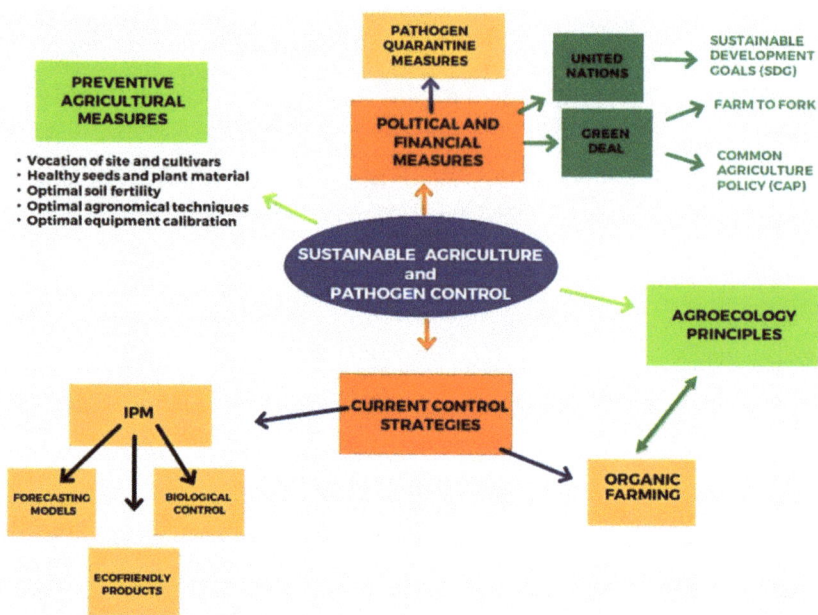

Figure 1. Synoptic panel that shows the current strategies and policies related to the achievement of sustainable disease control in horticulture.

2. The Basis for a Sustainable Disease Control: The Preventive Measures

2.1. Suitability and Selection of the Site and Cultivars

In addition to the economic aspects and infrastructural facilities, the climatic factors characterising an area must be considered for the choice of the crop to be cultivated. At present, this issue is relevant because of climatic changes that affect most areas of the world. Climate change can result in the adoption of different pathogen control strategies and agronomical techniques, owing to the possible adaptation of new pathogens to the new climatic scenario. For example, many areas with a Mediterranean climate that are traditionally free from freezing events either in winter or early spring have recently faced relevant frost damage during such periods, which seriously threatens the economic profit of crops [15,16]. In addition, for other areas such as Central and Eastern Asia, Central–North America, Northern India, Australia, and the Mediterranean Basin, the occurrence of "hot spots" (i.e., temperature > 40 °C for many consecutive days, accompanied by the absence of rainfall) during summer pose a risk to wheat cultivation [17] and can cause severe damage to heat-tolerant crops such as olive [18]. Edaphic (i.e., soil fertility, texture, and porosity) and biotic (i.e., occurrence of bees, pests, and pathogens) factors must also be considered to avoid future problems due to climate change.

In addition to area suitability, the right choice of cultivars is another basic element that can allow the success of the crop according to sustainability criteria. For woody species, the choice is of basic importance and, according to soil characteristics, should also consider the choice of rootstock. The right choice is even more critical, particularly when the crop reaches a new cultivation area [19]. The cultivar choice for herbaceous crops is also important in the context of climate change, as shown by an extensive survey performed in Germany with cereal producers. Farmers judged eco-stability, grain yield performance, and steadiness as being the most important cultivar requirements [20].

2.2. Healthy Seeds and Plant Material

The healthy phytosanitary status of seeds, tubers, plantlets, potted plants, and propagative material is a fundamental prerequisite for initiating cultural cycles. At present, this aspect is particularly important considering the extensive global circulation of these commodities. In recent years, the relevant increase in global plant circulation regarding the agricultural and forest trade has dramatically increased the possibility of pests and pathogens to rapidly reach new countries and, consequently, to colonise and infect new crops and the same crops cultivated on another continent [21]. Once introduced in a new area, phytopathogens can become part of the new environment(s) depending on a series of factors, such as the number of introduction events, the transmission rate of the pathogen, the density and spatial variation of the susceptible host, the favourability of the climatic conditions, the synchronicity between host susceptibility, and the pathogen life cycle [22]. An efficient surveillance system at the border should be developed in each country to rapidly intercept new threats before they can be established in a new territory. This issue is particularly important for countries that have not yet developed a phytosanitary regulation system based on quarantine principles. In contrast, seed companies and plant nurseries should efficiently implement all preventive measures that can reduce the colonisation of plant material (i.e., effective pathogen control strategies during plant growth and disinfection of plant material before shipping). In addition, farmers should carefully monitor crops, particularly during the first phases of growth, to observe and eliminate potential diseases.

2.3. Optimal Soil Fertility and Agronomical Techniques

One of the pillars of the European Common Agricultural Policy (CAP) is the maintenance and enhancement of soil fertility; correct soil management is one of the fundamental prerequisites for sustainability in agriculture [23]. According to agroecological principles, some effective practices can be applied to herbaceous and woody crops to maintain and augment soil fertility. Crop rotation with leguminous species and the planting of cover crops between tree rows are methods that can ensure, over a long-term period, the maintenance of natural soil fertility [24,25]. Crop rotation can also result in a better control of some soil-borne diseases, such as in potatoes affected by *Rhizoctonia solani* and *Streptomyces scabies* [26]. In addition, the application of plant growth-promoting rhizobacteria (PGPR), biofertilisers, composts, mycorrhiza, biochars, and humic and fulvic acids can augment nutrient acquisition and assimilation, improve soil texture and plant growth, and induce systemic resistance to biotic and abiotic stresses [26–32]. For example, the distribution to the soil of a biofertiliser that contained a mixed fungal and bacterial microflora induced conferred protection against Fusarium wilt of banana caused by *Fusarium oxysporum* f. sp. *cubense* after three years [33]. *Paecilomyces variotii*, a fungus obtained from agro-industrial compost, showed efficacy in the control of Fusarium wilt of melon caused by *Fusarium oxysporum* f. sp. *melonis* [34]. However, care should be given to the correct choice of compounds released into the soil to increase their overall fertility. In some circumstances, organic matter, particularly animal manure, can release antibiotics that can perturb native microflora, causing adverse effects on the crop [35]. It should be noted that balanced crop nutrition is an essential component of any integrative program for crop protection [36].

Given that the application of compounds for crop protection aims to deposit the highest amount of the active ingredient on the target plant part (i.e., buds, leaves, and canopy) in which the pathogen resides, an effective and desirable reduction of the spread of any compound in the environment can be achieved by the appropriate calibration of the sprayers. At present, it is possible to adjust the sprayer nozzles to achieve the intended target (i.e., the plant part that shows symptoms of disease), which also reduces water utilisation [37]. Soil solarisation is a well-known technique that, when properly applied, can effectively control important soil-borne pathogens. However, the utilisation of plastic covers poses a relevant concern for their subsequent removal and disposal [38]. Organic farming largely benefits from such preventive measures to obtain effective pathogen control, particularly soilborne pathogens [39].

3. Sustainable Agriculture and Pathogen Control

3.1. The Basis for an Effective Sustainable Pathogen Control

Knowledge of the genomic structure, virulence factors, and epidemiology of pathogens is the basis for developing fine-tuned strategies for the effective control of biotic diseases in crops. Selected biocontrol agents or compounds with potential curative effects should be tested against different strains of pathogens that represent the entire population structure. In addition, the disease cycle of the target pathogen should be fully understood to precisely apply the biocontrol agent/compound before and during plant colonisation or the internal multiplication of the microorganism. Some examples of the basic studies that link crop and pathogen epidemiology to the selection of biocontrol agents and fine-tune the spread of active ingredients for disease control are as follows [40–46]. Knowledge of the pathogen cycle of diseases is also the basis for the development and implementation of disease forecasting [47,48]. Moreover, the sustainability of modern pathogen control should be considered in addition to crop productivity, the ecological function of the crop, and the social acceptance of the strategy [49]. A report that concerns either the effectiveness or the social impact of different strategies to control fire blight of apple, caused by *Erwinia amylovora*, in Switzerland has been prepared. A thorough investigation performed by interviewing experts and a literature data search revealed that biological control performed with *Aureobasidium pullulans* is either effective or widely accepted in rural areas because of its feasibility, durability, low impact on animals, biodiversity, soil and water habitation, low cost, and acceptance by consumers [50]. In this case, the majority of the inhabitants of an area are aware of the importance and efficacy of sustainable agriculture for the maintenance and improvement of their lifestyle and environmental safety.

3.2. Current Control Strategies

Integrated pest management (IPM) is the current strategy that allows for the effective control of many plant pathogens in many cases. According to the European Union Framework Directive on the sustainable use of pesticides (Directive 2009/128/EC), IPM "means careful consideration of all available plant protection methods and subsequent integration of appropriate measures that discourage the development of populations of harmful organisms and keep the use of plant protection products and other forms of intervention to levels that are economically and ecologically justified and reduce or minimise risks to human health and the environment. IPM emphasises the growth of a healthy crop with the least possible disruption to agro-ecosystems and encourages natural pest control mechanisms". In addition, the FAO IPM programme involves three large areas located in Asia, the Near East, and West Africa to improve farming skills, raise the awareness of smallholder farmers of the risks posed by traditional agrochemicals, and promote sustainable agriculture (https://www.fao.org/agriculture/crops/core-themes/theme/pests/ipm/en/, accessed on 22 May 2022).

However, given the diversity and complexity of agricultural scenarios, IPM can differ significantly among countries, and each crop of a definite area of cultivation should apply the IPM criteria according to the local reality and a holistic approach; thus, the combination of control tactics into a planned strategy can provide more effective and sustainable results than the single-tactic approach [51,52]. The development and utilisation of ad hoc web-based platforms illustrating control thresholds, cultural practices that can influence disease attack, pathogen virulence, and fungicide efficacy can help farmers, advisors, and researchers to better plan the control strategy according to a real-time assessment of the environmental conditions of the area [53].

The application of the IPM strategy on a large scale would benefit from ad hoc studies that provide updated information on the current control strategies applied to any crop in an area in order to identify the lack of knowledge in the field, to be resolved through future studies. These systematic maps of knowledge have proven useful in applying IPM to arable crops (i.e., wheat, barley, oat, potato, sugar beet, and oilseed rape) cultivated in large areas in Sweden [54]. A refined IPM strategy which combines all the validated methods for monitoring and reducing the impact of the diseases would allow researchers to control apple scab and apple powdery mildew, caused by *Venturia inaequalis* and *Podosphaera leucotricha*, respectively. This strategy includes disease monitoring and forecasting, ecofriendly fungicides, adequate orchard sanitation, biological control, and insect control through mating disruption. It is comparable with the results obtained by conventional pest management methods [55]. IPM is also the basis for a transition from a chemical to a biological control strategy in Canada regarding greenhouse vegetable crops [56].

In addition, IPM strategies largely benefit from the cultivation of resistant/tolerant cultivars, as observed for the more effective application of biocontrol agents such as *Bacillus mycoides* to a sugar beet cultivar that is more tolerant to *Cercospora beticola*, or *Bacillus subtilis* towards chickpeas infected by *Fusarium oxysporum* f. sp. *ciceris* [57]. In the Netherlands and Ireland, the utilisation of novel potato cultivars resistant to late blight, caused by *Phytophthora infestans*, in combination with real-time pathogen population monitoring and checking of its genetic structure allowed for a reduction of 80–90% of fungicide use (https://www.wur.nl/en/newsarticle/more-sustainable-potato-production-through-extended-ipm-for-late-blight.htm, accessed on 22 May 2022). The protection of crops starting from the seed is another relevant option for better management of diseases. *Trichoderma gamsii*, applied to maize kernels, has been proven effective in reducing pink ear rot caused by *Fusarium verticillioides* and root infection [58].

Despite the higher incidence of pathogens in the crop, the application of organic farming principles for some years can also support, in many circumstances, the effective biological control of pathogens [59]. The observed increase in ecological intensifications of the agro-ecosystem promotes a higher occurrence of beneficial microorganisms in the crop [59]. The synergism between IPM strategies and organic farming principles in relation to pathogen control could provide benefits for improving environmental quality, farm economic viability, and soil and human health [60].

At present, the success of IPM and organic farming in relation to pathogen control is based on three research sectors that are closely related to each other: disease forecasting models, biological control, and environmentally friendly natural products or compounds.

3.2.1. Disease-Forecasting Models

Disease forecasting is based on mechanical models designed with the input of climatic data and the pathogen cycle of disease to alert the grower on whether, when, and how to apply an agrochemical or a biocontrol agent to protect crops. Such models are dynamic because they analyse the changes in the components of an epidemic over time according to external variables (i.e., climatic data, pathogen multiplication, and plant growth stage in relation to disease development) [47] (Figure 2). An effective example of a forecasting model that allows a relevant reduction in pesticide distribution in the environment is vite.net® [61]. Based on a decision support system that calculates vineyard parameters (i.e., air, soil, plant, pathogen, and disease development) and a web-based tool that analyses such data, vite.net® provides information for *Plasmopara viticola* management in the vineyard. The system is flexible and can be tailored to a single vineyard or an area characterised by high similarity. This tool was largely utilised by grape growers on more than 17,000 h in 2017 in Italy, Spain, Portugal, Greece, Romania, and the United Kingdom and allowed for a reduction of approximately 50% in pesticide utilisation [47,61].

Another effective and used disease-forecasting model is Brassica$_{spot}$™ that is applied to manage *Albugo candida*, a causal agent of white blister of *Brassica* crops (i.e., radish and broccoli) in Australia [62]. The application of the model allowed for a reduction of more than 80% of the disease and reduced the number of pesticide sprays from fourteen to one or two per year. In addition, the introduction of a resistant cultivar and the simple change in the time of irrigation from 2000 h to 4000 h also decreased disease incidence [62].

Similarly, in Florida, the Strawberry Advisory System (SAS) based on local weather data allows growers to reduce the use of chemical sprays for controlling anthracnose, caused by *Colletotrichum* spp., and grey mould, caused by *Botrytis cinerea* [63], by 50%. In South Korea, the EPIRICE model was developed to assess the daily risk for the occurrence of rice blast caused by *Magnaporthe oryzae* [48]. The model utilises some climatic data linked to fungus multiplication, such as air relative humidity, temperature, and precipitation, and can be used to predict the risk of disease at an early stage [48].

Figure 2. Scheme of a modern decision support system based on a forecasting model for plant disease management. Information about crop-specific characteristics (**A1**), environmental conditions (**A2**), crop and plant status (**A3**), and agricultural operations (**A4**) flows from the crop to a remote server (**B1**), and it is stored in database (**B2**). This information is then used as an input for running mathematical models and decision algorithms (**B3**), which generate decision supports and alerts to the grower for deciding when and how apply a protective agent (**C1**). (Reproduced from [47]).

3.2.2. Biological Control

Biological control agents for plant diseases are defined as naturally occurring microorganisms capable of suppressing the growth and proliferation of a target pathogen by different mechanisms of action (i.e., competition for space and nutrients, antibiosis, predation, induced host resistance, and lytic enzymes). In addition to living microorganisms, they sometimes utilise metabolite(s) that can be sprayed directly onto crops [64]. Beneficial microorganisms are registered as plant protection products, and they are usually applied to crops at a high density once or several times during the growing season. In the United States and Canada, government agencies are responsible for confirming the biosafety of the biocontrol agents (i.e., the Environmental Protection Agency (EPA) in the United States, and the Pest Management Regulatory Agency (ARLA) in Canada). In Europe, according to Regulation 1107/2009 for plant protection products, the authorisation for commercialising biocontrol agents is obtained through some related steps: (a) the bioactive microorganism should be approved at the European level by the European Food Safety Authority (EFSA) according to the physiochemical properties of the substance, its risk profile for human health, and its risk profile for the environment; (b) formulated products should be authorised at the member state level; (c) further scrutiny with regard to organic agriculture requirements. In addition, the Directorate General for Health and Food Safety (DG SANTE), the Standing Committee on Plants, Animal, Food, and Feed (PAFF Committee), and the Rapporteur Member State are involved in the decision according to Directive 91/414, which states that any active substance should be included in an approved EU list (Annex 1), and its further application must be authorised by member states. These procedures take a long time, thus creating an overall slowness for final approval [65].

The causal agents of plant diseases can have either a worldwide occurrence in specific crops or local distribution. For the first case, the application of the Nagoya protocol of October 2014, for the "Access and the Fair and Equitable Benefit-sharing of Genetic Resources" that restricts the international exchange of biological material, can result in a limitation on the circulation and use of biological control agents that are selected in a different geographical area [66]. Consequently, given the increase in organic food demand and the current Green Deal policy, the selection of native local biocontrol agents is of paramount importance [11]. It should be noted that this selection largely depends on the specific pathosystem under study. Two approaches illustrate how it is possible to proceed: (a) selection through consecutive screenings for testing the effectiveness of the biocontrol agent (i.e., in vitro assays for antibiosis, lytic enzymes, and antimicrobial metabolites, in planta assays for colonisation, control performance, and induced resistance); (b) selection through the assistance of genetic/genomic studies (i.e., the use of genetic markers for finding single biocontrol traits, genome-wide DNA markers for selecting complex traits) [11,67]. The durability of biocontrol agents is another important prerequisite for their long-standing efficacy. Indeed, there are documented reports of a significant reduction in the control effectiveness of *Botrytis cinerea* in *Astilbe* hybrids, as shown by *Pseudomonas chlororaphis* after eight treatments [68].

In some circumstances, natural selection yields biocontrol agents that are capable of displaying long-term positive effects on some diseases, such as for soybean root rot caused by *Fusarium* spp., *Pythium* spp., and *Rhizoctonia solani* in Northeast China. In this case, a naturally occurring suppressive soil was analysed, and *Trichoderma harzianum*, *Pochonia clamydosporia*, *Paecilomyces lilacinus*, and *Pseudomonas fluorescens* strains have been found to inhibit the fungal pathogens of soybean roots [69]. Another well-known example is the natural occurrence of mycoviruses of the family *Hypoviridae*, which infect *Cryphonectria parasitica*, the causal agent of chestnut blight. Upon infection, the virus incites hypovirulence in the fungus by reducing its parasitic growth and sporulation capacity. The virus can be isolated from chestnut cankers and utilised as a biocontrol agent to cure trees by inhibiting further canker development. In addition, the virus is capable of spreading naturally in the forest and reaching other infected trees [70].

Fungi, bacteria, and yeast are the most widely used biocontrol agents. The following species are among the most versatile: *Trichoderma harzianum*, *Trichoderma viride*, *Bacillus subtilis*, and *Pseudomonas fluorescens*. All of them, indeed, have shown control activity towards some common fungal pathogens such as *Botrytis cinerea*, *Monilinia fructicola*, *Plasmopara viticola*, *Puccinia graminis*, and *Erisiphe* spp. [71]. *Rhizobium* (*Agrobacterium*) *radiobacter* strain K84 is among the most known biocontrol agents used for many years to control crown gall caused by *Agrobacterium tumefaciens* [72]. A stand-alone effective biocontrol agent has been also selected for apple scab caused by *Venturia inaequalis*, namely *Cladosporium cladosporioides* H39, that, over a wide range of environment, showed a high control level and appears effective even when applied after some days from the infection event [73]. Similarly, *Bacillus amyloliquefaciens* strains are capable of effectively controlling *Fusarium equiseti* in broad bean cultivation [74]. In some circumstances, a single biocontrol agent is capable of effectively reducing two diseases caused by distantly related microorganisms, as was seen for *Paecilomyces variotii*, which showed control activity towards both the bacterium *Xanthomonas vesicatoria*, the causal agent of bacterial spot of tomato, and *Fusarium oxysporum* f. sp. *melonis*, the causal agent of Fusarium wilt of melon [34].

Trichoderma spp. are among the most studied biocontrol agents in agriculture. Through different metabolic pathways, the induction of systemic resistance to the plant, multiple adaptive mechanisms, antimicrobial molecules, and antagonistic behaviour, *Trichoderma* strains can either promote plant growth or act as effective biocontrol agents against fungal species under numerous agricultural conditions, including in greenhouses and nurseries [75,76]. These strains account for the greatest proportion of fungal biocontrol agents against the phytopathogenic microorganisms investigated, and many commercial formulations that contain a single *Trichoderma* strain or a mixture of different *Trichoderma* strains are available [75,77]. A series of common and widespread phytopathogenic fungi, such as *Rhizoctonia solani*, *Fusarium oxysporum*, *Botrytis cinerea*, *Pythium* spp., *Sclerotinia* spp., *Verticillium* spp., *Phytophthora* spp., and *Alternaria* spp. can be controlled by generalist *Trichoderma* spp. strains [75,78]. Other fungal genera that also show antagonistic activity toward phytopathogenic fungi are *Alternaria*, *Aspergillus*, *Penicillium*, *Pichia*, *Candida*, *Talaromyces*, and nonpathogenic *Fusarium*, *Pythium*, and *Verticillium* [79]. *Pichia anomala* is effective in controlling postharvest crown rot of banana caused by *Colletotrichum musae*, *Fusarium verticillioides*, and *Lasidiodiplodia theobromae* [80]. In addition, *Ampelomyces quisqualis* is commercially available for the preventive control of powdery mildew fungi in different crops, such as eggplant, cucumber, tomato, and strawberry.

Pseudomonad strains provide a very large amount of potential biocontrol agents that can be found within the following species and species complexes: *Pseudomonas fluorescens*, *P. chlororaphis*, *P. putida*, *P. syringae*, *P. aureofaciens*, *P. protegens*, *P. mandelii*, *P. corrugata*, *P. koreensis*, and *P. gessardii* [81]. However, very few commercially available products are currently available: *P. fluorescens* for *Erwinia amylovora*; nonpathogenic *Pseudomonas syringae* for postharvest disease of fruits, potato, and sweet potato; *Pseudomonas chlororaphis* for fungal diseases of ornamental crops and turf grass; *Pseudomonas aureofaciens* for lawn and grass management against soil-borne fungi. Genetic instability and poor shelf life are among the main causes of the limited registration of pseudomonads as biocontrol agents [81]. *Bacillus* species have a higher shelf life due to their possibility to form endospores, and have many potential biocontrol agents, stemming from their ample antagonism mechanisms (i.e., antibiosis, enzymes, lipopeptides, competition for space, and nutrients), are also present in this genus, with *Bacillus subtilis*, *B. amyloliquefaciens*, and *B. polymixa* being the richest in providing biocontrol effectiveness [82].

Among the yeasts, five species are most widely used as biocontrol agents, namely *Candida oleophila*, *Metschnikowia fructicola*, *Cryptococcus albidus*, *Saccharomyces cerevisiae*, and the yeast-like fungus *Aureobasidium pullulans*. The low production of toxic secondary metabolites by yeast is an important prerequisite that raises fewer biosafety concerns and increases their possibility of utilisation [83]. *Candida oleophila* is utilised as a postharvest biocontrol agent for banana, apple, and citrus, towards *Colletotrichum musae*, *Botrytis cinerea*, and *Penicillium* spp., whereas *Aureobasidium pullulans* is utilised as a biocontrol agent towards fire blight of apple and pear, caused by *Erwinia amylovora*, and for apple postharvest diseases [84]. Moreover, a satisfactory control has been also achieved in strawberry cultivated in greenhouses towards *Phytophthtora cactorum* and *Botrytis cinerea*, the causal agents of crown and root rot and grey mould, respectively [85].

In some cases, a combination of more biocontrol agents provides better control than the application of a single agent. The sprays of *Bacillus mycoides* and the yeast *Pichia guillermondii* on strawberry leaves provided a better performance for controlling *Botrytis cinerea* than the application of a single biocontrol agent [86]. In addition, to induce stimuli for the plant growth, plant growth-promoting rhizobacteria (i.e., *Bacillus subtilis* and *Curtobacterium flaccumfaciens*) can also act as effective biocontrol agents as the case of a multiple effectiveness against three cucumber pathogens, namely *Colletotrichum orbiculare*, *Pseudomonas syringae* pv. *Lachrymans*, and *Erwinia tracheiphila* [87]. However, the compatibility between different control agents should always be assessed because there are also cases of reduced control activity when two agents are applied without a prior compatibility assessment [88,89]. In the case of proven compatibility, the two biocontrol agents can provide satisfactory disease control activity through a multitrophic approach [90]. The addition of specific substances that promote the growth of biocontrol agents on the plant's surface can also enhance the performance of the biocontrol agents, as in the case of lactic acid, which is used as a biocontrol agent for *Erwinia amylovora* [91].

3.2.3. Natural Products and Compounds

A natural product with a potential use for controlling plant diseases can be defined as a physiologically active chemical that is synthesised by plants, microorganisms, or animals. These products can act as antimicrobials or inducers of systemic resistance, are usually easily biodegradable, and do not persist in the environment [92]. Some of them can act as templates for chemical pesticides (i.e., synthetic analogues), such as the fungicide strobilurin, which was named with reference to *Strobilurus tenacellus*, a wood-rotting mushroom. Chitosan and its derivatives, alkaloids, flavonoids, terpenes, proteins, and phenolic compounds are the most widely studied and used dried materials [93]. These products are the result of their coevolution with the biotic environment; thus, many of them are defence compounds toward other organisms and can possess potent bioactivities and selectivity [94]. Within this category could be included also antibiotics. However, in many countries, these products are banned for their utilisation in agriculture as protectants because of the possibility of transmission of genetic traits from bacterial phytopathogens to microbes of human and animal importance, which could confer resistance to a single or multiple antibiotics [95,96]. However, these are not discussed in this paper.

Harpin, a proteinaceous elicitor of the self-protective hypersensitivity reaction in plants against pathogens, obtained from the pathogen *Erwinia amylovora* [97], was among the first natural products to be commercially utilised in plant protection. Protection obtained through harpin was observed for harvested apples infected by blue mould, which is caused by *Penicillium expansum* [98]. Currently, it is included among the biostimulants. Another well-known natural fungicide was obtained from *Reynoutria sachalinensis* to control powdery mildew of cucumber and tomato in glasshouse cultivations, as well as downy mildew of grapevine [99,100]. The extract induces plant defence responses such as callose papillae and an increase in salicylic acid and caffeic acid [101]. Chitosan is a derivative polymer of chitin, the primary component of the cell wall of fungi and the exoskeleton of insects and crustaceans, and is usually extracted from shrimp shells, mud crabs, and fungi for utilisation in medical and chemical science, as well as in agriculture. Chitosan-based compounds are traditionally utilised postharvest to protect fruit decay [102]. Due to the very high economic losses in crops caused by phytopathogens that show pesticide resistance [103], the discovery, validation, and registration of new natural products that are characterised by different and novel modes of action are required, and DNA sequence technologies can help in identifying gene(s) or clusters of genes of potential interest for pathogen control [94].

4. Developing Control Strategies

A synoptic panel concerning developing control strategies is shown in Figure 3.

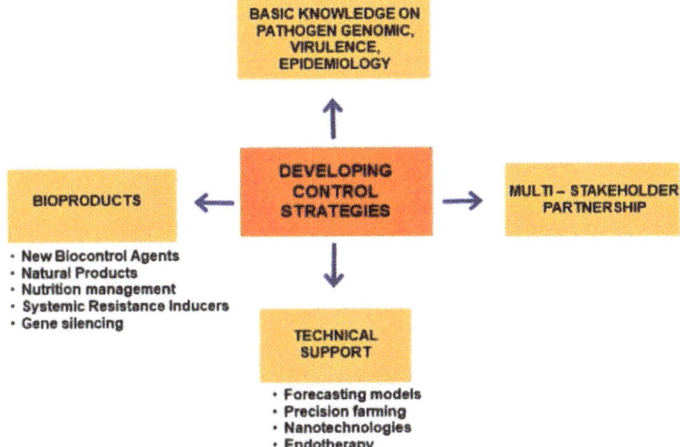

Figure 3. Synoptic panel that illustrates the developing disease control strategies for sustainable agriculture.

4.1. Technical Support

4.1.1. Precise Timing Decision for Pathogen Control

Modern forecasting models for plant disease should be characterised by (a) flexibility and accuracy; (b) differential interactions between the analysed data and statistical representation; (c) different thresholds of infection risks in relationships with possible mixed infections and changes in crop susceptibility during the season; (d) modelling the disease dynamics, taking into account the long-distance interactions of biological parameters (e.g., spore dispersal through wind and deposition via rainfall events); (e) defining the precise timing for applying biopesticides, taking into account the different scales of climatic and epidemiological data [104]. Knowledge of the pathogenic genetic structure is also important for including such variability into the model according to a fine-tuned pathogen disease cycle. In addition, the assessment of field infection over consecutive years can also be used to calibrate and compare the current-year prediction range [104]. Because some models of the past are hindered by sparse spatial data and a limited use of field monitoring technology, predictive models for pathogen infection based on weather forecasting can now largely benefit from recent advances in the large-scale monitoring of the space provided by satellites [104]. Consequently, a further improvement could be obtained by incorporating data besides the meteorological parameters, obtained by using remote sensing, into the model, such as the leaf reflectance of the red band obtained during the crop growing season [105]. In some circumstances, remote sensing analyses allow for the detection of pre-symptomatic outbreaks and distinguish symptoms caused by different causal agents, such as *Phytophthora infestans* and *Alternaria solani* [106]. The final validation of the models in the field and their utilisation by farmers is considered a mandatory step for including the model in an effective IPM strategy [12], and the direct involvement of farmers (i.e., the end-users) in developing such tools is important [107].

Climate change projections should also be considered, so that modern forecasting models also aim to predict possible scenarios of pathogen outbreaks and provide a fine-tuned timing for applying preventive and curative biocontrol agents, or natural products, for plant protection [108]. Through different global circulation models, it is now possible to generate climatic projections of diseases in different geographical areas. Several key plant pathogens, including *Fusarium* spp., *Puccinia recondita*, *Pyrenophora teres*, *Magnaporthe oryzae*, *Plasmopara viticola*, *Phytophthora infestans*, and *Xanthomonas oryzae* pv. *oryzae* have been assessed for their potential impact on climate change [108]. These models can also predict the impact of a pathogen outbreak at the landscape level when introducing a new crop adapted to the changing climate or in the case of a widely distributed crop facing a new pathogen introduced from abroad [108].

4.1.2. Precision Farming and Pathogen Control

Strictly connected with the forecasting models, precision farming can provide methods that can be utilised at a single-farm level in order to predict pathogen outbreaks in real-time and site-specifically. The geographical information system (GIS), in combination with on-site monitoring platforms for continuously assessing pedo-climatic data, are the basis for providing alerts to the farmers in relation to the pathogen outbreak [109]. A so-called phenomic approach can assist to establish the precise time for performing the preventive or curative treatment to the crop. The measurements of crop traits such as growth and performance during the season obtained through noninvasive techniques can detect pre-symptomatic events of disease-related changes in the crop that are not visually apparent [110]. In these cases, the "Internet-of-Things" monitoring platform utilises an artificial intelligence algorithm for emulating the decision-making ability of humans regarding the choice of the precise timing for controlling the pathogen [111]. An agro-weather station is installed in the field for measuring many climatic data and additional data such as leaf wetness, soil temperature and water content, and solar radiation in order to detect the early stages of the pathogen infection and to provide an alert to the farmer for applying control measures [111]. Another advantage of precision farming for pathogen control is the management of the disease according to its possible occurrence only in some area of the farm, this resulting in a reduction of pesticide distribution. The monitoring platforms can also detect soilborne diseases by recording the changes in the foliage characteristics [109]. An on-going application for monitoring disease spread both on a large scale and at the single-farm level is the agricultural research outcomes system (AgCROS), developed in Florida to manage the citrus greening disease of citrus crops, caused by the non-cultivable bacterium Candidatus *Liberibacter asiaticus*. AgCROS is a GIS-based monitoring platform for sharing research data between farmers and researchers and to provide decisions for a better management of the disease [112]. One of the most important outcomes of the platform is the early detection of trees infected by the pathogen. This result can be achieved by analysing, in each farm, some plant physiological indexes such as NDVI and NDRE, shown by the trees during the season and captured by remote sensing through unmanned aerial vehicles (UAV) [112]. Moreover, to reduce the cost of the analyses for large areas, data obtained from satellite can be also utilised [112].

Another application of precision farming coupled with remote sensing data analyses is currently performed to control cotton root rot, caused by the soilborne fungus *Phymatotrichopsis omnivora*, in Texas [113]. Field multispectral images are acquired though satellite sensors, and field maps are created to monitor the crop. A classification map distinguishes infected field parts from the healthy ones based on different colors pointing out the diseased plants. Consequently, different rates of fungicides and fertilisers are applied to any single farm according to the precise occurrence of diseased plants in a specific area, thus allowing for a relevant reduction of pesticide spread [113].

4.1.3. Nanotechnology

Nanomaterials released within a plant through nanotechnology can provide a relevant contribution to the reduction of agrochemicals spread in the environment. However, we must first assess the interactions of the nanoparticles with the plant tissue [114], as well as the bioavailability and durability of nanoparticles [115], their potential ecotoxicological risks and accumulation in food [116], and the relative costs for their application [116]. In addition, nano-biopesticides should have refined technical properties such as a high solubility of low-solubility active ingredients, a slow targeted release, and the non-premature degradation of active ingredients [117]. Additionally, some traditional pesticides have been re-formulated by developing nanoparticles to reduce the dispersal of the active ingredients in the environment [116]. Among the preventive measures for pathogen control to be applied in sustainable agriculture, copper and zinc-based nanoparticles could be used as seed coating substances or in foliar applications to improve the overall growth of plantlets [118]. The nanomaterials that show potential for application as nanopesticides include silver, copper, zinc, carbon, magnesium, manganese, silicon, calcium carbonate, and chitosan [114]. Bioactive products can also be encapsulated within biodegradable nanomaterials or, alternatively, loaded with plant extracts possessing antimicrobial activities. These nanocarriers can release the active product selectively to the plant. Among these nanocarriers, chitosan oligomers and methacrylated lignin could be effectively utilised [119]. PGPR strains could also be utilised as nano-fertilisers and/or inducers of systemic resistance to phytopathogens [120].

Greenhouse studies have shown the relevant activity of copper-based nanoparticles towards *Fusarium oxysporum* f. sp. *niveum*, the causal agent of watermelon root rot [121]. Similarly, sprayed copper-based nanoparticles showed relevant antifungal activities in eggplant grown in a glasshouse on a soilless medium and infected with *Verticillium dahliae*, whereas copper, zinc, and manganese nanoparticles significantly reduced the pathogenic activity of *Fusarium oxysporum* f. sp. *lycopersici* in tomato grown on a soilless medium [122]. Nanoparticles obtained from crude extracts of *Chaetomium cochlioides*, a fungus that possesses antimicrobial compounds, reduced the severity of rice blast caused by *Magnaporthe oryzae* by 60% [123]. It should be stressed that, to date, most applications of nanoparticles have been restricted to basic laboratory studies and some field applications [114].

Due to the low degradability of nanoparticles in the natural environment [49], the full development of this sector still requires in-depth studies on the impacts of the different active ingredients used as nanoparticles to determine the long-term effects of nanoparticles on the environment and their residual content in food. The field applications of silver nanoparticles retained as effective antimicrobials are the most striking example of this cautious approach [124]. The need to formulate precise guidelines for nanoparticle utilisation in the field is also required [114,125]. Notably, however, chitosan-derived nanoparticles offer a good avenue for assessing the biosafety risks, since these nanoparticles are created from biodegradable bioactive polymers, are not toxic to humans and animals, and offer good antimicrobial activity towards both fungi and bacteria [117,126].

Antimicrobial peptides obtained from natural compounds or that are designed and synthetically obtained de novo are small peptides that show the potential capacity for controlling plant diseases, as replacements for traditional pesticides once we have verified their impacts on the ecosystem (i.e., the epiphytic microbiota of leaves and fruit) [127]. A current limitation of the wide utilisation of such peptides is the high cost of field treatments. These peptides offer the possibility to control plant diseases by either inactivating the pathogenicity of the target phytopathogen or inciting plant defense mechanisms in a multitarget approach. This possibility was verified by applying a bifunctional synthetic peptide to tomato plants and observing its control capacity towards *Pseudomonas syringae* pv. *tomato*, the causal agent of tomato bacterial speck; *Xanthomonas campestris* pv. *vesicatoria*, the causal agent of tomato bacterial spot; *Botrytis cinerea*, the causal agent of tomato grey mould [128].

4.1.4. Endotherapy

The release of specific products within the xylem tissue of trees through trunk injection or trunk infusion is another technique that could allow for a notable reduction of agrochemical dispersal in the environment, reduce the risk of toxicity for farmers, and possibly reduce the overall cost of the protective treatment [129]. The products to be released through endotherapy should be preliminarily verified for their suitability in apoplastic transport within the plant, for the absence of phytotoxic effects [130], and to determine the most suitable plant port for effective injection [131]. There is also a need to reduce the trunk wounds that are caused by devices utilised for the injection [132–134]. Apple trees infected with *Erwinia amylovora*, the causal agent of fire blight, and injected with products that incite the induction of systemic acquired resistance, such as acibenzolar-S-methyl and potassium phosphite, were found to significantly reduce both blossom and shoot blight. This injection also induced the expression of some proteins related to plant defense [133]. Similarly, a study in Apulia (Italy) on evaluating the possibility of controlling *Xylella fastidiosa* subsp. *pauca*, the causal agent of olive quick decline syndrome, showed that the trunk injection of a curative biocomplex containing zinc, copper, and citric acid incited an increase in some plant defense-related metabolites, such as oleuropein and polyphenols, as well as a simultaneous decrease in other metabolites related to the disease, such as quinic acid and mannitol [135]. Endotherapy performed with glutaraldehyde on grapevine cuttings provided a satisfactory control of *Phaeoacremonium minimum*, one of the fungi involved in the esca disease complex [136]. In addition, endotherapy performed on grapevine plants showing symptoms of esca disease complex using chitosan oligomers as nanocarriers loaded with the extracts of some plants offering antimicrobial properties, such as *Rubia tinctorum*, *Equisetum arvensis*, *Urtica dioica*, and *Silybum marianum*, yielded a significant reduction of foliar symptoms [119].

4.2. New Bioproducts and Sustainability

4.2.1. Biocontrol Agents

The potential use of biocontrol agents in sustainable plant protection is currently limited by factors such as the fragmentation of biocontrol sub-disciplines and crop site-specific factors; unwieldy regulatory processes; increasingly bureaucratic barriers to access biocontrol agents; insufficient engagement and communication with the public, stakeholders, growers, and politicians in regards to the considerable economic benefits of biocontrol; relatively high costs [137,138]. For biosafety reasons, any microorganism released into the environment should be carefully assessed via the post-release field monitoring of putative negative environmental impacts, and well-designed ecological monitoring programs will provide data that can help regulators [139]. The cost-efficiency of the product plays a relevant role in farmer choice, and considerable attention should be devoted by the selling companies to the methods of formulation, storage, and delivery [32]. Additionally, the cost of registration for the biocontrol agents is very high, and some companies register their bioproducts as biofertilisers [140].

Apart from bureaucratic and regulatory barriers, some technical aspects for identifying potential biocontrol agents could also be improved, and the need for new high-throughput screening systems was previously suggested [141]. In particular, marker-based screening that can test, in vitro, many enzymes and/or metabolites linked to the antagonistic activity of the microorganism(s) and perform genomic-based searches for genetic traits that show antagonistic activity should be applied to reveal potential beneficial strains to be further assessed through in planta tests [11,141].

One branch of development for obtaining biocontrol agents is based on bacteriophages and phages, the viruses of bacteria. Apart from their stability in the environment [142], one of the most critical points for the utilisation of biocontrol agents is the definition of their host range. This aspect is critical since the potential dispersal of such agents into the environment should not create problems for beneficial bacterial microflora [143,144]. Moreover, phages should not interfere with the genetic material of the target plant by r eleasing traits through transduction that could be incorporated into the plant cell [143]. There are many studies demonstrating the in vitro effectiveness of phages and phage cocktails in infecting and killing plant pathogenic bacteria, even though relatively few strains have reached the commercial phase [142–144]. This lack of commercial applications could be due to the intrinsic difficulty in establishing a direct correlation between laboratory and field conditions [145].

In the U.S.A., *Xanthomonas vesicatoria*, the causal agent of bacterial spot of tomato and pepper, *Pseudomonas syringae* pv. *tomato*, the causal agent of bacterial speck of tomato, *Erwinia amylovora*, the causal agent of fire blight of apple and pear, and *Xylella fastidiosa* subsp. *fastidiosa*, the causal agent of Pierce's disease of grapevine are examples of bacterial pathogens that can be controlled or mitigated using commercially available products based on phages. In Europe, *Enterobacteriaceae* (*Pectobacterium*, *Dickeya*), which affects potato during post-harvest, and *Erwinia amylovora*, which is under strict regulation during spring, are the sole pathogens for which commercially distributed phages are allowed [143,144]. Soil-borne pathogens and seeds are other targets for biocontrol through phage utilisation [143,144]. In addition, phage-derived proteins, such as endolysin for Gram-positive species, can be exploited as potential antimicrobial agents [146]. In the future, precision farming through sensor-based technology could effectively assist and improve upon the field utilisation of phages. The detection of early stages of the disease coupled with the best time to avoid leaf desiccation and U.V. stress could facilitate the success of phage activity in the field [144].

Usually, the rhizosphere and phyllosphere are the most commonly exploited resources for isolating potential bacterial strains for biological control. Plant-growth-promoting rhizobacteria are well-known active agents that enhance the plant defense system, thus conferring a general healthy status to the crop. *Acinetobacter*, *Arthrobacter*, *Azospirillum*, *Bacillus*, *Bradyrhizobium*, *Rhizobium*, *Sphingomonas*, *Serratia*, and the nonpathogenic *Agrobacterium*, *Burkholderia*, and *Pseudomonas* are the most common isolated genera that show growth-promoting activities [147]. *Bacillus* spp. and *Pseudomonas* spp. isolated from the rhizosphere and/or soil are among the bacterial genera most widely utilized as biocontrol agents for soil-borne fungal and bacterial phytopathogens, whereas the genus *Streptomyces* yielded a relevant number of strains with potential as biocontrol agents, especially towards fungal species [148]. In the phyllosphere, *Pseudomonas*, *Bacillus*, and *Pantoea* are the predominant genera that can be isolated as potential biocontrol agents [149]. Preliminary screening to reveal the occurrence of antimicrobial compounds and antagonistic activities (e.g., hydrolytic enzymes, ammonia, and antibiosis) related to phytopathogen biocontrol and the promotion of plant growth should be performed on a large collection of isolates to highlight the most promising strains [150]. Other important features of successful biocontrol agents include niche-adaptability, competition for nutrients, and colonisation ability [151,152]. Moreover, some commercial products based on strains of the aforementioned bacterial genera are currently used as biocontrol agents for crop protection, including soil-borne diseases.

For example, *Pantoea vagans* and *Pseudomonas fluorescens* provide many strains with antagonistic activities to both phytopathogenic bacterial and fungal species. Some strains of both species show control capacity for *Erwinia amylovora*, the causal agent of fire blight of apple and pear [153]. Additionally, commercial compounds have been developed as potential substitutes for streptomycin application in disease control. *Pseudomonas fluorescens* is among the most versatile bacterial species and can act as either a plant-growth-promoting or biocontrol agent for plant diseases, especially soil-borne diseases [154]. Similarly, *Pseudomonas chlororaphis* strains have been registered in commercial formulations to control leaf blight and grey mould in tomato, which are caused by *Alternaria alternata* and *Botrytis cinerea*, respectively [155].

Bacillus subtilis, *Bacillus amyloliquefaciens*, and *Bacillus pumilus* are largely employed as biocontrol agents for citrus diseases both in the field and during storage—namely, *Colletothricum acutatum* and *Colletothricum gloeosporioides*, the causal agents of anthracnose; *Alternaria citri*, the causal agent of black rot; *Phytophthora citrophthora*, the causal agent of root rot; *Plenodomus thacheiphilus*, the causal agent of mal secco; *Penicillium digitatum*, *Penicillium italicum*, and *Geotrichum candidum*, which are involved in fruit decay during storage [156,157]. To improve the effectiveness of these biocontrol agents during fruit storage, additional compounds or physical treatments (i.e., copper hydroxide, sodium carbonate, tea saponins, and hot water) are applied to provide more integrated and effective management [156]. The salt-tolerant *Bacillus velezensis* showed significant biocontrol activity towards Verticillium wilt of olive [158] (Figure 4).

Diseased non-treated control   **Diseased *B. velezensis* XT1-treated**

Figure 4. Olive Verticillium wilt, caused by *Verticillim dahliae*, treated with *Bacillus velezensis*: olive trees in the field experiment before and after the treatment with *B. velezensis* strain XT1. (Reproduced from [158]).

Bacillus strains could also be effectively used in mixed compounds to exploit the synergistic effects of their diverse antagonistic modes of action towards target pathogenic bacteria and fungi [159]. Potato is another crop that largely benefits from the utilisation of biocontrol agents to prevent tuber soft rotting caused by *Pectobacterium* spp., *Dickeya* spp., and *Clostridium* spp. [160]. The overall strategy includes soaking potato tubers prior to sowing in a solution containing *Pseudomonas fluorescens* strains to prevent future colonisation by pathogens. Similarly, the soil can also be inoculated with antagonistic bacteria such as *Bacillus*, *Pantoea*, *Pseudomonas*, and *Streptomyces* to colonize the root zone where phytopathogens usually start plant colonisation.

Finally, cocktails of biocontrol agents in the form of a solution or powder are used to protect harvested tubers during storage. In this case, the tubers are soaked in the solution prior to storage [160]. A significant reduction in economic losses caused by rice blast incited by *Magnaporthe oryzae* is expected to be obtained through the application of *Streptomyces* strains as biocontrol agents [161,162]. Some lactic acid bacteria, which are usually used as bioprotective agents for foodborne pathogens and spoilage microorganisms, have also shown good potential to act as biocontrol agents for plant pathogens. Due to its antimicrobial metabolites, *Lactobacillus plantarum* shows broad-spectrum antagonism towards some bacterial phytopathogens such as *Pseudomonas syringae* pv. *actinidiae*, the causal agent of kiwifruit bacterial canker; *Xanthomonas arboricola* pv. *pruni*, the causal agent of bacterial spot of stone fruit, and *Xanthomonas fragariae*, the causal agent of angular leaf spot of strawberry [163] (Figure 5). A consortium of endophytic lactobacilli composed of *Weissella cibaria* and *Lactococcus lactis* enabled the control of papaya dieback caused by *Erwinia mallotivora* [164].

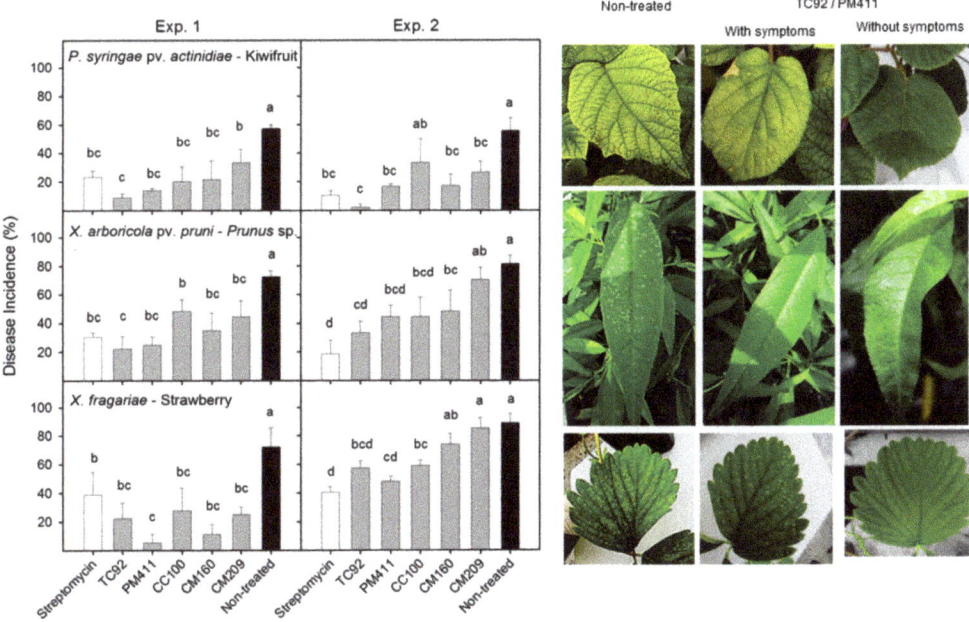

Figure 5. Effect of the treatments with lactobacilli strains (grey bars) on *Pseudomonas syringae* pv. *actinidiae*, *Xanthomonas arboricola* pv. *pruni*, and *Xanthomonas fragariae* infections in kiwifruit, Prunus, and strawberry plants, respectively, under greenhouse conditions. The effect of strains on disease incidence (%) was compared with streptomycin (white bars) and a non-treated control (black bars). Different letters indicate statistical significance (Reproduced from [163]).

The versatile behavior of *Trichoderma* spp. can also be exploited to provide effective control through the colonisation of additional niches besides those that are usually targeted, such as in the case of *T. gamsii*, which is commonly utilised to protect wheat spikes from *Fusarium graminearum*, the causal agent of Fusarium head blight of wheat. The addition of fungus in the soil or during sowing can enhance the overall biocontrol activity [165]. Similarly, the addition of *Trichoderma polysporum* spores to seeds in combination with liquid compost rich in organic matter supplied in fertigation during plant growth enhanced overall bioactivity towards melon wilt caused by *Fusarium oxysporum* f. sp. *melonis* under semiarid conditions [166]. A combination of *Trichoderma* strains with nanoparticles could also provide augmented control effectiveness [167].

The search for new beneficial traits within *Trichoderma* spp. can now be assisted through genomic prediction tools that reveal the occurrence of a single gene or cluster of genes that produce bioactive product(s) either *per se* or under some stimuli [168,169]. Several predicted terpenes and phytotoxins with potential bioactive behavior have been found in the genome by genomically screening publicly available *Trichoderma* genomes [169]. A reduction in production costs is also an important issue to resolve for this important sector of biocontrol agents, and new types of formulations exploiting low-cost substrates for producing pure fungal spores should be implemented [75].

Consortia of bacteria and fungi, so-called multi-strain biological control agents, or microbial synthetic communities, can also be utilised for the management of soil-borne diseases caused by bacteria, fungi, and oomycetes [170] or key fruit pathogens, such *Botrytis cinerea*, by developing niche-specific microbial interactions [171]. A microbial consortium based on *Trichoderma atroviride*, *Aureobasidium pullulans*, and *Bacillus subtilis* that colonises different ecological niches of the bunch during the season was proven to be effective in significantly reducing the activity of *B. cinerea*, the causal agent of grey mould of grapevine [171]. According to this strategy, *Trichoderma atroviride*, a good coloniser of dead plant tissue, provides protection at the bunch closure stage. Additionally, *Aureobasidium pullulans* can compete for sugar utilisation, with the fungus occurring on cracks in berries, whereas *Bacillus subtilis*, which produces antagonistic metabolites, should be sprayed close to harvest time [171]. Commercially available formulations contain *Trichoderma* strains and other bioactive species such as *Pseudomonas fluorescens*, *Pseudomonas aureofaciens*, *Bacillus* spp., and *Streptomyces* spp. [75]. Consortia of different *Streptomyces* strains showed the potential capability to reduce the severity of some diseases caused by *Fusarium oxysporum* in vegetables [172].

Non-harmful endophytic microorganisms are another source of potential biocontrol agents currently being exploited. Generally, these microbes live within plant tissues and provide useful metabolites for plant growth and tolerance to stress, including plant pathogens [13,173]. Volatile organic compounds could also be used to control plant diseases [174]. The genus *Trichoderma* also includes endophytic species that can be exploited as mycoparasites, as in the case of coffee leaf rust caused by *Hemileia vastatrix* [175]. Other fungal genera that have shown potential beneficial activities towards fungal phytopathogens include *Heteroconium*; *Ramularia*; *Xylaria*; *Candida*; and nonpathogenic strains of *Fusarium oxysporum*, *Cladosporium*, *Colletotrichum*, *Alternaria*, *Phoma*, *Pestalotiopsis*, and *Botryosphaeria* [13,173]. The genus *Streptomyces* is another source of beneficial endophytes that show antagonistic activities towards soilborne fungi such as *Fusarium*, *Pythium*, *Verticillium*, *Rhizoctonia*, and *Phytophthora* (directly to the phytopathogens or mediated through their metabolites). Some commercial products are also available [176]. *Bacillus*, *Paenibacillus* spp., and *Rhodococcus* are other good candidates for obtaining biocontrol agents [177].

4.2.2. Natural Products

Natural products that show biocidal activities are commonly obtained from plant or animal extracts or microbial metabolites. Essential oils, chitosan, some plant extracts (i.e., *Yucca schidigera*, *Equisetum arvense*, *Punica granatum*, *Allium cepa*, *Urtica dioica*, and

Camellia sinenis), alkaloids, and bacterial lipopeptides are currently being studied, and some are already available as commercial biopesticides [178]. Seaweeds represent another potential source for obtaining biopesticides [179], especially when utilised in a well-planned IPM strategy [180]. Essential oils (i.e., mixtures of terpenes hydrocarbons, alcohols, and phenols) extracted from many plant species have been studied in detail as products with potential biocidal activities towards phytopathogenic fungi, oomycetes, and bacteria, without inciting pathogen resistance [181,182]. However, the stability and persistence of essential oils in the environment, as well as their high cost of authorisation and regulatory barriers, remain obstacles for their wide utilisation in sustainable agriculture [181,182]. Essential oils obtained from *Mentha arvensis*, *Mentha spicata*, *Juniperus mexicana*, *Citrus x sinensis*, *Persicaria odorata*, *Piper nigrum*, *Canarium commune*, *Cinnamomum zeylanicum*, *Boswellia carterii*, *Cymbopogon flexuosus*, *Litsea cubeba*, *Artemisia alba*, *Cistus ladaniferus*, *Copaifera* tree, *Ferula galbaniflua*, *Citrus aurantium*, and *Schinus terebinthifolius* are registered in Europe by the European Chemical Agency (Homepage-ECHA (europa.eu)) as being suitable for use in agriculture, with some approved for their biocidal activities. In addition, when applied as seed protectants against cabbage black rot caused by *Xanthomonas campestris* pv. *campestris*, essential oils obtained from *Zataria multiflora* yielded a significant reduction in the further occurrence of disease in the field [183], whereas clove oil obtained from *Eugenia caryophyllata* was observed to control pomegranate bacterial blight caused by *Xanthomonas axonopodis* pv. *punicae* [184]. Essential oils obtained from cumin, basil, and geranium, applied either as seed protectants or directly in the soil, showed effectiveness towards cumin root rot caused by *Fusarium* spp. [185]. New techniques such as emulsion and the encapsulation of essential oils are currently being studied to enhance the stability and persistence of such oils in the environment [181].

Chitosan is currently being studied both as a plant defense inducer and for its direct involvement in disease control [186,187]. One of chitosan's properties is its film-forming abilities, which can be exploited to protect the surface of plants, thus avoiding colonization by pathogens. This feature can protect fruits and vegetables under postharvest conditions, and many crops benefit from treatments for controlling fungi involved in decay. These crops include table grapes, strawberry, pear, apple, citrus, peach, sweet cherry, plum, and mango [188]. The film formed by chitosan can also reduce both the incidence and severity of *Pseudomonas syringae* pv. *actinidiae*, the causal agent of kiwifruit bacterial canker [189]. Significant activity was also observed in rice infected with bacterial leaf blight caused by *Xanthomonas oryzae* pv. *oryzae* [190]. Seed dressing is another potential application of chitosan to reduce disease [191]. Interestingly, chitosan showed a synergistic effect with some biocontrol agents such as *Trichoderma* spp., indicating both castor seed protection [192] and a increased reduction in the pathogenicity of *Cercospora beticola* and *Fusarium oxysporum* [193]. Chitosan should also be added to the soil to reduce the severity of tomato crown and root rot caused by *Fusarium oxysporum* f. sp. *radicis lycopersici* [194]. Similarly, chitosan should be applied together with essential oils to reduce the volatility of such oils and to enhance their effectiveness in the control of *Aspergillus flavus* on dates [195].

In a previous study, *Yucca schidigera* extract protected sorghum seeds through the pathogenic activity of *Phoma sorghina*, *Curvularia lunata*, *Cladosporium* spp., and *Fusarium* spp. [196]. The macerate of *Equisetum arvense* (common horsetail) presented control activities that were very similar to those of different copper compounds towards *Phytophthora infestans*, the causal agent of tomato late blight; *Puccinia triticina*, the causal agent of durum wheat brown rust; *Fusarium graminearum*, the causal agent of Fusarium head blight of durum wheat [197]. Pomegranate (*Punica granatum*) peel extract is another natural product that shows potential activities for the control of phytopathogens. Pomegranate fruit peel extract showed a significant reduction in the severity and incidence of disease for *Pseudomonas syringae* pv. *tomato*, the causal agent of tomato bacterial speck, and tomato damping-off caused by *Fusarium oxysporum* f. sp. *lycopersici* [198,199]. Pomegranate (*Punica granatum*) fruit peel extract also showed interesting activities for the postharvest control of *Botrytis cinerea*, *Penicillium expansum*, and *Penicillium digitatum* on sweet cherry and citrus fruits [200]. *Camellia sinensis* extracts presented antibacterial activities towards *Pseudomonas syringae* pv. *actinidiae*, the causal agent of kiwifruit bacterial canker [201].

Tetrahydro-β-carboline alkaloids, which naturally occur in fruits, showed potential inhibitory activities towards some emergent and dangerous phytopathogenic bacteria such as *Xanthomonas axonopodis* pv. *citri*, the causal agent of Asiatic citrus canker; *Xanthomonas oryzae* pv. *oryzae*, the causal agent of rice bacterial blight; *Pseudomonas syringae* pv. *actinidiae*, the causal agent of kiwifruit bacterial canker [202].

4.2.3. Nutrition Management

Apart from playing fundamental roles in plant growth and development, nutrients are also involved in plant pathogenesis and disease control [203]. Moreover, for pathogen control in sustainable agriculture, preventive balanced nutrition coupled with rational agronomical techniques may be more cost effective and environmentally friendly than the application of any pesticide [36,204]. However, each pathosystem (i.e., the interaction between one single pathogen and one single crop in a certain environment) has its own peculiarities that should be investigated in order to verify the relationships between the contents of each nutrient within the plant and the development of disease over the season [36]. Among the various macronutrients, nitrogen can have either a beneficial or negative effect when applied to a crop to contain disease. These differing behaviors seem to be influenced by the overall characteristics of the pathogen (i.e., an increase in disease severity with obligate pathogens and a decrease in disease severity with facultative pathogens) [205].

The role of phosphorus in crop protection appears to be inconsistent and unpredictable. Relatively high potassium content generally has positive effects in reducing disease severity [202]. In particular, potassium phosphate was found to significantly reduce the severity of barley powdery mildew [206]. After establishing that a specific dose does not reach the limit imposed for the maximum residue level of phosphites, potassium phosphite can also contribute to disease reduction, as in the case of *Alternaria solani* on potato and *Alternaria alternata* on citrus fruit [207,208]. Calcium appears to be especially effective for post-harvest treatments against gray mould caused by *Botrytis cinerea* [209,210], whereas magnesium showed inconsistent results in controlling diseases [211]. Further ad hoc studies should be performed to determine the potential beneficial effects [212].

Among the various micronutrients, copper has been widely used as a fungicide and bactericide for a long period of time, although its excessive utilisation can result in the accumulation of copper in the soil, which poses risks for the microbial microflora [213]. Zinc and manganese are microelements that show clear beneficial activities against plant diseases [205]. Applying a supply of manure rich in zinc prior to planting winter wheat was found to significantly reduce the severity of spring blight caused by *Rhizoctonia cerealis* [211]. Moreover, the higher uptake of manganese in paddy-grown rice when compared to the low uptake in upland rice cultivations yielded a greater resistance to *Magnaporthe oryzae*, the causal agent of rice blast [214]. The wheat root rot disease caused by *Gaeumannomyces graminis* can be effectively managed through the nutrition of the host plant, especially through the supply of manganese [205]. Iron can also enhance the virulence of bacteria and fungi, and soilborne pathogens can be limited by adding rhizosphere-beneficial microorganisms that, through the activation of siderophores, reduce the iron content in the soil [36].

An in-depth study to determine which nutrients can impede the multiplication of pathogens showed that the direct supply of nutrients can help directly manage disease in some circumstances [215]. Another study explored the application of a biocomplex fertiliser containing low doses of zinc, copper, and citric acid to the crowns of infected trees, providing a striking example of the importance of plant nutrition as a strategy for significantly reducing infection in olive groves with symptoms of olive quick decline syndrome caused by *Xylella fastidiosa* subsp. *pauca* [216,217] (Figure 6). In this case, zinc and copper ions, which also showed direct bactericidal activities [215], increased in olive trees grown in soils that were characterised by low contents of such ions [218]. The increases in zinc and copper content within the plant that were caused by supplying this biocomplex over a few years allowed the severely infected trees to achieve good yields [214]. Additionally, supplying some nutrients as a foliar spray—in this case, calcium and boron—enabled the better control of gray mould in strawberry, caused by *Botrytis cinerea* [219].

Figure 6. Two olive groves in Nardò (Lecce province, Apulia, Italy): the trees on the right were abandoned upon the infection by *Xylella fastidiosa* subsp. *pauca*, whereas the trees on the left were treated with the zinc–copper–citric acid biocomplex described in [216,217]. The treated trees continue to yield despite the occurrence of the pathogen in the surrounding areas.

4.2.4. Systemic Resistance Inducers

The inducers of plant systemic resistance to pathogens show potential applications in IPM strategies. Such resistance can be mediated via (a) salicylic acid or its precursors, which incite systemic acquired resistance (SAR), or by (b) PGPR, which incites induced systemic resistance (ISR) by modulating the jasmonate and ethylene pathways [220]. Such inducers activate signals related to the plant defense mechanisms that are able to counteract pathogen colonisation. Acibenzolar-S-methyl, harpin, chitosan, and extracts of *Reynoutria sachalinensis* and *Saccharomyces cerevisiae* are commercially available inducers of resistance, whereas β-aminobutyric acid (BABA), probenazole, saccharin, phosphites, biochar, mycorrizhal fungi, endophytes, and algal extracts are among the most commonly studied future candidates for new commercial products [220]. Acibenzolar-S-methyl is able to induce resistance towards different kinds of pathogens in different crops, including the postharvest decay fungi of mango species such as *Colletotrichum gloeosporioides* [221]. Moreover, fava bean was found to be protected from *Uromyces viciae-fabae* and *Ascochyta rabiei* after acibenzolar-S-methyl treatments [222]. Within an IPM strategy, acibenzolar-S-methyl, in combination with copper treatments, reduced the severity of kiwifruit bacterial canker caused by *Pseudomonas syringae* pv. *actinidiae* [223].

Besides plant growth, PGPR mixtures showed the ability to protect crops, as in the case of pepper infected with *Xanthomonas vesicatoria*, the causal agent of bacterial spot [224]. Similarly, *Azospirillum brasilense* protected strawberry infected with *Colletotrichum acutatum*, the causal agent of anthracnose fruit rot [225]. Preventive BABA treatments protected lettuce and grapevine from downy mildew caused by *Bremia lactucae* and *Plasmopora viticola*, respectively [226,227]. As an inducer, biochar (i.e., a product of biomass pyrolysis obtained in the absence of oxygen) showed broad-spectrum activities when applied to the soil as an amendment, yielding protection against *Botrytis cinerea*, *Colletotrichum acutatum*, and

Podosphaera aphanis in strawberry [228]. This product, obtained from olive pruning, also showed inhibitory activities towards tomato spotted wilt virus, a systemic viral agent, without negatively affecting beneficial soil biocontrol agents such as *Bacillus* spp. and *Trichoderma* spp. [229]. Marine algae also showed potential broad-spectrum activities towards plant pathogens. Extracts from *Ulva fasciata* reduced the activities of *Colletotrichum lindemuthianum*, the causal agent of anthracnose of bean [230]. Additionally, extracts of *Ulva armoricana* showed activities towards the powdery mildew of grapevine, bean, and cucumber [231].

4.2.5. Gene Silencing

Using RNA interference technology to achieve host-induced silencing of the pathogen gene(s) involved in mechanisms of pathogenicity and virulence is another strategy that is currently being studied to provide sustainable tools for plant disease control, including the control of viruses [232–234]. Gene silencing through RNA interference is a common plant strategy aimed at blocking the activities of pathogens. Phytopathogens attempt to overcome this measure through anti-silencing mechanisms aimed at inactivating the host plant's RNA interference machinery to start the infection [232]. This technology is based on establishing small interfering host-induced (or more recently, exogenous (i.e., sprayed)) non-coding RNAs that, upon their uptake into the plant, can silence genes that are fundamental for the pathogen life cycle and/or infection (i.e., cross-kingdom RNA interference) [232–234]. The pathogen target genes used in developing the gene-silencing approach are numerous [234]. However, the success of this strategy largely depends on the efficiency of the pathogen in RNA uptake [235]. Some difficulties in the application of gene silencing include the degradation and uptake of RNA into the plant due to physical or biological interference, the lack of transformability among various crops, and the potential absence of genetic stability among silencing traits [236,237].

The pathogenicity of *Fusarium graminearum* was significantly reduced on barley leaves by using a spray that released double-stranded RNA, targeting the *FgAGO* and *FgDCL* genes in the RNA interference machinery of the fungus [238]. The pathogenic activity of *Phytophthora infestans*, the causal agent of potato late blight, was significantly reduced by gene-silencing targeting of the *hp-PiGPB1* gene, coding for the G protein β-subunit [239]. *Botrytis cinerea*, the causal agent of grey mould on many crops, is potentially manageable using gene-silencing technology. Spraying fruits, foliage, and flowers with exogenous RNAs targeted to the *DCL1* and *DCL2* genes, coding for the fungus' effectors involved in pathogenicity, can obtain a significant reduction of the disease on tomato, lettuce, onion, strawberry, and grapevine [240]. In addition, spray-induced gene silencing has been effectively applied to reduce the virulence of the fungus on grapevine, both on potted plants and on harvested bunches, by targeting the *erg11* gene of the cytochrome P450 monooxygenase. Double-stranded RNA was sprayed under high pressure on the leaves, absorbed by the petiole, and sprayed on the bunch at postharvest [241].

5. Concluding Remarks and Perspectives

The successful control of plant diseases in horticulture has long been a pillar of both food production and the conservation of agroecosystems. At present, these two goals are becoming increasingly urgent due to the rapid increase in human population and the threats posed by climate change [242]. The global circulation of plant material, with the potential introduction in new areas of dangerous phytopathogens, poses additional risks to safeguarding crop longevity and, in some circumstances, concerns the whole landscape of a territory. Within this context, high levels of crop productivity, conservation, and improvements to agroecosystem fertility should not be viewed as contradictory, and the development and utilisation of effective environmentally sustainable strategies should be applied more consistently to integrate and substitute for chemical pesticides in the near future [243]. Achieving good quality fruits and vegetables during the postharvest period is a complementary and relevant goal to achieve [244]. This scenario is largely fostered by supranational policies that are supporting projects and developing strategies to increase either the achievement of scientific and technical results or the wide acceptance of a "greener" agriculture. Many scientists involved in various aspects of sustainable agriculture complain about delays in achieving the regulation and commercial utilisation of known and novel biopesticides and/or products that can control phytopathogens. The safety and quality checks of these formulations are compulsory, but more frequent and initiative-taking activities among regulatory institutions are required to promote the spread of new active ingredients.

The conservation of agroecosystems according to sustainability principles greatly depends on the development and application of environmentally friendly strategies on a large scale. The "multi-stakeholder partnership" is an interesting strategy that could support the spread and acceptance of sustainable agriculture. This concept highlights the possibility that diverse groups in the agro-food chain can have common problems or aspirations despite having different interests [245]. A striking example of such a strategy for sustainable agriculture came from the agreement between Barilla, a well-known Italian brand of pasta, and some private companies that produce and supply ingredients such as tomato, sugar beet, and cereals; these organisations had no previous contact. These partners, including the farmers of each single company, share mutual and diversified interests in supporting their planned activities. The network also includes agronomists, research centers, and universities. Single farms are directed to adopt crop rotation, nitrogen fixer crops or green manure, and IPM in their agronomical practices. However, these practices are ruled out in most common methods of production. Farms receive a supply contract from Barilla with some guarantees concerning the price and quantity of the product over many years [246]. These initiatives appear to be positive for improving the circulation of sustainable principles in agriculture and should be expanded to promote sustainable agriculture.

Funding: This research received no external funding.

Informed Consent Statement: Not applicable.

Data Availability Statement: Data sharing not applicable.

Conflicts of Interest: The author declares no conflict of interest.

References

1. Siebrecht, N. Sustainable agriculture and its implementation gap-overcoming obstacles to implementation. *Sustainability* **2021**, *12*, 3853. [CrossRef]
2. Carlisle, L.; Montenegro de Wit, M.; Delonge, M.S.; Iles, A.; Calo, A.; Getz, C.; Ory, J.; Munden-Dixon, K.; Galt, R.; Melone, B.; et al. Transitioning to sustainable agriculture requires growing and sustaining an ecologically skilled workforce. *Front. Sustain. Food Syst.* **2019**, *3*, 96. [CrossRef]
3. McNeill, D. The contested discourse of sustainable agriculture. *Glob. Policy* **2019**, *10* (Suppl. 1), 16–27. [CrossRef]
4. Hertoge, K. Mals/Malles Venosta Referendum. 2014. Available online: http://www.marcozullo.it/wp-content/uploads/Malles-Venosta-Referendum.pdf (accessed on 22 May 2022).
5. Damalas, C.A.; Koutroubas, S.D. Current status and recent development in biopesticides use. *Agriculture* **2018**, *8*, 13. [CrossRef]
6. Fenibo, E.O.; Ijoma, G.N.; Matambo, T. Biopesticides in sustainable agriculture: A critical sustainable development driver governed by green chemistry principles. *Front. Sust. Food Syst.* **2021**, *5*, 619058. [CrossRef]
7. Garrett, K.A. Climate change and plant disease risk. In *Global Climate Change and Extreme Weather Events: Understanding the Contributions to Infectious Disease Emergence*; Relman, D.A., Hamburg, M.A., Choffnes, E.R., Mack, A., Eds.; National Academies Press: Washington, DC, USA, 2008; pp. 143–155.
8. Juroszek, P.; Von Tiedemann, A. Potential strategies and future requirements for plant disease management under a changing climate. *Plant Pathol.* **2011**, *60*, 100–112. [CrossRef]
9. Soussana, J.-F.; Graux, A.-I.; Tubiello, F.N. Improving the use of modelling for projections of climate change impacts on crops and pastures. *J. Exp. Bot.* **2010**, *61*, 2217–2228. [CrossRef]
10. Zhan, J.; Thrall, P.H.; Burdon, J.J. Achieving sustainable plant disease management through evolutionary principles. *Trends Plant Sci.* **2014**, *19*, 570–575. [CrossRef]
11. Leung, K.; Ras, E.; Ferguson, K.B.; Ariens, S.; Babendreier, D.; Bijma, P.; Bourtzis, K.; Brodeur, J.; Bruins, M.A.; Centurion, A.; et al. Next-generation biological control: The need for integrating genetics and genomics. *Biol. Rev.* **2020**, *95*, 1838–1854. [CrossRef]

12. Fenu, G.; Malloci, F.M. Forecasting plant and crop disease: An explorative study on current algorithms. *Big Data Cogn. Comput.* **2021**, *5*, 2. [CrossRef]
13. Fontana, D.C.; De Paula, S.; Torres, A.G.; Moura de Souza, V.H.; Pascholati, S.F.; Schmidt, D.; Dourado Neto, D. Endophytic fungi: Biological control and induced resistance to phytopathogens and abiotic stresses. *Pathogens* **2021**, *10*, 570. [CrossRef]
14. He, D.-C.; He, M.-H.; Amalin, D.M.; Liu, W.; Alvindia, D.G.; Zhan, J. Biological control of plant diseases: An evolutionary and eco-economics consideration. *Pathogens* **2021**, *10*, 1311. [CrossRef]
15. Ferrante, P.; Scortichini, M. Frost promotes the pathogenicity of *Pseudomonas syringae* pv. *actinidiae* in *Actinidia chinensis* and *A. deliciosa* plants. *Plant Pathol.* **2014**, *63*, 12–19.
16. Nuttall, J.G.; Perry, E.M.; Delahunty, A.J.; O'Leary, G.J.; Barlow, K.M.; Wallace, A.J. Frost response in wheat and early detection using proximal sensors. *J. Agron. Crop Sci.* **2019**, *205*, 220–234. [CrossRef]
17. Barlow, K.M.; Christy, B.P.; O'Leary, G.J.; Riffkin, P.A.; Nuttall, J.G. Simulating the impact of heat and frost events on wheat crop production: A review. *Field Crop Res.* **2015**, *171*, 109–119. [CrossRef]
18. Haworth, M.; Marino, G.; Brunetti, C.; Killi, D.; De Carlo, A.; Centritto, M. The impact of the heat stress and water deficit on the photosynthetic and stomatal physiology of olive (*Olea europea* L.)—A case study of the 2017 heat wave. *Plants* **2018**, *7*, 76. [CrossRef]
19. Silvestri, C.; Bacchetta, L.; Bellincontro, A.; Cristofori, V. Advances in cultivar choice, hazelnut orchard management, and nut storage to enhance product quality and safety: An overview. *Sci. Food Agric.* **2021**, *101*, 27–43. [CrossRef]
20. Macholdt, J.; Honermeier, B. Importance of variety choice: Adapting to climate change in organic and conventional farming system in Germany. *Outlook Agric.* **2017**, *46*, 178–184. [CrossRef]
21. Hulme, P.E. Unwelcome exchange: International trade as direct and indirect driver of biological invasions worldwide. *One Earth* **2021**, *4*, 666–679. [CrossRef]
22. Garbelotto, M.; Pautasso, M. Impacts of exotic forest pathogens on Mediterranean ecosystems: Four case studies. *Eur. J. Plant Pathol.* **2012**, *133*, 101–116. [CrossRef]
23. European Commission. Proposal for a regulation of the european parliament and of the council establishing rules on support for strategic plans to be drawn up by Member States under the Common agricultural policy (CAP Strategic Plans) and financed by the European Agricultural Guarantee Fund (EAGF) and by the European Agricultural Fund for Rural Development (EAFRD) and repealing Regulation (EU) No 1305/2013 of the European Parliament and of the Council and Regulation (EU) No 1307/2013 of the European Parliament and of the Council; COM/2018/392 final—2018/0216 (COD). 2018.
24. Wezel, A.; Casagrande, M.; Celette, F.; Vian, J.-F.; Ferrer, A.; Peigné, J. Agroecological practices for sustainable agriculture. A review. *Agron. Sustain. Dev.* **2014**, *34*, 1–20. [CrossRef]
25. Peeters, A.; Ambhul, E.; Barberi, P.; Migliorini, P.; Ostermann, O.; Goris, M.; Donham, J.; Wezel, A.; Batello, C. Integrating Agroecology into European Agricultural Policies. Position Paper and Recommendations to the European Commission on Eco-Schemes. 2021. Available online: https://www.agroecology-europe.org/wp-content/uploads/2021/07/AEEU_Positionpaper_Ecoschemes_FINAL_english.pdf (accessed on 22 May 2022).
26. Larkin, R.P.; Lynch, R.P. Use and effects of different Brassica and other rotation crops on soilborne diseases and yield of potato. *Horticulturae* **2018**, *4*, 37. [CrossRef]
27. Muscolo, A.; Sidari, M.; Nardi, S. Humic substance: Relationships between structure and activity. Deeper information suggests univocal findings. *J. Geochem. Explor.* **2013**, *129*, 57–63. [CrossRef]
28. Rillig, M.C.; Sosa-Hernandez, M.A.; Roy, J.; Aguilar-Trigueros, C.A.; Valyi, K.; Lehmann, A. Towards an integrated mycorrhizal technology: Harnessing mycorrhiza for sustainable intensification in agriculture. *Front. Plant Sci.* **2016**, *7*, 1625. [CrossRef]
29. Backer, R.; Rokem, J.S.; Ilangumaran, G.; Lamont, J.; Praslikova, D.; Ricci, E.; Subramanian, S.; Smith, D.L. Plant growth-promoting rhizobacteria: Context, mechanism of action, and roadmap to commercialization of biostimulant for sustainable agriculture. *Front. Plant Sci.* **2018**, *9*, 1473. [CrossRef]
30. Semida, W.M.; Beheiry, H.R.; Sétamou, M.; Simpson, C.R.; Abd El-Mageed, T.A.; Rady, M.M.; Nelson, S.D. Biochar implications for sustainable agriculture and environment: A review. *S. Afr. J. Bot.* **2019**, *127*, 333–347. [CrossRef]
31. Seenivasagan, R.; Babalola, O.O. Utilization of microbial consortia as biofertilizers and biopesticides for the production of feasible agricultural product. *Biology* **2021**, *10*, 1111. [CrossRef]
32. Elnahal, A.S.M.; El-Saadony, M.T.; Saad, A.M.; Desoky, E.M.; El-Tahan, A.M.; Rady, M.M.; AbuQamar, S.F.; El-Tarabily, K.A. The use of microbial inoculants for biological control, plant growth promotion, and sustainable agriculture: A review. *Eur. J. Plant Pathol.* **2022**, *162*, 759–792. [CrossRef]
33. Fu, L.; Penton, C.R.; Ruan, Y.; Shen, Z.; Xue, C.; Li, R.; Shen, Q. Inducing the rhizosphere microbiome by biofertilizer application to suppress banana Fusarium wilt disease. *Soil Biol. Biochem.* **2017**, *104*, 39–48. [CrossRef]
34. Suarez-Estrella, F.; Arcos-Nievas, M.A.; Lopez, M.J.; Vargas-Garcia, M.C.; Moreno, J. Biological control of plant pathogens by microorganisms isolated from agro-industrial composts. *Biol. Control* **2013**, *67*, 509–515. [CrossRef]
35. Zhou, X.; Wang, J.; Lu, C.; Liao, Q.; Gudda, F.O.; Ling, W. Antibiotics in animal manure and manure-based fertilizers: Occurrence and ecological risk assessment. *Chemosphere* **2020**, *255*, 127006. [CrossRef] [PubMed]
36. Dordas, C. Role of nutrients in controlling plant diseases in sustainable agriculture. A review. *Agron. Sustain. Dev.* **2008**, *28*, 33–46. [CrossRef]

37. Michael, C.; Gil, E.; Gallart, M.; Stavrinides, M.C. Influence of spray technology and application rate on leaf deposit and ground losses in mountain viticulture. *Agriculture* **2020**, *10*, 615. [CrossRef]
38. Gamliel, A. Application of soil solarization in the open field. In *Soil Solarization THEORY and Practice*; Gamliel, A., Katan, J., Eds.; American Phytopathological Society: St. Paul, MN, USA, 2017; pp. 175–180.
39. Van Bruggen, A.H.C.; Gamliel, A.; FinckH, M.R. Plant disease management in organic farming systems. *Pest Sci. Manag.* **2016**, *72*, 30–44. [CrossRef]
40. Hansen, Z.R.; Everts, K.L.; Fry, W.E.; Gevens, A.J.; Grunwald, N.J.; Gugino, B.K.; Johnson, D.A.; Johnson, S.B.; Knaus, B.J.; McGrath, M.T.; et al. Genetic variation within clonal lineages of *Phytophthora infestans* revealed through genotyping-by-sequencing, and implications for late blight epidemiology. *PLoS ONE* **2016**, *11*, e065690. [CrossRef]
41. Cohen, Y.; Ben Naim, Y.; Falach, L.; Rubin, A.E. Epidemiology of basil downy mildew. *Phytopathology* **2017**, *107*, 1149–1160. [CrossRef]
42. Newberry, A.A.; Babu, B.; Roberts, P.D.; Dufault, N.S.; Goss, E.M.; Jones, J.B.; Paret, M.L. Molecular epidemiology of *Pseudomonas syringae* pv. *syringae* causing bacterial leaf spot of watermelon and squash in Florida. *Plant Dis.* **2018**, *102*, 511–518.
43. Pegg, K.G.; Coates, L.M.; O'Neill, W.T.; Turner, D.W. The epidemiology of Fusarium wilt of banana. *Front. Plant Sci.* **2019**, *10*, 1395. [CrossRef]
44. Pruvost, O.; Boyer, K.; Ravigné, V.; Richard, D.; Vernière, C. Deciphering how plant pathogenic bacteria disperse and meet: Molecular epidemiology of *Xanthomonas citri* pv. *citri* at a microgeographic scales in a tropical area of Asiatic citrus canker endemicity. *Evol. Appl.* **2019**, *12*, 1523–1538.
45. Donati, I.; Cellini, A.; Sangiorgio, D.; Vanneste, J.L.; Scortichini, M.; Balestra, G.M.; Spinelli, F. *Pseudomonas syringae* pv. *actinidiae*: Ecology, infection dynamics and disease epidemiology. *Environ. Microbiol.* **2020**, *80*, 81–102.
46. Ostos, E.; Garcia-Lopez, M.T.; Porras, R.; Lopez-Escudero, F.J.; Trapero-Casas, A.; Michaelides, T.J.; Moral, J. Effect of cultivar resistance and soil management on spatial-temporal development of *Verticillium* wilt of olive: A long-term study. *Front. Plant Sci.* **2020**, *11*, 584496. [CrossRef]
47. Caffi, T.; Rossi, V. Fungicide models are components of multiple modeling approaches for decision-making in crop protection. *Phytopathol. Medit.* **2018**, *57*, 153–169.
48. Kim, K.-H.; Jung, I. Development of a daily epidemiological model for rice blast tailored for seasonal disease early warning in South Korea. *Plant Pathol. J.* **2020**, *36*, 406–417. [CrossRef]
49. He, X.; Fu, P.; Aker, W.G.; Hwang, H.-M. Toxicity of engineered nanomaterials mediated by nano–bio–eco interactions. *J. Environ. Sci. Health Part C* **2018**, *36*, 21–42. [CrossRef]
50. Gusberti, M.; Klemm, U.; Meier, M.S.; Maurhofer, M.; Hunger-Glaser, I. Fire blight control: The struggle goes on. A comparison of different fire blight control methods in Switzerland with respect to biosafety, efficacy and durability. *Int. J. Environ. Res. Public Health* **2015**, *12*, 11422–11447. [CrossRef]
51. Moser, R.; Pertot, I.; Elad, Y.; Raffaelli, R. Farmers' attitudes toward the use of biocontrol agents in IPM strawberry production in three countries. *Biol. Control* **2008**, *47*, 125–132. [CrossRef]
52. Barzmann, M.; Bàrberi, P.; Birch, A.N.E.; Boonekamp, P.; Dachbrodt-Saaydeh, S.; Graf, B.; Hommel, B.; Jensen, J.E.; Kiss, J.; Kudsk, P.; et al. Eight principles of integrated pest management. *Agron. Sustain. Dev.* **2015**, *35*, 1199–1215. [CrossRef]
53. Jorgensen, L.N.; Hovmoller, M.S.; Hansen, J.G.; Lassen, P.; Clark, B.; Bayles, R.; Rodemann, B.; Flath, K.; Jahn, M.; Goral, T.; et al. IPM strategies and their dilemmas including an introduction to www.eurowheat.org. *J. Integr. Agric.* **2014**, *13*, 265–281. [CrossRef]
54. Berlin, A.; Nordström Kälström, H.; Lindgren, A.; Olson, A. Scientific evidence for sustainable plant disease protection strategies for the main arable crops in Sweden. A systematic map protocol. *Environ. Evid.* **2018**, *7*, 31. [CrossRef]
55. Holb, I.J.; Abpnyi, F.; Bourma, J.; Heijne, B. On-farm and on-station evaluations of three orchard management approaches against apple scab and apple powdery mildew. *Crop Prot.* **2017**, *97*, 109–118. [CrossRef]
56. Shipp, L.; Elliott, D.; Gillespie, D.; Brodeur, J. From chemical to biological control in Canadian greenhouse crops. In *Biological Control: A Global Perspective*; Vincent, C., Goettel, M.S., Lazarovitis, C., Eds.; CABI: Wallingford, UK, 2007; pp. 118–127.
57. Jacobsen, B.J.; Zidack, N.K.; Larson, B.J. The role of *Bacillus*-based biological control agents in integrated pest management systems: Plant diseases. *Phytopathology* **2004**, *94*, 1272–1275. [CrossRef]
58. Galletti, S.; Paris, R.; Cianchetta, S. Selected isolates of *Trichoderma gamsii* induce different pathways of systemic resistance in maize upon *Fusarium verticillioides* challenge. *Microbiol. Res.* **2020**, *233*, 126406. [CrossRef]
59. Muneret, L.; Mitchell, M.; Seufert, V.; Aviron, S.; Djoud, E.A.; Pétillon, J.; Plantegenest, M.; Thiéry, D.; Rusch, A. Evidence that organic farming promotes pest control. *Nat. Sustain.* **2018**, *1*, 361–368. [CrossRef]
60. Baker, B.P.; Green, T.A.; Cooley, D.; Futrell, S.; Garling, L.; Gershuny, G.; Moyer, J.; Rajotte, E.G.; Seaman, A.J.; Young, S.L. Organic Agriculture and Integrated Pest Management: A Synergistic Partnership to Improve Sustainable Agriculture and Food Systems. 2015. Available online: https://organicipmwg.files.wordpress.com/2015/07/white-paper.pdf (accessed on 22 May 2022).
61. Rossi, V.; Salinari, F.; Poni, S.; Caffi, T.; Bettati, T. Addressing the implementation problem in decision support systems: The example of vite.net®. *Comput. Electron. Agric.* **2014**, *100*, 88–99. [CrossRef]
62. Minchinton, E.J.; Auer, D.P.F.; Thomson, F.M.; Trapnell, L.; Petkowski, J.E.; Galea, V.J.; Faggian, R.; Kita, N.; Murdoch, C.; Kennedy, R. Evaluation of the efficacy and economics of irrigation management, plant resistance and Brassica$_{spot}$™ models for management of white blister on *Brassica* crops. *Australasian Plant Pathol.* **2013**, *42*, 169–178. [CrossRef]

63. Pavan, W.; Fraisse, C.W.; Peres, N.A. Development of a web-based disease forecasting system for strawberry. *Comput. Electron. Agric.* **2011**, *75*, 169–175. [CrossRef]
64. Köhl, J.; Kolnaar, R.; Ravensberg, W.J. Mode of action of biological control agents against plant diseases: Relevance beyond efficacy. *Front. Plant Sci.* **2019**, *10*, 845. [CrossRef] [PubMed]
65. Sund, I.; Eilenberg, J. Why has the authorization of microbial biocontrol agents been slower in the EU than in comparable jurisdictions? *Pest Manag. Sci.* **2021**, *77*, 2170–2178. [CrossRef] [PubMed]
66. Smith, D.; Hinz, H.; Mulema, J.; Weyl, P.; Ryan, M.J. Biological control and the Nagoya Protocol on access and benefit sharing-a case of effective due diligence. *Biocontrol Sci. Technol.* **2018**, *28*, 914–926. [CrossRef]
67. Pliego, C.; Ramos, C.; De Vicente, A.; Cazorla, F.M. Screening for candidate bacterial biocontrol agents against soilborne fungal pathogens. *Plant Soil* **2011**, *340*, 505–520. [CrossRef]
68. Ajouz, S.; Nicot, P.C.; Bardin, M. Adaptation to pyrrolnitrin of *Botrytis cinerea* and cost of resistance. *Plant Pathol.* **2010**, *59*, 556–566. [CrossRef]
69. Wei, W.; Xu, Y.; Li, S.; Zhu, L.; Song, J. Developing suppressive soil for root disease of soybean with continuous long-term cropping of soybean in black soils of Northeast China. *Acta Agric. Scand. B Soil Plant Sci.* **2015**, *65*, 279–285. [CrossRef]
70. Milgroom, M.G.; Cortesi, P. Biological control of chestnut blight with hypovirulence: A critical analysis. *Annu. Rev. Phytopathol.* **2004**, *42*, 311–338. [CrossRef] [PubMed]
71. Nicot, P.C.; Bardin, M.; Alabouvette, C.; Köhl, J.; Ruocco, M. Potential of biological control based on published research. 1. Protection against plant pathogens of selected crops. In *Classical and Augmentative Biological Control against Diseases and Pests: Critical Status Analysis and Review of Factors Influencing Their Success*; Nicot, P.C., Ed.; IOBC/WPRS, 2011; pp. 1–11.
72. Moore, L.W.; Warren, G. *Agrobacterium radiobacter* strain K84 and biological control of crown gall. *Annu. Rev. Phytopathol.* **1979**, *17*, 163–179. [CrossRef]
73. Köhl, J.; Scheer, C.; Holb, I.J.; Masny, S.; Molhoek, W.M.L. Toward an integrated use of biological control of *Cladosporium cladosporioides* H39 in apple scab (*Venturia inaequalis*) management. *Plant Dis.* **2015**, *99*, 535–543. [CrossRef]
74. Haddoudi, I.; Cabrefiga, J.; Mora, I.; Mhadhbi, H.; Montesinos, E.; Mrabet, M. Biological control of fusarium wilt caused by *Fusarium equiseti* in *Vicia faba* with broad spectrum antifungal plant-associated *Bacillus* spp. *Biol. Control* **2021**, *160*, 104671. [CrossRef]
75. Woo, S.L.; Ruocco, M.; Vinale, F.; Nigro, M.; Marra, R.; Lombardi, N.; Pascale, A.; Lanzuise, S.; Manganiello, G.; Lorito, M. *Trichoderma*-based products and their widespread use in agriculture. *Open Mycol. J.* **2014**, *8*, 71–126. [CrossRef]
76. Sood, M.; Kapoor, D.; Sheteiwy, M.; Ramakhrisnan, M.; Landi, M.; Araniti, F.; Sharma, A. *Trichoderma*: The "secrets" of a multitalented biocontrol agent. *Plants* **2020**, *9*, 762. [CrossRef]
77. Hermosa, R.; Viterbo, A.; Chet, I.; Monte, E. Plant-beneficial effects of *Trichoderma* and of its genes. *Microbiology* **2012**, *158*, 17–25. [CrossRef]
78. Zin, N.A.; Badaluddin, N.A. Biological functions of *Trichoderma* spp. for agriculture applications. *Ann. Agric. Sci.* **2020**, *65*, 168–178. [CrossRef]
79. Thambugala, K.M.; Daranagama, D.A.; Phillips, A.J.L.; Kannangara, S.D.; Promputtha, I. Fungi vs. fungi in biocontrol: An overview of fungal antagonists applied against fungal plant pathogens. *Front. Cell Infect. Microbiol.* **2020**, *10*, 604293. [CrossRef]
80. Wisniewski, M.; Droby, S.; Norelli, J.; Liu, J.; Schena, L. Alternative management technologies for postharvest disease control: The journey from simplicity to complexity. *Postharvest Biol. Technol.* **2016**, *122*, 3–10. [CrossRef]
81. Höfte, M. The use of *Pseudomonas* spp. as bacterial biocontrol agents to control plant diseases. In *Microbial Bioprotectants for Plant Disease Management*; Köhl, J., Ravensberg, V., Eds.; Burleigh Dodds Science Publishing: Cambridge, UK, 2021; p. 74.
82. Shafi, J.; Tian, H.; Ji, M. *Bacillus* species as versatile weapon for plant pathogens: A review. *Biotechnol. Biotechnol. Equip.* **2017**, *31*, 3. [CrossRef]
83. Freimoser, F.M.; Rueda-Mejia, M.P.; Tilocca, B.; Migheli, Q. Biocontrol yeasts: Mechanisms and applications. *World J. Microbiol. Biotechnol.* **2019**, *35*, 154. [CrossRef]
84. Kunz, S.; Schmitt, A.; Haug, P. Field testing of strategies for fire blight control in organic fruit growing. *Acta Hortic.* **2011**, *896*, 431–436. [CrossRef]
85. Iqbal, M.; Jamshaid, M.; Zahid, M.A.; Andreasson, E.; Vetukuri, R.R.; Stenberg, J. Biological control of strawberry crown rot, root rot and grey mould by the beneficial fungus *Aureubasidium Pullulans*. *BioControl* **2021**, *66*, 535–545. [CrossRef]
86. Guetsky, R.; Shtienberg, D.; Elad, Y.; Dinoor, A. Combining biocontrol agents to reduce the variability of biological control. *Phytopathology* **2001**, *91*, 621–627. [CrossRef]
87. Raupach, G.S.; Kloepper, J.W. Mixtures of plant growth-promoting rhizobacteria enhance biological control of multiple cucumber pathogens. *Phytopathology* **1998**, *88*, 1158–1164. [CrossRef]
88. Leibinger, W.; Breuker, B.; Hahn, M.; Mendgen, K. Control of postharvest pathogens and colonization of the apple surface by antagonistic microorganisms in the field. *Phytopathology* **1997**, *87*, 1103–1110. [CrossRef]
89. Xu, X.-M.; Jeffries, P.; Pautasso, M.; Jeger, M.J. Combined use of biocontrol agents to manage plant disease in theory and practice. *Phytopathology* **2011**, *101*, 1024–1031. [CrossRef]
90. Sarrocco, S.; Valenti, F.; Manfredini, S.; Esteban, P.; Bernardi, R.; Puntoni, G.; Baroncelli, R.; Haidukowski, M.; Moretti, A.; Vannacci, G. Is exploitation competition involved in a multitrophic strategy for the biocontrol of Fusarium head blight? *Phytopathology* **2019**, *109*, 560–570. [CrossRef]

91. Roselló, G.; Francés, J.; Daranas, N.; Montesinos, E.; Bonaterra, A. Control of fire blight of pear trees with mixed inocula of two *Lactobacillus plantarum* strains and lactic acid. *J. Plant Pathol.* **2017**, *99*, 111–120.
92. Slusarenko, A.J.; Patel, A.; Portz, D. Control of plant diseases by natural products: Allicin from garlic as a case study. *Eur. J. Plant Pathol.* **2008**, *121*, 313–322. [CrossRef]
93. Singh, S.; Kumar, V.; Datta, S.; Dhangial, D.S.; Singh, J. Plant disease management by bioactive natural products. In *Natural Bioactive Products in Sustainable Agriculture*; Singh, J.I., Yadav, A.l., Eds.; Springer: Singapore, 2022; pp. 15–19.
94. Yan, Y.; Liu, Q.; Jacobsen, S.E.; Tang, Y. The impact and prospect of natural product discovery in agriculture. *EMBO Rep.* **2018**, *19*, e46824. [CrossRef]
95. Sundin, G.W.; Wang, N. Antibiotic resistance in plant pathogenic bacteria. *Annu. Rev. Phytopathol.* **2018**, *56*, 161–180. [CrossRef]
96. Taylor, P.; Reeder, R. Antibiotics use on crops in low and middle-income countries based on recommendations made by agricultural advisors. *CABI Agric. Biosci.* **2020**, *1*, 1. [CrossRef]
97. Wei, Z.-M.; Beer, S. Harpin from *Erwinia amylovora* induces plant resistance. *Acta Hortic.* **1996**, *411*, 223–226. [CrossRef]
98. De Capdeville, G.; Beer, S.V.; Wilson, C.L.; Aist, J.R. Alternative disease control agents induce resistance to blue mold in harvested "Red Delicious" apple fruit. *Phytopathology* **2002**, *92*, 900–908. [CrossRef] [PubMed]
99. Daayf, F.; Schmitt, A.; Bèlanger, R.R. The effects of plant extracts of *Reynouria sachalinensis* on powdery mildew development and leaf physiology of long English cucumber. *Plant Dis.* **1995**, *79*, 577–580. [CrossRef]
100. Trottin-Caudal, Y.; Fournier, C.; Leyre, J.-M.; Decognet, V.; Nicot, P.C.; Bardin, M. *Efficiency of Plant Extract from Reynoutria Sachalinensis (Milsana) to Control Powdery Mildew on Tomato (Oidium Neolycopersici)*; ffhal-02764311f; Colloque International: Avignon, France, 2003.
101. Margaritopoulos, T.; Toufexi, E.; Kizis, D.; Balayiannis, G.; Anagnostopoulos, C.; Theocharis, A.; Rempelos, L.; Troyanos, Y.; Leifert, C.; Markellou, E. *Reynoutria sachalinensis* elicits SA-dependent defense responses in courgetti genotypes against powdery mildew caused by *Podosphaera xanthii*. *Sci. Rep.* **2020**, *10*, 3354. [CrossRef] [PubMed]
102. El Ghaouth, A.; Arul, J.; Asselin, A.; Benhamou, N. Antifungal activity of chitosan on post-harvest pathogens: Induction of morphological and cytological alterations in *Rhizopus stolonifer*. *Mycol Res.* **1992**, *96*, 769–779. [CrossRef]
103. Gould, F.; Brown, Z.S.; Kuzma, J. Wicked evolution: Can we address the sociobiological dilemma of pesticide resistance? *Science* **2018**, *360*, 728–732. [CrossRef]
104. Newlands, N.K. Model-based forecasting of agricultural crop disease risk at the regional scale, integrating airborne inoculum, environmental, and satellite-based monitoring data. *Front. Environ. Sci.* **2018**, *6*, 63. [CrossRef]
105. Zhang, J.; Pu, R.; Yuan, L.; Huang, W.; Nie, C.; Yang, G. Integrating remotely sensed and meteorological observations to forecast wheat powdery mildew at a regional scale. *IEEE J. Sel. Top. Appl. Earth Obs. Remote Sens.* **2014**, *7*, 4328–4339. [CrossRef]
106. Gold, K.M.; Townsend, P.A.; Chlus, A.; Hermann, I.; Couture, J.J.; Larson, E.R.; Gevens, A.J. Hyperspectral measurements enable pre-symptomatic detection and differentiation of contrasting physiological effects of late blight and early blight in potato. *Remote Sens.* **2020**, *12*, 286. [CrossRef]
107. Lindblom, J.; Lundstrom, C.; Ljung, J.; Jonsson, A. Promoting sustainable intensification in precision agriculture: Review of decision support systems development and strategies. *Precis. Agric.* **2017**, *18*, 309–331. [CrossRef]
108. Newbery, F.; Qi, A.; Fitt, B.D.L. Modelling impacts of climate change on arable crop diseases: Progress, challenge and applications. *Curr. Opin. Plant Biol.* **2016**, *32*, 101–109. [CrossRef]
109. Roberts, D.P.; Short, N.M., Jr.; Sill, J.; Lakshman, D.K.; Hu, X.; Buser, M. Precision agriculture and geospatial techniques for sustainable disease control. *Ind. Phytopathol.* **2021**, *74*, 287–305. [CrossRef]
110. Simko, I.; Jimenez-Berni, J.A.; Sirault, X.R.R. Phenomic approach and tool for phytopathologists. *Phytopathology* **2017**, *107*, 6–17. [CrossRef]
111. Khattab, A.; Habib, S.E.D.; Ismail, H.; Zayan, S.; Fahmy, Y.; Khairy, M.M. An IoT-based cognitive monitoring system for early plant disease forecast. *Comput. Electron. Agric.* **2019**, *166*, 105028. [CrossRef]
112. Lu, X.; Lee, W.; Minzan, L.; Ehsani, R.; Mishra, A.; Yang, C.; Mangan, R. Feasibility study on huanglongbing (citrus greening) detection based on worldview-2 satellite imagery. *Biosyst. Eng.* **2015**, *132*, 28–38.
113. Yang, C. Remote sensing and precision agriculture technologies for crop disease detection and management with a practical application example. *Engineering* **2020**, *6*, 528–532. [CrossRef]
114. Mittal, D.; Kaur, G.; Singh, P.; Yadav, K.; Ali, S.A. Nanoparticle-based sustainable agriculture and food science: Recent advances and future overlook. *Front. Nanotechnol.* **2020**, *2*, 579954. [CrossRef]
115. Kah, M.; Hofmann, T. Nanopesticide research: Current trend and future priority. *Environ. Int.* **2014**, *63*, 224–235. [CrossRef] [PubMed]
116. Gilbertson, L.M.; Pourzahedi, L.; Laughton, S.; Gao, X.; Zimmermann, J.B.; Theis, T.L.; Westerhoff, P.; Lowry, G.V. Guiding the design space for nanotechnology to advance sustainable crop production. *Nature Nanotechnol.* **2020**, *15*, 801–810. [CrossRef] [PubMed]
117. Fortunati, E.; Mazzaglia, A.; Balestra, G.M. Sustainable control strategies for plant protection and food packaging sectors by natural substances and novel nanotechnologies approaches. *J. Sci. Food Agric.* **2019**, *99*, 986–1000. [CrossRef]
118. De Oliveira, J.L. Nano-biopesticides: Present concepts and future perspectives in integrated pest management. In *Advances in Nano-Fertilizers and Nano-Pesticides in Agriculture. A Smart Delivery System for Crop Improvement*; Jogaiah, S., Singh, H.B., Fraceto, L.F., De Lima, R., Eds.; Elsevier Science Publishing: Amsterdam, The Netherlands, 2021; pp. 1–27.

119. Sanchez-Hernandez, E.; Langa-Lomba, N.; Gonzalez-Garcia, V.; Casanova-Gascon, J.; Martin-Gil, J.; Santiago-Aliste, A.; Torres-Sanchez, S.; Martin-Ramos, P. Lignin-chitosan nanocarriers for the delivery of bioactive natural products against wood-decay phytopathogens. *Agronomy* **2022**, *12*, 461. [CrossRef]
120. Kalia, A.; Kaur, H. Nano-biofertilizers: Harnessing dual benefits of nano-nutrient and bio-fertilizers for enhanced nutrient use efficiency and sustainable productivity. In *Nanoscience for Sustainable Agriculture*; Pudake, R., Chauhan, N., Kole, C., Eds.; Springer: New York, NY, USA, 2019; pp. 51–73.
121. Borgatta, J.; Ma, C.; Hudson-Smith, N.; Elmer, W.; Plaza Peréz, C.D.; De La Torre-Roche, R.; Zuverza-Mena, N.; Haynes, C.L.; White, J.C.; Hamers, R.J. Copper based nanomaterials suppress root fungal disease in watermelon (*Citrullus lanatus*): Role of particle morphology, composition and dissolution behaviour. *ACS Sustain. Chem. Eng.* **2018**, *6*, 11. [CrossRef]
122. Elmer, W.H.; White, J.C. The use of metallic oxide nanoparticles to enhance growth of tomatoes and eggplants in diseased infected soil or soilless medium. *Environ. Sci. Nano* **2016**, *3*, 1072–1079. [CrossRef]
123. Song, J.J.; Soytong, K.; Kanokmedhakul, S.; Poeaim, S. Natural products of nanoparticles constructed from *Chaetomium* spp. to control rice blast disease caused by *Magnaporthe oryzae*. *Int. J. Agron. Biol.* **2020**, *23*, 1013–1020.
124. Sharma, J.; Singh, V.K.; Kumar, A.; Shankarayan, R.; Mallubhotla, S. Role of silver nanoparticles in treatment of plant diseases. In *Microbial Biotechnology*; Patra, J., Das, G., Shin, H.S., Eds.; Springer: Singapore, 2018; pp. 435–454.
125. Kookana, R.S.; Boxall, A.B.A.; Reeves, P.T.; Ashauer, R.; Beulke, S.; Chaudry, K.; Cornelis, G.; Fernandes, T.F.; Gan, J.; Kah, M.; et al. Nanopesticides: Guiding principles for regulatory evaluation of environmental risk. *J. Agric. Food Chem.* **2014**, *62*, 4227–4240. [CrossRef]
126. Freepons, D. Chitosan, does it have a place in agriculture? *Proc. Plant Growth Regul. Soc. Am.* **1991**, *10*, 11–19.
127. Montesinos, E.; Badosa, E.; Cabrefiga, J.; Planas, M.; Feliu, L.; Bardají, E. Antimicrobial peptides for plant disease control. From discovery to application. In *Small Wonders: Peptides for Disease Control*; Rajasekaran, K., Cary, J., Jaynes, J., Montesinos, E., Eds.; Blackwell: Oxford, UK, 2012; pp. 235–261.
128. Montesinos, L.; Gascon, B.; Ruz, L.; Badosa, E.; Planas, M.; Feliu, L.; Montesinos, E. A bifunctional synthetic peptide with antimicrobial and plant elicitation properties that protect tomato plants from bacterial and fungal infections. *Front. Plant Sci.* **2021**, *12*, 756357. [CrossRef]
129. Berger, C.; Laurent, F. Trunk injection of plant protection products to protect trees from pests and pathogens. *Crop Prot.* **2019**, *124*, 104831. [CrossRef]
130. Akinsamni, O.A.; Drenth, A. Phosphite and metalaxyl rejuvenate macadamia trees in decline caused by *Phytophtora cinnamomi*. *Crop Prot.* **2013**, *53*, 29–36. [CrossRef]
131. Archer, L.; Albrecht, U.; Crane, J. Trunk injection to deliver crop protection materials: An overview of basic principles and practical considerations. University of Florida, Horticultural Science Department, IFAS Extension. *EDIS* **2021**, *221*, HFS1426.
132. Montecchio, L. A Venturi effect can help cure our trees. *J. Vis. Exp.* **2013**, *80*, e51199. [CrossRef]
133. Acimovic, S.G.; Cregg, B.M.; Sundin, G.; Wise, J.C. Comparison of drill- and needle-based tree injection technologies in healing of trunk injection ports on apple trees. *Urban For. Urban Green* **2016**, *19*, 151–157. [CrossRef]
134. Scortichini, M.; Loreti, S.; Pucci, N.; Scala, V.; Tatulli, G.; Verweire, D.; Oehl, M.; Widmer, U.; Massana Codina, J.; Hertl, P.; et al. Progress towards a sustainable control of *Xylella fastidiosa* subsp. *pauca* in olive groves of Salento (Apulia, Italy). *Pathogens* **2021**, *10*, 668. [CrossRef]
135. Girelli, C.R.; Hussain, M.; Verweire, D.; Oehl, M.; Massana-Codina, J.; Avendano, M.S.; Migoni, D.; Scortichini, M.; Fanizzi, F.P. Agro-active endo-therapy treated *Xylella fastidiosa* subsp. *pauca*-infected olive trees assessed by the first 1H-NMR-based metabolomic study. *Sci. Rep.* **2022**, *12*, 5973.
136. Del Frari, G.; Costa, J.; Oliveira, H.; Boavida Ferreira, R. Endotherapy of infected grapevine cuttings for the control of *Phaeomoniella chlamydospora* and *Phaeoacremonium minimum*. *Phytopathol. Medit.* **2018**, *57*, 439–445.
137. Barratt, B.I.P.; Moran, V.C.; Bigler, F.; Van Lenteren, J.C. The status of biological control and recommendations for improving uptake for the future. *BioControl* **2017**, *63*, 155–167. [CrossRef]
138. Lamichhane, J.R.; Bischoff-Schaefer, M.; Bluemel, S.; Dachbrodt-Saaydeh, S.; Dreux, L.; Jansen, J.-P.; Kiss, J.; Kohl, J.; Kudsk, P.; Malausa, T.; et al. Identifying obstacles and ranking common biological research priorities for Europe to manage most economically important pests in arable, vegetable and perennial crops. *Pest Manag. Sci.* **2017**, *73*, 14–21. [CrossRef] [PubMed]
139. Bonaterra, A.; Badosa, E.; Cabriefiga, J.; Francés, J.; Montesinos, E. Prospects and limitations of microbial pesticides for control of bacterial and fungal pomefruit tree diseases. *Trees* **2012**, *26*, 215–226. [CrossRef] [PubMed]
140. Velivelli, S.L.S.; De Vos, P.; Kromann, P.; Declerck, S.; Prestwich, B.D. Biological control agents: From field to market, problems, and challenges. *Trends Biotechnol.* **2014**, *32*, 493–496. [CrossRef]
141. Raymaekers, K.; Ponet, L.; Holtappels, D.; Berckmans, B.; Cammue, B.P.A. Screening for new biocontrol agents applicable in plant disease management—A review. *Biol. Cont.* **2020**, *144*, 104240. [CrossRef]
142. Stefani, E.; Obradovic, A.; Gasic, K.; Altin, I.; Nagy, I.K.; Kovacs, T. Bacteriophage-mediated control of phytopathogenic xanthomonads: A promising green solution for the future. *Microorganisms* **2021**, *9*, 1056. [CrossRef]
143. Buttimer, C.; McAuliffe, O.; Ross, R.P.; Hill, C.; O'Mahony, J.; Coffey, A. Bacteriophages and bacterial plant diseases. *Front. Microbiol.* **2017**, *8*, 34. [CrossRef]
144. Holtappels, D.; Fortuna, K.; Lavigne, R.; Wagemans, J. The future of phage biocontrol in integrated plant protection for sustainable crop production. *Curr. Opin. Biotechnol.* **2021**, *68*, 60–71. [CrossRef]

145. Balogh, B.; Nga, N.T.T.; Jones, J.B. Relative level of bacteriophage multiplication in vitro or phyllosphere may not predict in planta efficacy for controlling bacterial leaf spot on tomato caused by *Xanthomonas perforans*. *Front. Microbiol.* **2018**, *9*, 2176. [CrossRef]
146. Vu, N.T.; Oh, C.-S. Bacteriophage usage for bacterial disease management and diagnosis in plant. *Plant Pathol. J.* **2020**, *36*, 204–217. [CrossRef]
147. Goswami, D.; Thakker, J.N.; Dhandhukia, P.C. Portraying mechanics of plant growth promoting rhizobacteria (PGPR): A review. *Cogent Food Agric.* **2016**, *2*, 1127500. [CrossRef]
148. Wang, X.; Wang, C.; Li, Q.; Zhang, J.; Ji, C.; Sui, J.; Liu, Z.; Song, X.; Liu, X. Isolation and characterization of antagonistic bacteria with the potential for biocontrol of soil-borne wheat disease. *Plant Pathol.* **2018**, *125*, 1868–1880.
149. Legein, M.; Smets, W.; Vandenheuvel, D.; Eilers, T.; Muyshondt, B.; Prinsen, E.; Samson, R.; Leeber, S. Modes of action of microbial biocontrol in the phyllosphere. *Front. Microbiol.* **2020**, *11*, 1619. [CrossRef]
150. Mota, M.S.; Gomes, C.B.; Souza, I.T., Jr.; Bittencourt Moura, A. Bacterial selection for biological control of plant disease: Criterion determination and validation. *Braz. J. Microbiol.* **2017**, *48*, 62–70. [CrossRef]
151. Gomez-Lama Cabanas, C.; Legarda, G.; Ruano-Rosa, D.; Pizarro-Tobias, P.; Valverde-Corredor, A.; Niqui, J.L.; Trivino, J.C.; Roca, A.; Mercado-Blanco, J. Indigenous *Pseudomonas* spp. strains from the olive (*Olea europea* L.) rhizosphere as effective biocontrol agents against *Verticillium dahliae*: From the host roots to the bacterial genome. *Front. Microbiol.* **2018**, *9*, 277. [CrossRef]
152. Salvatierra-Martinez, R.; Arancibia, W.; Araya, M.; Aguilera, S.; Olalde, V.; Bravo, J.; Stoll, A. Colonization ability as an indicator of enhanced biocontrol capacity—An example using two *Bacillus amyloliquefaciens* strains and *Botrytis cinerea* infection of tomatoes. *J. Phytopathol.* **2018**, *166*, 601–612. [CrossRef]
153. Stockwell, V.O.; Johnson, K.B.; Sugar, D.; Loper, J.E. Control of fire blight by *Pseudomonas fluorescens* A506 and *Pantoea vagans* C9-1 applied as single strains and mixed inocula. *Phytopathology* **2010**, *100*, 1330–1339. [CrossRef]
154. Couillerot, O.; Prigent-Combaret, C.; Caballero-Mellado, J.; Moenne-Loccoz, Y. *Pseudomonas fluorescens* and closely related fluorescent pseudomonads as biocontrol agents of soil-borne phytopathogens. *Lett. Appl. Microbiol.* **2009**, *48*, 505–512. [CrossRef]
155. Nam, H.S.; Anderson, A.J.; Kim, Y.C. Biocontrol efficacy of formulated *Pseudomonas chlororaphis* O6 against plant diseases and root-knot nematodes. *Plant Pathol. J.* **2018**, *34*, 241–249. [CrossRef]
156. Chen, K.; Tian, Z.; He, H.; Long, C.; Jiang, F. *Bacillus* species as potential biocontrol agents against citrus diseases. *Biol. Cont.* **2020**, *151*, 104419. [CrossRef]
157. Aiello, D.; Leonardi, G.R.; Di Petro, C.; Vitale, A.; Polizzi, G. A new strategy to improve management of Citrus mal secco disease using bioformulates based on *Bacillus amyloliquefaciens* strains. *Plants* **2022**, *11*, 446. [CrossRef] [PubMed]
158. Castro, D.; Torres, M.; Sampedro, I.; Martinez-Checa, F.; Torres, B.; Béjar, V. Biological control of *Verticillium* wilt on olive trees by the salt-tolerant strain *Bacillus velezensis* XT1. *Microorganisms* **2020**, *8*, 1080. [CrossRef] [PubMed]
159. Fira, D.; Dimkic, I.; Beric, T.; Lozo, J.; Stankovic, S. Biological control of plant pathogens by *Bacillus* species. *J. Biotechnol.* **2018**, *285*, 44–55. [CrossRef] [PubMed]
160. Osei, R.; Yang, C.; Cui, L.; Wei, L.; Jin, M.; Wei, X. Antagonistic bioagent mechanisms of controlling potato soft rot. *Plant Prot. Sci.* **2022**, *58*, 18–30. [CrossRef]
161. Law, J.W.-F.; Ser, H.-L.; Khan, T.M.; Chuah, L.-H.; Pusparajah, P.; Chan, K.-G.; Goh, B.-H.; Lee, L.-H. The potential of *Streptomyces* as biocontrol agents against the rice blast fungus, *Magnaporthe oryzae* (*Pyricularia oryzae*). *Front. Microbiol.* **2017**, *8*, 3. [CrossRef]
162. Chaiharn, M.; Theantana, T.; Pathom-aree, W. Evaluation of biocontrol activities of *Streptomyces* spp. against rice blast disease fungi. *Pathogens* **2020**, *9*, 126. [CrossRef]
163. Daranas, N.; Rossello, G.; Cabrefiga, J.; Donati, I.; Francés, J.; Badosa, E.; Spinelli, F.; Montesinos, E.; Bonaterra, A. Biological control of bacterial plant diseases with *Lactobacillus plantarum* strains selected for their broad spectrum activity. *Ann. Appl. Biol.* **2020**, *174*, 92–105. [CrossRef]
164. Mohd Taha, M.D.; Mohd Jain, M.F.; Saidi, N.B.; Abdul Rahim, R.; Shah, U.K.; Mohd Hashim, A. Biological control of *Erwinia mallotivora*, the causal agent of papaya dieback disease by indigenous seed-borne endophytic lactic acid bacteria consortium. *PLoS ONE* **2019**, *14*, e0224431. [CrossRef]
165. Sarrocco, S.; Esteban, P.; Vicente, I.; Bernardi, R.; Plainchamp, T.; Domenichini, S.; Puntoni, G.; Baroncelli, R.; Vannacci, G.; Dufresne, M. Straw competition and wheat root endophytism of *Trichoderma gamsii* T6085 as useful traits in the biological control of Fusarium head blight. *Phytopathology* **2021**, *111*, 1129–1136. [CrossRef]
166. Tuão, C.A.; Pinto, J.M. Biocontrol of melon wilt caused by *Fusarium oxysporum* Schlect f. sp. *melonis* using seed treatment with *Trichoderma* spp. and liquid compost. *Biol. Control* **2016**, *97*, 13–20.
167. Lahuf, A.A.; Kareem, A.A.; Al-Sweedi, T.M.; Alfarttoosi, H.A. Evaluation the potential of indigenous biocontrol agent *Trichoderma harzianum* and its interactive effect with nanosized ZnO particles against the sunflower damping-off pathogen, Rhizoctonia Solani. *IOP Conf. Ser. Earth Environ. Sci.* **2019**, *365*, 012033. [CrossRef]
168. Fanelli, F.; Liuzzi, V.C.; Logrieco, A.F.; Altomare, C. Genomic characterization of *Trichoderma atrobrunneum* (*T. harzianum* species complex) ITEM 908: Insight into genetic endowment of a multi-target biocontrol strain. *BMC Genom.* **2018**, *19*, 662. [CrossRef]
169. Rush, T.A.; Shersthra, H.K.; Meena, M.G.; Spangler, M.K.; Ellis, J.C.; Labbè, J.L.; Abraham, P.E. Bioprospecting *Trichoderma*: A systematic roadmap to screen genomes and natural products for biocontrol applications. *Front. Fung. Biol.* **2021**, *2*, 716511. [CrossRef]
170. Niu, B.; Wang, W.; Yuan, Z.; Sederoff, R.R.; Chaing, V.L.; Borriss, R. Microbial interactions within multiple-strain biological control agents impact soil-borne plant disease. *Front. Microbiol.* **2020**, *11*, 585404. [CrossRef]

171. Pertot, I.; Giovannini, O.; Benanchi, M.; Caffi, T.; Rossi, V.; Mugnai, L. Combining biocontrol agents with different mechanism of action in a strategy to control *Botrytis cinerea* on grapevine. *Crop Prot.* **2017**, *97*, 85–93. [CrossRef]
172. Bubici, G. *Streptomyces* spp. as biocontrol agents against *Fusarium* species. *CAB Rev.* **2018**, *13*, 050. [CrossRef]
173. De Silva, N.I.; Brooks, S.; Lumyong, S.; Hyde, K.D. Use of endophytes as biocontrol agents. *Fungal Biol. Rev.* **2019**, *33*, 133–148. [CrossRef]
174. Sdiri, Y.; Lopes, T.; Rodrigues, N.; Silva, K.; Rodrigues, I.; Pereira, J.A.; Baptista, P. Biocontrol ability and production of volatile organic compounds as a potential mechanism of action of olive endophytes against *Colletotrichum Acutatum*. *Microorganisms* **2022**, *10*, 571. [CrossRef]
175. Del Carmen Rodríguez, H.; Evans, H.C.; De Abreu, L.M.; De Macedo, D.M.; Ndacnou, M.K.; Bekele, K.B.; Barreto, R.W. New species and records of *Trichoderma* isolated as mycoparasites and endophytes from cultivated and wild coffee in Africa. *Sci. Rep.* **2021**, *11*, 5671.
176. Vurucunda, S.S.K.P.; Giovanardi, D.; Stefani, E. Plant growth promoting and biocontrol activity of *Streptomyces* spp. as endophytes. *Int. J. Mol. Sci.* **2018**, *19*, 952. [CrossRef] [PubMed]
177. Hong, C.E.; Park, J.M. Endophytic bacteria as biocontrol agents against plant pathogens: Current state-of-the-art. *Plant Biotechnol. Rep.* **2016**, *10*, 353–357. [CrossRef]
178. Andrivon, D.; Bardin, M.; Bertrand, C.; Brun, L.; Daire, X.; Fabre, F.; Gary, C.; Montarry, J.; Nicot, P.; Reignault, P.; et al. Can organic agriculture cope without copper for disease control? In *Synthesis of the Collective Scientific Assessment Report*; INRA: Paris, France, 2018; p. 64.
179. Ali, O.; Ramsubhag, A.; Jayaraman, J. Biostimulant properties of seaweed extracts in plants: Implications towards sustainable crop protection. *Plants* **2021**, *10*, 531. [CrossRef]
180. Chinnadurai, C.; Ramkissoon, A.; Rajendran, R.; DeAspa, S.; Ramsubhag, A.; Jayaraj, J. Integrated disease management in pumpkin in the southern Caribbean. *Trop. Agric. Univ. West Indies* **2018**, *95*, 132–140.
181. Raveau, R.; Fontaine, J.; Lounès-Hadj Sahraoui, A. Essential oils as potential alternative biocontrol products against plant pathogens and weeds: A review. *Foods* **2020**, *9*, 365. [CrossRef] [PubMed]
182. Alonso-Gato, M.; Astray, G.; Mejuto, J.C.; Simal-Gandara, J. Essential oils as antimicrobials in crop protection. *Antibiotics* **2021**, *10*, 34. [CrossRef] [PubMed]
183. Amini, L.; Soudi, M.R.; Saboora, A.; Mobasheri, H. Effect of essential oil from *Zataria multiflora* on local strains of *Xanthomonas campestris*: An efficient antimicrobial agent for decontamination of seeds of *Brassica oleracea* var. *capitata*. *Sci. Hort.* **2018**, *236*, 256–264. [CrossRef]
184. Kumar, P.; Lokesh, V.; Doddaraju, P.; Kumari, A.; Singh, P.; Meti, B.S.; Sharma, J.; Gupta, K.J.; Manjunatha, G. Greenhouse and field experiments revealed that clove oil can effectively reduce bacterial blight and increase yield in pomegranate. *Food Energy Secur.* **2021**, *10*, e305. [CrossRef]
185. Hashem, M.; Moharam, A.M.; Zaied, A.A.; Saleh, F.E.M. Efficacy of essential oils in the control of cumin root rot disease caused by *Fusarium* spp. *Crop Protect.* **2010**, *29*, 1111–1117. [CrossRef]
186. Katijar, D.; Hemantaranjan, A.; Singh, B.; Nishant Bhanu, A. A future perspective in crop protection: Chitosan and its oligosaccharides. *Adv. Plants Agric. Res.* **2014**, *1*, 23–30.
187. Xing, K.; Zhu, X.; Peng, X.; Qin, S. Chitosan antimicrobial and eliciting properties for pest control in agriculture: A review. *Agron. Sustain. Dev.* **2015**, *35*, 569–588. [CrossRef]
188. Romanazzi, G.; Feliziani, E.; Sivakumar, D. Chitosan, a biopolymer with triple action on postharvest decay of fruit and vegetables: Eliciting, antimicrobial and film-forming properties. *Front. Microbiol.* **2018**, *9*, 2745. [CrossRef]
189. Scortichini, M. Field efficacy of chitosan to control *Pseudomonas syringae* pv. *actinidiae*, the causal agent of kiwifruit bacterial canker. *Eur. J. Plant Pathol.* **2014**, *140*, 887–892.
190. Stanley-Raja, V.; Senthil-Nathan, S.; Chantini, K.M.P.; Sivanesh, H.; Ramasubramanian, R.; Karthi, S.; Shyam-Sundar, N.; Vasantha-Srinivasan, P.; Kalaivani, K. Biological activity of chitosan inducing resistance efficiency of rice (*Oryza sativa* L.) after treatment with fungal based chitosan. *Sci. Rep.* **2021**, *11*, 20488. [CrossRef]
191. Nandeeshkumar, P.; Sudisha, J.; Ramachandra, K.K.; Prakash, H.S.; Niranjana, S.R. Shekar, S.H. Chitosan induced resistance to downy mildew in sunflower caused by *Plasmopara halstedii*. *Physiological. Mol. Plant Pathol.* **2008**, *72*, 188–194. [CrossRef]
192. Chandrika, K.S.V.P.; Prasad, R.D.; Godbole, V. Development of chitosan-PEG blended films using *Trichoderma*: Enhancement of antimicrobial activity and seed quality. *Int. J. Biol. Macromol.* **2019**, *126*, 282–290. [CrossRef]
193. Kappel, L.; Kosa, N.; Gruber, S. The multilateral efficacy of chitosan and *Trichoderma* on sugar beet. *J. Fungi* **2022**, *8*, 137. [CrossRef]
194. Lafontaine, J.P.; Benhamou, N. Chitosan treatment: An emerging strategy for enhancing resistance of greenhouse tomato plants to infection by *Fusarium oxysporum* f. sp. *radicis lycopersici*. *Biocontrol Sci. Technol.* **1996**, *6*, 111–124. [CrossRef]
195. Aloui, H.; Khwaldia, K.; Licciardello, F.; Mazzaglia, A.; Muratore, G.; Hamdi, M.; Restuccia, C. Efficacy of the combined application of chitosan and locust bean gum with different citrus essential oils to control postharvest spoilage caused by *Aspergillus flavus* in dates. *Int. J. Food Microbiol.* **2014**, *170*, 21–28. [CrossRef]
196. Wullf, E.G.; Zida, E.; Torp, J.; Lund, O.S. *Yucca schidigera* extract: A potential biofungicide against seedborne pathogens of sorghum. *Plant Pathol.* **2012**, *61*, 331–338. [CrossRef]

197. Trebbi, G.; Negri, L.; Bosi, S.; Dinelli, G.; Cozzo, R.; Marotti, I. Evaluation of *Equisetum arvense* (horsetail macerate) as a copper substitute for pathogen management in field-grown organic tomato and durum wheat cultivation. *Agriculture* **2021**, *11*, 5. [CrossRef]
198. Quattrucci, A.; Ovidi, E.; Tiezzi, A.; Vinciguerra, V.; Balestra, G.M. Biological control of tomato bacterial speck using *Punica granatum* fruit peel extract. *Crop Prot.* **2013**, *46*, 18–22. [CrossRef]
199. Mohamad, T.G.; Khalil, A.A. Effect of agriculture waste: Pomegranate (*Punica granatum* L.) fruits peel on some important phytopathogenic fungi and control of tomato damping-off. *J. Appl. Life Sci. Int.* **2015**, *3*, 103–113. [CrossRef] [PubMed]
200. Li Destri Nicosia, M.G.; Pangallo, S.; Raphael, G.; Romeo, F.V.; Strano, M.C.; Rapisarda, P.; Droby, S.; Schena, L. Control of postharvest fungal rots on citrus fruit and sweet cherries using a pomegranate peel extract. *Postharvest Biol. Technol.* **2016**, *114*, 54–61. [CrossRef]
201. Lovato, A.; Pignatti, A.; Vitulo, N.; Vandelle, E.; Polverari, A. Inhibition of virulence-related traits in *Pseudomonas syringae* pv. *actinidiae* by gunpowder green tea extracts. *Front. Microbiol.* **2019**, *10*, 2362. [PubMed]
202. Liu, H.-W.; Ji, Q.-T.; Ren, G.-G.; Wang, F.; Su, F.; Wang, P.-Y.; Zhou, X.; Wu, Z.-B.; Li, Z.; Yang, S. Antibacterial functions and proposed modes of action of novel 1,2,3,4-tetrahydro-β-carboline derivatives that possess an attractive 1,3-diaminopropan-2-ol pattern against rice bacterial blight, kiwifruit bacterial canker, and citrus bacterial canker. *J. Agric. Food. Chem.* **2020**, *68*, 12558–12568. [CrossRef] [PubMed]
203. Agrios, N.G. *Plant Pathology*, 5th ed.; Elsevier-Academic Press: Amsterdam, The Netherlands, 2005; p. 635.
204. Huber, D.M.; Haneklaus, S. Managing nutrition to control plant diseases. *Landbauforsch. Völkenrode* **2007**, *57*, 313–322.
205. Gupta, N.; Debhnat, S.; Sharma, S.; Sharma, P.; Purohit, J. Role of nutrients in controlling the plant diseases in sustainable agriculture. In *Agriculturally Important Microbes for Sustainable Agriculture*; Meena, V.S., Mishra, P.K., Bisht, J.K., Pattanayak, A., Eds.; Springer Nature: Singapore, 2017; pp. 217–261.
206. Mitchell, A.F.; Walters, D.R. Potassium phosphate induces systemic protection in barley to powdery mildew infection. *Pest Manag. Sci.* **2004**, *60*, 126–134. [CrossRef] [PubMed]
207. Yogev, E.; Sadowsky, A.; Solei, Z.; Oren, Y.; Orbach, Y. The performance of potassium phosphite for controlling Alternaria brown spot of citrus fruit. *J. Plant Dis. Prot.* **2006**, *113*, 207–213. [CrossRef]
208. Liljeroth, E.; Lankinen, A.; Andreasson, E.; Alexandersson, E. Phosphite integrated the late blight treatment strategies in starch potato does not cause residue in the starch product. *Plant Dis.* **2020**, *104*, 3026–3032. [CrossRef]
209. Manganaris, G.A.; Vasilakakis, M.; Diamantidis, G.; Migani, I. The effect of postharvest calcium application on tissue calcium concentration, quality attributes, incidence of flesh browning and cell wall physicochemical aspects of peach fruits. *Food Chem.* **2007**, *100*, 1385–1392. [CrossRef]
210. Sun, C.; Zhu, C.; Tang, Y.; Ren, D.; Cai, Y.; Zhou, G.; Wang, Y.; Xu, L.; Zhu, P. Inhibition of *Botrytis cinerea* and control of gray mold of table grapes by calcium propionate. *Food Qual. Saf.* **2021**, *5*, fyab016. [CrossRef]
211. Huber, D.M.; Jones, J.B. The role of magnesium in plant disease. *Plant Soil* **2013**, *368*, 73–85. [CrossRef]
212. Moreira, W.R.; Bispo, W.M.S.; Rios, J.A.; Debona, D.; Nascimento, C.W.A.; Rodrigues, F.A. Magnesium induced alterations in the photosynthetic performance and resistance of the plants infected with *Bipolaris oryzae*. *Sci. Agric.* **2015**, *72*, 328–333. [CrossRef]
213. Fagnano, M.; Agrelli, D.; Pascale, A.; Adamo, P.; Fiorentino, N.; Rocco, C.; Pepe, O.; Ventorino, V. Copper accumulation in agricultural soils: Risks for the food chain and the soil microbial populations. *Sci. Total Environ.* **2020**, *734*, 139434. [CrossRef]
214. Pearson, C.J.; Jacobs, B.C. Elongation and retarded growth of rice during short-term submergence in three stages of development. *Field Crops Res.* **1986**, *13*, 331–344. [CrossRef]
215. Navarrete, F.; De La Fuente, L. Zinc detoxification is required for full virulence and modification of the host leaf ionome by *Xylella fastidiosa*. *Mol. Plant Microbe Interact.* **2015**, *28*, 497–507. [CrossRef]
216. Scortichini, M.; Chen, J.; De Caroli, M.; Dalessandro, G.; Pucci, N.; Modesti, V.; L'Aurora, A.; Petriccione, M.; Zampella, L.; Mastrobuoni, F.; et al. A zinc, copper and citric acid biocomplex shows promise for control of *Xylella fastidiosa* subsp. *pauca* in olive trees of Apulia region (southern Italy). *Phytopathol. Medit.* **2018**, *57*, 48–72.
217. Tatulli, G.; Modesti, V.; Pucci, N.; Scala, V.; L'Aurora, A.; Lucchesi, S.; Salustri, M.; Scortichini, M.; Loreti, S. Further *in vitro* assesment and mid-term evaluation of control strategy of *Xylella fastidiosa* subsp. *pauca* in olive groves of Salento (Apulia, Italy). *Pathogens* **2021**, *10*, 85. [CrossRef]
218. Del Coco, L.; Migoni, D.; Girelli, C.R.; Angilè, F.; Scortichini, M.; Fanizzi, F.P. Soil and leaf ionome heterogeneity in *Xylella fastidiosa* subsp. *pauca*-infected, non infected and treated olive groves in Italy. *Plants* **2020**, *9*, 760. [CrossRef]
219. Singh, R.; Sharma, R.R.; Tyagi, S.K. Pre-harvest foliar application of calcium and boron influences physiological disorders, fruit yield and quality of strawberry (*Fragaria* × *ananassa* Duch.). *Sci. Hort.* **2007**, *112*, 215–220. [CrossRef]
220. Walters, D.R.; Ratsep, J.; Havis, N.D. Controlling crop diseases using induced resistance: Challenges for the future. *J. Exp. Bot.* **2013**, *64*, 1263–1280. [CrossRef] [PubMed]
221. Lin, J.; Gong, D.; Zhu, S.; Zhang, L.; Zhang, L. Expression of PPO and POD genes and contents of polyphenolic compounds in harvested mango fruits in relation to benzathiadiazole-induced defense against anthracnose. *Sci. Hort.* **2011**, *130*, 85–89. [CrossRef]
222. Sillero, J.C.; Rojas-Molina, M.M.; Avila, C.M.; Rubiales, D. Induction of systemic acquired resistance against rust, ascochyta blight and broomrape in faba bean by exogenous application of salicylic acid and benzothiadiazole. *Crop Prot.* **2012**, *34*, 65–69. [CrossRef]

223. Monchiero, M.; Gullino, M.L.; Pugliese, M.; Spadaro, D.; Garibaldi, A. Efficacy of different chemical and biological products in the control of *Pseudomonas syringae* pv. *actinidiae* on kiwifruit. *Eur. J. Plant Pathol.* **2015**, *44*, 13–23.
224. Hahm, M.-S.; Sumayo, M.; Hwang, Y.-J.; Jeon, S.-A.; Park, S.-J.; Lee, J.Y.; Ahn, J.-H.; Kim, B.-S.; Ryu, C.-M.; Ghim, S.-Y. Biological control and plant growth promoting capacity of rhizobacteria on pepper under greenhouse and field conditions. *J. Microbiol.* **2012**, *50*, 380–385. [CrossRef]
225. Tortora, M.L.; Díaz-Ricci, J.C.; Pedraza, R.O. Protection of strawberry plants (*Fragaria ananassa* Duch.) against anthracnose disease induced by *Azospirillum brasilense*. *Plant Soil* **2012**, *356*, 279–290. [CrossRef]
226. Cohen, Y.; Rubin, A.E.; Kilfin, G. Mechanisms of induced resistance in lettuce against *Bremia lactucae* by DL-β-amino-butyric acid (BABA). *Eur. J. Plant Pathol.* **2010**, *126*, 553–573. [CrossRef]
227. Tamm, L.; Thürig, B.; Fliessbach, A.; Goltlieb, A.E.; Karavani, S.; Cohen, Y. Elicitors and soil management to induce resistance against fungal plant diseases. *NJAS Wagening. J. Life Sci.* **2011**, *58*, 131–137. [CrossRef]
228. Harel, Y.M.; Elad, Y.; Rav-David, D.; Borenstein, M.; Shulchani, R.; Lew, B.; Graber, E.R. Biochar mediates systemic response of strawberry to foliar fungal pathogens. *Plant Soil* **2012**, *357*, 245–257. [CrossRef]
229. Faggioli, F.; Luigi, M.; Mangili, A.; Dragone, I.; Antonelli, M.G.; Contarini, M.; Speranza, S.; Bertin, S.; Tiberini, A.; Gentili, A. Effects of biochar on the growth and development of tomato seedlings, and on the response of tomato plants to the infection of systemic viral agents. *Front. Microbiol.* **2022**, *13*, 862075.
230. Paulert, R.; Talamini, V.; Cassolato, J.E.F.; Duarte, M.E.R.; Noseda, M.D.; Smania, A.; Stadnik, M.J., Jr. Effects of sulphated polysaccharide and alcoholic extracts from green seaweed *Ulva fasciata* on anthracnose severity and growth of common bean (*Phaseolus vulgaris* L.). *J. Plant Dis. Protect.* **2009**, *116*, 263–270. [CrossRef]
231. Jaulneau, V.; Lafitte, C.; Corio-Costet, M.-F.; Stadnik, M.J.; Salamagne, S.; Briand, X.; Esquerré-Tugayé, M.-T.; Dumas, B. An *Ulva armoricana* extract protects plants against three powdery mildew pathogens. *Eur. J. Plant Pathol.* **2011**, *131*, 393–401. [CrossRef]
232. Koch, A.; Kogel, K.-H. New wind in the sails: Improving the agronomic value of crop plants through RNAi-mediated gene silencing. *Plant Biotechnol. J.* **2014**, *12*, 821–831. [CrossRef]
233. Muhammad, T.; Zhang, F.; Zhang, Y.; Liang, Y. RNA interference: A natural immune system of plants to counteract biotic stressors. *Cells* **2019**, *8*, 38. [CrossRef]
234. Gebremichael, D.E.; Mehari Haile, Z.; Negrini, F.; Sabbadini, S.; Capriotti, L.; Mezzetti, B.; Baraldi, E. RNA interference strategies for future management of plant pathogenic fungi: Prospects and challenges. *Plants* **2021**, *10*, 650. [CrossRef]
235. Qiao, L.; Lan, C.; Capriotti, L.; Ah-Fong, A.; Sanchez, J.N.; Hamby, R.; Heller, J.; Zhao, H.; Glass, N.L.; Judelson, H.S.; et al. Spray-induced gene silencing for disease control is dependent on the efficiency of the pathogen RNA uptake. *Plant Biotechnol. J.* **2021**, *19*, 1756–1768. [CrossRef]
236. Koch, A.; Biedenkopf, D.; Furch, A.; Weber, L.; Rossbach, O.; Abteilatef, E.; Linicus, L.; Johannsmeier, J.; Jelonek, L.; Goesmann, A.; et al. An RNAi-based control of *Fusarium graminearum* infections through spraying of long dsRNAs involves a plant passage and is controlled by the fungal silencing machinery. *PLoS Pathog.* **2016**, *12*, e1005901. [CrossRef]
237. Goodfellow, S.; Zhang, D.; Wang, M.-B.; Zhang, R. Bacterium-mediated RNA interference: Potential application in plant protection. *Plants* **2019**, *8*, 572. [CrossRef]
238. Werner, B.T.; Gaffar, F.Y.; Schuemann, J.; Biedenkopf, D.; Koch, A.M. RNA-spray-mediated silencing of *Fusarium graminearum AGO* and *DCL* genes improve barley disease resistance. *Front. Plant Sci.* **2020**, *11*, 476. [CrossRef]
239. Jahan, S.N.; Asman, A.K.M.; Corcoran, P.; Fogelqvist, J.; Vetukuri, R.R.; Dixelius, C. Plant-mediated gene silencing restricts growth of the potato late blight pathogen *Phytophthora infestans*. *J. Exp. Bot.* **2015**, *66*, 2785–2794. [CrossRef]
240. Wang, M.; Weiberg, A.; Lin, F.-M.; Thomma, B.; Huang, H.-D.; Jin, H. Bidirection cross-kingdom RNAi and fungal uptake of external RNAs confer plant protection. *Nat. Plants* **2016**, *2*, 16151. [CrossRef]
241. Nerva, L.; Sandrini, M.; Gambino, G.; Chitarra, W. Double-stranded RNAs (dsRNAs) as a sustainable tool against gray mold (*Botrytis cinerea*) in grapevine: Effectiveness of different application methods in an open-air environment. *Biomolecules* **2020**, *10*, 200. [CrossRef]
242. Legreve, A.; Duveiller, E. Preventing potential diseases and pest epidemics under a changing climate. In *Climate Change and Crop Production*; Reynolds, M.P., Ed.; CABI Publishing: Wallingford, UK, 2010; pp. 50–70.
243. Rippa, M.; Battaglia, V.; Cermola, M.; Sicignano, M.; Lahoz, E.; Mormile, P. Monitoring the copper persistence on plant leaves using pulsed thermography. *Environ. Minit. Assess.* **2022**, *194*, 160. [CrossRef] [PubMed]
244. Wisniewsky, M.; Wilson, C.; Droby, S.; Chalutz, E.; El-Ghaouth, A.; Stevens, C. Postharvest biocontrol: New concepts and applications. In *Biological Control: A Global Perspective*; Vincent, C., Goettel, M.S., Lazarovitis, C., Eds.; CABI: Wallingford, UK, 2007; pp. 262–273.
245. Brouwer, H.; Woodhill, J.; Hemmati, M.; Verhoosel, K.; Van Vugt, S. *The MSP Guides: How to Design and Facilitate Multistakeholder Partnerships*; Centre for Development and Innovation: Wageningen, The Netherlands, 2015.
246. Pancino, B.; Blasi, E.; Rappoldt, A.; Pascucci, S.; Ruini, L.; Ronchi, C. Partnering for sustainability in agri-food supply chains: The case of Barilla sustainable farming in the Po valley. *Agric. Food. Econ.* **2019**, *7*, 13. [CrossRef]

MDPI
St. Alban-Anlage 66
4052 Basel
Switzerland
www.mdpi.com

Horticulturae Editorial Office
E-mail: horticulturae@mdpi.com
www.mdpi.com/journal/horticulturae

Disclaimer/Publisher's Note: The statements, opinions and data contained in all publications are solely those of the individual author(s) and contributor(s) and not of MDPI and/or the editor(s). MDPI and/or the editor(s) disclaim responsibility for any injury to people or property resulting from any ideas, methods, instructions or products referred to in the content.

www.ingramcontent.com/pod-product-compliance
Lightning Source LLC
LaVergne TN
LVHW070444100526
838202LV00014B/1666